MANAGING OUR WILDLIFE RESOURCES

MANAGING OUR WILDLIFE RESOURCES

Third Edition

STANLEY H. ANDERSON

Wyoming Cooperative Fish
and Wildlife Research Unit
University of Wyoming

 PRENTICE HALL, Upper Saddle River, New Jersey 07458

Library of Congress Cataloging-in-Publication Data

Anderson, Stanley H.
   Managing our wildlife resources / Stanley H. Anderson. —3rd ed.
     p. cm.
   Includes bibliographical references and index.
   ISBN 0-13-901232-X
   1. Wildlife management—United States. I. Title.
SK361.A74 1999
  639.9—dc21   98-5859
                                      CIP

Acquisitions editor: Charles Stewart
Director of Production and Manufacturing: Bruce Johnson
Managing Editor: Mary Carnis
Editorial/production supervision and
   interior design: Tally Morgan, WordCrafters Editorial Services, Inc.
Cover design: Marianne Frasco
Cover photo: LuRay Parker, © Wyoming
   Game and Fish Department
Manufacturing buyer: Marc Bove
Marketing manager: Melissa Bruner

© 1999, 1991, 1985 by Prentice-Hall, Inc.
Simon & Schuster/A Viacom Company
Upper Saddle River, New Jersey 07458

Printed in the United States of America

10 9 8 7 6 5 4 3 2 1

ISBN 0-13-901232-X

PRENTICE-HALL INTERNATIONAL (UK) LIMITED, *London*
PRENTICE-HALL OF AUSTRALIA PTY. LIMITED, *Sydney*
PRENTICE-HALL CANADA INC., *Toronto*
PRENTICE-HALL HISPANOAMERICANA, S.A., *Mexico*
PRENTICE-HALL OF INDIA PRIVATE LIMITED, *New Delhi*
PRENTICE-HALL OF JAPAN, INC., *Tokyo*
SIMON & SCHUSTER ASIA PTE. LTD., *Singapore*
EDITORA PRENTICE-HALL DO BRASIL, LTDA., *Rio de Janeiro*

For
Donna, Becky, and Greg

CONTENTS

PREFACE

Because wildlife management takes on many roles in the United States, the wildlife manager must confront many complicated questions. For example, does the public have the right to hunt, fish, and look for wildlife on private land? How does wildlife management relate to the objectives of the private landowner? Who should manage wildlife resources? Although much of the land in the eastern, central, and southern United States is privately owned, a large proportion of land in the Rocky Mountains, the northwest, and, to a lesser extent, the southwest belongs to the public. Wildlife management on these publicly owned lands involves cooperation among private, state, and federal agency personnel. The wildlife manager must also work with politicians to resolve pressing environmental and biological problems. Because of increased public interest in wildlife resources, politicians have begun to play an even more important role in wildlife management decisions. However, sometimes these decisions ignore important biological questions.

Although wildlife legislation has evolved to settle some disputes, defining wildlife and wildlife management often creates gray areas. As a consequence, new subdivisions are forming in wildlife management. For example, specialties in animal damage control have been created to reduce the impact of wildlife on humans and human activity. Other wildlife management disciplines include endangered species biologists, nongame biologists, urban wildlife managers, and raptor specialists.

Wildlife management not only involves the direct manipulation of wildlife populations and their habitats, but consists of educational programs. These programs help private landowners to achieve their management objectives. Education also cultivates public awareness of such issues as wildlife conservation and the use of wildlife by hunters of big game, small game, or waterfowl; fishermen; photographers; and wildlife watchers.

In 1935, Aldo Leopold published the text *Game Management* because he recognized the need for providing academic training for wildlife managers. Throughout his life, he stressed the importance of a holistic approach toward managing wildlife. This approach has continued in recent years with developments in ecology and wildlife management philosophies and techniques. We now study communities and ecosystems, realizing that game animals as well as all other wildlife are integral to these units—no one part exists or can be adequately managed without the other parts.

Training for wildlife managers has evolved from on-the-job field experience to include academic courses in areas such as biology, physiology, botany, forestry, and ecology. The applied situations in *Managing Our Wildlife Resources* pull many of these areas together and provide undergraduate students with techniques on how these disciplines are used to get the job done.

Training today in all wildlife disciplines requires a strong foundation in the dynamics of wildlife populations and their habitat needs. Most curricula provide such a foundation. The wildlife manager must understand the evolution of wildlife management, laws governing management, and the impact of politics on management.

Academic training, however, has not replaced on-the-job training. Knowledge gained in the classroom can only provide a foundation for a better understanding of, and for answers to, the complex questions and problems encountered in the field. *Managing Our Wildlife Resources* bridges the gap between the academic arena and the field.

The purpose of this book is to relate biological concepts to wildlife management and to present management techniques that can be used at the different levels of wildlife management—field, regional, national, and international. The tools for and constraints on wildlife management and the wildlife manager are addressed throughout the book. Selected examples are used in each chapter.

Managing Our Wildlife Resources was written to meet the needs of students preparing for a career in wildlife management and people interested in our wildlife resources. What do wildlife resource managers need to know? They must understand how populations grow and interact with each other and with the natural system; how habitats support species and communities; and how to control the populations and habitats of not just one or a few wildlife species, but of the great diversity of species, each with different needs. Managers must also understand the management principles, planning processes, impact predictions, and possible results of different techniques.

Part 1 describes the meaning of wildlife management. A brief history of the subject is presented and attitudes toward wildlife are discussed. In Part 2, principles of population ecology are discussed. There are discussions of birth and death rates, genetic composition, growth, and the interaction, regulation, and movement of populations. Methods of measuring the characteristics of populations, manipulation techniques, and the purpose and meaning of population models are presented.

Wildlife habitat is the subject of Part 3. There we discuss the habitat needs of species and the ways in which managers can meet those needs. Techniques for managing habitats and the impact of environmental changes on wildlife habitats are included.

In Part 4, we examine the background the manager needs to set goals. This examination covers the legislative context, the relationship between the planning process and the administrative role of managers, and the methods of evaluating public desires. We develop a personality profile and describe the background and skills of wildlife managers.

Management techniques for different groups of animals receive attention in Part 5. Included is information on the biology, behavior, population dynamics, and habitats of these groups. Chapters are devoted to big-game and nongame animals, small mammals, waterfowl, shore and upland birds, fish, endangered species, and damage due to animals. Techniques involving the manipulation of populations and habitats are treated as they apply to each group. Our examples show how management makes an impact on species. In each chapter of Part 5, a group of species is selected to show how management efforts can affect those animals. The species are selected to represent an array of wildlife across the North American continent and to cover examples from different taxa in nongame species.

Each chapter in the book includes a summary and a series of questions. The questions are helpful in highlighting some of the many facets of wildlife management and relating management practices to principles of ecology. The student should develop an understanding of wildlife populations and habitat manipulation techniques that can be applied to different management needs. Most of all, the student should see how effective management involves not only biological knowledge, but effective communication and public relations skills.

The gathering and sorting of data and material for the text, tables, and figures in this book have involved many people. I would like to thank some of the major contributors to this effort: John Cook assisted with library searches and data assimilation; Bob Lanka and Cathy Raley assisted with data collection; Kevin Gutzwiller, Doug Inkley, George Menkens, Wayne Hubert, Willie Suchy, Chris Maser, and Donna Anderson provided highly useful comments on the manuscript; and Becky Anderson, Nellore Collins, and Angela Brummond assisted in preparing the manuscript. I also wish to thank my colleagues in the U.S. Fish and Wildlife Service, the Wyoming Game and Fish Department, and the University of Wyoming Zoology Department for their helpful suggestions and assistance. Donald Van Meter (Ball State University), Samuel J. Mazzer (Kent State University), and James S. Wakeley (Pennsylvania State University) provided helpful reviews of the first edition of this text. Doug Crowe and Archie Reeve provided excellent comprehensive reviews of the entire text.

Stanley H. Anderson

PREFACE TO THE THIRD EDITION

There have been many changes in the world since *Managing Our Wildlife Resources* was first published in 1985, due primarily to ever-increasing numbers of people. These changes have significantly affected our wildlife resources, although the principles of management have remained the same: the control of populations, the maintenance of habitat, and the management of people in order to sustain wildlife. I have tried to show how the changes have affected wildlife management in this edition. All chapters have been updated. Changes in legislation, new and updated techniques, public perceptions, and the current status of all species of wildlife are discussed. The differences between private and public land management for wildlife are included. To each chapter has been added boxed examples of how the concepts in the chapter have been applied.

I am grateful for the help and input of many people in preparing this revision. The users of the earlier editions had many helpful suggestions. I very much appreciate the great ideas, rewarding discussions, and research results from the graduate students and research associates at the Wyoming Cooperative Fish and Wildlife Research Unit. Many people provided data and information for this revision, including Mark McKinstry, Greg Anderson, Fred Lindzey, Wayne Hubert, Dave McDonald, Loren Ayers, and Kira Young. My thanks to all of them. Christine Waters and Linda Ohler were of immense help in preparing the manuscript. A very special thanks to my wife, Donna, who aided with all aspects of the project.

Part 1
MANAGING OUR WILDLIFE RESOURCES

*1 What Is Wildlife
   Management?*

River otter are found in many waterways of North America.

Part 1 offers an insight into the history of wildlife management. We see which factors influence management and learn of the many people affected by or involved in management. This part sets the stage for the rest of the book.

1

WHAT IS WILDLIFE MANAGEMENT?

Wildlife populations are found in areas where their basic needs—shelter, reproduction, food and water, and movement—are satisfied. We call the area of a particular population its **habitat.** *Wildlife management* is the art and science of manipulating **populations** and habitats for the animals and for human benefit.

According to the dictionary, *wild* means living in a state of nature—not tamed or domesticated. Management is the act of controlling or directing; husbandry, on the other hand, is the management of domestic animals. Today, many efforts that supposedly fall to the wildlife manager are questionable wildlife efforts. When pheasant farms produce birds and release them for hunts the day before hunting season, are we truly managing wildlife? Releasing fish raised in hatcheries is considered management, but is it? Raising species in captivity for return to the wild is an important part of endangered species research. We must, however, keep in mind that wildlife management means controlling untamed animals.

We can often assert such control by means of (1) managing habitats, (2) managing people, and (3) managing the individuals in a population by allowing them to increase, decrease, or remain constant. The principles of wildlife management are based on these three methods. We manage a habitat for a population or a community of populations. When we manage the habitat for all the populations in the community, we are managing for biodiversity. We manage people by doing such things as controlling the number of hunters or wildlife viewers, controlling access to sensitive populations or habitats, and educating people about wildlife. The combination of the three methods should be kept in mind throughout this book as we look at managing our wildlife resources.

As the human population increases, many more issues face wildlife managers. Everyone from urbanites to ranchers to farmers views various kinds of wildlife differently. Species considered to be pests by some are much admired by others. Ranchers and farmers see their private land impinged on by wildlife legislation involving endangered species, wetlands habitats, and migratory birds. Businesspeople sometimes feel that wildlife hinders their jobs. For instance, mitigating the destruction of habitats resulting from mining processes can be expensive. Other groups want to maintain the diversity of wildlife. How do we apply the principles of wildlife management with different people's interests and attitudes? There is no one answer.

Many people have an interest in wildlife management, and many individuals and organizations seek to impose their own values on wildlife policies. Among the complex of individuals and groups that can benefit from wildlife management are hunters, fishing enthusiasts, photographers, bird-watchers, tourists, and landowners (Figure 1–1). Additionally, entrepreneurs are involved, including some who may never see the open country or many forms of wildlife; among these are outfitters and businesspeople catering to tourists, such as hotel owners and restauranteurs. This wide array of people and the great diversity of interests they express greatly complicate the wildlife-management process.

Wildlife—terrestrial and aquatic—is an important component of the natural system, a renewable resource when proper habitat and population management procedures are followed. Because of its complexity and because the plants, animals, and physical resources that make up the natural system are so interrelated and interdependent, most people understand only a small portion of the natural system's structure. Thus, frequently unwittingly, people can impose themselves on the system in such a way that

Figure 1–1 Black bear are found in North America. They are a hunted species, yet a popular species to watch and photograph. In addition, they cause damage to some human resources. (Courtesy Fred Lindzey, U.S. Geological Survey.)

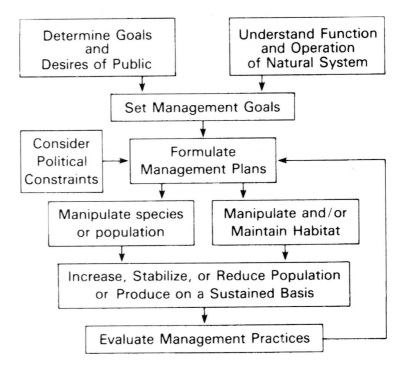

Figure 1–2 Process of wildlife management.

resources are depleted or destroyed. Subsystems such as forests and lakes are often changed. Animal interactions may change as livestock is introduced, forests are cut, dams are constructed, or human population increases. Wildlife management is, therefore, a complex procedure of inventorying and evaluating habitats and populations, determining people's goals, and superimposing those goals on the natural system (Figure 1–2).

APPROACHES

Various approaches can be used in wildlife management. **Preservation,** letting the natural system alone, is the controlling philosophy in some national parks. Some people feel that all wildlife management should use this approach. It should be remembered, however, that an undisturbed system is not a stable one: Natural changes constantly create different habitats for wildlife. Populations interacting in each area can so change the system that the environment may not remain suitable for their own continued existence.

Conservation is the effort to maintain and use natural resources wisely. Forms of conservation range from active managing efforts, such as the manipulation of habitats and the introduction of species, to a complete "let alone" attitude (preservation). But conservation in general means attempting to save resources for future generations.

Management is the manipulation of populations or habitats to achieve desired goals. These goals may include any of the following: (1) to *increase* the size of a population; (2) to *remove* individuals from a population on a continuing basis (take a sustained yield), which requires that enough individuals be left to reproduce and so replace

Habitat Preservation

A fire in the summer of 1988 showed the effects of preservation in a natural system. As our national parks have been managed by a "let alone," or preservation, policy for many years, a great deal of downed and woody material has accumulated in the forests of Yellowstone and Grand Teton National Parks, as well as in the John D. Rockerfeller Memorial Parkway between the parks. Fires started by careless campers or lightning were immediately put out. In these areas, fires undoubtedly occurred naturally at intervals prior to the establishment of the park system.

In 1988, a series of events created a massive wildfire in these public areas. During the preceding winter, the snowpack was low. Spring was dry and river waters low. In July, dry, hot days with afternoon lightning storms resulted in a fire. Because the fire was in the wilderness, it was described as a prescribed burn. The dry conditions and wind caused the fire to spread, at which point it was declared a wildfire.

The fire continued out of control throughout July and August, with several additional fires being started by lightning. Some of the fires came together. In late September, snowfall helped firefighters control the blazes. All in all, some 370,000 hectares (925,000 acres) burned.

The fires made national news for months. Public focus was on the number of acres destroyed and the firefighting effort. What did not get reported were the changes following the fires. The next spring, grasses and wildflowers covered soils in the burned areas. Aspen sprouts increased within a few years, resulting in browsing by elks. Bison moved into the sagebrush areas that were burned.

Preservation is, therefore, a form of management in which change occurs because of natural events. Whether this is good or bad depends on one's perspective. Preservation creates conditions for other change, in this case a fire, which results in different habitats and changes in wildlife.

those removed; (3) to *stabilize* or reduce the size of a population[1]. Combinations of these goals can be used for either a single species or a number of species.

Ira Gabrielson, former director of the U.S. Bureau of Biological Survey (now the U.S. Fish and Wildlife Service) and author of *Wildlife Conservation,* writes that "wildlife management recognizes the reality and operation of ecological communities and that man's activities often greatly disrupt them, thence that it is often desirable from the human viewpoint to work with these communities and attempt to modify or manage them in man's interests"[2].

Management generally involves manipulating numbers of individuals, increasing or decreasing the birthrate, increasing or decreasing the death rate, or manipulating the habitat to change the distribution or density of species. Management can also be passive. If the goal is to allow nature to manage itself, the manager may not need to do anything to the habitat, because the natural process is dynamic; however, even under a "let alone" policy wildlife will change over time.

Species management has been used for many years. Game, endangered species, and control of nuisance wildlife all require some form of species management. In the first comprehensive textbook on wildlife management in the United States, Aldo Leopold, considered the father of the discipline in the United States, writes that game management "is the art of making land produce sustained annual crops of wild game for recreational use." He includes the following practices under game management[3]:

1. Restriction of hunting
2. Control of predators
3. Reservations of game lands (such as parks, forests, and refuges)
4. Artificial replenishment of wildlife (such as restocking and game farming)
5. Environment controls (such as control of food, special factors, and disease)

Leopold was influenced by his fieldwork as a forester in New Mexico, where he saw declines in major big-game species and the impact of predators that could "hunt 365 days of the year." His later experiences caused him to extend his concern to all wildlife.

Since the early 1970s, more and more people have come to realize that a variety of wildlife species is a sign of a healthy community or natural system. Many techniques are used to secure diversity of species. Some think that a variation of the single-species (indicator) approach, in which, by managing for one species, "the others will follow in a healthy environment," is best. Other approaches include maintaining a *diversity* of plant communities to bring about diversity of wildlife (Figure 1–3).

Since 1960, wildlife management has also applied the concept of multiple use of resources. Thus, agencies and companies, as well as individuals responsible for forests, ranges, lakes, or other natural resources, are trying to establish wildlife as a secondary management objective. Part of the reason for this approach is legislation that requires such federal land agencies as the Forest Service, Bureau of Land Management, and Department of Defense to manage for wildlife in addition to managing timber, grazing, and land units. Special federal legislative acts applicable to the Forest Service (the Multiple Use Sustained Yield Act and the National Forest Management Act) and the Bureau of Land Management (the Federal Land Policy and Management Act) provide that wildlife must be considered in the multiple-use approach to natural-resource management.

The diversity and multiple-use approaches to wildlife management have come about to a large extent because of public pressure. Astute politicians realize that wildlife management is not just for hunters and fishing enthusiasts. While these sportsmen and -women have contributed considerably to the goals and financing of management, photographers, bird-watchers, hikers, and nature lovers are now demanding management for their interests. For the most part, these groups of **nonconsumptive users** are willing to pay for their interests. They provide a great deal of revenue to motel and restaurant owners and to outdoor equipment suppliers. In fact, nonconsumptive users have caused industrial changes. Nonconsumptive wildlife users charter boats to see and pho-

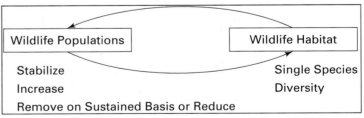

Figure 1–3 Approaches to maintaining diversity of wildlife. Manipulation of the habitat influences the diversity of the wildlife population.

tograph seabirds and sea mammals. Some boat operators regularly schedule wildlife tours; others have converted completely to wildlife trips. Motels in popular bird-watching areas cater to the interests of these enthusiasts. There are now tours organized specifically to observe individual species or the diversity of wildlife throughout the world. Thus, public desires are changing. It is not that hunting and fishing are unimportant components of wildlife management, but that other demands are now also considered. Economic considerations accompany these new, mostly nonconsumptive uses of wildlife.

Wildlife management today goes beyond the effort of field people to manipulate habitats and populations. Conservation organizations exert political pressure on elected state and federal officials to attain the goals of their members. Such organizations often maintain staffs in Washington, DC, where they lobby Congress and members of the executive branch for their ideas. More and more, these organizations use the legal system to achieve their goals.

Conservation organizations were working hard in the late 1980s and 1990s to increase national interest in biological diversity. Their efforts included a major campaign to lobby Congress. Such actions have been a key step in increasing public awareness of the interrelationships among our natural systems. The actions included a national policy statement on the conservation of biological diversity and management actions leading toward increased biological diversity.

The aims of conservation organizations vary, of course. Some wish to preserve all natural systems, others seek to protect the rights of hunters or fishers, and still others lobby for the preservation of endangered species. Professional wildlife groups such as the Wildlife Society are also strong public-interest groups. They all lobby for legislation to protect wildlife, and they employ professionals to conduct their fight.

WILDLIFE MANAGERS

While many people become interested in wildlife management as a career for the outdoor experiences it offers, working with wildlife requires meticulous planning, data collection, analysis, and evaluation. The professional wildlife manager must have a broad background in the biology of wildlife populations, as well as their habitats, life histories, and behavioral interactions. In addition, managers must be good communicators: They must have the ability to understand and be understood by other people. Persuasive and political skills are also essential characteristics of a good manager. The skill to convince others without creating animosity is important. An effective manager, in other words, combines many skills with an in-depth scientific background. Chapter 11 is devoted to the activities and background of the wildlife manager.

People interested in wildlife can make management a vocation or retrain their interest as an avocation. Professional managers work for public agencies, corporations, conservation groups, or special environmental companies. These professionals might manage public or private lands, work in the field, or plan and supervise others' management efforts in the field on a local, regional, or national basis.

There is a great difference between public and private wildlife management, yet animals do not recognize boundaries. Management of wildlife on public lands is determined largely by the public agency's goals and public desires. Management of wildlife on private land is often determined by profit-related goals. The legal bases of management also differ: Public lands often have legislative mandates specifying management approaches, while private lands are usually affected only by harvest legislation. Some legislation, of course, such as that relating to endangered species, applies to both public and private lands.

Private landowners are now trying to make more income from wildlife on their lands. Access fees for hunting and fishing vary considerably, from a few dollars to thousands. Large landowners sometimes form cooperative hunting or fishing clubs that charge annual fees. Others are establishing package tours for sportsmen, including guides, room, board, and entertainment, for a set fee. In all cases, state hunting or fishing licenses are required; however, pressure is placed on some state wildlife agencies to let landowners control the issue of licenses for their property.

EVOLUTION OF WILDLIFE MANAGEMENT IN THE UNITED STATES

The European Legacy

In early English history, game species were considered the property of no one, much like the air and oceans. However, unlike the air and oceans, wildlife could become the property of anyone who captured it. This attitude can be traced back to the Roman Empire. Apparently, the only relevant provision in early Roman law was the exclusive right of a private owner to possess and kill the wildlife on his property.

As the land became subdivided among feudal lords, attitudes toward wildlife changed. In England, royal forests were designated as places where the king and others of his choosing might engage in the chase. Actually, the king had title to game species on all the land, and one way he had of rewarding his favorite nobles was by conferring franchises over game on their property. Thus, land for nobility and royal forests constituted most of the wildlife habitat, and the general public had little or no access to either fish or wildlife. The domination of hunting rights for wildlife by a few became such an acute problem, that in 1215 the Magna Carta, which resulted from a rebellion against the dictatorial power of the king, included wildlife-related clauses. It directed that private veers or docks in rivers and streams be removed because they were becoming a hazard to navigation[4]. These veers had been placed as a means of access to the fishery resources by nobility. A later amendment to the Magna Carta barred the king from granting private fishery rights.

In Great Britain, royal power over wildlife gradually gave way to Parliament, which, however, still maintained wildlife for a chosen few. Statutes prohibited unqualified people from taking wildlife, and "qualified" usually meant "wealthy" or "prestigious." Parliament parceled out wildlife land rights and determined what rights people had with respect to taking wildlife. To the people, wildlife management meant, for the most part, controlling the removal of species. So even though the king's powers were reduced, the change meant little for the people.

The thrust of game laws that initially prevailed in the United States was that Parliament owned resources on the land. Permission to use the resources could be obtained from Parliament or its designee. The colonies—later the states—assumed the transfer of parliamentary power for the resources and thus owned the game within their boundaries.

The states' ownership of wildlife dominated the way the legislation evolved for wildlife management and protection in the United States until the late eighteenth century, when the federal government passed the Lacey Act to assist the states with importation and interstate commerce of wildlife. (See Chapter 10.)

Land and People

Early visitors to the New World reported on the abundance of wildlife. John Cabot noted the abundance of fish in the New World before 1500[5]. His son, Sebastian Cabot, later commented that fish were so plentiful along the coastal waters, that they could slow the progress of a ship[6]. Such reports brought many fishing ships from European nations to the coastal waters of the Americas. It was reported that 350 to 400 vessels visited the area each year by the mid-16th century[5].

Explorers on both the east and west coasts of America were amazed at the number and variety of wildlife species. It must be remembered that the early explorers came from lands where human population was increasing and food was becoming scarce. The long history of dense human habitation on the European continent had changed wildlife populations there. Species that could not adapt to human activity had disappeared from the fields and oceans. Species that could coexist with the expanding human population flourished. Thus, the bison and wild goats of the Mediterranean vanished, while species such as starlings, house sparrows, and rats increased[5].

North American wildlife had not been without human impact, of course. Wildlife had been used for food, clothing, and daily living for thousands of years. But the native Indians had apparently practiced a conservation that meant minimum interference with nature, despite occasional lapses from grace. For instance, an account by Colonel R. I. Dodge, who lived and fought among the plains Indians, describes their fall feasts, at which only a small part of each of hundreds of bison was consumed[7,8].

Still, Indians of the plains had little impact on the great bison herds that provided them with food, clothes, and shelter[9] (Figure 1–4). One of the reasons was the constant warfare between tribes, which kept their numbers down. Even with the introduction of guns to the Indians, wildlife—particularly bison—could sustain their populations. It was the opening of the trade markets and invasion by white people that changed the picture. The whites gave no thought to replacing individuals in the wildlife population.

Indians of the eastern forests practiced a form of habitat management. They cleared and often burned extensive areas around their villages for cultivation, protection, and attraction of some wildlife[9]. Thus, a diversity of habitats evolved. Grasslands created by fires attracted elk and bison.

Early white settlers of the Americas came to an area with an abundance of natural resources, including wildlife. Certainly, some colonists were interested in protecting the

Figure 1–4 Indians relied on bison for food, clothing, and shelter. (Courtesy of the Minnesota Historical Society.)

wildlife and so respected the resources[5]. They found abundant food and clothes obtainable from wildlife (Figure 1–5). But for others, trade with the Indians was a very profitable venture. Beaver were eliminated from many ponds and streams of the northeast. Too, agricultural practices changed the habitat of many species of animals.

Uses and Exploitation

While wildlife in the European countries had been taken chiefly for sport, some was used for food and clothing. In the New World, there were three major uses of wildlife—two of them, unfortunately, exploitative. First, the stories of wildlife abundance led to commercial ventures, particularly the fur-trading business. Second, groups of people moving westward and people getting their food from the land were sustained by the removal of wildlife. These people often exploited the populations and altered the habitat. Finally, somewhat later, the attitude developed that wildlife should be managed for sport hunting and fishing.

Figure 1–5 Like the Indians, early settlers depended on wildlife for food and clothing. (Courtesy of the Missouri Historical Society, St. Louis.)

Commercial Use. As wildlife became a source of revenue, excessive exploitation occurred. Early explorers found the native Indians eager to trade furs and bird plumes for shiny trinkets and small knives. Pelts and plumes brought high prices on the European markets, setting the stage for vast commercial ventures that would exploit the wildlife resources of the New World. Enterprising individuals established commercial organizations that bought from or traded with Indians and trappers and sold on the European markets. The largest of these organizations eventually became the Hudson Bay Company, chartered by the British Crown. The company was granted a full monopoly over all streams entering Hudson Bay and all traffic in adjacent lands[9].

Throughout the 18th and 19th centuries, the fur business expanded on the North American continent. Trappers became independent businesspeople traveling alone or in small groups to western streams and taking native animals, mostly beaver, for the pelts (Figure 1–6). These people lived rough, adventuresome lives, deriving their sustenance from the land. They often trapped one area until their take declined and then moved to a new area. Periodically, they would go east to sell their pelts and then return to the west to collect more. Their stories of the west were part of the motivation for homesteaders to expand westward.

The impact of the removal of wildlife for commercial use resulted in the extinction or near extinction of some species. Passenger pigeons in concentrated flocks from Canada to Mexico were easy to exploit (Figure 1–7). Flocks numbering more than 1 trillion birds were reported in Kentucky, Ohio, and Michigan around 1800. Markets for the birds were found in eastern city restaurants, where the "pigeoners," as they were called, sent the birds by the tons. The birds roosted on tree branches and nested in colonies. As people settled these areas where pigeons nested, they changed the habitat through the settlement and the accompanying removal of trees. Still, people apparently assumed that resources like the pigeon were inexhaustible. In 1857, the Ohio Senate quickly voted down a bill that would have offered protection to the passenger pigeon. Members of the Senate noted that, the birds being wonderfully prolific, no ordinary

Figure 1–6 Fur trader sorting pelts. (Courtesy of the State Historical Society of Wisconsin.)

Figure 1–7 Passenger pigeon. (Courtesy of the State Historical Society of Wisconsin.)

destruction could affect the numbers that were produced each year[10]! By the early 20th century, the species had been all but exterminated. The last passenger pigeon, Martha, died in the Cincinnati Zoo on September 1, 1919[9].

Meanwhile, exploitation of the oceans and coastal areas continued at a relentless pace. Eggs and flesh from easily captured seabirds that nested on rocky shores and cliffs were in demand by Europeans. Gannets, puffins, and auks were slaughtered. Fishing boats brought back the products of whales, walrus, polar bears, and seals (Figure 1–8).

Figure 1–8 Early whaling expedition. (Courtesy of the State Historical Society of Wisconsin.)

Means of Sustenance. The westward movement of people in the United States brought a new dimension to exploitation. Exploratory trips to the west were organized following reports from trappers and explorers. The Lewis and Clark expedition (1803–1806) brought back information on wildlife. Then came wagon trains and the railroad.

Conservation-minded people realized that animals like the bison could not withstand the slaughter that was occurring. So a number of noted people began a campaign to increase the public's awareness of wildlife conservation. John J. Audubon (1785–1851) and a number of other writers and painters eulogized American wildlife (Figure 1–9). But although they helped create an interest in, and awareness of, wildlife, their efforts did little to help the plight of the bison.

White people had begun to hunt bison as early as 1540. Still, in 1871 the animal was fairly abundant in most parts of the west. But the railroad fragmented the herds and brought large numbers of people, including teams of hunters with a variety of new hunting gear. Too, soldiers living in forts in the west utilized bison for food and sport. Some military people tried to reduce the herds as a means of depriving Indians of their food supply. It is estimated that with the railroad and heavy use of the prairie by travelers and settlers, nearly 3,700,000 bison were slaughtered in the period 1872–1874[6]. While bison once numbered almost 60 million, they were reduced to around a million and a half by 1800, and expansion of the railroad system in the 1880s caused a further decline.

In the east, farming was the dominant method of earning a living. Forests were cleared and replaced by food plants (Figure 1–10). Europeans brought wheat, oats, barley, onions, and sugarcane to the New World. Domestic animals, including horses, dogs, pigs, cattle, chickens, sheep, and goats, were also brought from Europe. The New World contributed potatoes, maize, beans, squash, avocados, pineapples, tomatoes, and rubber to the European markets[11].

Farming in the New World had a major impact on wildlife resources. Forest clearing was extensive, reducing the habitat for many native species, such as migratory songbirds, and increasing it for others, such as white-tailed deer. Removal of forests for fuel and construction also affected native species. Meanwhile, livestock owners found

Figure 1–9 Hunters shooting bison from a train ca. 1875. (Courtesy of the Edward E. Ayer Collection, the Newberry Library.)

Figure 1–10   Ohio settlers clearing forest for a cabin. (Courtesy of the Everett D. Graff Collection, the Newberry Library.)

wolves, cougars, and bear attacking their animals. Bounties were established, but the predator population persisted on the east coast for some hundred years after the first settlements[9].

Control of predators was important in the early development of wildlife management. Wildlife was usually classified as good or bad. Farmers and ranchers had lists of wildlife that were harmful to their operations. Bounties were established as a means of controlling wildlife suspected of killing commercial, game, or desirable nongame species. Bounties are still seen by some as a means of controlling predators. In a number of counties of western states, predator control boards institute bounty payments during periods when complaints about predators are high.

Sports Hunting and Fishing.   The third use of wildlife, sports hunting and fishing, started to develop in the mid-19th century. In the east, the wealth of natural resources produced a group of people who, in various commercial projects, accumulated great wealth. With the money came leisure time. Hunting and fishing for sport were among the activities pursued by this new group of wealthy businesspeople and landowners. Growing aware of the decline in wildlife and the loss of habitats and access to land that supported wildlife, sports hunting and fishing groups initiated efforts to conserve and manage wildlife. Some formed clubs that purchased vast tracts of land which were then made available to their members for hunting. Clubs such as the New York Sportsman Club, established in 1844, were influential in protecting the wildlife resources from exploitation by sports enthusiasts.

EVOLUTION OF AMERICAN WILDLIFE CONSERVATION

A group of people interested in nature, the outdoors, and wildlife emerged at about the same time as the clubs. This unique combination of people included foresters, most of whom had been trained in European schools, writers, artists, and businesspeople.

The extinction of some species and the discovery of others on western exploration trips prompted writers such as Henry David Thoreau and Alexander Wilson to write about wildlife, describing its wonders and sometimes its slaughter. Magazine articles on wildlife became popular. *Forest and Stream,* a sportsman's journal established in 1873, was a strong advocate of wildlife conservation. Artists painted pictures of new and declining wildlife. Alexander Wilson and Audubon became noted for their wildlife paintings (Figure 1–11).

Federal Government

Other forms of natural resource conservation greatly assisted the wildlife-conservation movement in the late 19th and early 20th centuries. The establishment of land as a trust for all the people came in 1872, when President Grant set aside 8,671 square kilometers (km^2) [3,348 square miles (sq mi)] of land that would eventually become Yellowstone National Park (Figure 1–12). The park area was considered a refuge, but law enforcement efforts were minimal during the early years, and poachers took a heavy toll of the wildlife[9]. By the late 19th century, a number of forest reserves had been established to protect and manage America's timber resources. These reserves, the forerunners of the national forests, became important means of maintaining habitats for wildlife.

The federal government set up a wildlife agency in 1885, when funds were allocated to establish the predecessor of the Biological Survey, now the Fish and Wildlife Service. C. Hart Merriam was the first head of the Division of Economic Ornithology and Mammalogy, as it was originally known. This agency was established by Congress as a result of a resolution adopted at the first annual meeting of the American Ornithologists' Union in New York City.

Figure 1–11 Pigeon hawk painted by Audubon. (From John J. Audubon, *Birds of America* [New York: Macmillan, 1937].)

Figure 1–12 Elk in early Yellowstone National Park. (Courtesy of the Wyoming Game and Fish Department.)

By the end of the 19th century, the federal government had a fragmented conservation program resulting from the pressure of writers, artists, and the beginning interests of states in maintaining their wildlife species. A few areas with historic or aesthetic values, such as Yellowstone, had been set aside as public land.

President Theodore Roosevelt proved to be one of the most important people in the conservation movement. Much national wildlife-management legislation, as well as action, can be traced to the Roosevelt years. Aiding Roosevelt in these efforts was Gifford Pinchot, who proposed the philosophy of sustained-yield forestry (Figure 1–13). This was, in effect, the beginning of looking at natural resources as renewable. Theoretically, if properly managed, some could be utilized forever.

A good deal of land was put under federal management during the Roosevelt years. It was Roosevelt who initiated the national wildlife refuge system by issuing an executive order setting aside Pelican Island in Florida as a federal bird refuge. During the Roosevelt administration, the cumulative area of the national forest system increased from 17 million hectares (42 million acres) in 1902 to 70 million hectares (172 million acres) in 1909. The program of charging people for running their livestock on federal land was initiated. Timber sales and regulations were supervised by forest rangers.

In 1908, a White House conference on conservation advocated prudent use of natural resources, without waste and tempered by reason and consideration for the basic supply[9]. As a result of the conference, a national conservation commission, with Pinchot

Figure 1–13 Gifford Pinchot. (Courtesy of the U.S. Forest Service.)

as chairman, was appointed. This commission was responsible for making a detailed inventory of the nation's natural resources, including water, forests, land, and minerals, and for staging conferences to discuss methods of improving natural-resource management. Even though wildlife was not the major topic in these conferences, the secondary effects on wildlife were considerable. It was during the conferences that a bond was forged between the federal government and states to manage wildlife resources.

In 1940, the Fish and Wildlife Service was formed from the Bureau of Biological Survey and the Bureau of Fisheries and made a part of the Department of the Interior. In 1956 the consolidation was undone by the Fish and Wildlife Act, which created a Bureau of Sport Fisheries and Wildlife and a Bureau of Commercial Fisheries. In 1970, the Bureau of Commercial Fisheries was transferred to the Department of Commerce and renamed the National Marine Fisheries Service. Meanwhile, the Bureau of Sport Fisheries and Wildlife remained a branch of the Department of the Interior and was called the Fish and Wildlife Service.

State Governments

Between 1865 and 1900, a number of states established game-management agencies. In 1852, Maine appointed a person in each of its counties to enforce deer- and moose-hunting regulations. A bag limit on deer was imposed in 1873, and a game and fish commission was established in 1800[9]. Although the procedures varied from state to state, the states began to recognize their role in protecting wildlife resources. Most early agencies were funded by appropriations from the general treasury, but in 1895 North Dakota passed a law requiring hunters to purchase licenses. The proceeds were used to help run the game agency (Figure 1–14).

Conservation organizations at the federal, state, and local levels frequently had to give top priority to predator control in their attempts to preserve wildlife. When states

Figure 1–14 Hunting camp. (Courtesy of the U.S. Soil Conservation Service.)

established their wildlife agencies, they usually gave them responsibility for administering the bounty programs.

Private Organizations

In the early 1920s, wildlife management became an accepted means of maintaining wildlife resources. Although national land was available for wildlife and a wildlife refuge system was well under way, the pressure for the program came at this time from the hunting and fishing communities, which were interested in increasing the nation's stock of game and fish. Part of the movement was designed to provide clean waters and to restore fish and wildlife resources. The Issac Walton League of America, established primarily as a result of hunting and fishing activity, emphasized pollution control and prudent use of natural resources. The League became one of the first organizations to lobby for wildlife laws and a conservation ethic.

One of the more influential sportsmen's clubs was the Boone and Crockett Club, founded in 1887 at a dinner given by Theodore Roosevelt for selected friends. The earliest members included Henry Cabot Lodge, Francis Parkman, Senator George G. Vest, D. G. Elliott, Colonel R. I. Dodge, Gifford Pinchot, and J. P. Morgan[10]. The club, which promoted hunts throughout the world, made it a prerequisite for membership that one have killed with a rifle an American big-game animal.

Nongame interests were also organizing. The American Ornithologists' Union was established in 1883 in New York City. Modeled after the British Ornithological Union, its members included some of the best known naturalists and ornithologists in America. Among its first efforts was the proposal of a "model law" to be considered by state legislatures. The purpose of the law was to protect nongame birds and their eggs and roosts, with collection for scientific purposes allowed only after careful review by the state. Several states, including New York and Pennsylvania, adopted the law before 1900.

Nonconsumptive interests in wildlife continued with the formation of the Audubon Society in 1886. Until 1889, this society, formed by the publishers of *Forest and Stream,* provided free membership to persons who pledged to prevent, to the extent possible, the killing of wild birds not used for food, the destruction of nests of eggs of wild birds, and the wearing of feathers for dress. By 1887, membership numbered 37,400. Local Audubon clubs continued to develop, primarily in response to feather fashion. In 1905, the various local chapters combined to form the National Association of Audubon Societies[10].

Aldo Leopold

During the early part of the 20th century, there were no courses in wildlife management: Game wardens were trained on the job. Aldo Leopold changed that. Born in Iowa in1887, he grew up in a large home overlooking the Mississippi River. Living near bottomlands and in the migratory paths of ducks and geese, Leopold developed an early interest in wildlife. Later employed by the U.S. Forest Service in New Mexico, he was responsible for working with the state of New Mexico in enforcing game laws. Many of Leopold's ideas on game management were forged in those years (Figure 1–15).

Having moved to Wisconsin, Leopold was later appointed to the faculty of the University of Wisconsin, where he became a one-person department of wildlife management and set up a graduate program for managers. During that period, he wrote his text *Game Management.* His famous *Sand County Almanac,* an account of land–wildlife interactions, was published posthumously. Leopold stressed the importance of ecological principles in developing wildlife-management techniques. He emphasized the need for cooperative integration of land use, including farming, forests, wildlife, and recreation favorable to local conditions on both public and private lands[12].

J. N. "Ding" Darling

In early 1934, Leopold and cartoonist-conservationist J. N. "Ding" Darling (Figure 1–16) served on the President's Committee on Wildlife Restoration. Darling was a Pulitzer prize–winning political cartoonist for the *Des Moines Register* with a degree in

Figure 1–15 Aldo Leopold. (Courtesy of the State Historical Society of Wisconsin.)

Figure 1–16 J. N. "Ding" Darling and one of his environmental cartoons. (Photo courtesy of *The Des Moines Register*. Cartoon reprinted with permission of the J. N. (Ding) Darling Foundation, Inc.)

Part I / Managing Our Wildlife Resources

biology. He had poked barbs at the federal conservation program and its father, President Franklin D. Roosevelt. But in 1934, President Roosevelt appointed Darling chief of the Bureau of Biological Survey. Darling had just designed the first migratory bird-hunting stamp, under the act that went into effect in 1934, to accumulate funds to purchase waterfowl refuges. Instrumental in bringing new life to a federal wildlife-management agency, Darling realized that there were not enough people to staff basic research, management, and administrative posts. In 1934, he invited a group of industrialists to meet with him in the Waldorf–Astoria Hotel in New York. As a result of this meeting, a number of major conservation organizations that were to influence the direction of wildlife management were established. One further outcome was the Cooperative Wildlife Research Unit program, a cooperative program between the U.S. Fish and Wildlife Service (U.S. Geological Survey as of 1996), state game and fish management agencies, the Wildlife Management Institute, and universities. The American Wildlife Management Institute, the North American Wildlife Foundation, the National Wildlife Federation, and the North American Wildlife and Natural Resources conferences also began at this time. These organizations are now major components of wildlife conservation, maintaining research facilities for wildlife. After only 18 months, Darling left government service to become the first president of the National Wildlife Federation. Ira N. Gabrielson, an equally strong wildlife administrator, succeeded him.

ATTITUDES TOWARD WILDLIFE

Attitudes toward wildlife and wildlife management in the United States have been cyclic in nature. The abundance of wildlife found by early settlers encouraged exploitation both by individuals wanting to remove wildlife and by commercial interests, which removed animals at a rate sometimes causing the extinction of their species. Early conservation movements responded to these destructive trends. Conservation efforts were aided by mass media and both aided and deterred by political interests. At times, divisions within the conservation ranks have impeded progress.

Wildlife management began with interested conservation and sports groups. This initial interest resulted in a variety of laws controlling the removal of wildlife and eventually led to the establishment of areas where wildlife could live without human encroachment. As more people moved into areas inhabited primarily by wildlife, competition for these lands became keener. Thus, wildlife management came to involve not only fieldwork, but politics as well.

Wildlife resource management received a great boost during the administration of President Franklin D. Roosevelt, who established the Civilian Conservation Corps to provide employment for many young people entering the job market during the Great Depression of the 1930s (Figure 1–17). During these years, many wildlife refuges had water impoundments constructed to provide habitats for fish and waterfowl. Trees were planted and trails constructed in the national forests and parks. Visitor centers were built, so that many could learn about and enjoy wildlife. The soil conservation principles developed and implemented at this time helped maintain wildlife habitats.

A major surge in wildlife conservation occurred in the late 1960s and early 1970s. Spurred on by the proconservation policies of the Kennedy and Johnson administrations, Earth Day created in the American public a major awareness of our natural resources. Wildlife was one of the major benefactors. New forms of protective legislation were passed. Curricula changed in schools: *Ecology* became a word known by all. The introduction of wildlife conservation into the public education system signaled a new

Figure 1–17 Civilian Conservation Corps workers banding ducks. (Courtesy of the National Archives.)

milestone in public awareness. By 1975, surveys of public attitudes showed a major concern for conserving our nation's wildlife resources.

The political indifference—even antagonism—to wildlife management of the late 1920s and early 1930s during the Coolidge, Harding, and Hoover administrations resurfaced during the early 1980s and continued through the 1990s as the political climate in Washington became anti-natural-resource management. It is worth noting of the latter decline in support for wildlife management that, while some influential politicians appeared to be indifferent to maintaining our national resources, polls indicated that public interest in conservation was high. Private conservation groups received more money than ever before, and the number of people interested in various aspects of wildlife and natural resource management was higher than it had ever been.

This separation of political and popular interests is disturbing. Today, very few wildlife decisions are based primarily on the biological knowledge of wildlife managers in the field. Most decisions involve trade-offs and political pressures, which often are not related to managing wildlife at all and frequently have no biological basis. Ding Darling once said that the greatest threats to wildlife are the Republican and Democratic parties. Indeed, the problems of wildlife management would be more easily solved without the intervention of politicians, who have little knowledge or understanding of wildlife interactions. The cold fact is, however, that the political arena is where decisions are being made for wildlife. So it is this arena that must be understood by the wildlife-management profession if it is going to maintain wildlife resources in the nation.

Another complication for management is that wildlife is not confined to political boundaries. Thus, cooperation among different government subdivisions is necessary. Early in U.S. history states had control of wildlife, but now many federal acts allow U.S. government involvement on federal land and in relation to migratory and endangered species. In addition, international treaties are currently the basis of international cooperation in some areas of wildlife management. Lobby interests for wildlife in Washington, DC, and most states had taken a businesslike approach by the end of the 1980s.

Efforts to promote biological diversity, the preservation of endangered species, and nonconsumptive wildlife are now actively undertaken. Funding for agencies involved in wildlife management receives intense scrutiny and comment from conservation organizations, which have large memberships that write letters to support their positions.

The public plays a critical role in the future of wildlife. Therefore, it is essential that people be educated and informed in wildlife management. Some wildlife biologists refer to the concept of *wildlife acceptance capacity,* or the wildlife population level acceptable to people in an area[13]. Where the reintroduction of wolves is discussed, the ranchers think not of ecosystems, but of their cattle. When the number of deer–vehicle collisions increases, people are concerned about both accidents and conservation. Increases in the number of hunted species on private lands can bring claims of damage from the landowners, and decreases in animals can bring complaints from hunters. Lack of birds in wetlands arouses the concern of bird-watchers and their organizations. Wildlife agencies must balance these concerns with those of managing the diversity of the natural system, so that all species can survive.

Today, the wildlife manager is in a crucial and delicate position. The public is attuned to the need for wildlife conservation as never before. History tells us that most of the major steps in legislation for managing wildlife have resulted from public attitudes converted into the political pressure needed to maintain our wildlife resources.

Wildlife management is forever changing. No longer can we manage only ducks, big game, or hunters. Managers need to be open to new ideas and continue their

---

Land Ethics

"All ethics so far evolved rest upon a single premise; that the individual is a member of a community of interdependent parts. His instincts prompt him to compete for his place in that community, but his ethics prompt him to also cooperate (perhaps in order that there may be a place to compete for).

The land ethic simply enlarges the boundaries of the community to include soils, waters, plants, and animals, or collectively, the land.

This sounds simple: do we not already sing our love for and obligation to the land of the free and the home of the brave? Yes, but just what and whom do we love? Certainly not the soil, which we are sending helter-skelter downriver. Certainly not the waters, which we assume have no function except to turn turbines, float barges, and carry off sewage. Certainly not the plants, of which we exterminate whole communities without batting an eye. Certainly not the animals, of which we have already extirpated many of the largest and most beautiful species. A land ethic of course, cannot prevent the alteration, management, and use of these "resources," but it does affirm their right to continued existence, and, at least in spots, their continued existence in a natural state.

In short, a land ethic changes the role of *Homo sapiens* from conqueror of the land-community to plain member and citizen of it. It implies respect for his fellow members, and also respect for the community as such."

Leopold, Aldo. 1949. *A Sand County Almanac.* Oxford University Press.

education past formal training, as well as integrating many disciplines into their plans[14]. Managing our wildlife resources involves everyone working together so that we can maintain this renewable resource for the future.

SUMMARY

Wildlife management involves the manipulation of populations or habitats to achieve established goals. The attitudes of people have always been the major influence in wildlife-management decisions. While many of our attitudes toward wildlife and approaches to wildlife management stem from European customs, influential Americans have helped shape management direction in the United States. Artists and writers such as Audubon and Thoreau created public awareness of wildlife during the early 19th century. In the early 1900s, President Theodore Roosevelt was important in the movement for conservation of our natural resources.

During the first half of the 20th century, the public attitude toward wildlife management was shaped largely by such figures as Gifford Pinchot, Aldo Leopold, and J. N. "Ding" Darling, as well as by President Franklin D. Roosevelt. Leopold, considered by many to be the father of wildlife management, was influential in starting university programs in the discipline and wrote the first textbook on the subject, *Game Management.*

Today, there is a great public awareness of the importance of wildlife and wildlife conservation. Not only are hunters and fishing enthusiasts interested in management, but many nonconsumptive users are making their wishes known. But governmental attitudes toward conservation of natural resources in general are ambiguous at best, and wildlife professionals must try to relate to both the public and political forces in developing wildlife policy.

DISCUSSION QUESTIONS

1. Discuss the role of Aldo Leopold in the evolution of wildlife management in the United States.
2. Distinguish between wildlife management and conservation.
3. Do you agree that sports clubs and conservation organizations are essential for managing wildlife resources? Why or why not?
4. Wildlife managers are often frustrated during the first few years on the job. Can you speculate why? Can you suggest methods to reduce the frustration?
5. What major forces shaped the evolution of wildlife management in the United States?
6. What factors contribute to the decision-making process in wildlife management today?
7. Describe different approaches to wildlife management.
8. Discuss the role of politics in wildlife management.
9. How have hunters and fishing enthusiasts been involved in formulating wildlife policy?
10. What people are responsible for setting the goals of wildlife management?

LITERATURE CITED

1. Caughley, G. 1977. *Analysis of Vertebrate Populations.* New York: Wiley.

2. Gabrielson, I. N. 1963. *Wildlife Conservation.* New York: Macmillan.

3. Leopold, A. 1933. *Game Management.* New York: Scribner's.

4. Bean, M. J. 1983. *The Evolution of National Wildlife Law.* New York: Praeger.

5. Kimball, T. L., and R. E. Johnson. 1978. The Richness of American Wildlife. In H. P. Brokaw (Ed.), *Wildlife in America* (pp. 3–17). Washington, DC: Council on Environmental Quality.

6. Morison, S. E. 1971. *The European Discovery of America: The Northern Voyages,* A.D. *500–1600.* New York: Oxford University Press.

7. Allen, D. L. 1954. *Our Wildlife Legacy.* New York: Funk & Wagnalls.

8. Dodge, R. I. 1882. *Our Wild Indians: Thirty-Three Years Personal Experience among Red Men of the Great West.* Hartford, CT: Worthington.

9. Trefethen, J. B. 1975. *An American Crusade for Wildlife.* New York: Winchester Press and the Boone and Crockett Club.

10. Tober, J. A. 1981. *Who Owns Wildlife?* Westport, CT: Greenwood Press.

11. Crosby, A. W. 1972. *The Columbian Exchange.* Westport, CT: Greenwood Press.

12. Flader, S. L. 1974. *Thinking like a Mountain.* Columbia, MO: University of Missouri Press.

13. Decker, D. J. and K. G. Purdy. 1988. Toward a Concept of Acceptance Capacity in Wildlife Management. *Wildlife Society Bulletin* 16:53–57.

14. Jacobson, S. 1995. New directions in education for natural resources. In R. L. Knight and S. F. Bates, *A New Century for Natural Resources Management* pp. 297–311. Washington, DC: Island Press.

Part 2
WILDLIFE POPULATIONS

Royal tern colony, North Carolina. (Courtesy, U.S. Fish and Wildlife Service.)

Population control is a key element in managing wildlife. Through population control, either by removal or sterilization of a population or by habitat enhancement, managers can accomplish their goals. There are many components involved in creating successful management programs.

The next four chapters present factors that influence populations. The evaluation of population growth, interactions among species, and population numbers help managers determine the optimal number of individuals for the good of the population and the habitat. This section should give the student a good understanding of how the principles of population management are used by wildlife managers.

26

2

CHARACTERISTICS OF WILDLIFE POPULATIONS

The study of wildlife populations is basic to an understanding of wildlife management. In this chapter, we examine some characteristics of populations: birth and death rates, density, age structure, and genetic makeup. We also give some attention to the balance between habitats and populations and to various functions of management in that relationship.

POPULATIONS, COMMUNITIES, AND ECOSYSTEMS

The word *population* has a number of meanings. Most population biologists define a **population** as *a group of organisms of a single species that interact and interbreed in a common place at a given time.* A species may have many members and occupy a large range; an example is the North American mule deer (Figure 2–1).

Wildlife managers often deal with herds (animals of the same species that travel or feed together) or smaller populations. From a manager's perspective, a population is the smallest subunit in which a group of animals of the same species is self-sufficient. This group of animals has the necessary seasonal ranges, access to habitats, and genetic integrity and thus is the management unit. The North American mule deer population, for example, has many individuals and occupies a large range. Managers, however, may manage herds that have summer and winter ranges occupying an area smaller than 250 km^2 (100 sq mi). In some cases a population, such as the Kirtland's warbler, may have very few individuals and occupy a small area for breeding. Other populations, such as

Figure 2–1 Mule deer range through most of the western United States, but overlap with white-tailed deer in some areas.

migratory waterfowl, may require large areas on a seasonal basis in order to meet all their requirements. Local populations are separated by physical and social barriers. A population has characteristics—for example, its birthrate, death rate, growth potential, density, age structure, dispersion, and genetic composition—that differ from characteristics of individuals within the population. Population characteristics are usually measured statistically, and populations can be identified by these characteristics or descriptors[1].

Normally, populations of different species live and interact in an area. A group of populations (plants and animals) that live within a particular area is called a **community.** Some biologists subdivide communities into plants, birds, big game, and other kinds of wildlife. When the populations in a community interact with their surrounding physical environment by food, energy flow, and mineral exchange, we call the unit interactions an *ecosystem*. The community forms the living or *biotic* part of the ecosystem; energy, minerals, nutrients, and water form the nonliving, or *abiotic,* component (Figure 2–2). Because ecosystems have arbitrary boundaries, an ocean, a forest, a watershed, and a basin may each be defined as an ecosystem. A pond is a good example of an ecosystem. These systems are not closed, but have both living and nonliving material interacting with surrounding systems.

The limits of some populations are difficult to define. Although snowshoe hare live and breed throughout a large portion of the northern United States and Canada (Figure 2–3), it is unlikely that those found in Maine will breed with those in Oregon. Although the hare's range must be viewed very broadly, local conditions influence the animals, and local populations can be delineated. In other cases, geographic distance separates populations. Burrowing owls found in Florida are separated from those found in the western United States, and the two are considered separate populations. When we try to reconcile the concept of population with various taxonomic structures, we

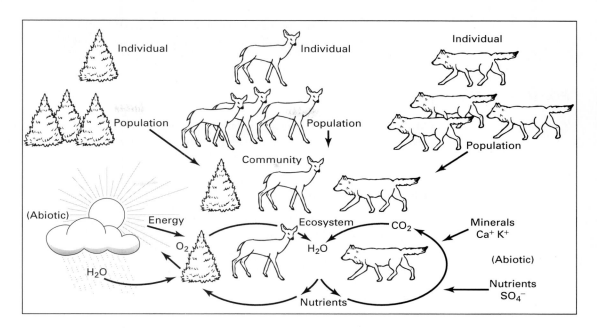

Figure 2–2 Components of an ecosystem.

also run into trouble. Taxonomists maintain that some separate populations are the same species or subspecies, while others are not. The general criterion for taxonomy is the ability to breed, but this is often difficult to determine.

Fish have been established in many lakes and streams around the country; biologists, however, often consider the fish species in a particular lake as one population. In the case of anadromous fish, such as the coho salmon, which migrates from streams out

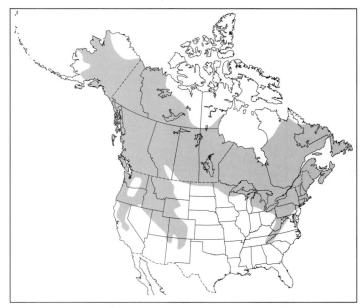

Figure 2–3 Range of snowshoe hare in North America.

to the ocean and returns, population ecologists view a population of salmon as one coming from the same stream and returning to that area (Figure 2–4).

POPULATION CHARACTERISTICS

Mechanisms for managing a population are frequently based on one or more of its characteristics. We will discuss these characteristics and some other considerations in this section and look at methods of measuring the characteristics in Chapter 4.

Birth or Natality Rate

The **birthrate** or **natality rate,** is the number of offspring a population produces in a unit of time, usually one year. The rate is often expressed as the average number of female births per female in the population. If the maximum natality rate were achieved, there would be no population control, and the population would increase dramatically. Charles Darwin once calculated that if elephants, with a gestation period of 600 to 630

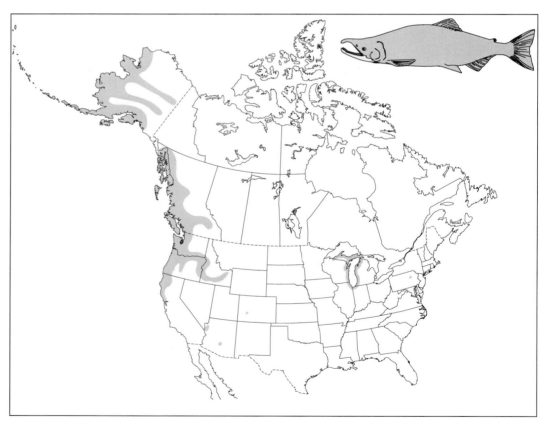

Figure 2–4 Coho salmon distribution in North America. (From D. S. Lee, *Atlas of North American Freshwater Fish* (Raleigh, NC: North Carolina State Museum of Natural History, 1980).)

days, reproduced at their maximum rate, a single pair would have 19 million descendants after 750 years.

The natality rate varies with species and even within species. For example, the average number of young raised by prairie falcons can be 2.5 in the northwest and 0.5 in the Rocky Mountains in the same year, due to differences in the availability of food. Fish commonly produce a great number of eggs, but few survive to become adults. Deer have one or two young per year; however, only one commonly survives. If more than two young survive in the lifetime of the doe—which may be five or more years—the population will increase, unless some other mortality factor, such as a harvest of the animal or disease, occurs. Generally, the longer an adult cares for its young, the fewer offspring it produces.

Reproduction can be measured as a rate, often as the **crude birthrate,** which is expressed in relation to the size of a population; for example, 100 births per 1,000 individuals per year is a crude birthrate.

Natality may be treated in several ways. Sometimes, management is based on natality rates among different age groups. For example, investigators wanted to know the age-specific natality between minimum and maximum breeding ages of Dall sheep in the Yukon. Sheep were captured and marked distinctively, and ages were determined from horn annuli, or rings. It was found that there were 58 live births in 1971 and 50 in 1972, the first and second years after marking. These data revealed that sheep began reproducing at four years of age[2] (Figure 2–5).

At other times, it is appropriate to look at natality for the whole population. A study in Yellowstone National Park was conducted to identify factors affecting the reproductive success of ospreys. Biologists found that from 5 to 41 occupied nests each year between 1972 and 1977. The number of birds produced ranged from 6 to 30 each year.

It is important to distinguish between *natality* and *growth rate*. *Natality* means that the number of births, usually per unit time, such as one year, and can be a positive value or zero. *Growth rate* means net increase or decrease and can be positive, zero, or negative.

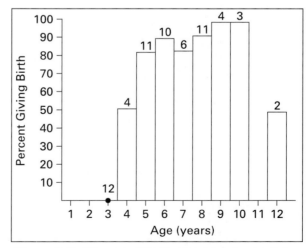

Figure 2–5 Percent of Dall's sheep giving birth in each age class. Numbers above histogram are sample size. (From [2].)

Actually, a variety of environmental factors affect a population's maximum birthrate, thereby reducing it to the *realized birthrate*. This term can be somewhat misleading: It can mean a reduction either in the number of young born or in the number of young that survive. With either definition, however, to the manager, it means fewer animals. For example, waterfowl produce a large number of young; but because of hunters, predators, and disease, many do not survive to reproduce.

Wildlife managers obviously need to know a population's natality rate if they want to increase or decrease its population. The manager can influence the realized natality by manipulating factors that affect it. An increase in the number of does taken by hunters will presumably keep the realized natality rate of the deer population low. Complete restriction of hunting against females will presumably increase natality. Biologists also examine the ability of the female to rear young. Endangered-species biologists find that, although whooping cranes often lay two eggs, it is common for only one to survive. Thus, they can remove one egg and rear the young in captivity without interfering with the effect on the natural population.

Death or Mortality Rate

The **mortality rate,** or **death rate,** of a population is defined as the number of individuals dying as a proportion of the total population during a time unit—again, ordinarily one year. Death rates are used by managers to determine the proportion of individuals in different age classes of a population. Managers can observe whether harsh weather or environmental changes alter the death rate.

Duckling Morality

A study of spectacled eider ducks in Alaska indicated that only 34 percent of the hatched ducklings survived to 30 days of age. During the first 10 days of life, 74 percent of the duck mortality occurred mostly because of adverse weather and predators. Half of the females lost their entire brood 30 days after hatching. Thus, in constructing a life table or looking at the survival of spectacled eider ducks, this crucial period must be considered. It could be a major controlling factor in the population. Harsh storms could destroy the entire cohort of young in some years.

Flint, P. L., and J. B. Grant. 1997. Survival of Spectacled Eider Adult Females and Ducklings during Brood Rearing. *Journal of Wildlife Management* 61:217–221.

As with natality, the mortality rate can be expressed for the whole population or for different age classes. Since different environmental conditions contribute to the death rate, data can be collected to show how factors such as climate, food, predators, hunters, and disease affect mortality. For example, whooping crane nests in Wood Buffalo National Park in northern Alberta, near the Northwest Territory, are normally isolated by water and marshes, which prevent wolves from disrupting nesting (Figure 2–6). But 1982 was a dry year, so the nesting area became accessible by land. In that year, 10 of the 16 whooping crane chicks apparently fell prey to wolves or other predators before they could fly.

Figure 2–6 Whooping cranes generally lay two eggs; however, only one normally hatches.

Life Tables

To use mortality rates, it is necessary to quantify data and develop meaningful comparisons. One method is to examine the life history of the population and develop a table showing how many animals die during each period of time. These tables are called **life tables.**

Life tables were developed by students of human populations for use by life insurance companies to determine death rates at different ages. The data in the tables indicate death rates by age and sex for human populations in different parts of the country and so project life expectancy. This information is used to determine rates charged by life insurance companies. Obviously, life tables can vary in usefulness. For a big game herd with a relatively closed population, the life table is relatively accurate. If migration between areas is common, the life table will be less accurate.

A life table generally begins with a standard number of individuals, commonly 1,000. It is assumed that these individuals are part of a stationary population. The table is divided into five columns for both males and females. (See Table 2–1.) In the first column, the age classes of the animals are listed (x). The next column (l_x) indicates the number of individuals out of the original 1,000 that are alive at the beginning of the age interval. The third column shows the number of individuals dying in that age interval out of the original 1,000 (d_x). Mortality per 1,000—that is, the *proportion* of individuals that die during that interval—is indicated in the fourth column (q_x). The last column (e_x) is the calculation of the life expectancy or mean lifetime remaining for individuals that have reached that age. The group of 1,000 individuals that start in each age group is referred to as a *cohort.*

There are two kinds of life tables. One is the **cohort,** or **dynamic,** life table, which shows the status of a group of animals, all born at the same time. This table is often difficult to construct, because it is hard for managers to obtain a large enough cohort to follow through a period of time. Immigration and emigration also affect this table. It is necessary to mark each animal and know the exact time of its death. For this reason, cohort life tables are most useful for laboratory animals, where researchers can determine the exact fate of each animal. Cohort life tables are also used in bird banding or other field studies involving tag returns.

The second type of life table is the **static,** or **time-specific,** life table, in which the mortality of each age group is recorded over a period of time, usually a year. This

TABLE 2-1

Life Table for Isle Royale Moose

Age (years)	Males l_x	Males d_x	Males q_x	Males e_x	Females l_x	Females d_x	Females q_x	Females e_x	Total l_x	Total d_x	Total q_x	Total e_x
1–2	1,000	108	0.108	7.02	1000	74	0.074	7.83	1000	88	0.088	7.29
2–3	892	62	0.070	6.81	926	42	0.045	7.41	912	51	0.056	6.94
3–4	830	46	0.055	6.28	884	56	0.063	6.73	861	55	0.064	6.32
4–5	784	46	0.059	5.62	828	37	0.045	6.15	806	41	0.051	5.71
5–6	738	50	0.068	4.93	791	32	0.040	5.41	765	49	0.064	5.00
6–7	688	46	0.067	4.26	759	97	0.128	4.62	716	70	0.098	4.31
7–8	642	77	0.120	3.53	662	69	0.104	4.22	646	85	0.132	3.72
8–9	565	116	0.205	2.94	593	69	0.116	3.66	561	94	0.168	3.21
9–10	449	85	0.189	2.56	524	88	0.168	3.07	467	86	0.184	2.75
10–11	364	100	0.275	2.05	436	97	0.222	2.59	381	102	0.268	2.26
11–12	264	112	0.424	1.64	339	93	0.274	2.19	279	98	0.351	1.90
12–13	152	58	0.382	1.47	246	106	0.431	1.83	181	73	0.403	1.66
13–14	94	54	0.574	1.07	140	69	0.493	1.83	108	56	0.519	1.44
14–15	40	27	0.675	0.85	71	28	0.394	2.11	52	28	0.538	1.44
15–16	13	12	–	0.54	43	14	–	2.16	24	13	0.542	1.54
16–17	–	–	–	–	29	9	–	1.97	11	4	–	1.64
17–18	–	–	–	–	20	9	–	1.60	7	4	–	1.29
18–19	–	–	–	–	11	–	–	1.45	3	–	–	1.33
19–20	–	–	–	–	11	9	–	0.55	3	4	–	0.25

table is constructed from a sample of animals of each age class, taken in proportion to the total number of individuals in that age class. It assumes that the birth and death rates are constant within the class and that there is no movement from the area. This is the kind of life table that insurance companies often develop. You can see such a table does not reflect changes that occur in the health pattern of the population which might reduce the birthrate rate or change the death rate.

In wildlife work, life tables may not reflect true field conditions[3]. For example populations that reproduce year-round would not be well reflected in tables that assume one period of births per year. Furthermore, the mathematics of modeling with life tables assumes that the deaths occurring each year do so right after the pulse of birth. Field biologists, therefore, generally work with only the female portion of the population. One measure of recruitment at the time of census is the number of daughters produced by females of age X, divided by the total females of age X[3]. This allows us to determine the number of females in each age class[4].

Life tables, although somewhat arbitrary, do indicate time frames in which the greatest number of deaths occur. For example, insurance companies note an increase in the number of deaths in the human population during the teens and early twenties. Since further analysis of the data indicates that this increase is the result of automobile accidents, auto insurance rates are higher for that age group.

It is important to recognize that life tables are subject to bias if not all age classes are sampled adequately. Still, a manager can utilize life tables to determine changes in mortality rates over a period of time. They also can be useful in determining the population growth rate, sustainable harvest levels, and population projections[3].

Survivorship Curves

To portray a life table, biologists often draw a **survivorship curve,** with data from the l_x column on the abscissa and age on the ordinate. By viewing these curves, biologists can determine at what age the greatest changes occur in the population. Survivorship curves take three basic forms (Figure 2–7). The curve of some populations dips slightly after birth and levels off for a time, before dropping dramatically toward the post-reproductive period. This is called a *type I* curve. The moose survivorship curve drawn from life table data conforms closely to this type (Figure 2–8). Many large vertebrate populations, as well as the human population, display the type I curve. If mortality is relatively constant in each age group of the population, the survivorship curve will be a diagonal line plotted on semilog paper (*type II*). Such a curve is characteristic of rodents and some bird and invertebrate populations. If mortality rates are extremely high during early life, as with fish, the survivorship curve drops dramatically during early stages and then levels off (*type III*). This curve is characteristic of most populations of animals that do not take care of their young. These hypothetical curves are modified, of course, when one looks at individual data found in the field, so that managers should use them as reference points for field populations.

Let us consider for a moment survivorship data for a specific population. The major cause of death among moose on Isle Royale National Park in Lake Superior is predation by wolves. Between 37 and 72 percent of the calves are killed by wolves. Converting the data for adult moose into a life table (Table 2–1), we see that moose surviving their first year of life have a mean life expectancy of 7.3 years. Between the ages of one and two years, annual mortality rates are about 10 percent or less, while mortality rates increase steadily thereafter. Most males have died by age 15.5 and females by age 19.5[5].

From the data in the table, a survivorship curve (Figure 2–9) shows the age selectivity of wolf predation. These data can be graphed in a different manner from the data previously described by plotting morality rate (q_x) against age (Figure 2–10). The resulting U-shaped curve has been reported for many species of mammals. It shows the death rate of the young, a stable population, and a die-off of older individuals. These configurations indicate long-term stability in the moose population. Average adult

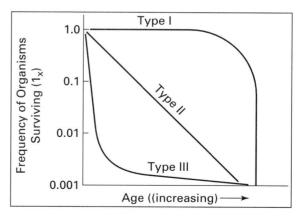

Figure 2–7 Three forms of survivorship curves.

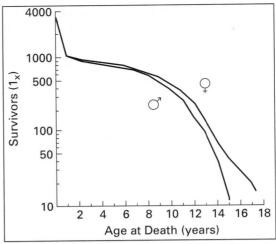

Figure 2–8 Survivorship curves for male and female adult moose on Isle Royale.

mortality is calculated at 13 percent, while yearling recruitment is 11 to 13 percent. Data are also used to examine differences in survival of the sexes. Males of a given age tend to have a slightly higher mortality rate than females of the same age (Table 2–1). Bull moose apparently have a higher incidence of malnutrition and arthritis.

Survivorship curves of males and females in the moose population show how the sexes differ in age at death (Figures 2–10.) The sex ratio remains essentially even from ages 1 to 8. After age 8, the proportion of males drops gradually for several years, with

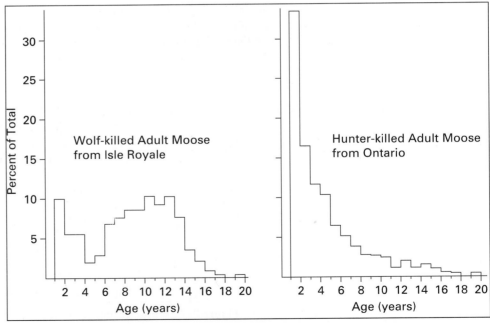

Figure 2–9 Comparison of age-specific mortality in moose due to hunters and wolves.

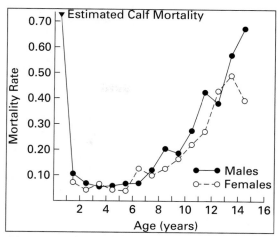

Figure 2–10 Mortality versus age of Isle Royale moose.

the rate of decline accelerating in the oldest age classes. The proportion of males does not fall below 45 percent until after 10 years of age. Since only about 12 percent of the population is calculated to be older than 18 years, differential mortality in males and females has little effect on the sex ratio of the entire population.

Higher mortality rates have been observed in males of some ungulate species, such as deer, moose, and elk. The reasons suggested for disproportionate male mortality include stress during the rut, when male activity is greatest, take by hunters, and, in some cases, greater susceptibility to disease. These factors may explain the greater incidence of malnutrition among bull moose on Isle Royale. Wolves appear capable of detecting slight incapacities and so many attack arthritic males, thereby shortening the average male life span[5]. By using life tables and survivorship curves with separate data for males and females, the manager can detect differential mortality between the sexes that may affect the harvest rate (Figure 2–11).

Productivity

Life tables can be utilized to calculate *population fertility.* For this, a new column, m_x, the *age-specific birthrate,* or the number of offspring per female produced at each age interval, is added to the table. This fertility schedules gives the average number of female offspring per female. When $m = 0$, no births occur, a common situation in the first year of life for many long-lived animals (Table 2–2). The fertility schedule can be plotted, producing a fertility curve (Figure 2–12), which indicates the age at which the fertility rate increases and when it levels off. We can multiply columns l_x and m_x and obtain the sum of the values for the different age classes and the net reproductive rate:

$$R_o = \frac{\Sigma \, l_x m_x}{1,000}$$

In most cases, l_x refers to females only. R_o is the average number of female offspring produced during a lifetime and is a useful figure for computing population growth rates. In the case of species with nonoverlapping generations, R_o is the exact amount by which the population increases in each generation. The life table can be used to

Figure 2–11 Bull moose generally have a shorter life span than females do. (Courtesy of the U.S. Fish and Wildlife Service; photo by J. M. Greany.)

TABLE 2–2

Fertility Table

Age Group	l_x	m_x	$l_x m_x$
0	1,000	0	0
1	800	0.3	240
2	700	0.4	280
3	600	0.3	180
4	400	0.1	40
5	100	0	0

$$R_o = 740/1{,}000 = 0.74$$

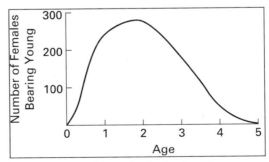

Figure 2–12 Hypothetical fertility curve based on data from Table 2–2.

calculate R_o. For example, if we determine that during the first year ($x = 0$) there are no births ($m = 0$) and assume that all females survive ($l_o = 1.0$), then $l_o m_o = 1.0 \times 0 = 0$. At the end of the first year, 50 percent of the females survive ($l_1 = 0.5$). Each gives birth, on an average, to two female offspring ($m_1 = 2$); hence, $l_1 m_1 = 0.5 \times 2 = 1.0$. At the end of the second year, 20 percent of the original females survive ($l_2 = 0.2$), and each has given birth by that time, on average, to four female offspring ($m_2 = 4$). Hence, $l_2 m_2 = 0.2 \times 4 = 0.8$. No females live into the third year ($l_3 = 0, l_3 m_3 = 0$). The net reproductive rate is the sum of all the $l_x m_x$ values just obtained, or 1.8[6].

Assuming that l_x relates to females only, an R_o of 1 indicates a stable population, one that is just replacing itself. Values of R_o above 1 indicate an increasing population, while R_o values below 1 indicate a declining population. In Table 2–2, the R_o of 0.74 would indicate a declining population. In other words, 740 animals replaced 1,000.

Density

The density of a population is the number of individuals per unit area—for example, the number of deer per hectare or fish in a pond. Most populations are not evenly spread throughout an area, so that wildlife managers need to develop techniques to measure these mobile animals. Data on density are used with information on the availability of food to determine whether a population can be supported by an area. Density estimators are discussed in Chapter 4.

Age Structure

The age structure of a population—the number of individuals in each age category—is very important in determining management potential. A population consisting primarily of old individuals obviously will soon die off. In contrast, a population that consists primarily of young individuals is likely to increase rapidly in size, and managers need to institute methods to contain their numbers. Knowledge of age structure can also be used to help solve some wildlife problems. For example, in areas where sheep are being taken by mountain lions, reproduction may be high in the lion population, and the young may be dispersing from the area where the parents live. As the young disperse, they find sheep a good food.

From life-table data on males and females, *age–sex-distribution pyramids* can be drawn. Managers divide the population structure into prereproductive, reproductive, and postreproductive sections. Age–sex-distribution pyramids display three basic patterns, shown in Figure 2–13. If the mortality rate equals the birthrate and is distributed through all age classes of the population, the age distribution will remain the same and the population will be stationary. This is the case with the moose. When more individuals are surviving than dying for each age class, there is a young population. Populations that are growing display this pattern. If fewer individuals are found at the base of the pyramid, indicating a low birthrate, there is a declining population. Thus, life tables portray the type of population that exists and so help the manager determine where management efforts are needed. It must be emphasized that age-distribution pyramids are only indicators of trends; the effective manager will use the table in deciding whether a more intensive data-collection effort is needed.

Genetic Composition

Each living organism that engages in sexual reproduction possesses unique genetic material, half of which combines with that of its mate and is passed on to their offspring. Within a population, the total genetic material constitutes the *genetic composition* of

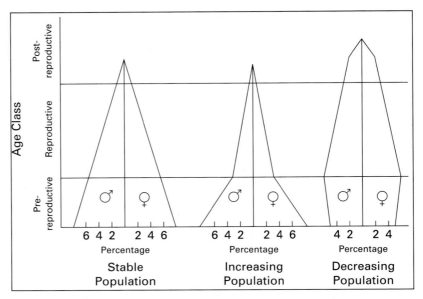

Figure 2–13 Age–sex-distribution pyramid for populations.

that population. Of course, individuals in the population vary in specific genetic makeup. Thus, members will have different coloration, and some will be swifter runners than others, more adept at digesting food, or better camouflaged, and so on.

Members of a population with a certain combination of genetic material may be better able than others to survive environmental change or to move to a new area. Similarly, changes in the genetic material, called mutations, may make it easier for some members to survive changing conditions or produce more young than other members. If all the young survive, those that produce more young will be contributing more genetic material to the population.

A study of the genetics and population characteristics of the yellow-cheeked pocket gopher gives some insight into the impact of isolated populations and population genetics[7]. The population studied was in an area near Lubbock, Texas. The gophers were relatively isolated, and the females had a number of small litters throughout a relatively long breeding period. At least 25 percent of the females survived through two breeding seasons. This was in contrast to female gophers in many other parts of the country, which had litters only during one breeding season. The study showed that female gophers lived an average of 56 weeks, with 25 percent still alive 86 weeks after entering the trappable population. Males, on the other hand, had a life expectancy of only 31 weeks.

The initial thinking when the population was studied was that the relatively small, isolated population would suffer an increased probability of extinction because of inbreeding. However, it was found that the low reproductive output through a long breeding season produced a population with a complex age structure that apparently buffered random environmental catastrophes, such as flooding, fire, and droughts. The animals that bred several times during the field season provided more males to interact and breed with females, thus increasing the genetic diversity of the population. It was felt that a catastrophe which reduced or inhibited the reproduction of animals with a highly synchronized breeding season could disrupt the age structure enough to cause it to become extinct. The encouragement of more random male–female matings may produce many offspring that are the products of a variety of males. This would result in a continued diversity in the population, rather than a genetic composition produced by one or a few males.

Much research has been done in the genetics of population interaction, particularly in endangered species. All organisms need to be able to find mates; when the population decreases to such an extent that mates cannot be found—as had occurred in whales—the species is obviously in serious jeopardy. Larger populations have larger gene pools and so can tolerate more change. Constant inbreeding so reduces the viability of the population, that it is often unable to withstand environmental change.

Mating of close relatives in natural populations increases the likelihood of offspring with genetic defects. As a result, populations with relatively low numbers may be affected by *genetic abnormalities.* Thus, there is a minimum number (often difficult to define) for populations below which extinction may occur due to genetic effects. This minimum number is called the *minimum viable population.* Both the genetic makeup of a population and the environment influence the minimum viable population.

Biologists have been unable to agree on an actual number that constitutes a minimum viable population for each endangered species. They do, however, feel that this is an important concept and must be considered when populations reach very low numbers. Although few genetic studies have been made of bird and big-game populations, it can be assumed that these species develop differing genetic characteristics when they isolate themselves into breeding groups, herds, or flocks (Figure 2–14).

Minimal Viable Population

As animal populations decline in numbers, we often hear of genetic viability. This term means that a population of animals has the diversity of genetic material needed to respond to adverse conditions such as weather or disease. If some animals die, there will be enough left with genetic material to enable the population to reproduce and survive. As bison were removed from Yellowstone Park, people asked what the minimum population size was to retain genetic viability in the herd.

There has been no clear answer, although two biologists, O. H. Frankel and Michael Soule, stated that they believed 500 animals to be the minimum viable population. They indicated that many factors contributed to their estimate, so this number could vary in different populations. Other biologists have held that 500 breeding females was a more accurate estimate.

In the case of Yellowstone bison, the herd started from a small population numbering in the thirties. Thus, the minimum viable population might not be a reliable concept.

Frankel, O. H., and M. E. Soule. 1981. *Conservation and Evolution* (Cambridge, U.K.: Cambridge University Press, p. 321).

Figure 2–14 Cutthroat trout in a lake can be genetically isolated.

Metapopulations

In some cases, populations of species can be separated from one another, usually geographically. As a result, there are a number of isolated local populations, each of which may have its own extinction and recolonization rates. The occasional dispersion from one population to another links the species together and allows the flow of genetic material if breeding occurs. Local populations can be examined together as a single large population, or *metapopulation*. If movement among the locations is very infrequent or migration routes are blocked, local extinction can occur. If many of the local population are destroyed, the whole population or species can be lost.

Traditionally, wildlife managers have looked at large, visible populations of big game and waterfowl. Many mammal, bird, reptile, amphibian, and invertebrate populations could function with a series of local populations making up one large metapopulation. For example, prairie rattlesnakes occupy dens in rocks. There are usually clusters of dens around rock outcrops. Most snakes return to the same den each fall; however, some snakes do migrate to other clusters and den there. These clusters of dens are often isolated from other clusters. Each cluster could act as a local population for the whole population or metapopulation. If snakes are collected, so that a number of the clusters have no snakes, then the whole population could be harmed. When wildlife biologists issue permits for taking animals, they must consider the concept of metapopulations and how the take would affect the whole population.

WHERE ANIMALS ARE FOUND

Habitat

Populations are found in areas where all their needs can be met. A population's basic needs are for shelter, reproduction, food and water, and movement. An area that fills all these requirements is called a **habitat.** Even though all the requisite needs of a population are found in a particular habitat, a population may not exist in it. If a species is to be introduced or reintroduced into an area, its requirements should be known and the habitat evaluated before action is taken. Thus, in introducing white-tailed deer into a new area, one would study their current habitat, usually open fields near brushy edges with sparse trees, and try to match that habitat as much as possible. If the new area were heavily forested, it would be appropriate to create small openings or edges.

Niche

A population's **niche** is its role in the habitat[8]. What does it do? Does it eat plants or animals? What time does it feed? What minerals does it use and cycle? Some species occupy broad ecological niches by feeding on many kinds of food. Others have very specialized niches and feed on only one or a few food items.

A wide variety of animals can occupy similar niches. To identify the niche of a population, its behavior and physical factors, as well as food, must be considered. Thus, large ungulates and insects may occupy similar niches, since both use grass or herbs as their diet. This concept is very important when one looks at land utilization. Often, domestic cattle, small rodents, insects, and ungulates may occupy similar niches.

An understanding of a population's habitat and niche are very important when managing wildlife species. Any time a change in the number of species occurs, both the habitat and niche are affected. Most important is the reintroduction of new species; when the bison were removed from an area, its niche was not left unoccupied. The role that the bison played in the community was assumed in part by one or more different organisms, including insects, other grazing mammals, and birds. The reintroduction of bison into the area would again shift the population structure.

EVOLUTION AND ADAPTATIONS

Every population has evolved biological characteristics with physical and chemical makeup through a selection process known as **evolution.** Most characteristics have developed because they made the population better able to survive in a particular environment. The explanation is fairly simple, but the process is very complex and usually slow. Individuals with characteristics that make them better fit to survive produce more young that survive and contribute more genetic material to future generations of the population. As water, sunlight, or other environmental factors change, the advantage may shift to other members of a population. Subtle **adaptations** (changes that make a population better suited to its environment) can develop in relatively small groups of animals, such as a herd of deer, a population of cutthroat trout in one lake, or the grizzly bears of Yellowstone Park. Human-induced habitat changes, such as altering the temperature of a lake a few degrees or removing a physical barrier between two herds, can cause a major change in populations and even destroy a population that cannot adapt rapidly enough.

Natural Balance

One of the difficult phenomena to measure is the *natural balance,* or interrelationships that develop among different populations in a community and in an ecosystem. People who study natural balance point out that living systems, including groups of organisms and individuals living together in the same environment, have regulating mechanisms, that work through a feedback system. **Feedback** is the return of output or part of the output to a system's input. Information from the interaction of parts of a system is returned, which causes change in a particular state. The mechanism is like that of a thermostat. When you set the thermostat at a desired temperature, this information is fed to a temperature-sensing device. If the temperature of the room in which the thermostat is located is lower than the temperature you indicate, the furnace starts up; if it is higher, the furnace shuts off. The temperature of the room is continually monitored as signals are sent to the furnace. When the temperature falls and more heat is supplied, the negative or reverse relationship between the input of information and the response, **negative feedback,** occurs. When a change in the system in one direction is converted to a command to continue the system in the same direction, **positive feedback** occurs. Generally speaking, negative feedback keeps the system in equilibrium and positive feedback disrupts the equilibrium, causing the system to become unstable.

Negative and positive feedback tend to maintain a population in **homeostasis** (equilibrium) with its environment. A population can then grow until it reaches the limits of the support area. Beyond that point, habitat factors act as a negative feedback mechanism, reducing the number of individuals that can survive or that are born. When the decreasing population reaches a level the habitat can support, the inherent drive for reproduction, which is a positive-feedback mechanism, again takes over, and the population increases until a negative force is triggered once more. This mechanism tends to occur in a cyclic fashion. The cycles, however, are not always uniform. Some ecologists feel that the negative trigger in the cycle occurs just below the limits of the habitats, so that populations tend to be maintained in a homeostatic or equilibrium state within their habitats. When disturbances develop within a community, changes occur in the population. Any change in habitat, such as the introduction of fire, logging, or water pollution, tends to operate as a disruptive mechanism to the natural equilibrium. Populations that expand beyond the limits of their environment become their own destructive mechanisms. However, that does not mean that this is bad for the community. New balances are established.

Most management activities alter natural homeostatic mechanisms, thus causing changes in the natural cycles of populations. In some cases, the disruption of the natural cycle can mean that food is not available for predators which normally prey on the population. The predatory animals may then need alternative sources of food. For eagles, coyotes, and mountain lions, one new source can be domestic animals. When setting hunting quotas, the manager is substituting hunting for natural mortality factors. A hunting program which removes animals that would die from disease and starvation is an artificial destructive mechanism that acts to maintain the natural balance of the system.

Disruption of the Natural Balance of Yellowstone Lake

The introduction of a different species often causes a disturbance in natural systems. Such introductions can harm native species and disturb other components of the animal community. One example is the introduction of lake trout into Yellowstone Lake in Yellowstone National Park. On July 30, 1994, lake trout were discovered in Yellowstone Lake, the core of the remaining undisturbed natural habitat for native cutthroat trout. Data suggest that lake trout, which eat the cutthroat trout, may have been in the lake since 1989, when they were probably illegally introduced. Biologists estimate that lake trout could reduce the number of cutthroat trout by more than 90 percent in the next 20 to 100 years. Since cutthroat trout spawn in the streams around the lake, they are a popular sport fish in both the lake and its tributaries. In addition, they are an important food item for grizzly bear, ospreys, white pelicans, river otter, and other animals. Unfortunately, biologists see little chance of removing the lake trout from Yellowstone Lake.

Kaeding, L. R., G. D. Boltz, and D. G. Carty. 1996. Lake trout discovered in Yellowstone Lake threaten native cutthroat trout. *Fisheries* 21:16–20.

Management

To manage populations effectively, the wildlife manager must have a good understanding of population dynamics and population–habitat interactions. The manager can remove animals through harvesting them (a form of people management), which affects either the birthrate or death rate, or can increase or decrease the area available for a suitable habitat (Chapter 1). Features of the habitat necessary for survival, such as nesting areas or spawning grounds, can be increased. Physical and chemical changes can be made to encourage a population to live in a specific habitat or discourage it from living there. Other plant or animal populations can be increased or decreased to change the biotic environment. Any changes that occur are likely to affect all populations.

All population management must be based on the available habitats. Any habitat can support just so many animals until the available resources that satisfy their needs are used up. Removal of animals from a population before it saturates the habitat is one management device. By harvesting animals on a planned, continuous basis, managers can obtain a sustained yield. More animals can be removed if the manager wants to decrease the population.

Several studies show the impact of harvesting on species. For example, in Newfoundland, if 20 to 25 percent of the beaver population were harvested, the population would remain stable. In the absence of harvesting, some compensatory natural mortality occurs[9]. In this case, the habitat could support fewer animals than are produced. In planning removal programs, however, it is important to recognize that other populations may be affected by the program. Thus, predators or prey might increase, or new animals might move into an area.

Of course, managers may not want to remove or decrease a population, but rather, may desire to increase the variety of animals. Habitat manipulation appears to be the key to such a goal. Two of the approaches to wildlife management involve **featured species** (also known as indicator, key, and single species) and **species richness**[8].

In the featured-species approach, the goal is to produce selected species at desired numbers and specific locations. This can be done by adjusting vegetation and increasing food, cover, water, and nesting areas for the featured species. Game, endangered species, or species with aesthetic value are frequently selected. The idea is that these featured species will indicate a healthy environment for other desired species. The overall species richness of an area would be maintained.

There are several guidelines in managing indicator species. First, one must be careful to pick a species that is indeed an indicator of the general habitat. It would be inappropriate to pick all big game, raptors, or all small mammals. The roles of species in community competition or energy flow through predation must be considered. The status of the population in terms of genetic isolation from other populations should be evaluated. Second, it is important to pick species that are indicative of the particular habitat, including the amount of moisture, vegetation, and other factors. The great blue heron is common in saltwater marshes of southern Florida. It could be used as an indicator species there.

Decisions as to how many species to list as indicator species for a particular area must be based on several considerations. The risk of not maintaining all species will vary inversely with the proportion and number of species addressed. If the species selected have broad requirements, this approach may not be practical for indicating the welfare of species with specific requirements. Species with very specialized needs may be very expensive to manage.

In the process of selecting an indicator species, the effect of the habitat on that species needs to be considered. For example, features that can maintain elk and deer in the Blue Mountains of Oregon are known[10]. Forage, water, and cover are the primary habitat-limiting factors. Elk presumably will be indicators of these features. The habitat can be manipulated and human activities in the forest controlled in such a way as to reduce disturbance and maintain the features. Specific methods of controlling logging and burning and optimizing the winter and summer range can be used to assist the elk[10]. When these factors are altered, the number of elk changes.

The concept of indicator species must be used cautiously. No one species can be an indicator of the whole system. Some management agencies use this concept in an attempt to simplify a complex system. This only misleads us into a false approach to management[11].

The species richness or diversity approach consists of providing as varied a habitat as possible to support as many populations as possible. This approach to management has been tried in a number of areas with success. In the Blue Mountains of Oregon, species richness was attributed strictly to stand size and could be approached with an average habitat block of 24 hectares (58 acres)[10]. This involves the size of the habitat, which we talked about earlier. Species richness and indicator species are discussed in relation to nongame wildlife in Chapter 20.

SUMMARY

A population is a group of organisms that live within a particular geographic area and usually interact behaviorally and interbreed. Populations have characteristics, including birth and death rates, age structure, and genetic composition, that individual members do not have. Life tables and survivorship curves can be drawn to reveal at what age mortality affects the population. Age-distribution pyramids provide a visual display of the population's growth characteristics and probable future productivity.

Animals are found in habitats where all their needs are met. Within its habitat, each population has a unique functional role or niche. The population evolves in its niche through the process of selection. Within a group of populations living in the same area (community) and interacting with the physical environment (the ecosystem), a series of regulatory mechanisms (feedback) evolves to allow each population to achieve a balance with its environment (homeostasis).

Management of a population can take the form of removal of some of its members to decrease the size of the population or maintenance of the habitat and proper selection of individuals to produce a sustained yield. Habitat management can be implemented by indicator species or diversity of species.

DISCUSSION QUESTIONS

1. Define *population*.
2. Does a natural balance really exist? Give an example.
3. When would a life table be used to manage a population? Give examples and tell what data would be derived from the tables.
4. What information is provided by a survivorship curve, and how is this information used?
5. How can a manager calculate natality and mortality from population size and age data?
6. Why would it be desirable to have population productivity data before setting hunting or fishing seasons?

7. What is the relevance of the genetic makeup of populations to management decisions?
8. Why can a habitat exist without a population?
9. How do you see feedback controlling natality and mortality?
10. What approaches can be taken in managing a population? A community?

LITERATURE CITED

1. Solomon, M. E. 1976. *Population Dynamics.* London: Edward Arnold.
2. Bunnell, F. L., and N. A. Olsen. 1981. Age Specific Natality in Dall's Sheep. *Journal of Mammalogy* 62:379–80.
3. Menkens, G. E., and S. H. Anderson. 1989. Temporal-Spatial Variation in White-Tailed Prairie Dog Demography and Life Histories in Wyoming. *Canadian Journal of Zoology* 67:343–49.
4. Taylor, M., and J. S. Carley. 1988. Life Table Analysis of Age Structure Populations in Seasonal Environment. *Journal of Wildlife Management* 52:366–73.
5. Peterson, R. O. 1977. *Wolf Ecology and Prey Relationships on Isle Royale.* Scientific Monograph Series 11. Washington, DC: National Park Service.
6. Wilson, E. O., and W. H. Bossert. 1971. *A Primer of Population Biology.* Sunderland, MA: Sinauer Associates.
7. Smolen, M. J., H. H. Genoways, and R. J. Baker. 1980. Demographic and Reproductive Parameters of the Yellow-Cheeked Pocket Gopher. *Journal of Mammalogy* 61:224–36.
8. Odum, E. P. 1983. *Basic Ecology.* Philadelphia: W. B. Saunders.
9. Payne, N. F. 1984. Mortality Rates of Beaver in Newfoundland. *Journal of Wildlife Management* 48:117–26.
10. Thomas, J. W. (Ed.). 1979. *Wildlife Habitats on Managed Forests of the Blue Mountains of Oregon and Washington.* Agricultural Handbook 553. Washington, DC: USDA Forest Service.
11. Graul, W. D., and G. C. Miller, 1984. Strengthening Ecosystem Management Approaches. *Wildlife Society Bulletin* 12:282–88.

3

POPULATION GROWTH AND INTERACTIONS

Most populations exhibit forms of growth that result from the populations' genetic makeup and interactions with their habitat. Within the habitat, populations interact in a variety of ways. The presence of some populations limits the growth of others, but sometimes the presence of one population accelerates the growth of another. Some interactions cause the members of one population to move; other interactions alter the habitat.

In this chapter we discuss population growth and interaction in the community, as well as factors that cause interactions and their impact on management. As we have noted, any change in a homeostatic balance can have an impact on the system. We examine both competition and predation in relation to animal numbers and energy flow and the impact of wildlife movement patterns on population dynamics. Nowadays, attempts to manage communities of populations in a holistic fashion are being tried. We discuss some of these approaches, all of which require knowledge of interactions within and between populations, communities, and ecosystems.

POPULATION GROWTH

Forms of Growth

In a population that is increasing, we can discover the patterns of growth by plotting the numbers of individuals at various instants of time and then connecting the numbers in linear fashion. Most populations exhibit a variation of one of two basic growth patterns: **exponential** or **sigmoid** (logistic) (Figure 3–1).

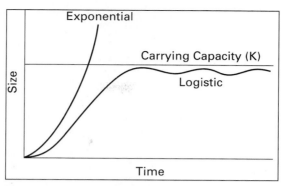

Figure 3–1 Exponential and logistic growth curves. Notice how the logistic curve approaches the carrying capacity as environmental resistance acts.

In a population experiencing exponential growth, the growth curve starts out flat and becomes increasingly steep until the individuals saturate the habitat. At that point the population decreases very rapidly. Any population that continues to grow at an exponential rate eventually reaches a level where the environment can no longer support it, and the number of individuals in the population then declines rapidly. This type of growth occurs with lemmings in northern Europe: When natural controls are removed, the population grows until environmental destruction occurs. Commonly, exponential growth is exhibited when organisms colonize new areas or predators are removed. Populations that regularly show exponential growth are usually found in unstable environments.

Sigmoid, or logistic, growth curves also start out flat, but ascend sharply almost immediately. They then flatten as the population reaches the number of individuals the area can support. Populations that exhibit logistic or sigmoid growth generally grow to a fairly constant size of approximately the number that can be supported by the habitat. Such populations are often found in areas where environmental disturbances are few.

But populations cannot grow indefinitely. When a population reaches the number of individuals the area can support, either it produces fewer young or fewer young survive. This number, called the **carrying capacity** (Figure 3–2), is the equilibrium between animal and habitat[1]. In populations exhibiting a logistic growth pattern, the mean number fluctuates around or just below the carrying capacity. In populations exhibiting exponential growth, the population size often extends well beyond the carrying capacity for a short time and then declines well below the carrying capacity for a short time, only to increase again later. The carrying capacity of a habitat for one population can be difficult to ascertain, since all other populations using the habitat alter the area. Although we can project the carrying capacity of an environment theoretically, we must remember that it can and does change, as when a change in the amount of moisture available in grasslands during a particular season influences the number of herbivores that can graze there.

Carrying capacity changes with many factors, such as climate, the presence or absence of other animals, and disturbance. We cannot always clearly define a number, yet we can use this concept to help us plan. At the same time, we must not become so attached to the concept as to allow it to be the only driving force in management. Many populations are managed for numbers negotiated by different interest groups, such as landowners, hunters, and conservation groups. Accordingly, the carrying capacity is a number based on socioeconomic, not biological, considerations (Figure 3–2).

Figure 3–2 The presence of cattle can influence the growth of ungulate populations. (Courtesy M. McKinstry.)

Growth Equations

Equations to predict the size of populations exhibiting exponential or logistic growth will become increasingly helpful as computer facilities are used to a greater extent by the wildlife profession. Here we discuss the basic growth equations; some of their uses in models are discussed in Chapter 5.

Exponential Growth.   To simplify our discussion, we will talk about females in a population. (We assume that the males are produced at the same rate and in the same numbers.) To calculate the rate of increase of the population, biologists must determine the reproductive rate of the population. If each female produces two female offspring that survive, the population will double each generation: If we start with 20 individuals, we will have 40 in the next generation, and so on.

Suppose that we let R_0 stand for the net replacement rate per generation. ($R_0 = 2$ when each female produces an average of two female offspring in her lifetime.) If we let N stand for the size of the population (the number of individuals in the breeding generation), N_0 for the initial number of individuals, and t for the number of generations from the time we begin our calculations to the time we end, we can write the formula

$$N_t = R_0^t N_0 \tag{1}$$

If we now use the example in which $R_0 = 2$ and start with 500 individuals, we calculate that 4,000 individuals will be alive after three generations:

$$N = 2^3 \times 500 = 4,000$$

This equation assumes that animals breed only in one season and do not have overlapping generations. We can extend the equation to calculate the rate of growth of a population when breeding goes on all the time. Then a change in the population (as long as there is no emigration or immigration) is accounted for by birth and death rates. In other words, the change in the size of the population during a time unit is the difference between natality and mortality, that is,

$$\frac{\text{change in numbers}}{\text{change in time}} = \text{constant} \times \text{number in population}$$

or

$$\frac{dN}{dT} = rN$$

This equation describes the growth rate, or change in the number of individuals (increase or decrease), of a population that are breeding all the time with constant birth and death rates. It can be rewritten

$$\frac{dN}{dt} = (b_0 - d_0)N$$

where N = number of individuals in the population at a given moment

t = time unit

r = constant, called the *intrinsic rate of increase,* which is equal to the birthrate minus the death rate

b_0 = birthrate, or the number of offspring one individual will have, on average, per unit of time

d_0 = death rate—that is, the number of individuals dying per unit of time

This *basic exponential growth equation* indicates the rate at which a population is increasing or decreasing. It states that the rate at which a population grows per unit of time is a constant multiplied by the number of individuals present. The equation is an idealistic population growth descriptor. The constant r, the intrinsic rate of increase, can vary in the same species that has members in different areas. For example, a mule deer herd in one valley may have an intrinsic rate of increase different from that of a herd in the next valley because of differences in food availability.

The r value of a population can also change as a function of the age when the population starts to reproduce. In an elk population in south-central Washington, r was 0.20 for 1975–1986; however, between 1982 and 1986, the r value was 0.30. While yearlings do not always reproduce, a value of 0.30 in the population would indicate that they did[2]. Undoubtedly, the forage conditions were of such good quality, that the yearling elk could reproduce.

If there are very few or no restraints on population growth, the intrinsic rate of increase can become larger—that is, there may be a greater increase in births than in deaths. The maximum value of r for a population is referred to as r_{max}, the *maximum intrinsic rate of increase.* The realized intrinsic rate of increase is usually lower than r_{max}. The ability of a population to exhibit exponential growth at its r_{max} value is called its *biotic potential.* Environmental constraints reduce the biotic potential of a population by causing deaths and preventing births.

Wildlife managers rewrite Equation (1) to obtain a generalized growth equation, $N = N_0 e^{rt}$, in which they define N_0 as the number of organisms in a population at the moment when observations begin, t as the amount of time elapsed after observations begin, and e as a constant, 2.7183. This equation allows managers to start with N_0 organisms and (assuming exponential growth) determine how many there were in the past or will be after a specific time (weeks, years, generations).

Logistic Growth. We can now consider what happens to a population as it begins to reach the carrying capacity of its habitat. Our original growth equation, $dN/dt = rN$, which describes a population experiencing exponential growth, is the starting point. No population can continue to grow without restraints. Environmental factors (**environmental resistance**) that slow the population growth combine to help determine the carrying capacity. We designate the carrying capacity as K. In most cases, environmental resistance or factors in a population itself prevent it from exceeding its carrying capacity.

Factors in the environment that become limiting decrease the birthrate, increase the mortality rate, or both. Since the exponential equation represents the growth of a population in an environment with little or no environmental resistance, we must add factors to indicate how the environment changes the growth rate of the population. The exponential growth equation can be multiplied by the term $(K - N)/K$ to show the impact of environmental resistance. The formula

$$\frac{dN}{dt} = rN\left(\frac{K - N}{K}\right)$$

becomes the logistic or sigmoid growth equation.

When N is close to zero or the population is just starting to grow, dN/dt comes very close to equaling rN: The growth is nearly exponential. As the population increases, the factor $(K - N)/K$ slows the growth rate. If N exceeds K (the population exceeds the capacity of the environment), the factor $(K - N)/K$ becomes negative, and the population returns to within the carrying capacity from a value higher than K. In fact, any impact on population size from K affects the rate of growth so as to return the population to its equilibrium size. K is what mathematicians call a *stable,* or *persistent, equilibrium.*

The carrying capacity K is useful in setting quotas. But as we indicated earlier, it is difficult to assess its value, so indirect methods of determining carrying capacity are used. Carrying capacity is not an absolute value, but an approximation, and managers may have trouble measuring K in field situations. Thus, if observations indicate that X antelope are dying after each successive day in which snow covers the ground, the number can assist in determining K.

The logistic equation is a simplified model of population growth that assists us in developing a number of other models useful in determining the size of a population. Discussions of derivations and implications of population growth regulations are available in a number of population biology texts[3,4].

r and K Strategy

The values of r and K for a population in a particular habitat are the result of the genetic makeup of that population. These parameters are formed during the evolution of the population. They can be viewed as the population's adaptability to a particular environment. Populations that live in relatively unpredictable or unstable environments need to be capable of reproducing rapidly and using resources before their competitors do. They must also be able to disperse and occupy new habitats quickly. These species, which rely on high r values and are therefore referred to as "r strategists," are able to make use of resources quickly. The r strategy allows them to make full use of the habitat, which, because of its shifting nature, keeps many other populations on a lower part of the logistic growth curve. Under such circumstances, the populations that can reproduce rapidly—have a high r value or r selection—are favored. Thus, in a disturbed habitat, *r* strategists are the first to appear. Populations with high *r* values lose this advantage in a stable habitat, since they lack the ability to compete in crowded circumstances.

The *K* strategists, or *K*-selected populations, have the advantage of being able to interact at the carrying-capacity density level. Here, being able to reproduce rapidly is not as important as being able to utilize the habitat and extract food from it while competing with other animals. These organisms may be able to defend an area against others, keep them out, or prevent them from obtaining some required resource.

K strategists and *r* strategists are not mutually exclusive, but can be viewed as opposite poles of a continuum. Most animals exhibit some combination of intermediate characteristics: They can still reproduce if the environment is disturbed and also compete to some degree with other animals in a relatively stable environment. The extent to which a particular population leans toward the *K* or *r* pole will help a manager make certain decisions. Hunting will not have as large an impact on a population exhibiting an *r* strategy as it will on one having a *K* strategy. Reintroduction of populations into an area is more likely to succeed with a population exhibiting an *r* strategy and will require less work on the part of the manager. Population-reduction measures involving a population exhibiting an *r* strategy may mean reducing the number of habitats in which the population can grow rapidly. On the other hand, population-reduction measures for populations with a *K* strategy may include removing individuals from the population (Figure 3–3).

The use of an *r* or *K* strategy in field situations is very complex, because all the species in a community exhibit variation. A species' strategy may be influenced by the environment, disturbance of its habitat, or interaction with other animals. Thus, managers must be aware of the growth potential of the species and how it interacts with others when proposing wildlife management activities.

Figure 3–3 Many raptors are near the top of the food chain and exhibit a *K* strategy. The ferruginous hawk is an example.

Competition

Competition may be defined as *the active demand by two or more organisms for a limited resource.* When the resource is not found in sufficient amounts to meet the requirements of all organisms in a population, it becomes a limiting factor on the growth of the population[3]. When two or more populations require the same limited resource, competition results. One of two things can happen to populations that are competing. First, they can develop an equilibrium, in which both survive, usually by some subdivision of the habitat. Second, one population can outcompete the other, causing it to become extinct or move out of the area. This happens when one population has a competitive advantage under the existing environmental conditions.

Competition can occur among members of the same species (**intraspecific competition**) or between individuals belonging to different species (**interspecific competition**). Most of our discussion will concern interspecific competition, centering on competitive mechanisms that result in the exclusion of a species and those that allow them to coexist. Theories of competition help us understand how populations regulate the growth and presence of other populations.

The effect of interspecific competition can be indicated by a mathematical model, starting with the logistic growth equation. The rate of population growth of either one of two populations living in the same area will affect the other, as the second population uses some of the resources. In other words, each population reduces the habitat's carrying capacity (K) for the other. The interaction that results between two or more populations is the competition. We designate the two populations as N_1 and N_2. Their rates of increase will be r_1 and r_2, respectively. We then need to consider the carrying capacity of their habitats, which will be K_1 and K_2, respectively.

Assuming that the impact of one population on another increases as their numbers increase, the growth of the first population (N_1) has an inhibiting effect of $1/k_1$ on one individual of its own population. The inhibitory effect of an individual of N_1 on the other population (N_2) can be written as B/K_2, where B is the *competitive coefficient* of the second population. Individuals in N_1 reduce the carrying capacity of N_2 in proportion to the number of individuals in N_1 by B/N_1. The effect of one individual of N_2 on the growth of population N_1 is K_1. The presence of N_2 will influence N_1 by the factor α/N_2, where α is the competitive coefficient of the first population. As a result, we can modify the logistic growth equation.

The modified growth equation for N_1 when competition with N_2 occurs is

$$\frac{dN_1}{dt} = r_1 N_1 \frac{K_1 - N_1 - \alpha N_2}{K_1}$$

The growth equation for N_2 when competition with N_1 occurs is

$$\frac{dN_2}{dt} = r_2 N_2 \frac{K_2 - N_2 - B N_1}{K_2}$$

These populations will be in equilibrium when

$$\frac{dN_1}{dt} = \frac{dN_2}{dt} = 0$$

We can connect the values on a graph, where each population has a zero growth curve (Figure 3–4). Above the line, dN/dt, N_1 can decrease; below the line, N_1 can increase. As a result, there are four possible outcomes of competition between the two populations[5] (Figure 3–5):

1. Population N_1 is the only survivor, since the zero growth curve of N_1 is above that of N_2. N_1 can utilize the habitat in a better manner than N_2, resulting in a higher B value.
2. Population N_2 is the only survivor, since the zero growth curve of N_2 is above that of N_1.
3. Either population N_1 or N_2 alone can survive; which one does depends on the number of individuals with which each population starts. If there are more N_1 individuals initially, the N_1 population will survive; if there are more N_2 individuals initially, the N_2 population will survive.
4. The two species can coexist. As N_1 increases and N_2 decreases, the two reach a point where the process is reversed.

These equations, the Lotka–Volterra equations, developed by A. J. Lotka and V. Volterra a number of years ago, show why it is unlikely that two populations will continue to occupy the same ecological niche. How, then, can some form of stable co-existence come about, as in alternative 4? When two competing species continue to live together, they must have some kind of internal control that prompts them to stop increasing before enough individuals are produced to destroy the habitat of the other population. In other words, their internal density controls stop their growth before they eliminate the competition. It is also possible that, although the environment defining an ecological niche is normally continuous as a result of competition between species, the distribution can be discontinuous. As a population's numbers decrease in the presence of competing populations, it makes a finer subdivision of its niche. This fact led to the formation of the Gause hypothesis, or **competitive exclusion principle.** The principle states that no two species can occupy the same niche at the same time.

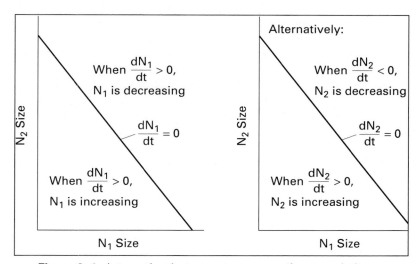

Figure 3–4 Interaction between two competing populations.

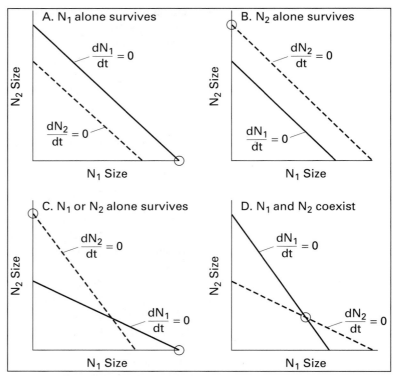

Figure 3–5 Possible outcome of competitive interactions between two populations.

Although competition is often difficult to observe in field situations, populations do interact with other populations and the physical factors in the system, resulting in a dynamic set of regulating mechanisms. Minor impacts on a community can establish regulating mechanisms through relatively subtle and immediate feedback. Major impacts are more visible and may need longer periods of time before "normal" population interactions are reestablished. The populations observed in the field generally have gone through competitive interactions and, as a result, have evolved morphological, physiological, or behavioral characteristics that have reduced competition.

Mechanisms that have evolved to reduce competition most often protect food or living space. Birds, for example, define territories to keep out other individuals of the same species. This is easily seen around marshes, where red-winged blackbird males vigorously chase others from territories they dominate. If other male red-winged blackbirds enter the area, they are attacked by the defending males (Figure 3–6).

The size of a territory varies with its characteristics as a habitat and the species' ability to defend the area. Territorial size for two populations of clay-colored sparrows in shrub land near the urban area of Manitoba, Canada, averaged less than for those in open country. Experimental removals of territorial males in mid-May revealed a surplus of nonterritorial birds that came to occupy the territory. Biologists found that nest predation increased as territories became smaller. The upper limits of territorial size were determined by nesting cover and pressure from adjacent territory holders, while predation determined the minimum size[6].

Figure 3–6 Like red-winged blackbirds, yellow-headed blackbirds vigorously defend territories in marshes. (Courtesy of the Wyoming Game and Fish Department.)

A study in Great Britain was conducted on great and blue tits, two bird species that coexisted in the same apparently preferred habitat. When investigators excluded the great tits from nest boxes in two study areas, the number of breeding blue tits increased significantly. In a controlled area where birds were not excluded from nest boxes, no increase was observed in the number of blue tits[7]. The presence of nest boxes therefore conferred enough competitive advantage on blue tits, that they could coexist with great tits.

The habitat relationships of four sympatric (living in the same area) species of meadow mice were studied in northwest Arizona. The data, which were gathered by snap trapping and live trapping on grids, were accompanied by surveys of the distribution of vegetation among microhabitats. The species' responses to vegetation provide an insight into their coexistence: The mice subdivided the habitat by types of vegetation. The researchers developed a three-dimensional structure of the habitat, including information that the species fed in various parts of the habitat at different times of day and selected different types and sizes of food[8] (Figure 3–7). Therefore, an alteration of the habitat would in all likelihood change the composition of the population of the four species.

In the central Great Basin of the United States, two species of chipmunks were found to coexist by reason of variation in vegetation. One species constantly chased the other. The species that was being chased, however, was more adept at climbing trees and moving from branch to branch. The more arboreal species coexisted where a

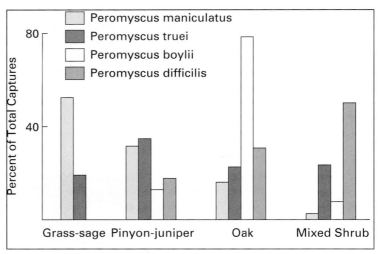

Figure 3–7 Comparison of habitats where four species of mice are found. (From Ref. 8.)

number of piñon and juniper trees grew close together. Where the trees were more scattered, the arboreal species lost its competitive advantage and could not subsist[9].

Studies of big game also show a variety of relationships that reduce competition. A number of big-game species interact and subdivide their habitat in accordance with type of forage, elevation, and availability of food[10]. When deer, elk, and other grazing species occur together, it is possible to determine which factors prevent interspecific competition (Table 3–1).

A study was made in Alberta, Canada, of competitive exclusion involving coyotes and wolves. Data on the movement patterns of the two species, collected by means of radio collars placed on the animals, indicated that coyotes lived primarily where the chance of encountering wolf packs was low. Dispersing juvenile coyotes generally skirted these areas. Coyotes that did not avoid areas of intensive wolf activity became prey. On the basis of this study, it is quite likely that the coyote population would expand into areas vacated by wolf populations[11]. In fact, the coyote population of the American west increased with the extirpation of the wolf.

Competition among cold-blooded vertebrates can also be studied. In the Great Smoky Mountains, two species of plethodontid (lungless) salamanders were considered competitors because when one was removed, there was a significant increase in the density of the second. Apparently, the species had been subdividing the habitat in some microfashion[12].

TABLE 3–1
Potential Means of Avoiding Interspecific Competition

Animal	Food Similarities	Seasonal Differences	Locational Differences	Ability to Shift to Alternative Food	Behavioral Differences
Mule deer	x				x
White-tailed deer	x	x	x	x	
Elk	x				

One result of competition can be a reduced range. When wolves and coyotes inhabit the same area, the coyotes are considerably more restricted than if wolves were not present. This type of side effect must be considered when planning for a population. Over a period of time, populations can develop special adaptive characteristics that reduce competition. In closely related species of birds, bills or feet have become modified to extract food from different areas of tree bark or the ground, thereby reducing competition. This is true for three species of nuthatches, which have overlapping distribution patterns in some parts of their ranges.

Predation

Like competition, predation is an interaction between two populations that can produce a reduced rate of increase on one or both populations. When a predator population depends on a single prey population as source of food, one population can strongly influence the rate of increase of the other. One of the mortality factors in a prey population is predation. Thus, in holistic management, the proportion of prey removed by the predator must be considered. If a manager allows increased hunting, there will be fewer animals available for the predators, which will then look for alternative sources of food. Too, predation is part of the energy-flow pattern from sunlight through plants and animals. Predation also serves as a population density regulation mechanism: When prey is high, the number of predators increases; when prey populations decline, the number of predators decreases.

In natural communities, predator–prey relationships evolve over a long period. As the ability of a predator to capture its prey improves, prey populations develop mechanisms to elude the predator. This is a continuous process. Thus, individual birds of prey that have better ability to see capture more food and have more young that survive because they are better fed.

The result of predation is a reduction in the rate of increase, r, in interacting populations. Although predation could threaten the survival of prey, this is an unlikely development, because the predator then would no longer have a source of food. As a result, predator–prey interactions tend to establish a balance.

The Lotka–Volterra equation can be applied to predator–prey interactions. Representing the size of the predator population as N_1 and that of the prey population as N_2, we can develop the following equations, in which B refers to the birthrate and D to the death rate:

$$\frac{dN_1}{dt} = B_1 N_1 N_2 - D_1 N_1 \quad \text{(predator growth)}$$

$$\frac{dN_2}{dt} = B_2 N_2 - D_2 N_1 N_2 \quad \text{(prey growth)}$$

In this case, the product $N_1 N_2$ determines the birthrate of the predator and the death rate of the prey. If both N_1 and N_2 are reduced proportionately, the effect will be much greater in the product than in N_1 or N_2 separately. Therefore, the predator's birth rate will be affected more drastically, and the prey will have its death rate reduced. If we graph the predator–prey interactions, we can see how both oscillate over a period of time (Figure 3–8). However, many factors influence these population cycles, and as a result, the fluctuations generally are greater than those of a neat cyclic pattern[3].

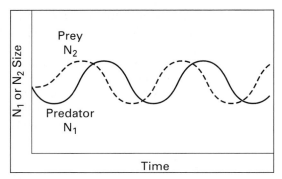

Figure 3–8 Predator–prey interactions.

Here are some observations on predator–prey interactions involving game animals:

1. Rarely have the predators been implicated as the sole or direct cause of population decline, except when people are one of the predators.
2. Predators sometimes appear to be the agents by which numbers are adjusted to the carrying capacity of the habitat.
3. Predators may increase the mortality of prey populations during severe weather and retard or prevent the recovery of populations at other times.
4. When other environmental conditions change, predator–prey interactions can change, accelerating increases or declines. For example, wolves may remove more ungulates as the habitat deteriorates.
5. Predation is sometimes considered a limiting factor. It can be when it involves the removal of young animals; however, other factors can also affect the survival of the young.
6. Habitat frequently is the ultimate limiting factor, but at times, predation can depress or maintain the numbers of animals below the habitat's carrying capacity[13].
7. Changes can occur in the age composition of the prey.
8. The availability of prey (food) can control a predator population.

A number of cases indicate that predators do not control ungulate populations. For example, golden eagles are widely reported to prey on ungulates, especially young fawns. Pronghorn antelopes also are leery of golden eagles, but no study has shown eagle predation to be serious. It has been found that predators more commonly take weak, maimed, or old animals and only occasionally take healthy individuals[13].

Data on the impact of predators on ungulates show that, up to a point, predation increases as the number of prey increases. In one study, wolves did not completely control white-tailed deer when the number of deer per wolf exceeded 150[13]. Wolf predation simply could not keep up with the higher number of deer. Predators serve to lessen prey population oscillations, but when the prey population increases dramatically, this controlling mechanism is ineffective.

Because predators are so important in energy flow, we need to note that there are often alternative forms of prey. A study of mountain lions in Arizona indicated that alternative prey in the form of livestock enabled the lions to maintain a higher density than would have been possible with native prey alone. In this situation, the availability of domestic calves was greatest in late spring, when deer numbers were lowest[12].

Coyotes generally rely on rodents and rabbits for food. This permits the maintenance of a higher coyote–ungulate ratio than if ungulates were the sole primary food of coyotes. When coyotes are numerous, they can inflict serious losses on ungulate populations in a short time if weather or other conditions temporarily increase ungulate vulnerability. On the other hand, an abundance of alternative prey conceivably could reduce or postpone coyote predation on ungulates. There is an implication here for sheep ranching. Sheep ranchers usually want to destroy the coyote population, but if they do, wild grazers may increase, thereby reducing the forage available for sheep. Because the number of alternative prey species can be substantial, it is necessary to explore such possibilities in research work.

How can predator control mechanisms be used to manage game species? To answer this question, it is necessary to look at the objectives of management and then to find out whether the habitat will support more game if control of predators occurs. Conclusions about using predators in management must be made on an individual basis, since studies show that some predators limit the number of their prey, while others have little effect. It appears that predator–prey interactions form a major component in energy flow, allowing energy to move between the two organisms. Thus, any type of management that involves the removal of one or the other species can affect predator–prey cycles. Predation is one form of population regulation that occurs in the coexistence of populations in a community.

Hunting and Fishing

When people hunt or fish, they act as predators, but they often use advanced techniques that have been developed so rapidly that the prey populations cannot respond. In ocean fishing, modern fleets, with their new methods of locating and capturing schools of fish, can so reduce a population that it cannot reproduce at its replacement rate (Figure 3–9).

Without this extreme human interference, most populations evolve toward a state of self-regulation that eliminates extreme fluctuations in their size through positive or negative feedback mechanisms. The growth of a prey population is a positive feedback to a predator population—more prey means more food for more predators. The growth of a predator population is a negative feedback mechanism to a prey population—more predators mean that more prey are eaten, leaving fewer to reproduce.

The impact of hunting is always a question. In a study conducted in Ohio, the impact of hunting on squirrels was found to be negligible. The squirrel population fluctuated in relation to seed productivity[14]. In another study, experiments were designed to determine the impact of hunting rock ptarmigan in the autumn and spring. Approximately 40 percent of the autumn population in a 10.4-km² area were removed by shooting in three consecutive years. During May of 1971 and 1972, 40 percent of the breeding population in a 7.5-km² area were removed. Comparison of annual breeding densities in the experimental and controlled areas suggested that neither the autumn nor spring removals at the 40-percent level depressed the rock ptarmigan population. There was some evidence that autumn removal may have contributed to a higher breeding density than did spring removal. In most cases, male ptarmigan taken in the spring were not replaced until the following year[15]. Other studies also show that as long as hunting is a control measure, it substitutes for mortality factors and therefore does not have an adverse impact on population[16].

When hunting *substitutes* for other forms of mortality, it is said to be **compensatory.** When hunting *adds* to total mortality (when it is another form of mortality) it

Wolf the Predator, Caribou and Moose the Prey

From 1987 to 1991, the Alaska Department of Fish and Game conducted an extensive study of wolves in northwestern Alaska. Results from the study helped us understand predator–prey interactions. The department radio-collared 86 wolves and followed them to find out what they captured and what killed the wolves. Wolves preyed primarily on caribou, killing 6 to 7 percent of the herd annually. In the winter, the caribou migrated away from the territory of wolf packs. At that time, the wolves changed prey, capturing resident moose primarily. The wolves removed 11 to 14 percent of the moose population annually. Because the moose were affected by severe weather, this mortality could have had an impact on the moose population in years of low numbers. Some moose were captured when caribou were present, but caribou were the preferred food.

The moose population was established in the area some 40 years before the study began, so biologists speculate that wolves used to leave the area and follow the caribou during their migration in order to obtain adequate food.

During the study, biologists noted the following causes of mortality on radio-collared wolves:

Alive at the end of the study	11%
Missing at the end of the study	28%
Killed by a hunter	42%
Starved	4%
Killed by another wolf	1%
Died of old age	1%
Died from rabies	13%

In all likelihood, the food available was not sufficient to support the wolf population. If hunter kills were reduced, other causes of mortality could increase (wolf kills, disease, starvation). Was this an example of compensatory or additive morality? The study also showed that wolf populations were likely at or above the level at which the habitat had food to support them.

From Ballard, W. B., L. A. Ayaares, P. R. Krausman, D. J. Reed, and S. G. Fancy. 1997. Ecology of Wolves in Relation to a Migratory Caribou Herd in Northwest Alaska. *Wildlife Monograph* 135:1–47.

is said to be **additive.** The basic idea underlying the compensatory mortality hypothesis is it that as hunting increases, other forms of mortality, such as disease and predation, decrease, keeping the survival rate for the population the same. At some size, however, the population could be reduced to such low numbers that it does not have enough reproductive animals to maintain a population. The additive hypothesis states that as hunting increases, the size of the population decreases in a linear manner. Hunting, then adds to all other mortality factors that are occurring. There is much debate concerning the impact of hunting on populations, particularly waterfowl.

Hunting also leaves some unanswered questions concerning natural selection. Many ungulate species are subject to hunting pressure directed heavily toward the larger, more robust males. Thus, people might have an adverse impact on these populations over time.

Figure 3–9 Fishing is a form of predation. (Courtesy S. McCutcheon.)

Symbiosis

When organisms of two or more species live in the same habitat, they have a **symbiotic** relationship. Some ecologists restrict the term to a living relationship beneficial to the species involved. We shall define it simply as living together, so that it includes parasitism, mutualism, and commensalism.

Parasitism

In predation, each population exists independently until the predator captures the prey. In **parasitism,** the parasite population locates itself in or on the prey (the host species). In some cases, one parasite uses more than one host population. Similarly, a host may have more than one parasite population.

Parasitism acts as a controlling or regulatory mechanism in some wildlife populations. For example, roundworms in white-tailed deer apparently have little effect on the size of the deer population, but do reduce moose numbers. When deer and moose are separated by deep snow, transmission of roundworms to moose is reduced. In two areas of northwestern Ontario, deer and moose overlap during the winter, when snowfall causes the moose population to descend to lower altitudes. The population of moose then declines because of infestation by roundworms. Too, comparisons of deer populations with and without roundworms showed no significant difference, but comparisons of moose populations that had and had not overlapped with deer infested with roundworms did show a significant difference. Obviously, the parasitic roundworm did have an impact on the moose population. In one Maine study, the prevalence of roundworms in moose was directly related to the density of the deer population[17].

Interaction involving a parasitic yeast occurs in the gut of the green frog. A study conducted in North Carolina found that a parasitic yeast enhanced the growth of tadpoles under certain conditions. Low concentrations of the yeast increased the growth rates of tadpoles, but not of siblings raised under identical conditions but without the yeast cells. Furthermore, the impact of a given concentration of yeast varied with the size of the tadpole: The yeast appeared to act as a stimulant for larger tadpoles and a parasite on smaller ones. Because larger tadpoles are more likely than their smaller siblings to metamorphose and reproduce, the benefit was to those individuals more likely to contribute to the continuation of the green frog population. For this reason, resistance to the parasite is not likely to develop. The parasite may act to regulate the size of the adult green frog population by altering the probability of metamorphosing for individual tadpoles[18].

In parasitic interactions, the host population usually has some form of density-dependent mechanism that prevents the parasite from increasing to a level where the host is destroyed. If the host dies, there obviously is no future for the parasites on that host! Through the process of selection, mechanisms have evolved to keep parasites in check. When a parasite invades a new host population, the result can be destruction for both the host and the parasite if there is no mechanism to regulate parasitic growth.

Malaria, caused by a parasite that inhabits the blood of some wildlife and cattle in areas of central Africa, is transmitted from one host to another by mosquitoes. When mosquitoes carrying the parasite bite people, severe illness and sometimes death follows. The parasite acts by destroying red blood cells in people. Some people have red blood cells that are deformed and look like sickles (Figure 3–10); these people have more resistance than others to the malaria parasite.

Mutualism

When two populations interact so that each benefits, the interaction is called **mutualism.** A good example of this is the relationship between nitrogen-fixing bacteria and their leguminous host. When nitrogen-fixing bacteria are not present in the soil, legumes grow slowly. When they are present, they assist in converting atmospheric nitrogen to a form that the legume is able to use (Figure 3–11). Another example is the relationship between tree roots and some fungi. Special species of fungi grow around some tree roots, particularly the very small rootlets. The fungi provide protection and some nutrients for the tree, and the tree returns the nutrient favor. Neither the fungi nor the tree can grow properly without the other[19]. And the cast of characters is sometimes larger. When soils are devoid of fungi spores, trees are unable to grow. In areas where clear-cutting occurs, small mammals such as red-backed voles eat fungal spores and deposit them in openings in the forest through fecal material. When there are no small-mammal populations, reestablishment of the forest in clear-cut areas may take much longer.

Commensalism

Commensalism is a relationship that benefits one population and has no effect on the other. For example, cattle egrets and cowbirds follow large grazing animals around. As the herbivores pull up grass and cause disturbances with their hooves, many insects are dislodged, making them an easy meal for the birds, but with no apparent benefit to the herbivores.

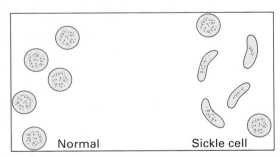

Figure 3–10 Schematic diagram of red blood cells affected by sickle cell anemia.

Figure 3–11 Root nodules on a legume. (Courtesy of the U.S. Soil Conservation Service.)

Disease

Disease also affects population levels. Disease occurs when one organism invades the body of another. In most cases, the disease organism disrupts the physiological functioning of the host. Unless the invaded organism builds up a defense to repel the invader, both organisms can die. Disease can often spread rapidly through a population, affecting its growth and reproduction.

Some forms of disease are caused by parasites, particularly parasites invading an abnormal host. Bacteria and viruses, the major causes of disease can be viewed as parasitic invaders, because the disease organism invades the host and affects it adversely. Sometimes, large numbers of the host population are destroyed by disease-causing organisms. If a few members of the host population have a resistance factor to the disease organism, their gene pool becomes the basis for reestablishing the population.

Sylvatic Plague and Prairie Dogs

Plague is a disease that affects many species of wildlife and can be passed on to humans. The disease is responsible for at least four great pandemics in world history. In the 14th century, it killed 25 million people, or one-fourth of the world's population. The causative agent is a bacteria (*Yersinia pestis*) found in some of species flea. When fleas bite an animal, the bacteria pass into the bloodstream. Since black-footed ferrets, an endangered species, feed primarily on prairie dogs, the prairie dog population is the focus of a number of studies.

There are five prairie dog species in the United States. Some are more susceptible to plague than others. When plague strikes a susceptible colony, its members can all die within two months. The population might recover due to immigration, within a period of two to seven years.

Plague is, therefore, a major form of population control in prairie dog colonies. The disease can sweep through colonies at intervals of 5 to 15 years. As some prairie dogs die due to plague and others immigrate into the area and establish themselves, we see many population parameters at work.

Disease, therefore, is a mortality factor in wildlife populations that should be evaluated whenever a manager is trying to change the population density of a wildlife species.

POPULATION CYCLES

Most populations vary seasonally, yearly, or over several years. Such changes must be considered when managers are determining the optimum size of a population or an area's carrying capacity. Irregular changes in the size of a population, referred to as **fluctuations,** may be the result of changes in weather or food supply or of the impact of human hunting. A regular pattern of change is called a **cycle.** The annual cycle of reproduction and subsequent decline in wildlife populations is an example. Some cycles extend over a period of years. Cycles can involve changes in the number of animals (predators) that eat others. In Canada the lynx preys on snowshoe hares. Data from more than a hundred years of the Hudson Bay Company's records show cyclic oscillations in lynx and hare populations, with densities peaking every 9 to 10 years (Figure 3–12). In North America, the snowy owl commonly has a 5-year-cycle, appearing in the northern part of the United States every fifth year, apparently as a result of a decrease in the rodent population in the arctic. Most of the snowy owls that appear in North America do not survive to return to the arctic. Thus, the snowy owl population crashes the year after it is found in the United States. The cycle starts again when the few surviving owls return to the arctic. A number of explanations have been proposed for cycles or the cyclic nature of populations. Three are considered here.

Stress

On the basis of a study of small mammals, J. J. Christian theorized that population cycles are caused by the stress inherent in large numbers and that declines in population density are associated with changes in some regulating hormones in the body[20,21]. Hormones are chemicals released by some body glands that control many physiological and behavioral activities in animals. Christian surmised that all mammals limit their own population densities by a combination of behavioral and physiological changes caused by the hormonal changes. His hypothesis is that the changes cause an

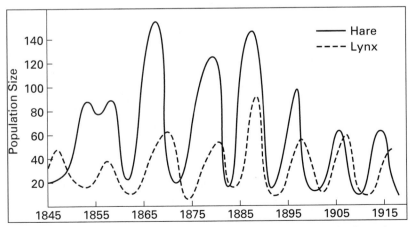

Figure 3–12 Cyclic nature of lynx and hare populations in Canada.

increase in the death rate and a reduction in reproductive activity[20]. If this hypothesis is correct, populations responding to stress tend to limit their numbers themselves[21] (Figure 3–13).

Food

Investigators who support the food hypothesis feel that changes in the size of a population are the result of changes in the quantity and quality of food in the habitat (Figure 3–14). Studies, primarily of small mammals, indicate that as the population increases, the impact on the grassy substrates, or ground cover, reduces the food supply and causes the population to decline. It is believed that this reaction occurs regardless of predation. Voles were used to test the hypothesis. High-quality supplemental food (rabbit pellets) was given to a population in a relatively unfavorable environment. Throughout the winter, the experimental population continued to show a higher breeding intensity, better adult survival, and heavier body weights than did a controlled population in a similar habitat, but without the supplemental food. The population with supplemental food reached densities 50 percent higher than those of the control population. Body growth, litter size, and density remained higher during the following spring and summer in the population receiving food supplements; however, their numbers declined in the late summer and autumn. Food appeared to influence the amplitude of the population cycle, but did not prevent the decline in numbers[22].

In another study, reindeer were introduced into an island off the coast of Alaska. The population increased from 29 animals to 6,000 between 1944 and 1963. In 1963, the population crashed, leaving fewer than 50 animals. During the time the population was increasing, there was a large quantity of high-quality forage on the island. As the density of reindeer on the island increased, the quality of the forage declined. The ratio of fawns to adult cows dropped: In 1944, 75 percent of the cows had calves; by 1963, the figure had declined to 60 percent. By 1963, lichen, a favorite food of reindeer, had been eliminated as a significant component of the winter diet. Harsh weather also occurred during the period. Sedges and grasses began expanding into sites previously occupied by lichen. Thus, the nature of the food supply appeared to have an impact on population growth[23].

In another cyclical change, wolf populations declined as white-tailed deer decreased in the Superior National Forest in northeastern Minnesota. Malnutrition and intraspecific strife increased. As wolf numbers began to decline, pup starvation became apparent,

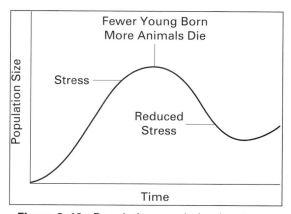

Figure 3–13 Population regulation by stress.

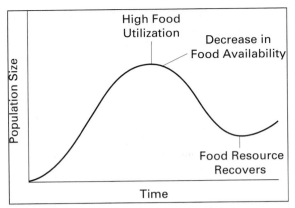

Figure 3–14 Population regulation by food.

followed by a lower production of pups and then by increased intraspecific strife. At higher densities, adult wolves were the most secure members of the population. But as the population declined, they became less secure because of the intraspecific strife[24].

Genetics

D. Chitty suggests that the antagonism associated with high breeding densities brings about a change in the genetic makeup of populations (Figure 3–15). When a population's density is high, animals produce offspring that are less likely to survive or reproduce. This could be caused by less energy being directed into reproduction or by changes in the genetic composition of the young. As a result, more aggressive individuals become dominant over more prolific ones. The process does not take place independently of the external environment, but rather is influenced by a variety of factors, such as weather. One study showed that as the population of voles increased, the effects of weather became more severe[25]. There is little direct evidence to test Chitty's theory because of the difficulties in evaluating genetic makeup. Animal population cycles controlled by genetic change would have density cycles similar to those created by stress factors.

Some biologists feel that the physical environment or ecosystem causes rhythmic fluctuations and that cycles are really irregular fluctuations. Others feel that cycles are

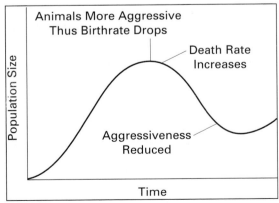

Figure 3–15 Population regulation by changes in the genetic makeup of young that survive.

Part 2 / Wildlife Populations

simply random changes caused by a variety of forces that operate haphazardly[26]. All we can say for sure is that the factors which cause populations to increase or decrease can be mechanisms or stimuli from outside the population such as predators and environmental change, or factors, within the population such as genetic composition. Of course both kinds of factors may work in combination.

Understanding the complexity of these interactions can help the wildlife manager plan. Causes may not always be apparent in field observations, but managers can make better decisions if they consider density data over extended periods. Cyclic changes in populations may need to be allowed for in hunting quotas or claims of damage to animals. Or an irregular fluctuation may result from environmental change, disease in the population, or a new population in the area.

POPULATION REGULATION

We have discussed population growth curves in populations exhibiting *r* and *K* selection. We have looked at possible causes of population cycles. It is apparent that various factors act independently or in combination to control the growth of populations. These factors are *population-regulating mechanisms*. The size of a population can be influenced by forces outside the population or within the population. Weather is an example of the former; genetics, stress, and changes in reproduction rates are examples of the latter.

There are different schools of thought as to what determines the number of organisms in a population. One group of biologists feels that population size is determined by factors external to the population itself and so is *density independent*. Another group feels that the number of individuals in a population is determined by *density-dependent mechanisms*.

Density Independence

Density-independent factors, such as climate, cannot actually regulate the growth of a population, because "regulation" connotes that a feedback mechanism is present. Climate-related events occur regardless of population numbers. No mechanism signals the population that changes are occurring. Still, density-independent mechanisms can influence the population growth. They can interact with density-dependent factors to cause a change in numbers, or they can cause a population to have so few members that density-dependent factors cannot come into play. Thus, density-independent factors affect a proportion of a population or the whole population, whether the population is large or small. If adverse weather conditions occur and a relatively small population contains 50 voles, 10 percent might die. The same percentage might die if the population were 500.

Density Dependence

As a population becomes more dense, density-dependent factors are brought to bear, and a larger number of the individuals in the population die or fewer individuals successfully reproduce. Density-dependent factors show an increase proportional to increases in population size. Thus, the three population-cycle hypotheses discussed earlier can be considered density-dependent factors, since higher proportions of animals are likely to die from stress, lack of food, or reduction in productivity due to genetic changes at higher population densities. Most mechanisms that regulate population are density dependent.

Density-dependent Growth in Gray Partridges

In Washington state, biologists working on the gray partridge examined the interaction between population size and density-dependent regulation. Using data collected from 1940 to 1992, they showed that the recruitment rate was low when the adult population density was high. Conversely, when the adult density was low, the recruitment rate of chicks was high. The biologists believed that an increase in the number of nesting birds caused an increase in nest predation. Ideal nest cover was difficult to find when there were a large number of nesting hens. In addition, predators developed a search image for nests and were able to locate them more easily when they were more abundant. As a result, there were fewer eggs left to hatch.

Rotella, J. J., J. T. Ratti, K. P. Reese, M. L. Taper, and B. Dennis. 1996. Long-term Population Analysis of Gray Partridge in Eastern Washington. *Journal of Wildlife Management* 60:817–25.

Population Constraints

Dispersal is one way in which density independence influences density dependence. Habitat limitation can control the level to which a population can increase. The limitation can be on an area or on requisite needs. The effect of forest manipulation on ruffed grouse is a good example. Ruffed grouse prefer areas of shrubs near the edges of forests. When transmission-line corridors are cut through forests, more area is created for grouse populations to increase. Cutting operations opened areas of 0.1 to 0.4 hectare (1/4 to 1 acre) in a 300-hectare (735-acre) pole timber forest, encouraging grouse broods to increase. By the seventh season, these openings, which were left undisturbed, were losing their shrub and other ground-level vegetation, and grouse broods began to decline[26]. After 10 years, the openings were filled in by a dense brush canopy and were of little value as brood feeding grounds. Thus, the grouse population density was a regulatory factor; however, habitat deterioration, a density-dependent factor, affected the ability of the habitat to support the grouse.

SUMMARY

Populations undergoing exponential growth generally increase rapidly, saturate their habitat, use the resources, and then decline. Logistic growth involves a rapid increase in size until the carrying capacity of the habitat (K) is approached. Populations then fluctuate around that level or remain below it. The carrying capacity itself can change as the characteristics of the habitat change.

Populations can interact in a variety of ways. When they seek a similar resource that is in short supply (i.e., when there is competition), the result is a negative impact on their growth rates. Parasitism is a form of competition in which an intimate relationship allows a parasite population to exist at the expense (energy use) of its host. The phenomenon of two populations that coexist to the benefit of both is called mutualism; coexistence that benefits only one population is called commensalism.

Populations can exhibit fluctuations (irregular changes in numbers) or cycles (regular changes in numbers). Cycles are thought to be caused by changes in stress felt by animals as their numbers increase, by a change in food supply, or by a change in the genetic makeup of the population as individuals that can survive better under crowded conditions remain when population densities are high.

Most population-regulating mechanisms are classified as density dependent. (Their impact on populations is more severe as the size of the population increases.) The impact of density-independent factors is on a constant proportion of the population, regardless of its size. Regulating mechanisms can be combinations of many population and habitat interactions.

DISCUSSION QUESTIONS

1. Why is it difficult to document competition in the field?
2. How are population cycles controlled by predators?
3. Discuss the role of hunting and fishing in population control.
4. What is the importance of competition, predation, and disease in hunting quotas?
5. How important is mutualism in population management?
6. Would inter- and intraspecific competition be likely to occur at the same time? Why or why not?
7. Two populations compete for food. One is always more abundant than the other in part of the common range, while the other is more abundant in the rest of the range. What will happen?
8. When can a predator act as a population-controlling mechanism for a prey population? When not?
9. What is carrying capacity, and why is it likely to change? How is carrying capacity related to environmental resistance?
10. Can you think of interactions between populations that have a negative impact on a population? A positive impact?

LITERATURE CITED

1. Caughley, G. 1979. What Is This Thing Called Carrying Capacity? In M. S. Boyce and L. D. Hayden-Wing (Eds.), *North American Elk: Ecology, Behavior, and Management,* pp. 1–8. Laramie, WY: University of Wyoming.
2. McCorquodale, S. M., L. L. Eberhardt, and L. E. Eberhardt. 1988. Dynamics of a Colonizing Elk Population. *Journal of Wildlife Management* 52:309–13.
3. Wilson, E. O., and W. H. Bossart. 1971. *A Primer of Population Biology.* Sunderland, MA: Sinauer Associates.
4. Emlen, J. M. 1984. *Population Biology.* New York: Macmillan.
5. Boughey, A. S. 1973. *Ecology of Populations.* New York: Macmillan.
6. Knapton, R. W. 1979. Optimal Size of Territory in Clay-Colored Sparrow. *Canadian Journal of Zoology* 57:1358–70.
7. Dhondt, A. A. 1980. Competition between the Great Tit and the Blue Tit outside the Breeding Season in Field Experiments. *Ecology* 61:129–96.
8. Holbrook, S. J. 1979. Vegetational Affinities, Arboreal Activity, and Coexistence of Four Species of Rodents. *Journal of Mammalogy* 60:528–42.
9. Brown, J. H. 1971. Mechanisms of Competitive Exclusion between Two Species of Chipmunks. *Ecology* 52:305–11.

10. Mackie, R. J. 1976. Interspecific Competition between Mule Deer, Other Game Animals, and Livestock. In G. W. Workman and J. B. Low (Eds.), *Mule Deer Decline in the West: A Symposium,* pp. 49–54. Logan, UT: Utah State University.

11. Fuller, T. K., and L. B. Keith. 1981. Non-overlapping Range of Coyotes and Wolves in Northeastern Alberta. *Journal of Mammalogy* 62:403–5.

12. Hairston, N. G. 1980. The Experimental Test of an Analysis of Field Distribution: Competition in Terrestrial Salamanders. *Ecology* 61:817–26.

13. Connolly, G. E. 1978. Predators and Predator Control. In J. L. Schmidt and D. L. Gilbert (Eds.), *Big Game of North America: Ecology and Management,* pp. 269–394. Harrisburg, PA: Stackpole Books.

14. Nixon, C. M., M. M. McClain, and R. W. Donohoe. 1975. Effects of Hunting and Mast Crop on a Squirrel Population. *Journal of Wildlife Management* 39:1–25.

15. McGowan, J. D. 1975. Effect of Autumn and Spring Hunting on Ptarmigan Population Trends. *Journal of Wildlife Management* 39:491–95.

16. Burnham, K. P., G. C. White, and D. R. Anderson. 1984. Estimating the Effect of Hunting on Annual Survival Rates of Adult Mallards. *Journal of Wildlife Management* 48:350–61.

17. Saunders, B. B. 1973. Meningeal Worm in White-Tailed Deer in Northwestern Ontario. *Journal of Wildlife Management* 37:327–30.

18. Steinwascher, K. 1979. Host–Parasite Interaction as a Potential Population Regulatory Mechanism. *Ecology* 60:884–90.

19. Maser, C., J. M. Trappe, and R. A. Nussbaum. 1978. Fungi–Small Mammal Interrelationships with Emphasis on Oregon Coniferous Forests. *Ecology* 59:799–809.

20. Christian, J. J. 1950. The Adreno-Pituitary System and Population Cycles in Mammals. *Journal of Mammalogy* 31:247–59.

21. Christian, J. J. 1971. Population Density and Reproductive Efficiency. *Biological Reproduction* 4:248–94.

22. Cole, F. R., and G. O. Gatzli. 1978. Influence of Supplemental Feeding on a Vole Population. *Journal of Mammalogy* 59:417–25.

23. Klein, D. R. 1968. The Introduction, Increase, and Crash of Reindeer on St. Matthew Island. *Journal of Wildlife Management* 32:350–67.

24. Mech, L. D. 1977. Productivity, Mortality, and Population Trends of Wolves in Northeastern Minnesota. *Journal of Mammalogy* 58:559–74.

25. Chitty, D. 1960. Population Processes in Vole and Their Relevance to General Theory. *Canadian Journal of Zoology* 38:99–113.

26. Wakeley, J. S. (Ed.) 1982. *Wildlife Population Ecology.* University Park, PA: Pennsylvania State University Press.

4
POPULATION MOVEMENTS AND MEASUREMENTS

Most animal populations engage in some form of movement—daily, seasonally, annually, or once in a lifetime. The nature and extent of this movement is important for a number of considerations, including how to estimate the number of individuals existing in an area. When the size of a population is to be estimated, it is necessary to know how many animals come into or leave the area. Some of the approaches to determining population numbers are the subject of this chapter.

POPULATION MOVEMENT

Movement of individuals in a population contributes to a change in the size of the population. In considering population growth or decline, managers must determine how many animals are added or removed through movement.

Migration

Many populations have periodic movement patterns. Regular movement of a population from a wintering area to a summer area and back again is called *migration*. Migration is common in many bird, mammal, and fish species. People usually associate migration with birds moving from the tropics up into North America to breed and then returning to the tropics. It is not as commonly recognized that elk, deer, antelope, and moose also migrate. Most ungulates divide into herds, which travel together. The trip may be only 32 km (20 mi) or less for a change in elevation. These patterns of movements have often been disrupted by human activities. Major highways, for example, have blocked the migration of many ungulates. Transmission-line rights-of-way, oil and gas development, and like disruptions can block critical migration routes, as well as destroy important summer or winter ranges (Figure 4–1).

Immigration and Emigration

It is important to have information about how many animals move in or out of a population. How do the young disperse? When animals leave their parental setting, they emigrate to settle another area. Emigration is common in many species , such as weasels, cougars, and many small mammals. When an animal settles in a new area, it has immigrated to that area.

It has been found that disruption of the social structure in animals will also disrupt their regular dispersal patterns. A study of the mountain lion population in Wyoming disclosed that when older males were removed, more younger males—which had formerly dispersed—remained, but maintained smaller territories. The number of females remained the same, but the number of males increased; thus, there were more mountain lions concentrated in the area. The extra prey needed to sustain the larger population was provided by domestic livestock, so that the disruption caused by harvesting old males actually increased the problem it was supposed to solve.

Evaluation of Movement

Telemetry involves placing small radio tracking units on animals and following them (Figure 4–2). Such units have been used in a study of whooping cranes, which migrate from Canada to Texas. Working with transmitters, biologists have been able to deter-

Figure 4–1 Migrating elk. The movement of elk, particularly in the fall, can be very spectacular. (Courtesy of the U.S. Fish and Wildlife Service Photo by B. Smith.)

Figure 4–2 Placing a collar around the neck of a moose. (Courtesy of the Wyoming Game and Fish Department. Photo by C. Anderson.)

mine the exact movement pattern, behavior, and preferred habitat of cranes. Whooping cranes apparently travel in small family groups of three or four individuals. Lakes and marshes along the way are very critical to the survival of this endangered species. Birds in general migrate each fall and spring and follow well-defined corridors. (See Chapter 19.)

Considerable information on movement patterns was gathered in a study of Nebraska coyotes by using radio telemetry. Coyote activity around the den sites increased during prebreeding, breeding, gestation, nursing, and early pup training, presumably because of strong social bonds between the mated pair. Because coyotes do not encroach on the territory used by other mated pairs, they require an area of at least 5 km^2[1] (1.9 sq mi). The coyotes were active at night and inactive or resting in the den during the day. As a result of changes in movement patterns during the different reproductive periods, the coyotes took different forms of prey. This type of study can provide information to aid in controlling coyote predation.

The area where animals move during their daily activity pattern is generally referred to as their **home range.** Home range differs from territory in that territory is a defended area and home range is an activity area[2].

In Idaho, surgical techniques were used to place small transmitters under the skin in four species of small mammals. Movement patterns of these animals were determined in relation to a radioactive waste disposal site planted with crested wheat grass. Results showed that the animals moved from sage desert habitat up to 200 meters (656 ft) to feed on the crested wheat grass and then returned to their burrows in the sage. Telemetry revealed that the home range increased when a new food supply appeared[3].

Dispersal from sites where animals were born has been documented by radio-tagging the animals. One study of mule deer in Utah in which more than 900 deer fawns were tagged provided a good deal of insight into the dispersal of the deer. Very few fawns moved far from their birthplace during the first summer. Yearlings tended to move a much greater distance. Sixty percent of the bucks and 35 percent of the does had left the area by the time they were 16 months of age. Thereafter, little dispersal occurred among the bucks, but the doe dispersal percentage and the distance between the tagged and kill sites increased for the next two to three years. This information can be related to carrying capacity. Generally, fawns killed by hunters were most commonly encountered about 1.1 km (0.7 mi) from the area in which they were born. Some of the bucks moved from 1.13 km (0.7 mi) up to more than 24 km (15 mi), with a mean distance of 7.25 km (4.25 mi), while the relatively few does moved as much as 24 km. The average distance that does moved was 5.4 km (3.67 mi).

The use of winter and breeding habitats was shown in Merriam's turkey in Wyoming. Using transmitters strapped on the turkeys' backs, biologists showed how turkeys dispersed from barnyard hay piles to dense forest 20 km away. Roost trees were identified so that the birds could be protected, thereby increasing the chances that the turkey population would grow.

Movement-pattern studies assisted an analysis of the social and spatial organization of lynx in northern Minnesota[4]. By tagging four males and 10 females, the investigator found that lynx tended to remain in a 50-km² (19.3-sq-mi) area. Females tended to overlap in range more than males did. Two females in the study shared part of the same area during one breeding season. Male ranges did not overlap, although they did abut one another. There is evidence that the spatial organization of the lynx is similar to that of the mountain lion. Male mountain lions tend to maintain home ranges that include one, two, or three females. The males breed with these females and can remain in the same area for a number of years if their social system is not disrupted.

Satellite Telemetry

A recent advancement in telemetry techniques incorporates the Global Positioning System (GPS) unit in a radio-telemetry collar that can be placed on an animal. A satellite-based navigation system, the GPS consists of more than 24 satellites circumnavigating the earth in four orbital planes. The telemetry unit, which can be in a collar or a hand-held GPS unit, picks up signals from the satellites and stores them for later downloading into a computer.

The accuracy of a GPS location depends on the number of satellite signals that reach the receiving unit. If data from four satellites are picked up, the elevation, which is a three-dimensional location, the latitude, and the longitude of an animal can be calculated to an accuracy of less than 10 meters. If data from only three satellites are picked up, no information on elevation is available. When fewer than three satellites provide signals, no locations can be determined[5].

Currently, biologists are investigating the effects of forests, vegetation, and topographic changes on the accuracy of GPS collar locations. Because animals spend time in different habitats with variable vegetation, biases can occur to influence the accuracy of the locations obtained. Biologists are trying to determine the magnitude of the biases so that corrections can be made for specific habitats. Because the GPS collars cost around $8,000 each, biologists are exploring new, less expensive, and more accurate technology. There are many advantages to these alternatives, since both nighttime and foul-weather locating are possible. These improvements can help biologists collect more accurate population estimates.

Marking Animals. Many devices are used to mark animals. Color neck markers around big-game species and some waterfowl are used to identify locations from which animals have come. Color dyes are used to mark feathers of bird and the fur of mammals.

Radioactive chemicals have been used successfully for movement studies. For example, [54]Cr was used to determine daily and seasonal movement of the tree frog, found in the pine barrens of New Jersey[6]. Selastic tubing containing radioactive wire was sewn around the frogs' bodies. Frogs could then be located with a hand-held radiation detector.

Determining Bear Numbers and Movement by Biomarkers

Biologists in Michigan and Minnesota marked bear to determine the size of their population by the mark–recapture technique. Tetracycline, a biomarker of bone and teeth that fluoresces under ultraviolet light, was placed in bear food at selected sites. Pieces of the bait were small enough to be eaten by individual bear. The bait sites were placed sparsely throughout both states. Biologists estimated the number of bear that ate baits by claw marks on the trees around the bait sites. This gave them an estimate of the number of marked bear in the area. When bear were killed by hunters or were reported dead, biologists could collect teeth from the dead animals and determine whether tetracycline was in them. As a result, the researchers had enough data to use mark–recapture methods to estimate the numbers of bear. They also received valuable data on individual bear movement.

From Garshelis, D. L., and L. G. Visser. 1997. Enumerating Metapopulations of Wild Bears with an Ingested Biomarker. *Journal of Wildlife Management* 61:466–480.

COLLECTING POPULATION DATA

Before collecting data, population managers need to define their objectives. They should have a clear understanding of how their data will be analyzed and used in decision making. Generally, there are two fundamental pieces of information needed by managers on any population: (1) How many animals are there? (2) Where are the animals? This information is basic to any other kinds of inquiry. Data can then be collected in a systematic manner to answer these questions. Too often, data are collected in a catch-as-catch-can manner in hopes that the information can help in management. To calculate growth, survival, natality, mortality, density, and other characteristics of a population, it is necessary to determine the number of individuals in the population at different points in time. Sometimes, desired data on sex and age may be collected at the same time as data on other population characteristics. If age and sex data are desired, proper techniques must be used. Determining the number of individuals in a population is probably one of the most important field tasks of the manager. These data are commonly used to calculate population characteristics and make management decisions involving removal, habitual enlargement, and multiple-use activities.

Since the population parameters of natality, mortality, dispersal, growth, and productivity are interrelated, information gathered on any one of them will be useful in determining another. Basically, we need to count the number of animals present in an area at one time. We can do this by total counts (counting all the animals) or sampling (counting some of the animals and assuming that they are some proportion of the total population).

In this section, we discuss some of the background needed to use selection techniques. We will look at several approaches and then relate them to some actual cases to provide an overview. The discussion is by no means comprehensive, but it should give some ideas of various methods of gathering information about populations. There are many individual papers that can be referred to for specific populations. In addition, sampling and statistics texts should be reviewed for data analysis.

Total Counts

When we want to know the numbers of animals present in an area, one way is to count all of them. But this method of conducting a census is used only in special circumstances. It is not often possible to count or to capture and record information about all the animals in a population, even though this is the way to get the most accurate information. **Total counts** are easiest when the population is grouped.

If the area where the population lives is very confined, such as fishtanks at hatcheries, total counts can be made at a number of points in time. This technique also works for nesting birds, such as raptors, with visible nests in a particular region. If all the nests can be located and are easily accessible, eggs and young can be counted in the nest.

Total counts can also be made if populations pass a particular area during the period when counts are desired. If entire big-game herds cross a roadway during their spring or fall migration, they can be counted. Waterfowl resting on lakes during their migratory flights or colonial nesting birds can be counted from photographs.

Sampling

More commonly, samples are taken of the total population. These samples are then projected into an estimate of the total number of individuals. Sampling reduces both the expense and the time necessary to determine the total size of a population. In most cases, samples provide all the information needed.

The first step in developing a sampling technique is to decide what kind of information is needed. After objectives are established, realistic programs can be set up based on the amount of time and money available. Most of the time we are interested in a **random sample,** which ensures that each individual in the population has a known probability of selection[3]. The chances of selection for each individual need not be equal in all the samples, but must be known. Populations are distributed in different ways. Animals may be clumped around water or a forest edge. Competition for human-made habitats, such as orchards, can encourage a uniform distribution (Figure 4–3). In any event, random samples are required to allow all organisms an equal probability of being counted.

By definition, a random sample is free from selection bias, which is a tendency (intentional or unintentional) to favor the selection of individuals that have certain characteristics. For example, if we are sampling trees in a forest, the temptation may be to select especially tall trees of large diameter. To avoid doing so, we need a technique that assures all trees an equal chance of being sampled. If we are interested in sampling the number of eggs found in raptor nests, we must be assured that *all* nests are part of the sampling procedures, not just those that are easy to reach. To avoid selection bias, then, we must develop a sampling scheme. There is no reason that we should select one raptor nest over another. Thus, in a simple random sample, we might give a number to each nest in an area occupied this year and pull the numbers blindly from a hat, giving all members an equal chance of being selected. A table of random numbers is often used to determine simple random samples. Such tables are found in most statistics textbooks.

In sampling for population size, the investigator wants to be sure that the population is not changing in a manner that might make the results of the sample poor indi-

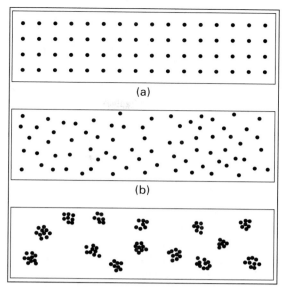

Figure 4–3 Potential distribution patterns: (a) uniform; (b) random; (c) clumped.

Line Transects

Line transects are used for sampling a number of animal populations. For example, we can count all birds seen along a transect. Traps for small mammal can be placed along a transect. It is important, however, to know how effective the transect is in determining the number of animals. Line transects have been used to sample harbor porpoises. Aircraft fly over prescribed transects and count the number of animals seen. Harbor porpoises, however, spend part of their time submerged and might not be detectable from the air. In Washington, a group of biologists tracked a group of porpoises to determine the time they spent at the surface. During seven days, 33 hours of observation were made in an area with a high density of porpoises. Observers saw that the group spent 23 percent of its time near the surface. Furthermore, of the porpoises observed from land observations, only 31 percent were seen from aircraft. Inexperienced observers saw even fewer animals. This experiment tells biologists that it is very important to understand the limits of any technique in determining population density.

Lakey, J. L., J. Calambokichis, S. S. Osmek, and D. J. Rugh. 1977. Probability of Detecting Harbor Porpoise from Aerial Surveys: Estimating of (O). *Journal of Wildlife Management*, 61:63–75.

cators of the population. Normally, we either assume that mortality and recruitment during the data-collection period are negligible or correct for them in the estimate. This can mean avoiding sampling during the breeding season, when the population is changing, and not sampling during migratory periods, when animals are moving in and out of an area.

In addition to making sure that the sample is random, we want to know how accurate and precise it is. **Accuracy** is the closeness of a sample to the true size of the population. For example, a method which shows that there are 100 animals when, in reality, there are 105 is said to be accurate within 5 animals. A sampling technique that does not take into account behavioral interactions, time of day, or correct habitat would be very inaccurate. In estimating a population of tree frogs, we must consider the time of day when they are most active, particularly if the technique involves listening to their calls. Designing a sample with great accuracy is important in getting a correct count of the animals present.

Precision means how close to the actual population size each sampling estimate comes. If the actual size of a population is 100 and we look at two sets of samples— 75, 100, 125, and 98, 100, 102—the latter set is more precise, although both give a mean value of 100 individuals. Repeating samples with great differences in number can usually give us a more accurate average. However, reconstructing the sampling technique may give a more precise answer. Thus, by clustering animals, sampling only in their habitat, and looking carefully at what hours or seasons we conduct sampling, we may come up with numbers that are consistently closer and more repeatable. We also may need to have a larger sample to obtain a more precise answer[7].

To estimate the size of a population accurately, we can use a number of approaches. As indicated earlier, we need to ask, first, why we want the information. Sometimes the mere determination that a species is present in an area is all that we need to know. This can be true of an endangered species. On the other hand, we may want a rough estimate of a population size for management purposes. Thus, if a certain number of eagle nests are observed from roadways, we can project what the minimum population is, so we will know that the population is viable.

Stratified Sampling.   Stratification is a precision-increasing device by which we divide a population into several subpopulations, or strata, each of which is separately sampled and the results combined to give estimates of the whole population. When population data are needed from large areas, investigators often stratify their sample. Stratified sampling allows heavier sampling in areas where animals are either known or expected to be concentrated. The stratified regions are then randomly sampled. One technique is to have, say, four out of five samples in areas where animals are known to be found and the other sample outside the area. If it is known that animals do not occur in certain habitats, naturally it would be improper to sample in those areas. Some areas warrant only minimum sampling. If, for example, investigators want to know how many antelope are found in a basin, they take into account the fact that the animals are commonly found on open prairie lands with relatively hilly topography. By aerial survey, they might mark areas with open prairie on a map and conduct a transect flight across the area to give an estimate of the population. They would spend minimal time around the edge of the woods and in forested areas, where antelope are not commonly found.

Cluster Sampling.   Another form of sampling involves cluster techniques. Although simpler than stratified sampling, cluster sampling is less precise and is commonly used when lack of funds or time precludes greater precision. Cluster sampling is simple random sampling applied to groups of a population, each group being considered as a single unit. The process consists of selecting one member of each cluster, making sure that no two members of the same cluster enter the sample. For example, we can first divide a population of, say, six into three groups of two members and then select one of these three-cluster groups by a simple random process. The group selected will constitute an unbiased sample. Refuge managers may be interested in determining the age and sex distribution of a population of mallards. If the refuges they manage have a number of impoundments, they may want to determine the age and sex distribution from several of the impoundments (clusters) and then form a conclusion about mallards' age–sex ratio within an entire refuge.

In clustering samples, maximum precision can be achieved by forming clusters such that individuals within a cluster vary as much as possible. In the example that we just used, it would be inappropriate, after observing mallards on ponds, to select ponds having an excessive number of males or females. Similarly, in cluster sampling antelope for age and sex information, it would be inappropriate to sample herds when the males and females were isolated.

Sample Size. How to determine the size of a sample is a problem confronting biologists and statisticians. The answers are not always straightforward, because many considerations are involved. We can say that, in general, the greater the variation in the characteristics to be measured in the population, the larger the sample needed.

Generally, managers want to ascertain characteristics such as the number of males or females, age class, or size. It is necessary to start with some idea of the variance that might be expected in the populations to be sampled. Thus, some initial fieldwork is required in several locations, to estimate the within-locality variation. Once these data are available, we can decide on the accuracy desired (how close to the true size of the population we want to come). Remember, the greater the accuracy, the more are the cost involved and the time required. We can then consult one of many statistics textbooks and use an appropriate formula to determine a sample size. In setting up samples, it is important to consult a statistician who is knowledgeable about wildlife statistics.

Indices. Population size, as well as age and sex, can be projected from animal signs, banded animals returned, calls, roadside observations, and other data. The results, or indices, do not give estimates of the total population size, but can indicate trends from year to year and from habitat to habitat. When trend data are all that are needed, costs are greatly reduced.

Trend data should not be used to project or even talk about the size of a population. Such data can indicate only whether populations are increasing, decreasing, or stable in a habitat. Some of the methods discussed in the following section, such as trapping and transects, can be used to develop a population index.

Methods

Once we have determined the type of sampling that is appropriate, we need to look at techniques that can be used with different species. A number of excellent books and papers discuss these techniques. The Wildlife Society publishes the *Wildlife Management Techniques Manual*[8], which discusses a variety of useful field applications. The *Handbook of Census Methods for Terrestrial Vertebrates*[9] also has a discussion of sampling problems. The Cooper Ornithological Society publication titled *Estimating Numbers of Terrestrial Birds*[10] provides information on techniques applicable in different seasons and at different times of the day. All these will help investigators find techniques for sampling specific organisms.

Aerial Surveys. Aerial surveys are commonly used to sample animals or animal signs visible from the air—for example, to count or check dens and rendezvous sites of wolves in northern Minnesota[11]. A number of aerial studies have been made

Swift Fox Survey Techniques

The small, nocturnal swift fox is found in grassland habitat in parts of the western United States. People now feel that this fox, which weighs about 2.2 kg (5 lb), is declining in numbers, although we have no accurate estimates of its population size. The information we do have comes from sightings recorded in trappers' records. Since the fox is small and nocturnal, people seldom see it. Biologists are examining three techniques for obtaining an indication of the animal's range and population size: spotlighting along roadways, counting scat, and using tracking plates. Surveys using spotlights involve driving secondary roads at night. Scat surveys are conducted while walking along roadways that swift foxes use as travel routes. Tracking-plate surveys involve placing .5 3 .5-m metal plates covered with talcum powder and baited with a mackerel scent along roads within the survey area. The foxes are lured onto the plates, on which they leave tracks in the talcum residue.

Spotlighting seemed to be the least accurate means of detecting foxes. Biologists did not detect many using this method. They also found fewer animals in the fall, when the young should be dispersing, than in the spring of the year. Scat, on the other hand, seemed to be a useful indirect method of detecting and estimating swift fox numbers. In some cases, it was difficult to distinguish between swift fox scat and that of other small carnivores in the areas. Tracking plates detected the most swift foxes. This method also seemed to be sensitive to seasonal differences in numbers.

to locate radio-tagged animals and determine their movement patterns. Typically, game and fish departments count animal populations by aerial surveys.

It takes some experience to determine the size of a population accurately. Unless the aerial sampling scheme is properly designed, it will not yield good results. In fact, most big-game aerial surveys yield only trends. But aerial surveys have been successful in locating raptor nests and determining the number of active nests (Figure 4–4).

Figure 4–4 Bald eaglet in its nest. (Courtesy L. Cuthbert-Millett.)

Data collected by aerial surveys can then be diagrammed on maps, and ground checks can be made to determine the productivity of the nests. Winter roosts of raptors, particularly bald eagles, have typically been located by aerial surveys.

When aerial surveys of mobile animals are made, boundaries should be used to provide a valid base for comparison. These boundaries can be roads, highways, or power lines, which are easily spotted from the air. Streams are not usually good boundaries, since big game may congregate along them, particularly in late summer. Boundaries should be used to set off a count block no larger than approximately 260 km² (100 sq mi), to allow a complete census in one flight, with little or no change of animals' moving into an adjacent area. The pilot and one observer usually make up the crew. The observer should be the only person counting animals.

The airplane should be flown in parallel strips approximately 0.8 km (0.5 mi) wide. The width of a strip can be judged by sightings on colored marks or streamers attached to wing struts. A large piece of paper may be taped to the observer's window to prevent counting animals outside the strip. All animals are counted that are in the field of vision immediately adjacent to the aircraft outward to where the line of vision reaches a predetermined point on the wing strut or edge of the paper (Figure 4–5).

The observer should count only on the side of the plane away from the sun. A straight course (transect) must be flown from north to south and south to north, so that the sun is directly behind the observer. The transects should be flown so that each successive one takes the aircraft further away from the sun, putting the observer in the best position to locate and count the animals.

No census should begin or continue unless there is unlimited visibility, with bright sunlight from approximately 20 degrees above the horizon. Counting should cease by the time the sun reaches 60 degrees above the horizon. Usually, this means that morning counts should begin 30 minutes after sunrise and afternoon counts should be completed by 30 minutes before sunset.

Before the flight begins, the observer should explain to the pilot how the transects are to be flown and what the pilot's responsibilities are. Doing so will prevent misunderstandings and improve the efficiency of the operation.

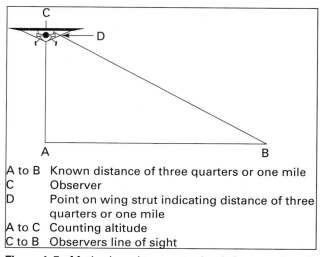

A to B	Known distance of three quarters or one mile
C	Observer
D	Point on wing strut indicating distance of three quarters or one mile
A to C	Counting altitude
C to B	Observers line of sight

Figure 4–5 Method used to count animals from an aircraft.

Trapping. There are many trapping techniques, some of which are discussed in detail in the texts mentioned earlier. Generally, some form of enclosure trap can be used for large mammals. Often a trap is set up in an area of migratory paths or around water holes. The trap may consist of large netting placed between poles. The animals can then wander in if the areas are baited or can be driven in by helicopter. A mechanism to close off the ends of these relatively large traps must be available. Once the animals are within the area, they need to be forced into a funnel trap so that they can be contained and counted.

Small animals are generally trapped in snap, or live traps, which must be placed in such a way that the population is sampled in a stratified or stratified random manner. When using a grid system, which in theory involves placing traps at prescribed intervals, trappers must consider the behavior patterns of small mammals; placing the traps near logs or holes can increase the traps' effectiveness. Some investigators place two or even three traps together to assure that if one trap is filled and another mammal comes by, it, too, will be captured. In a study conducted in South Carolina, investigators estimated the density of a population of small mammals by placing additional traps on a transect radiating from a trapping grid and on parallel census lines (Figure 4–6). They compared results using live trappings first and then removal trapping. The 12 × 12 grid with 144 traps consisted of eight transects radiating from the center. The census lines were crossed by the transect. The removal study was conducted with museum special traps (snap traps) placed at the stations[12] (Figure 4–6). These traps kill the animal, removing it from the population. The results indicated that live trapping yielded better density estimates than removal trapping. The results also indicated that parallel census lines are the most time- and cost-efficient methods of estimating density. If additional data on home range and dispersal are desired, a grid is recommended.

In another study using trapping, investigators were interested in finding how the distribution and movement pattern of a raccoon population changed in a residential community. An area in Ohio near the town of Glendale was divided into grids (Figure 4–8), and 15 traps were put out. The traps, baited with a mixture of sardines and dog food, were checked daily[13]. By marking captured animals, investigators determined the area of movement of the raccoons when the animals were caught again.

Transects. A transect is a straight line along which an observer moves or along which traps are placed. Transect counts are often used to count animals. It is possible to place traps in a transect through a habitat. Investigators can walk along transects and count the number of animals, particularly birds. When transects are used for observation, it is important for the observer to determine the size of the area in which the animals are detectable. To do this requires elaborate preparation. Observers must determine

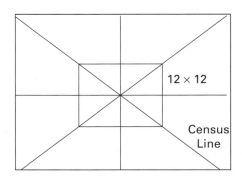

Figure 4–6 Trapping grids used for chipmunks. (From Ref. 12.)

Part 2 / Wildlife Populations

Figure 4–7 Museum special small-mammal trap.

at what distance they can see or hear animals. Some investigators have set up an index of conspicuousness, a scale of distance at which one can see various species[14]. Transect counting has been altered somewhat into what is referred to as the *variable circular plot* technique. Here, investigators walk along a transect, stop at intervals, and look for wildlife species, particularly birds. A species that is not easy to see will have a very short coefficient of detectability; one heard or seen at a greater distance will have a much longer coefficient of detectability—thus the name *variable circular plot*.

Mapping. Mapping can be combined with transect counts or used alone. In mapping, one draws a rough map of the area to be surveyed, including poles, rocks, buildings, fences, roads, and other noticeable landmarks. The location of each animal seen is marked on the map during field surveys. Mapping is particularly effective for animals, such as birds, that maintain territories during the breeding season.

Indirect Methods. Indirect methods of counting animals, used in a number of wildlife studies, involve projecting figures from animal signs or markings. For example, the number of beaver in a particular area can be calculated by counting the number of beaver houses along a stream if the investigator knows the average number of beaver per house. This technique can also be used for muskrats. Investigators frequently estimate bird populations during the breeding season by counting the number of singing males in a particular area. This number is projected into two birds for each male. Since birds defend territories, the technique is fairly accurate.

Owls and some other raptors swallow their food whole and regurgitate nondigestible parts, such as fur and bones. This regurgitated material is called a *pellet*. Pellet analysis can help determine what species of rodents, insects, and birds are in the area. Owl pellets can also indicate populations, since the pellets are usually found

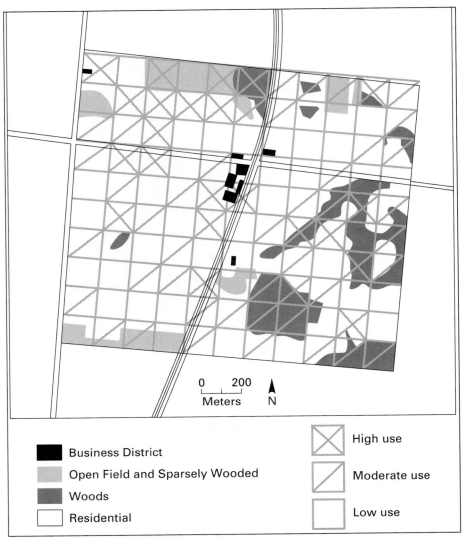

Figure 4–8 Grid showing distribution of the raccoon population in a residential area. (From Ref. 13.)

Business District

Open Field and Sparsely Wooded

Woods

Residential

High use

Moderate use

Low use

below roost sites. The estimate of rodent populations from owl pellets is complicated somewhat by the fact that some parts are unrecognizable. However, it is possible to determine what types of populations there are and the proportion of the owl's diet that comes from each of these populations. Of course, an accurate estimate of the population sizes is not possible, because the proportion of the population the owls have taken is unknown.

Some investigators use tracking methods to determine population sizes. If it is known that herds of large ungulates such as deer move across roadways during their migration period, these roadways can be prepared so that investigators can determine the number of deer crossing each morning by counting the hoof marks.

Several investigators have used scat analysis to get an indication of a population size. However, most of the data collected from this kind of analysis have furnished information only about the prey taken by animals such as wolves and areas in which those animals have fed[9]. Thus, individual behavior influences the results, as does digestability of the food eaten. In a study conducted in the Bridger Teton National Forest, scat analysis was used to ascertain the approximate number of black bear and their movement patterns. An important objective of the study was to determine variations in feeding habits during the various seasons. Investigators were then able to identify plants particularly important to the bear and so could recommend reduction in disturbances to areas in which those plants grew.

Indirect data have also been collected to determine the number of animals harvested in each age class to find the productivity of a population. One technique involves rings on the teeth (*cementum annuli*). The teeth of sea otter or other animals can be removed and the cementum annuli counted to obtain information on the age structure of the population[15]. The ages of many ungulates are determined by this method. Teeth can be removed at harvest check stations (set up by game wardens to examine how successful hunters have been) and sent to a laboratory for analysis. (See Chapter 15.)

Mark Recapture. The mark-recapture technique is used to estimate numbers of birds, small mammals, lizards, fish, and snakes. The technique, which consists of capturing an animal, uniquely marking it, and releasing it back into the population (Figure 4–9), is based on the fact that at any particular time after the initial animals have been marked, the population will consist of some marked and some unmarked animals. To utilize the technique, it is necessary to know the number of marked animals in the population and their proportion of the total population. A number of formulas are used to obtain an estimate of the population.

A study using mark recapture was made of chipmunks near Linfield, Pennsylvania[16]. Live traps were set up at three sites. Chipmunks were trapped and marked by toe clipping. Investigators clipped off a tip of a toe from each foot and then determined, based on the combination of the number of toe clips on the left foot versus the right foot, which animals had been marked previously.

A grid system with 20-m (66-ft) spacing was established to encompass the study area. One hundred ninety-four traps were set and baited with sunflower seeds. The grid was operated for eight days. Each chipmunk was marked. The chipmunk number and trap location were recorded for each captured individual, followed by release at the point of capture. The data were used to calculate the size of the population size by modified Lincoln–Peterson, Schumaker–Eschmeyer, and Schnabel methods.

Figure 4–9 Marking a field mouse with a ear tag before releasing the animal. (Courtesy D. Koehler.)

The *Lincoln–Peterson index* involves only two sampling periods, the first a capture, mark, and release period and the second a period of capture to check for marked animals. The second sample must be taken within a short time after the first, since the method assumes no recruitment of new animals. The Lincoln–Peterson index is expressed as

$$N = n_1 n_2 / m_2$$

where n_1 is the number of animals captured in the second trapping session, n_2 the number of animals marked in the first sample, and m_2 the number of marked animals caught in the second sample.

The *Schumaker–Eschmeyer method* uses the formula

$$N = \frac{\Sigma \, n_2^2 m_2}{\Sigma \, m_2 n_2}$$

The *Schnabel method* employs the equation

$$N = \frac{\Sigma \, n_1 n_2}{(\Sigma \, m_2)}$$

which is a weighted average of a number of Lincoln–Peterson estimates.

In the Schumaker–Eschmeyer and Schnabel methods, sampling can continue until stability is seen in successive population estimates. Thus, n_1 is the total number of animals caught in the individual sample time, n_2 is the total number of marked animals in the population, and m_2 is the number of marked animals caught in the particular trapping session. As censuses are repeated, the number of marked animals can decrease because of deaths and emigration and increase by the marking of new animals. These changes complicate the mark-recapture technique, in which it is assumed that (1) marked and unmarked animals are recaptured randomly, (2) all animals in the population are subject to the same rate of mortality, and (3) all marks are visible. These methods require closed populations.

Other mark-recapture techniques, such as the *Jolly–Sieber* and *capture* methods, can also be used[17]. In the capture program, managers must consider such population parameters as the timing of reproduction (e.g., once a year or all the time) and the behavior and movements of the animals. This elaborate approach involves the use of several mathematical equations. To obtain adequate results, large population samples are often necessary. The reader should consult a manual of techniques or details on these methods[8].

When multiple-sampling periods are used, we obtain an estimate of the number of marked animals in each successive sample. For example, suppose that after six sample periods marked animals are estimated at 400 and after seven sample periods at 300. Then the animals surviving between capture periods are

$$\frac{S_7}{S_6} = \frac{300}{400} = 75\%$$

The loss rate because of death and emigration is

$$100 - 75 = 25\%$$

An estimate can be made of population increase as a result of births and immigration during the period. Assume a total estimated population in sample period 6 of 800, of which 400 were marked. In sample period 7, the total population estimate was 700 and the marked animals 300. Projecting a survival rate of 75 percent, we can account for 600 animals (800 × 0.75 = 600). Thus, 100 animals were added to the population by birth and immigration between sampling periods 6 and 7.

Banding. Bands can be attached to the legs of many birds and bats and so are frequently used in mark-recapture work. Indeed, an elaborate system of banding birds has been set up by the U.S. Fish and Wildlife Service, which, through its Bird Banding Laboratory, coordinates a nationwide bird-banding effort. Aluminum bands are available to authorized banders (Figure 4–10). Banded birds that are recaptured or found dead provide information about their species' movement patterns and survival.

Investigators starting a project may obtain a great deal of useful information from the Bird Banding Laboratory. For example, a researcher studying a population of prairie falcons may want to know these birds' summer and winter ranges and migratory paths. By plotting data from the Bird Banding Laboratory on a map, the investigator will have information concerning the birds' movement patterns that will help in the design of a study plan. Range extensions, migratory behavior, and distribution patterns are types of important information gathered by the Bird Banding Laboratory. The laboratory's waterfowl-banding records have helped biologists establish hunting seasons. Data from these records are often used in life tables and in models that estimate population size[18].

Bands have been used to determine the survivorship of the gray bat throughout the southeastern United States[19]. Investigators censused bats from 1970 to 1976 by going into their roost sites, or hibernacula, most of them in caves. By banding bats while they were roosting, the researchers could get information of the bats' sex and age distribution. Continual banding and recapture over a period of years rendered information for a life table and gave investigators insight into the productivity of the population

Figure 4–10 U.S. Fish and Wildlife Service bands for banding prairie falcon chicks.

Monitoring Changes in Songbirds with Mist Nets

Small migratory birds are very difficult to count. Many factors, such as the availability of breeding and winter habitats, influence their populations. Generally, these short-lived birds are banded, but because they are so small, no significant numbers of banded birds are returning. In parts of North America, biologists are developing mist net programs along known migratory routes. Birds caught in the nets are identified with U.S. Fish and Wildlife leg bands. Volunteers operate the nets each spring in selected locations in hope of recapturing birds that return along the same route. Biologists would like to determine trends in small-bird populations by comparing the numbers of birds captured at each location year after year. These trends would be valuable indicators regarding population sizes, productivity, and the quality of the birds' habitat.

each year, as well as its survivorship and dispersal. Metal tags attached to the ears of animals or the gills of fish can be used to obtain comparable results.

Roadside and Call Counts. A number of national surveys, mostly of birds, are set up to determine population trends or distribution. One survey, in operation since 1965, is the Breeding Bird Survey, coordinated annually by the U.S. Fish and Wildlife Service. A sampling scheme based on blocks of 1 degree latitude and 1 degree longitude, or about 80 3 113 km (50 3 70 mi), was devised for selecting survey routes throughout North America. The number of routes varies with the availability of qualified personnel, but is uniform across a state or province[20]. There is one route per degree of latitude and longitude in most of the western states and Canadian provinces, two in the central and southern states, and four from Tennessee and Virginia northward. More intensive coverage is given to those states or provinces that have a high number of qualified birders. In such areas, the sampling density can be increased once all established routes are being run. The routes are randomly drawn, the starting point and direction of travel being picked from a table of random numbers. Some 2,300 routes have been drawn up in this way, and every effort is made to see that as many of them as possible are run each year. The routes are on secondary roads in order to minimize interference from traffic.

Qualified volunteers are recruited each year. Each observer starts one-half hour before local sunrise, counting and recording all the birds detected in a three-minute stop at the starting point. The count is repeated at 49 more stops 0.8 km (0.5 mi) apart. Only birds counted during the 50 three-minute stops are included in the total. Normally, a route takes from one to four and one-half hours to complete. The data are used to help the U.S. Fish and Wildlife Service and other interested people make decisions about various nongame species. The survey is biased, since it does not necessarily include all shorebirds, owls, or bird species active at different times of the day. It also is biased toward birds frequently found along roadways.

The Breeding Bird Survey has been very valuable in detecting changes in the ranges of species. For example, it was quite easy to show the invasion of southeastern states by the barn swallow (Figure 4–11). The movement of the cattle egret throughout the United States has also been tracked by the Breeding Bird Survey.

Figure 4–11 Changes in the distribution of barn swallows shown by the Breeding Bird Survey.

Data on trends in numbers of big game are also gathered by aerial survey. Yearly counts are compared to ascertain whether the numbers of animals seen are increasing or decreasing. (An increase or decrease in only one year does not signify any change in the population.) The Breeding Bird Survey is a good indicator of trends, since it combines data from many routes each year.

One of the better known national surveys of birds is the National Audubon Society's Christmas Bird Count, which is coordinated during the Christmas season throughout the United States[21]. The count is often made by local bird clubs, which establish sections of their areas for qualified observers to visit. The information garnered is tabulated and published annually in *American Birds*. Population trends can be developed from these results. Another bird survey is the Breeding Bird Census, which is also coordinated by the National Audubon Society. Plots are set up throughout the United States, and qualified observers count the number of birds in these plots every year. Numbers are tabulated and the results printed, also in *American Birds*. The data are valuable because correlations can be developed with the habitats in which the birds are found. The U.S. Fish and Wildlife Service also conducts a variety of waterfowl surveys, including some of winter waterfowl. Most of these are aerial surveys over preestablished routes. They provide an index of the number of waterfowl found in the selected areas.

Call counts are used to discern population trends of some upland game birds. The U.S. Fish and Wildlife Service coordinates annual woodcock and mourning dove call-count surveys. Some states conduct bobwhite, quail, and pheasant call-count surveys. Estimates of coyotes have also been made with this type of survey.

Census and survey techniques are useful in determining productivity. While we indicated earlier that census work should not be done when young are being born unless due allowance is made, information about the productivity of the population is often desirable. For example, raptor studies frequently involve counting the eggs laid in each nest in an area. Observers return to count the chicks that have hatched and those that have fledged.

Animal Welfare. The U.S. Congress and various state legislatures have passed legislation protecting the rights of animals. Most universities and public agencies have individuals or committees that review proposed projects in which animals are handled or marked. It is important to check on the legislation that applies to your area before beginning any project involving the handling of a wild animal.

Animal welfare legislation now governs most of the work biologists are involved in, including all field and laboratory studies, from the traditional to the nontraditional, as well as studies of farm animals. Specific regulations are issued by the Animal and Plant Health Inspection Service of the U.S. Department of Agriculture. In addition, the U.S. Department of Health and Human Services' National Institute of Health provides guidelines that cover areas such as project design, performance, and the relevance of the project to human health. When activities such as monitoring animals by implanted transmitters or experiments involving animals are undertaken, biologists must justify the appropriate use of the species and the number of animals to be used. They must show how undue stress or discomfort to the animal is avoided. If appropriate, sedation or anesthesia must be used. The end point of the project must be established. In addition, proper care and housing of the animals must be demonstrated, as must the use of qualified people to do the work[22].

SUMMARY

The characteristics and size of a population must be measured in order to develop management plans regarding the population. Migration, emigration, and immigration can all affect the number of animals and their habitat. Most movement patterns and habitat use are ascertained by radiotelemetry techniques. Some sampling techniques provide indirect information about animal movement.

Population size is normally estimated by sampling techniques, because total counts are too costly. Most sampling involves the use of a random sample, or a sample in which the chance of selection for each member of the population is known. Techniques used to sample a population of animals are as varied as the animals themselves. Transects can be used to sample birds, small mammals, and reptiles. Aerial counts are sometimes used for big game and trapping for a variety of wildlife. Indirect methods are useful for animals that are difficult to observe.

To get the best results, managers must carefully define the purpose of sampling. The techniques, design, and statistics that can best achieve the desired results must be selected. Decisions must be made as to accuracy, precision desired, and sample needed. The assumptions of the techniques chosen should be known. Knowledge of the animal's biology and behavior must be considered when establishing a sampling program.

DISCUSSION QUESTIONS

1. Define and distinguish migration, immigration, and emigration.
2. Describe the techniques that can be used to measure population movements.
3. What conditions should exist for the use of mark-recapture techniques to estimate population size?
4. What assumptions are made in sampling? When does one sample rather than census?

5. When might you not want a random sample?

6. When would you use a population index rather than a census?

7. What are the differences among random, stratified, and cluster sampling? When would you use each?

8. Of what value are data on population trends? How do you collect such data?

9. It is desirable to find out about black bear productivity in the upper peninsula of Michigan. Suggest a census approach.

10. Can you use signs, such as owl pellets, footprints, and scat, to determine the size of a population? What techniques and limitations are involved?

LITERATURE CITED

1. Andelt, W. F., D. P. Althoff, and P. S. Gibson. 1979. Movement of Breeding Coyotes with Emphasis on Den Size Relationships. *Journal of Mammalogy* 60:568–75.

2. Laundre, J. W., and B. L. Keller. 1984. Home Range Size of Coyotes: A Critical Review. *Journal of Wildlife Management* 48:127–39.

3. Koehler, D. K., T. D. Reynolds, and S. H. Anderson. 1987. Radio-Transmitter Implants in 4 Species of Small Mammals. *Journal of Wildlife Management* 51:105–8.

4. Mech, L. D. 1980. Age, Sex, Reproduction, and Spatial Organization of Lynx Colonizing Northeastern Minnesota. *Journal of Mammalogy* 61:261–67.

5. *Argos Newsletter,* 1988. Anon. Landover, MD.

6. Freda, J., and R. J. Gonzales. 1986. Daily Movements of the Treefrog. *Journal of Herpetology* 20:469–71.

7. Stuart, A. 1964. *Basic Ideas of Scientific Sampling.* London: Charles Griffin.

8. Bookhout, T. A. (Ed.). 1994. *Wildlife Management Techniques Manual.* Washington, DC: Wildlife Society.

9. Davis, D. E. 1981. *Handbook of Census Methods for Terrestrial Vertebrates.* Boca Raton, FL: CRC Press.

10. Ralph, C. J., and J. M. Scott. 1981. *Estimating Numbers of Terrestrial Birds.* Studies in Avian Biology 6. Lawrence, KS: The Cooper Ornithological Society.

11. Mech, L. D. 1977. Productivity, Mortality, and Population Trends of Wolves in Northeastern Minnesota. *Journal of Mammalogy* 58:559–74.

12. O'Farrell, M. J., D. W. Kaufman, and D. W. Lundahl. 1977. Use of Live Trapping with the Assessment Line Method for Density Estimation. *Journal of Mammalogy* 58:575–82.

13. Hoffmann, C. O., and J. L. Gottschange. 1977. Numbers, Distribution, and Movements of a Raccoon Population in a Suburban Residential Community. *Journal of Mammalogy* 58: 623–36.

14. Emlen, J. T. 1971. Population Densities of Birds Derived from Transect Counts. *Auk* 88:323–42.

15. Garsmelis, D. L. 1984. Age Estimation of Living Sea Otter. *Journal of Wildlife Management* 48:456–63.

16. Mares, M. A., K. E. Streilein, and M. R. Willig. 1981. Experimental Assessment of Several Population Estimation Techniques on an Introduced Population of Eastern Chipmunks. *Journal of Mammalogy* 62:315–28.

17. Menkens, G. E., and S. H. Anderson. 1988. Estimation of Small Mammal Population Size. *Ecology* 69:1952–60.

18. Brownie, C., D. R. Anderson, K. P. Burnham, and D. S. Robson. 1985. *Statistical Inference from Band Recovery Data: A Handbook.* Resource Publication 156. Washington, DC: U.S. Fish and Wildlife Service.

19. Stevenson, D. E., and M. D. Tuttle. 1981. Survivorship of the Endangered Gray Bat. *Journal of Mammalogy* 62:344–57.

20. Robbins, C. S., D. Bystrak, and P. H. Geissler. 1986. *The Breeding Bird Survey: Its First Fifteen Years, 1965–1979.* Resource Publication 157, Washington, DC: U.S. Fish and Wildlife Service.

21. Office of Migratory Bird Management. 1988. *Nongame Bird Strategies.* Washington DC: U.S. Fish and Wildlife Service.

22. Anon. 1996. *Guide for the Care and Use of Laboratory Animals.* National Research Council. Washington, DC.

5
POPULATION MODELING

MODELS AND WILDLIFE

Today, wildlife managers recognize that mathematical models can help them organize, manipulate, and analyze extensive data in ways useful in decision making. *Model building* involves the evaluation of changes in any living system, generally through mathematical equations. By varying the input on one side of the equation, we can predict the outcome symbolized by the other side. The process need not be only descriptive, but can be predictive as well. Until the late 1970s, most people working in wildlife management lacked the mathematical background to study populations through the modeling process. Statisticians, engineers, and physicists outside the field, aware of the vast amount of data that managers have accumulated, have attempted to do some modeling. Unfortunately, their lack of biological training has resulted in less-then-perfect models.

Since management involves living systems, it is a very complex process. In many cases, a manager does not have the data necessary to make decisions. Thus, data must be simulated and models based on data that the manager *feels* are correct. The increasing number of people with mathematical and statistical training who are entering the wildlife management field will assist greatly in the collection and analysis of data.

The use of desktop and hand-held computers has brought many changes to wildlife management, particularly in the area of planning. Most field offices now have computers. In some cases, data are entered in the field directly onto the computer. Statistical packages and previously developed models can be used by biologists to make predictions, and therefore decisions, about management almost immediately.

Definition of a Model

A **model** is a means by which we can examine changes in a system. A system is a collection of interacting parts that function as a unit. A **system** can be an animal, a population, a community, or the interaction between habitat and population. In addition, a system may be composed of subsystems. Thus, an animal, one system, includes a digestive system and a circulatory system.

Systems can be viewed as closed when they function without input from outside themselves. This is not generally true of wildlife systems, on which many outside forces impinge. Since most systems we examine are open, modelers must take external forces into consideration.

To characterize a system, it is necessary to describe the nature of its parts. Components that can change are called *variables*. Most complex systems have many variables, some of which may not be measurable. In wildlife management we generally deal with very complex systems. Managers want to know enough about how a system works to be able to predict events or manipulate variables to control events. To this end, observations are made of the real system and deductions made from them about the system's operation[1]. Constructing a model is one way of making the deductions.

Models can be very simple. An equation can model an increase in the growth of a potted plant with the addition of water when all other conditions are held constant. A more complex model could involve a series of equations showing how population growth in a community is controlled by interactions between the community and the ecosystem. Model building often incorporates the interactions as negative or positive feedback. (See Chapter 2.)

Development of a Model

In developing a model, the researcher must first examine the system to be modeled. The important components of the system must be identified. These entities must be selected so that they can be related to one another and are relevant to the problem to be solved. For example, if a model were to be used to predict the impact of stream channelization on cutthroat trout, the physical and biological habitat requirements would be important factors to ascertain, but fishers' success rates would not be relevant. Next, the relationships among the selected components must be understood[2]. What factors influence natality? What are the causes of mortality? These relationships must be represented accurately if the model is to be valid.

Modelers must clarify their thinking and make it specific. Often, they find it useful to write in longhand the information they want to evaluate in the model. A general description of the problem and then a list of input variables and their relationships will usually help. The next step involves drawing box diagrams showing how the input and output variables of the model interact. This process moves from a simple sketch to a drawing with increased details (Figure 5–1). Sometimes modelers need go no further than this descriptive phase. Simply by listing the various input and output components and varying them, they can come up with the desired predictions.

Following the development of a box diagram, the modeler can begin to develop mathematical relationships. He or she must put the pieces together to see whether the whole operates logically (Figure 5–1). Computers are a great help in developing the mathematical relationships, but the modeler must recognize that the computer is simply an accounting and calculating tool and that it is he or she who must find the insight or truths the model reveals[3]. The fact that a model has been run on a computer does

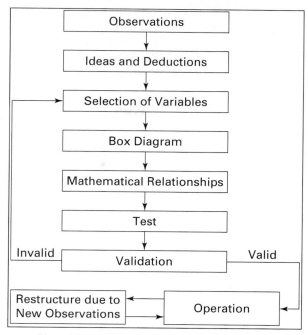

Figure 5–1 The model-building process.

not make its input data, assumptions, or conclusions more creditable. The worth of a program depends on what data are available, how valid the data are, and what assumptions have been made.

Some modelers like to have their models undergo *sensitivity analysis.* This is particularly valuable if field verification is difficult or impossible. Sensitivity analysis is the determination of the relative effects of changes in the model's parameters or variables—for instance, new natality and mortality data entered into a population growth model and the change in outcome observed. One method of performing sensitivity analysis is to alter the model's input variables one at a time while holding the other input variables constant. The output under the various combinations can then be evaluated.

This leads us to the final step in model building: validation[4]. Data from field situations are collected and used to test the projected model (Figure 5–1). Once a valid model has been developed, it can be refined as more data become available. A number of guidelines can be used in establishing a model, including the following:

1. Establish the boundaries, purpose, and time scale of the model.
2. Determine which variables to include and which to exclude from the model.
3. State how various quantities in the model are related to one another. (This is usually accomplished by block diagrams or flowcharts, with lines connecting the boxes that influence one another. It is helpful to write down the assumptions that are incorporated in the diagram.)
4. Describe the relationships used in the model in mathematical terms for processing on calculators or computers.
5. Evaluate, verify, and modify the running model.
6. Make simulated experiments.
7. Assess the results in the light of all assumptions made.

8. Present the data and model in such a way that other people can understand them. (It is important to remember that the less complicated the system and the more information available, the greater will be the likelihood of the model's success.)

Any real system can be looked at from many different points of view. Yet all of the known perspectives together do not equal the real system, because it is always possible to find an additional perspective. Each view does give some information about the system, and a collection of views permits a system concept to be formed. This concept is a function of the observer; the real system exists very well without it and regardless of whether it is right or wrong[5].

Model Characteristics

The manager must remember that *all models are abstract.* They simulate a particular system or event. The degree of abstraction to be used is a value judgment to be made in the light of the model's purpose, but the key to effective modeling is to strike a proper balance between *realism* and *abstraction.* Theoretically, a model is a representation of a real system and should reflect the real system as faithfully as possible. However, managers must understand the input data that went into a model and should never accept the output and develop concrete decisions on trust; they must understand how the model was developed.

Three features of models constructed for natural systems have been identified[6]: generality, precision, and realism. **Generality** is the applicability of a model to different conditions. **Precision** is the ability of a model to provide exactly the results asked for. **Realism** is how closely those results conform to the real system. It is very difficult to obtain all three of these attributes in any one model. One or more has to be sacrificed to obtain strong representation of the other. Different trade-offs have strengths and utility in particular applications[7]. When generality is sacrificed for realism and precision, as happens in many fishery studies, the parameters are reduced to those relevant to the short-term behavior of the organism. The resulting model can produce fairly accurate measurements of populations and can be worked on the computer, giving precise, testable, predictable results.

When realism is sacrificed for generality and precision, the equation often omits important components. This type of model has been relatively unsuccessful in wildlife work. Most models developed for wildlife management application have forgone generality[8]. Wildlife models usually require realism for credibility and precision for prediction and decision making.

Models can be demographic or functional. *Demographic models* are more commonly black boxes—systems with elements we do not fully understand, but know enough about to predict results. Thus, we may not know all the factors in reproductive success, but using average production values, we can develop a population model. In *functional models,* most of the components of the system are known. Predator–prey models are of this type[8].

Model building using mathematical formulas generally requires an understanding of matrix algebra, calculus, and systems design. The descriptions in this section will not provide a background in this area, and we make only limited references to these components; however, managers wishing to develop skill in modeling must have good skills in mathematics and statistics, as well as knowledge of the field in which they are modeling.

Types of Models

Two types of models, stochastic and deterministic, are generally used in wildlife work. When designing a project, we need to decide whether to use a stochastic or deterministic model. Although a deterministic model is less complicated and easier to interpret, if chance events or variability in the answers to questions are important, then a stochastic model should be used[9]. This kind of model enables us to deal with different probabilities. When many groups are likely to influence the results, we may want to present them as a possibility by using a stochastic model.

Stochastic. **Stochastic models** picture a random process of change. The objective is to determine the probability that an event will happen. Figure 5–2 is a stochastic model in which, after time T_p, population N_1 will achieve a size dependent on the probability of event *a*, *b*, or *c*, which can influence the birth or death rate of the population.

As another example, what is the probability that a population of 50 animals of an endangered species will survive for 50 years? We can set up a branch diagram, in which we start with the 50 individuals and then indicate a series of possible impacts on the population, with the degree of probability for each. Thus, that one set of probabilities will be as follows: The probability that the endangered-species population will produce one offspring per female member each year for 50 years may be 0.1, that the number of females will average 0.05 offspring per year for each year may be 0.4, and that the females will produce 0.25 young per year may be 0.5. From this type of branch diagram, we can draw conclusions as to the probability that the population will contain *x* individuals after 50 years.

In essence, the modeler indicates a series of probabilities, one through each branch in the diagram, allowing different results after a period of time, depending on different inputs. In stochastic modeling, these different probabilities may be based on

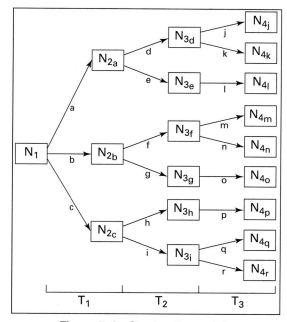

Figure 5–2 Stochastic model.

real data from the field or, since field data are not always available, on assumptions of the modeler. When we are trying to project the future, we obviously do not know what is going to happen, so in our stochastic model we can indicate only probabilities.

Deterministic. A **deterministic model** predicts a future event using a known set of initial conditions. This kind of model can generally be graphed as a curved line (Figure 5–3). The modeler indicates a variety of conditions that may exist at the outset of the modeling exercise. Thus, a population can be given a certain natality and mortality rate and biotic potential. Then the modeler can develop an equation to indicate how, given these values, the population will grow over a period of years and so determine the size of the population after year 1, year 2, year 3, and so on. To look at alter-

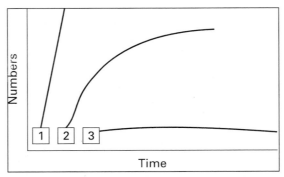

Figure 5–3 Deterministic model. Numbers in boxes refer to three different sets of initial conditions.

Deterministic models use unchanging, "average" data inputs and yield a single set of values as outputs. Most such models start with a group of animals to determine the future size of the population, based on the original size. A variation of this approach is to consider the reproductive value of different age classes in the population. Such a model exists to examine the population dynamics of birds. Four age classes are considered: fledglings, first-year birds, second-year birds, and third-year birds. We can determine the proportion of each age class that survives to become a member of the next age class. In this model, we assume that no birds survive to four years of age. With the help of a computer, we can calculate the population growth rate on the basis of the survival of members of each age class. The results show the number of individuals that should be in each age class and how they are distributed, as well as providing insight into the stability of the population being examined. If we calculated that a high proportion of the birds are in the oldest age class, the population could be decreasing. A major assumption of this deterministic model is that the "average" values used for the survival and fertility rates do not change with time. Stochastic models relax this assumption, so that the input values vary in response in order to simulate random fluctuations in parameters such as temperature or precipitation.

McDonald, D. B., and H. Caswell. 1993. Matrix Methods for Avian Demography. In D. Powers (Ed.), *Current Ornithology,* Vol 10, pp. 139–85. New York: Plenum Press.

native population growths, the modeler will vary the input data on natality, mortality, and biotic potentials. A new curve can be developed from each set of input data. A series of curves can then be plotted on a graph, giving a simulation model. In a deterministic model, conditions are set—if a system is to change from *a* to *b*, the probability is 1. In other words, the previous state of the system determines which system will follow: There is only one outcome that can be inferred from a particular state. In wildlife work, most models are deterministic.

Wildlife Modeling

Population Models. Modeling has been used for a variety of purposes in the biological sciences. Most of the models wildlife managers are concerned with look at how populations change in size. This is a reflection back to the life table. By inputting natality and mortality rates, and a population's age structure, managers can look at the growth rate and make determinations about environmental variables that may affect the population. Given a certain number of animals, for example, managers can determine what effects different rates of harvest may have on the population and what different environmental factors or types of changes in habitat might influence the population. Models are used extensively to evaluate survival based on mark recapture, including returns of banded animals, and other techniques described in Chapter 4. Some population models rely on matrix algebra, as matrices have been used to project population numbers over time[10,11].

A deterministic model can be built to find out how many animals in a mule deer herd are produced under certain temperature and moisture conditions. Mortality rates can be calculated from deer harvests and the information placed into a computer program, using a logistic growth equation, to indicate the number of fawns that will have survived at one-, two-, three-, or four-year intervals. Researchers have inferred populations sizes from returns of banded animals, using deterministic models[12].

Habitat Models. Habitat assessment is another area of modeling in wildlife work. A habitat can be a component of population modeling, entered as environmental resistance or limiting factors. Habitat modeling is a way of examining wildlife habitat requirements and assessing the suitability of a habitat. Once factors of the habitat necessary for the support of a population have been identified, communities can be classified as suitable or at least potentially suitable for desired species. Presumably, when these conditions can be found elsewhere or created elsewhere, a wildlife habitat can be effectively maintained or created. Although much effort has been expended on developing models that indicate types of habitat for different species or communities of wildlife, little effort has been made to verify these results under field conditions.

Planning Models. Wildlife-management agencies are now finding the modeling process useful in planning. By looking at recreation use and funds generated by wildlife-related activities, managers have been able to generate more dollars for management from the legislators.

Spatially Explicit Population Models. These models are becoming increasingly useful tools in managing wildlife. Models are spatially explicit when they combine a population simulation with landscape features of a given habitat. Spatially explicit population models incorporate birthrates, death rates, and population movements dynamically in the context of a habitat. The locations of different types of habitat and other conditions can be put into the model, and the effect of changing landscape features can be related to changes in the dynamics of a population. Spatially, models can be individually based, following the location of each animal across a landscape, or they can be population based, following the movements of a population over a season, a period of time, or the lifetime of a cohort. The models can incorporate the impact of natural and human causes to show how the population responds. Introduced species or an increase in predators can be simulated. The more we know about a species' life history and its response to habitat variables, the more effective our model will be.

Biologists have used spatially explicit population models to evaluate different conservation strategies on spotted owls[13], the introduction of wolves, the introduction of an endangered species, and how the effects of changes in grazing practices affects some large ungulates[14].

SOME MODELS USED IN WILDLIFE MANAGEMENT

Population Models

ONE POP. A model called ONE POP, developed by the Colorado Cooperative Wildlife Research Unit has been used to show the dynamics of animal populations, particularly big game and their associated food supplies. This model, which has been applied in the Rocky Mountain and south-central regions, is deterministic, being based on life-table information and the dynamics associated with sex and age structure, natural mortality, harvesting, and reproduction in ungulates[15]. In a study of an elk population in the Wichita Mountains National Wildlife Refuge in Oklahoma, ONE POP was used to validate population counts and develop hunting strategy. The three mortality periods of summer, winter, and the hunting season were used in examining the change in the elk population. The model simulated each mortality period by subtracting animals from each sex and age class according to specific natural mortality and harvest rates. Then, at the end of each simulated year, the model calculated an expected number of calves on the basis of specific reproduction rates and the number of breeding-age-females remaining in the population. The model then advanced all animals in each age class and proceeded through the next simulated year. The ONE POP model simulates a maximum of 28 age classes and 100 years[15].

Collecting data on the elk herd was a relatively easy task: Most of the data were available from the refuge's files. Simulation using the demographic data produced a population trend that constantly increased in numbers after the herd was established. This trend, similar to that of a real population, reached a level of approximately 300 elk by the mid-1930s[16]. By the 1940s, the simulated population consistently contained about 100 more elk than field counts indicated. The disparity became greater during the 1960s, when field estimates suggested a maximum population of about 700 elk, while the simulator produced a maximum population of nearly 4,000.

Refuge managers suspected that the refuge's elk population was higher than the field counts showed. The suspicion was confirmed and the model's credibility established during the early 1970s, when intensive sampling showed elk numbers to be greater than the field estimates and closer to the prediction of the model.

Turkey. Turkey is a model used to predict the influence of simulated harvests on Iowa wild turkey populations (Figure 5–4). Investigators developed the model to investigate the effects of variable harvests on the population. They wanted to determine the harvest mortality rate that would maintain a stationary population, given the average population parameters and assuming various combinations of hunting and nonhunting mortalities. The modelers divided the population into three age classes: poults, subadults (first-year birds), and adults (birds older than one year). Young birds were classified as subadults before the hunting season in October. Harvesting was incorporated by first calculating how many animals died because of nonhunting mortality before the hunting season, assuming that the annual nonhunting mortality rates applied

Figure 5–4 Wild turkeys are both a sight to see and a popular sports bird in many parts of the country. (Courtesy of the Wyoming Game and Fish Department.)

over the period. Next, hunting mortality was incorporated. Finally, the number that died during the remainder of the year was calculated by assuming that the remaining mortality was due to natural factors only (Figure 5–5).

The simulations were made by using data from each year separately, as well as average values for the four hunting years 1977–1981. Projections were then extended over a 10 year period. The results of these simulations were projected for males and females in the population. Although the parameters would not actually be constant for 10 years, these simulations illustrate the trends to be expected with the given combinations of reproduction and survival rates. Simulations using the parameter values for 1977–1978 resulted in populations that declined about 7 percent per year. The 1978–1979 estimate resulted in an increase of 40 percent per year. The 1979–1980 estimates resulted in an increase of about 2 percent per year and the 1980–1981 estimates in an increase of 6 percent per year. The four-year average increase was about 4 percent per year. In this case, the model was verified by track counts of 465 birds in 1977 and by incorporating the observed survival and fecundity rates for each successive year. No removal of birds for transplantation was incorporated. Winter population estimates derived from counts and tracks indicated that the real population grew about 14 percent (from 465 turkeys in 1977 to 600 in 1980) (Figure 5–6). Simulated populations did not reach as high a level as that observed in the field account, although the projected rapid growth of population was consistent with the field observation[17].

The results of this model show how precision is lost because of realism and generality. They also show that it is necessary to extract several years of data in order to get an accurate estimate of a population and its trends. Models based on one year's data obviously provide inaccurate projections for future populations. There is some indication that the use of only four years of data was inadequate for this study's population, which apparently was influenced by a great variety of environmental factors.

Fisher. A fisher is a large brown weasel-like North American animal. Fisher populations have been declining or are extinct in some areas of the country. Managers need to know the impact of trapping on this furbearer. To investigate the possible ef-

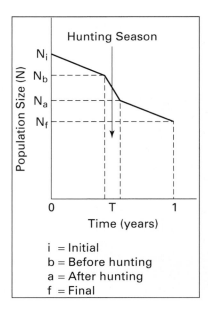

Figure 5–5 The impact of hunting on the Iowa turkey population. (From Ref. 17.)

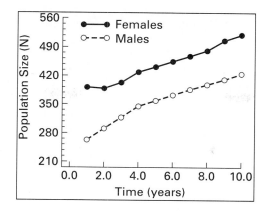

Figure 5–6 Projection of size of turkey population simulated by assuming constant average and fecundity survival rates over 10 years. (From Ref. 17.)

fects of trapping on fisher populations in the upper peninsula of Michigan, biologists used functional predator–prey models that were variants of the one discussed in Chapter 3. Data input included the fisher-population density, porcupine (prey) densities, and alternative-prey (snowshoe hare and white-tailed deer) densities. Rates of predation, mortality rates of fishers, and their feeding efficiencies were entered into the models. Utilizing these data, the models predicted a stable community, but suggested that even small increases in fisher mortality levels above the natural level might cause local extinction of fishers. The conclusion derived from these predator–prey models was that trapping could cause extermination of the fisher population[18].

Deer. Another model based on indirect methods was used in connection with the Colorado mule deer population[4] (Figure 5–7). In this case, preliminary information about the population dynamics of the deer, including population size, sex and age composition, birthrate, death rate, and exploitation rate, was used. Variables were indexed by age groups for each sex. In the diagram, boxes represent stated variables, five-sided boxes decision variables, circles auxiliary variables or functions that influence rates of flow, and cloudlike symbols sources or sinks. Each variable was increased or decreased by its associated rate of flow, indicated by a valvelike symbol. Dashed arrows represent information connectors, and solid arrows show flows of deer. Information flows between and interconnects many variables in the system. The transfer of an information flow does not affect the variable from which the information is taken.

This deer-population model was interlinked with an environmental submodel, a complex model indicating such factors as precipitation, mineral content of the soil, and fat reserves of the animal. This information, when collected from the field, provided an insight into the population density of deer. Predictions of the total number of fawns born, mortality rate, and population size were possible. A variety of harvest strategies were proposed. The model, somewhat more complex than the others discussed up to now, indicates the extent to which the modeling exercise can be taken. Again, data were available to managers, but had to be used with full knowledge of the nature of the information.

Eagle. Both stochastic and deterministic models were developed to show a series of outcomes for a population of bald eagles. The deterministic model used life-table data, and the stochastic model incorporated the eagle's chance for survival, mortality, reproduction, and sex of offspring through a series of random numbers. The life-table model showed that a population with a high survival rate increases at a rate of 11 percent per year, while a population with only a 10-percent lower survival rate decreases

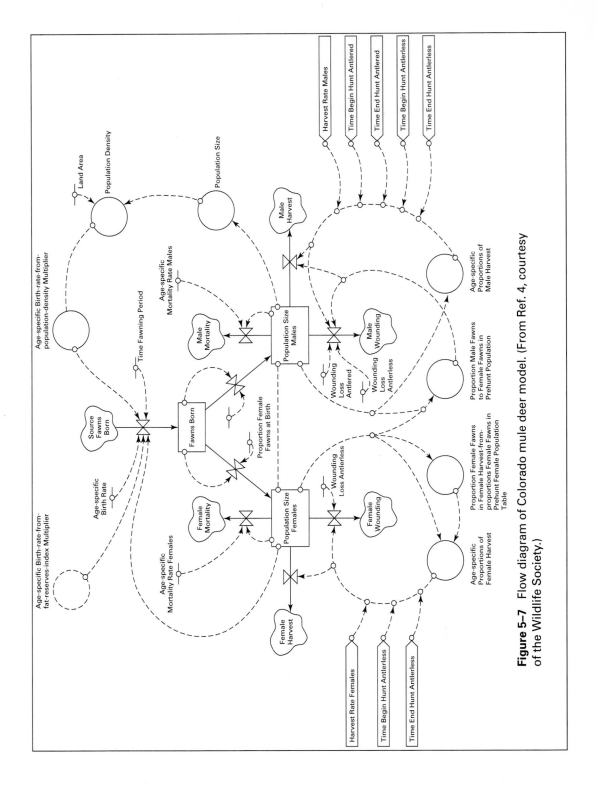

Figure 5–7 Flow diagram of Colorado mule deer model. (From Ref. 4, courtesy of the Wildlife Society.)

2 percent per year, both having the same rate of reproduction and essentially identical age ratios. The modeling also showed many possible outcomes.

The stochastic model showed similar results. Extinction of a population depended largely on its survival rate, partly on the initial population, and scarcely at all on the reproduction rate. Higher reproduction simply led to a larger number. Thus, according to the model, the population dynamics of bald eagles hinges mainly on survival, not reproduction[19].

Fish.   In a correlation model developed to look at fish populations of North American lakes, regression analysis was used to ascertain relationships between fish catches and both biotic and abiotic conditions (Figure 5–8). The fish catch went up as the fauna, or variety of invertebrate organisms and plant material on the bottom of the lake, increased in weight (Figure 5–9). It also increased with increases in the mean depth of the lake. This type of information helped managers determine productivity and so account for variations in fish catches. As the size and depth of a lake increases, the biotic bottom fauna tends to increase, and this causes the catch to go up[20].

Crane.   The impact of hunting on a population is the object of a model on the sandhill crane. The model is a simple deterministic system that embodies density-dependent rates of survival and recruitment. It employs four kinds of data: (1) the spring population of sandhill cranes, estimated from aerial surveys to be between 250,000 and 400,000 birds (Figure 5–10); (2) the age composition of the birds in the fall, estimated to attain 11.3 percent young cranes in 1974–1976; (3) the annual harvest of cranes, estimated from a variety of sources to be from 5 to 7 percent of the spring population; and (4) the age composition of harvested cranes, which, though difficult to estimate, suggests that immature birds were two to four times as vulnerable to hunting as adults.

Because the true nature of sandhill crane population dynamics is so poorly understood, it was necessary to try numerous (768 in all) combinations of survival and recruitment functions and to focus on the relatively few (37) that yielded population sizes and age structures comparable to those in the real population. Hunting was then applied to those simulated populations. In all combinations, hunting resulted in a lower crane population, the decline ranging from 5 to 54 percent. The median decline, 22 percent, suggests that a hunted sandhill crane population might be three-fourths as large as it would be if left unhunted[21].

Figure 5–8 Largemouth bass. (Courtesy of the Wyoming Game and Fish Department.)

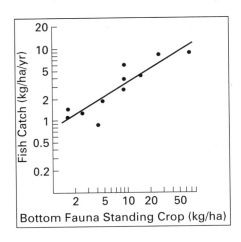

Figure 5–9 Comparison of fish catches and bottom-fauna weight on 11 lakes in the United States. (From Ref. 20.)

Kite. Two population models, one deterministic and the other stochastic, were developed to examine the life-history patterns of the Everglade kite, an endangered raptor species found in south Florida[22]. A life table was constructed from data derived from 161 birds fledged from a total of 183 nests. The data were used to develop a deterministic model. Some of the data needed, such as the proportion of the population that breeds each year, was not available, so estimates were made by field personnel. The model allowed biologists to determine the importance of females of each age to population growth. Figure 5–11 shows that at approximately age three, females begin to contribute greatly to population growth by their reproductive success. Younger females, if successful in rearing young, contribute more to population growth because the mothers' and offspring's fertility periods overlap for a longer period of time. (This principle is used in reverse by those wishing to control the human population. Postponing childbirth will mean that fewer generations will be alive at one time.)

A stochastic model was developed to look at different environmental factors that might affect the kite population (Figure 5–12). By establishing different environmental conditions, biologists could estimate what life-history patterns could cause the bird's extinction. The models helped managers evaluate management options. They showed that the kite apparently evolved a high adult survival rate in partial response to variable

Figure 5–10 Sandhill cranes roosting in the Platte River, Nebraska.

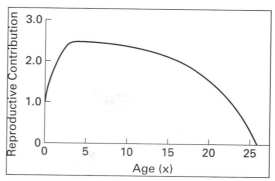

Figure 5–11 Reproductive contribution of each age class in the Everglade kite population. (From Ref. 22.)

reproductive output, resulting from changes in water level in the lower tip of Florida. During a series of dry years, reproductive success can be very low.

Disease. Disease can play a role in the population status of many species of wildlife. Biologists can often recommend corrective actions. A simulation model of Jackson Hole elk used reproduction rates from herds with and without brucellosis. The

Figure 5–12 Everglade kite. (Courtesy of the U.S. Fish and Wildlife Service.)

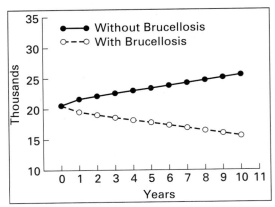

Figure 5–13 Jackson Hole elk population projections for herds with and without brucellosis.

predictions for a 10-year period indicated a decline in herds with brucellosis (Figure 5–13). Managers may therefore want to institute a program of vaccinating elk against brucellosis.

Habitat Models

Correlation. Habitat models can be related to population parameters. A study was made in a deciduous forest in Tennessee to determine what factors of the environment are correlated with populations of breeding birds. Data were collected on habitat variables in the areas populated by these birds. By comparing the populated areas with areas in which birds were not present, the researchers could develop a model of habitat requirements for the different species[23]. The data were subjected to discriminant function analysis and correlations between habitat and discriminant function variables for 13 abundant species identified. Downy woodpeckers were associated with areas where there were saplings, white-breasted nuthatches were commonly found in areas with short trees, cardinals were commonly associated with trees of middle height, and red-eyed vireos frequented taller trees.

This type of information can obviously be useful to managers. For instance, it helps to know that removing certain types of trees will affect some species of birds. In general, this sort of model is helpful in predicting what changes will occur with alterations in habitat.

Predictive. A forest simulation model was developed to predict the structure of habitats for nongame bird species[24]. The model was developed by inputting data on changes in the forest that might occur as a result of succession (natural changes in the community) over a 500-year period. From this model, simulations could be based on the habitat requirements of nongame birds. Thus, as trees change because of succession and timber harvesting, the bird species may change. Figure 5–14 shows how the percentage of habitats available for the ovenbird varies in disturbed and undisturbed forests over a 500-year period. In a 120-year cycle of timber harvesting, the variation in population size is more pronounced. These types of data can be combined for different simulations to indicate what changes should be effected as alterations in habitat occur.

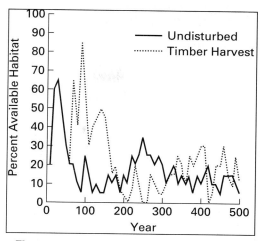

Figure 5–14 Available habitat for oven-birds over a 500-year period in disturbed and undisturbed forests.

Today, databases are commonly used to make predictions of changes in wildlife populations. Models were developed by the U.S. Fish and Wildlife Service that could use databases on habitat availability and nesting success to evaluate a mallard management plan in the prairie states. Individual treatments were proposed and included (1) the purchase of land for waterfowl production, (2) the purchase of wetland easements, (3) leasing of uplands for waterfowl management, (4) retirement of croplands, (5) the use of no-till winter wheat, (6) delayed cutting of alfalfa, (7) the use of nest boxes, and (8) the construction of islands of nests and the use of predator-resistant fencing.

Using simulation models with the databases, biologists predicted that nest boxes would increase the survival of young birds by 20 percent. No-till winter wheat would increase the survival of the young by 6 percent. The use of all of the mentioned methods would increase survival by 24 percent. In another simulation using this kind of model, nest baskets were found to be the most economical of the management practices for mallards[25].

SUMMARY

Models that predict changes in wildlife populations or habitat are important decision-making tools. Models are a means of evaluating many interacting variables in a system. They are built by making deductions from observations of systems and are validated by data collected in the field.

Models are abstractions of reality that should strike as close a balance as possible among generality, precision, and realism. The latter two qualities are the more important in wildlife work. Both stochastic models, which depend on random change, and deterministic models, which predict outcomes derived from known conditions, are used in wildlife management.

Wildlife biologists have used models to predict the impact of various harvest seasons and quotas on populations and survival rates for endangered species. Models are also being used to correlate habitat characteristics with wildlife and to predict the impact of changes in habitats. Thus wildlife planners also are finding models useful.

DISCUSSION QUESTIONS

1. What is a model?
2. How can models be validated? Give examples.
3. Describe a simulation experiment using a model.
4. How can managers use deterministic and stochastic models?
5. What should managers know about a model before accepting its results?
6. How can the precision of a model be increased? What is sacrificed in the process?
7. Give examples of how a habitat model might be used by a manager in a field situation.
8. Construct a simple model for a grouse population, showing ways to increase the harvest.
9. Suggest ways in which population models can help managers reduce predation.
10. All prospective wildlife managers should take courses in statistics, calculus, and systems biology. Why?

LITERATURE CITED

1. Berryman, A. A. 1981. *Population Systems.* New York: Plenum Press.
2. Dale, M. B. 1970. System Analysis and Ecology. *Ecology* 51:2–16.
3. Connolly, G. E. 1981. Assessing Populations. In O. C. Wallmo (Ed.), *Mule and Black-Tailed Deer of North America,* pp. 287–345. Lincoln, NE: Wildlife Management Institute and University of Nebraska Press.
4. Medin, D. E., and A. E. Anderson. 1979. *Modeling the Dynamics of a Colorado Mule Deer Population.* Wildlife Monograph 68. Washington, DC: Wildlife Society.
5. Pielou, E. C. 1978. *Population and Community Ecology.* New York: Gordon and Breach.
6. Levins, R. 1966. The Strategy of Model Building in Population Biology. *American Scientist* 53:421–31.
7. Fowler, C. W., and T. D. Smith. 1981. *Dynamics of Large Mammal Populations.* New York: Wiley.
8. Johnson, D. H. 1982. In G. G. Sanderson (Ed.), *Population Modeling for Furbearer Management,* pp. 25–35. Midwest Furbearers Management North Central Section, Central Montana and Plains Sections, and Kansas Chapter of the Wildlife Society, Lawrence. KS.
9. Starfield, M. M. 1997. A Pragmatic Approach to Modeling for Wildlife Management. *Journal of Wildlife Management* 61:261–70.
10. Leslie, P. H. 1945. The Use of Matrices in Certain Population Mathematics. *Biometrika* 33:183–212.
11. Leslie, P. H. 1948. Some Further Notes on the Use of Matrices in Population Mathematics. *Biometrika* 35:213–45.
12. Brownie, C., D. R. Anderson, K. P. Burnham, and D. S. Robson. 1985. *Statistical Inference from Band Recovery Data: A Handbook.* Resource Publication 156. Washington, DC: U.S. Fish and Wildlife Service.
13. Lamberson, R. H., et. al. 1994. Reserve Design for Terrestrial Species: The Effects of Patch Size and Spacing on the Viability of the Northern Spotted Owl. *Conservation Biology:* Vol. 8:185–95.
14. Dunning J. B., et. al. 1995. Spatially Explicit Population Models: Current Forms and Future Uses. *Ecological Applications* 5:3–11.
15. Gross, J. E., J. E. Roelle, and G. L. Williams. 1973. *Progress Report, Program ONE POP and Information Processor: A Systems Modeling and Communication Project.* Fort Collins, CO: Colorado Cooperative Wildlife Research Unit, Colorado State University.

16. Williams, G. L. 1981. An Example of Simulation Models as Decision Tools in Wildlife Management. *Wildlife Society Bulletin* 9(2):101–7, esp. Fig. 2, p. 104.

17. Suchy, W. J., W. R. Clark, and T. W. Little. 1983. Influence of Simulated Harvest on Iowa Wild Turkey Populations. *Proceedings of the Iowa Academy of Science* 90:98–102.

18. Powell, R. A. 1979. Fisher: Population Models and Trapping. *Wildlife Society Bulletin.* 7:149–54.

19. Grier, J. W. 1980. Modeling Approaches and Bald Eagle Population Dynamics. *Wildlife Society Bulletin* 8:316–22.

20. Matuszek, J. E. 1978. Improved Predation from Yields of Large North American Lakes. *Transactions of the American Fisheries Society* 107:385–94.

21. Johnson, D. H. 1979. *Modeling Sandhill Crane Population Dynamics.* Special Scientific Report 222, Washington, DC: U.S. Fish and Wildlife Service.

22. Nichols, J. D., G. L. Hensler, and P. W. Sykes, Jr. 1980. Demographics of the Everglades Kite: Implications for Population Management. *Ecological Modeling* 9:215–32.

23. Anderson, S. H., and H. H. Shugart. 1974. Habitat Selection of Breeding Birds in an East Tennessee Deciduous Forest. *Ecology* 55:828–37.

24. Smith, T., H. H. Shugart, and D. West. 1980. A Forest Simulation Model to Predict Habitat Structure for Nongame Birds. In D. E. Capern (Ed.), *The Use of Multivariate Statistics in Studies of Wildlife Habitat,* pp. 114–24. Washington, DC: USDA Forest Service.

25. Cowardin, L. M., D. H. Johnson, T. L. Shaffer, and D. W. Sparling. 1988. *Applications of a Simulation Model to Decisions in Mallard Management.* Fish and Wildlife Technical Report 17. Washington, DC: U.S. Fish and Wildlife Service.

Part 3
WILDLIFE HABITAT

The tropical palm rain forest in Puerto Rico provides a habitat for many species of wildlife. (Courtesy of the U.S. Department of Agriculture.)

We now understand the characteristics of populations, as well as the methods that can be used to determine present and future numbers. Next, we need to turn our attention to the habitats where the animals live. The next three chapters describe the needs of animals, how the habitat supplies those needs, and what happens when the habitat is altered. After reading this part, the student should see how habitat manipulation can influence natality and mortality, as well as how models can be used as predictors of changes in population resulting from a stable or an altered habitat. From this information, the most appropriate approaches to managing the habitat for populations and communities can be planned.

6

ENVIRONMENTAL CONDITIONS NECESSARY FOR SURVIVAL

Why are animals found where they are? Because their basic needs for food, water, and shelter are met there. Some animals require a very specialized habitat; for example, some species of fish can thrive only in water that varies just a few degrees. On the other end of the scale, English sparrows seem to be able to live in a multitude of habitats. If we wish to maintain desirable species, we must be certain that their basic needs continue to be met. This includes preventing impacts on the quality of a habitat from any type of pollution. Of course, pinpointing the quality of a habitat is somewhat difficult, because the same habitat that is undesirable for one species can be desirable for another. Also, animals desired by people may decline in numbers while others increase, making the general term *habitat quality* misleading. We need to know which species are desired before we can plan management programs.

Conditions necessary for the survival of species differ. The composition of each community is a result of evolutionary and recent history, as well as of human influences on the natural system. Every species has physical and biological needs that must be satisfied in the community or habitat in order for the species to survive. The more mobile species can use more than one type of habitat to supply these needs. Birds frequently nest in one habitat and feed in another. The great gray whale gives birth in warmer waters around Baja California, but migrates north during the summer months. Many large mammals move from harsher high altitudes to low-level meadows as their winter range. We are also familiar with the long migratory routes of many species of birds.

FACTORS AFFECTING POPULATION SUCCESS

The needs of a species can be divided into physical and biological needs. If all needs are met in the habitat, each population grows until it competes with other populations or uses all materials available for survival. For a population to survive, the minimal quantity of each physical and biological resource to satisfy each need must be available. When one physical constituent, such as a mineral necessary for the survival of a plant or animal, is not present in the soil in an amount sufficient to support a population, the population cannot exist in that habitat. A commodity that is present in limiting amounts will constrict the population's growth or distribution. This concept is called the **law of the minimum,** and the factor limiting the population growth is known as the **limiting factor.** The factor present in the smallest amount determines the limits of the population's growth.

Animals can be limited by food, water, temperature, humidity, barometric pressure, shelter, or reproductive sites. Biological needs must also be satisfied for a population to survive (Figure 6–1). In one case involving the dusky seaside sparrow, a shortage of mates became so limiting that the population could not recover when a fire destroyed most of the nesting females. Some whale populations are so small that males and females cannot find each other.

Limiting factors are, of course, a form of environmental resistance. They contribute to the K factor in population growth equations and therefore affect the habitat's carrying capacity. Changes in limiting factors can alter the carrying capacity. Thus, the addition of fertilizers can increase a habitat's carrying capacity for grazing animals. The addition of water impoundments to dry areas increases their carrying power for various species.

While the presence or absence of some factors can influence the populations of animals, so can the degree to which a factor is present. Water, sun, temperature, chemicals, and other physical factors have an acceptable range, called the *zone of tolerance,* in which an animal can survive and reproduce. Animals may also be able to survive in a *zone of stress,* beyond the tolerable range (Figure 6–2). Beyond the zone of stress, the animal dies. Thermal limits are best seen in fish, which cannot carry on normal activity in the zone of stress.

Physical Factors

A number of physical needs must be met in order for a population to survive. A few are described next.

Figure 6–1 Tiger salamander require moisture and heat for survival. (Courtesy of the Wyoming Game and Fish Department.)

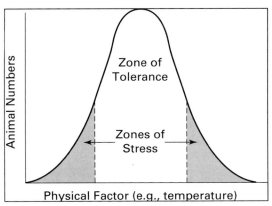

Figure 6–2 Zone of tolerance and zone of stress with respect to temperature for living organisms.

Water. Nothing can live without water or without the proper kind of water. The evolutionary invasion of land did not occur until plants and animals developed autonomous means of retaining water. Highly refined developments of this ability are found in desert life. Desert animals often avoid the heat of day by being more active at night. Cacti can retain water because of their tough outer coating. Many desert plants are physiologically adapted to remain as seeds until enough rain falls for them to grow, flower, and form more seeds. Some desert plants do not reproduce in a dry year, but do so several times during a wet year. Polar areas on the earth have their own type of desert: Although water exists in the form of ice, it is not readily available to support life.

The movement of water controls the distribution of aquatic life. Organisms in fast-flowing streams and rivers or tidal areas must attach themselves to a stable object or be able to move themselves so as not to be washed away. Sometimes this object turns out to be another organism. In the ocean, major upwellings and currents influence the distribution of life. Upwellings bring additional oxygen to deep water and additional nutrients to the surface. Ocean currents move marine organisms throughout the world.

Habitat improvement techniques for wildlife include an increase in the water supply. For example, managers found that the mule deer herd on Fort Stanton property in New Mexico responded to the development of permanent sources of water. In a five-year period, when the available water increased, the deer went from 14.7 to 19.2 in number per section. When the amount of water was decreased, the number of deer dropped to 9.4 per section. When water was made available again, the deer increased to 22.1 per section.[1] Observations showed no change in reproductive behavior—simply that deer moved to habitats with more water and away from those with less water.

Solar Radiation. The amount of solar radiation received on the earth varies according to the location, time of day, and time of year. Amounts can vary in an area because of trees and mountains. The solar radiation reaching the bottom of a forest is much less than that striking the canopy (Figure 6–3). A similar variation occurs in grasslands and shrub lands. Each habitat, then, has certain microclimatic conditions that favor different species. The microclimates may differ considerably when the composition and structure of the habitat create wide radiation and temperature ranges at different levels.

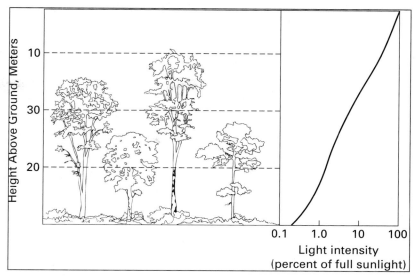

Figure 6–3 Less solar radiation is available at the bottom of the forest canopy than at the top, as vegetation filters out the sun's rays.

Temperature. The maximum yearly temperature, minimum daily temperature, and mean annual temperature of a habitat have a major effect on the distribution of organisms living there. Cold-blooded animals, such as snakes, lizards, and salamanders, are found primarily in the warmer parts of the world, because their bodies must obtain heat from the surrounding environment. Plants are also limited by temperature in their distribution.

The temperature tolerance of living organisms is relatively narrow. Human beings, for example, prefer to live in areas with temperatures around 20 to 22°C (70 to 72°F). Of course, we have found ways to extend the range of acceptable temperature. When the air becomes cooler, we wear heavier clothing; when it becomes warmer, we don lighter clothing. Still, we are restricted to a certain range. By reason of body physiology and enzyme-controlled reactions, animals are able to withstand greater drops than increases in temperature. Many live near their upper tolerance limits. Some specialized lower organisms can live in hot springs with temperatures of 85°C (185°F). The highest temperature encountered in spring water is generally about 36°C (97°F). In deserts, the temperature may reach a daily maximum of around 46°C (115°F) for as long as a month. That is when it is important for shaded areas to be available. The lower tolerance limit for temperatures is less critical for life, and very low temperatures may be tolerated for short periods of time, especially by organisms in a dormant condition. Water that falls below 0°C (32°F)—in the case of salt water, below −2.5°C (28°F)—freezes and, obviously, has a major impact on life.

Warm-blooded animals (**homeotherms**) are able to regulate their temperatures physiologically and therefore to survive in a broader range of temperatures than can cold-blooded animals (**poikilotherms**) (Figure 6–4). The poikilotherms tend to assume a temperature close to that of their environment. However, a number of behavior mechanisms allow them to regulate their temperature. For instance, they can move into or away from the sun or lie on the surface of soil or rocks to help maintain their temperature (Figure 6–5).

Figure 6–4 Avocets and black-necked stilts are warm-blooded, or homeothermic, shorebirds. (Courtesy of U.S. Fish and Wildlife Service; photo by L. G. Goldman.)

Aquatic life appears to have a very narrow temperature tolerance; thus, there is concern when industrial activities change the temperature of rivers and streams. It is believed that a change may cause some of the higher organisms living in the stream to die and other, less desirable biota to move in. Fish generally have a narrow temperature tolerance than terrestrial animals do. Fish species can survive only in waters whose temperatures are compatible with the fishes' internal tissues and chemicals[2] (Figure 6–2).

Different species of fish respond to different temperatures. For example, most brook trout are found in water at about 11°C (52°F). However, they can live in waters of from 2 to 18°C (35 to 64°F), indicating that this range is their zone of tolerance. Waters below 6°C (42°F) and above 14°C (57°F) are zones of stress, where the trout can survive for a time, but apparently cannot reproduce. Atlantic salmon, on the other hand, live in waters of from 6 to 20°C (43 to 68°F), but are most abundant at 14°C (57°F).[3]

Daylight. The length of the day, called the **photoperiod,** is detected in plants by chemical means. Plants control their flowering and sprouting by this chemical sensing of daylight. Many plants flower when the hours of daylight increase, signaling the arrival of warmer weather. Some tropical plants apparently have evolved a very sensitive response to the photoperiod; responding to changes of only a few minutes in the

Figure 6–5 The prairie rattlesnake is a cold-blooded animal (poikilotherm). (Courtesy of the Wyoming Game and Fish Department.)

length of the day. When plants are moved long distances, either north or south, their photoperiod mechanism frequently operates incorrectly, causing them to flower when weather conditions are not opportune. Photoperiod-sensitive differences also exist in plants at different altitudes on the same mountain slope.

Many forms of animal activity also are triggered by daylight. Birds, for example, have special photosensitive cells in their eyes (Figure 6–6). In many species, migration is thought to be triggered by the length of the day, as registered on these cells. In wildlife, the length of the day causes differences in the use or subdivision of habitats. It is quite common to see different animals at different times of day. When checking small-mammal traps, biologists find that some animals are caught during the evening, some at night, others in early morning, and still others during daylight hours.

In a study of small mammals in Idaho, ground squirrels were caught in morning and afternoon trap sessions. Kangaroo rats, pocket mice, and voles were active during the night. Birds are most active in the early morning hours, because most feed on insects active at that time. A study in the eastern deciduous forest showed that most birds sing from 6:00 to 10:30 A.M. After that time, the number of birds singing declines. Thus, if biologists want to census birds by listening for their singing, they need to finish before 10:30 A.M. each day for best results.

Disturbance of the habitat by intrusion during a particular time of day or night may alter the activity patterns of animals using the habitat. Animals that change their activity patterns to a different time of day can come in competition with those animals normally active during that time.

Air Currents. Air currents are important to many species of wildlife. Updrafts from valleys provide wind patterns along mountain ridges, which allow migratory birds, particularly raptors, to soar, thereby reducing their energy expenditure. Winds distribute plant seeds and some microorganisms, and local winds affect moisture levels in the soil and the relative humidity of the air. Winds also have many undesirable effects. Forest and range fires spread rapidly and may get out of control because of winds. High winds depress plant growth. Plants above the timberline are

Figure 6–6 Ruddy duck. Migration in many species of birds is triggered by the length of the day.

usually stunted and have spreading characteristics. Hurricanes or tornadoes force major changes in the community.

When the landscape is altered, the wind pattern can undergo a major change. Eagles are known to nest in canyons that protect them from the prevailing winds; however, when developments alter these areas, nesting raptors may no longer be able to utilize them.

Soil.   Soil influences the distribution of plant and animal life. The texture and composition of the soil provide the physical basis for plant growth. Some animals, such as badgers, make use of local physical characteristics in digging burrows. Generally, badgers prefer areas near rocks with enough clay to support the structure of the burrow. Other burrowing animals, such as prairie dogs, gophers, and ground squirrels, and secondary burrow users, such as burrowing owls, are also affected by the soil's physical characteristics.

Worldwide, the various types of soil influence the major types of vegetation, which in turn provide the structure for animal communities (Figure 6–7). Each of the major types of soil has characteristics that are related to the type of vegetation supported. There are a number of ways to classify soils, but the most important aspects for wildlife managers relate to how well the soils can retain water. To indicate this, we use drainage classes. Soil texture is also very important (Figure 6–8). Different soil characteristics can be found on soil maps, often developed for counties, watersheds, or management units such as national forests.

Most chemicals essential for plant growth, with the exception of oxygen and carbon dioxide, are taken from the soil. The quantities of these elements present in the soil indicate its fertility. Although the nutrient level is usually high in relation to plant needs, not all elements are in a readily usable form. Soils, then, are a medium from which the biological system extracts nutrients.

Many factors determine the nutrients that plants get from soils. The size and arrangement of soil particles affect water flow and storage, air movement, and the soil's ability to release nutrients to plants. Water infiltration into the soil is influenced by its characteristics. Sandy soils, which have larger particles, will take up more water than clay soils, which have smaller particles. Vegetation helps retain water and increase the infiltration rate. Ground cover can increase infiltration from three to seven times. Soil that has been cultivated for planting generally absorbs less water than does undisturbed soil with a vegetation covering.

Organisms also contribute to the fertility of the soil. Some are part of the decomposition cycle, participating in the breakdown and release of nutrients from dead organic matter. These organisms also play an important part in the formation of *humus,* which is decomposed organic material that contributes significantly to the texture, water-holding capacity, and in some cases, ion-exchanges capacity of the soil. Some organisms release minerals from soil particles, making them available to plants. These organisms are responsible for some of the soil's chemical reactions, and they may create by-products, such as acids, that change the physical makeup of the soil. Earthworms and other, larger organisms are important in maintaining aeration and texture.

Soils vary considerably from one ecosystem to another. Soils with the greatest potential for agricultural productivity, such as soils in the grasslands, generally have highly diverse biota, while poorer soils, such as those found in the northern coniferous forests, have fewer species. Unfortunately, many human activities decrease both the diversity and fertility of the soil. Although the application of chemical fertilizers in normal amounts is not generally detrimental to life, the use of excessive nitrogen fertilizers or various biocides may destroy whole groups of organisms.

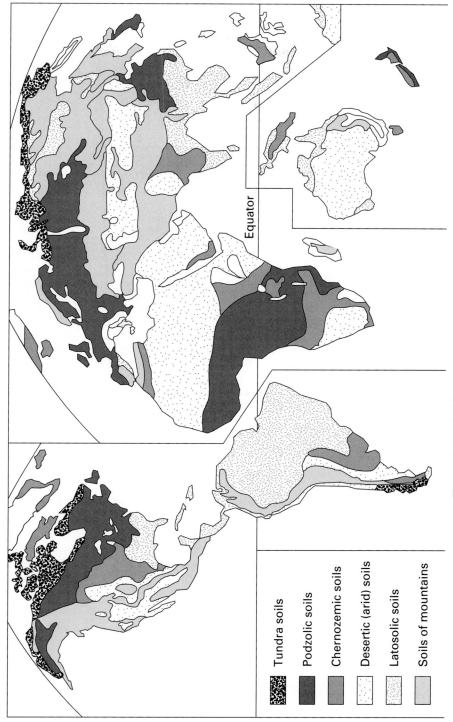

Figure 6–7 Worldwide distribution of types of soil.

Tundra soils

Podzolic soils

Chernozemic soils

Desertic (arid) soils

Latosolic soils

Soils of mountains

Equator

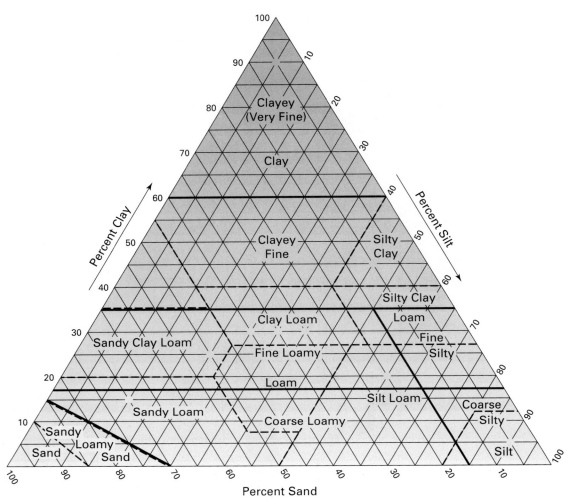

Figure 6–8 Soil texture scale.

Farmers have known for many centuries that fertilizers in the form of organic materials, including human and animal wastes or animal and plant parts, improve the structure of the soil and increase crop yields. Wildlife managers are beginning to recognize the potential for managing wildlife through habitat-improvement techniques using fertilizers. Some plants desired by wildlife respond to fertilizers; others do not. Nitrogen in the form of urea was applied to parts of the Fort Stanton Cooperative Range Research Center in south-central New Mexico. Even though the growth rate of mountain mahogany did not increase as a result, and wavy leaf oak and sagebrush showed no production response, the crude protein of oak leaves did increase, with fertilized plants containing 26 percent more crude protein than did unfertilized plants. Deer activity, as measured by pellets, was greater in the fertilized than in the non-fertilized area.[4] Studies have indicated that other factors, such as rainfall and minerals in the soil, may play a part in the improvement of deer browse by fertilizers. For example, nitrogen was absorbed by grasses in greater amounts when applied with lime in the Black Hills of South Dakota.[5] More deer were observed in the fertilized meadows.

Barometric Pressure. Atmospheric pressure is responsible in part for the fact that different forms of life are found at different altitudes. The higher the altitude, the lower is the pressure, making respiration more difficult for animals as altitude increases. Exploration of the highest altitudes by people is highly restricted without a means of obtaining additional oxygen. Those species of animals that live in oceanic depths show structural adaptations to pressure. Atmospheric pressure also indicates changes in the weather. Some animal species seem able to perceive certain atmospheric pressure changes that are indicative of weather changes.

Biological Factors

Food. All organisms require a source of energy. Whereas plants can use sunlight directly, animals must use other living matter—either plant or animal, depending on their position in the food chain. Some animals can use a variety of food sources. Black bear, for example, can eat berries, plants, fish, and even garbage. Other animals have highly restricted diets. The Everglade kite eats only the apple snail found in the south Florida marshes. Animals with specialized menus generally have restricted habitats, of course.

Most migration patterns result from the fact that food is seasonally more readily available in one place than in another. Migration from summer to winter ranges occurs because food is no longer available when snow covers the summer area. Often the migration of large game to their winter range becomes critical, however, because animals may be confined to a relatively small area with a limited source of food.

Social Structure. The social structure of a population is a component necessary for the population's survival. Animals often go through elaborate courtship rituals prior to mating. These rituals can be influenced by the type of habitat in which they take place. Some species remain isolated individually or in sex-specific groups during periods other than courtship. Environmental stimuli and compatible habitat must be available to allow these cycles to continue. Social animals such as prairie dogs have evolved very specific defense mechanisms upon which the whole group is dependent. We have discussed territory as a behavioral mechanism for preventing overpopulation of a habitat. Behavioral attributes have evolved with each population. They allow the population to survive and propagate.

Community. The community, which is composed of all living organisms in an area, influences each population. Competition among animals, the structure of vegetation, and different microclimates are all attributes that result from the community. The community of populations develops dynamic characteristics that provide for the needs of some animals.

Attributes of the community that can be measured and related to the presence of wildlife include the community's *pattern* or *distribution, structure, size, layering, edge,* and *diversity.*

Pattern. The distribution of plants in a community is generally influenced by soils, water, and exposure. This is dramatically illustrated by the difference in plant and animal species living on north- and south-facing slopes in eastern deciduous forests. Species also exhibit different distribution patterns. Statisticians show us that animals (or plants) can have *random, regular,* or *aggregate* distribution patterns. (See Chapter 4.)

Random distribution patterns would occur if the position of each animal or plant were independent of the positions of all the others. Although many sampling techniques assume some form of randomness, actual cases are very hard to find. "Regular distribution" means that there is some repeatable pattern. An example is the distribution of territorial birds, which defend areas against intrusion by others of the same species. Competition tends to disrupt regular distribution. Trees in an orchard are given a regular or uniform distribution pattern to minimize such competition.

Because many plants are clumped, animals tend to form aggregates. A concentration of animals around water sources and riparian habitats is common. Structures or elevated places in deserts and grasslands often have larger concentrations of wildlife than the surrounding land. Behavioral characteristics may also create aggregates of animals. Herds of deer, antelope, and elk are common. Prairie dogs live in colonies, and some species of fish live in schools.

Pattern is very important in wildlife management. Habitat-enhancement projects must take into consideration characteristics that attract wildlife. Additional water impoundments, rock piles, fence rows, poles, and downed logs are forms of habitat improvement that can encourage aggregations of wildlife.

Patterns must be considered, too, in designing and executing a sampling program. Consider the distribution pattern in Figure 6–9. Suppose that some form of quadrant sampling were initiated. In quadrant *A*, aggregation would probably not be detected, whereas quadrant *B*, used repeatedly, would probably provide evidence of aggregation. The results of transect, quadrant, or plotless sampling techniques would also be influenced by the distribution of animals.

Structure. Community structure consists of the physical makeup of the vegetation, topography, and remnant structures. Structure has an important influence on the wildlife community. Communities with scattered trees provide open areas for the movement of animals. Forests with very dense understories obstruct the movement of larger animals, but can provide shelter. In grasslands, ground-nesting birds, such as savannah sparrows and lark buntings, are found in the open. Scattered brush attracts nesting lark sparrows, and sage sparrows are found, as one would expect, in the sage. Sage grouse appear in scattered sage with open areas, but avoid dense sage.

Snags are standing dead trees or trees with dead branches that provide food, shelter, and perch sites for many species of plants and animals. In the Blue Mountains of Oregon, 139 bird and 23 mammal species were found to use snags for nesting or shelter. Some

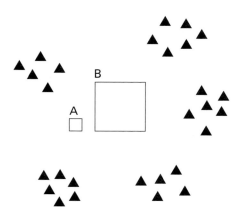

Figure 6–9 Sampling quadrants superimposed on animal distribution patterns.

animals excavate their own cavities (*primary-cavity nesters*); others use existing cavities (*secondary-cavity nesters*). Some bats and brown creepers use loose bark for roosts or nests. Many species of insects living in dead wood are food for birds and mammals.

Snags for wildlife can be hard or soft.[6] Foresters find the hard snags marketable, but not the soft snags. Different wildlife species show preferences for different types of snags, the amount of fungal decay apparently being important in their choices.[7]

Wildlife managers classify snags according to their state of decomposition, using the outline in Figure 6–10, which includes stages from live trees to stumps.[7] Each of these stages has value for some species of wildlife. Snag size (diameter and height) is related to the types of animals that use the snag. Thus, different diameters are selected by different species in Vermont[6] (Figure 6–11).

Decisions must often be made as to how many snags should be available for cavity-nesting bird species. One approach involves a determination of the territorial size from the literature of field studies. One can apply the formula

$$P = \frac{100 \text{ hectares}}{T}$$

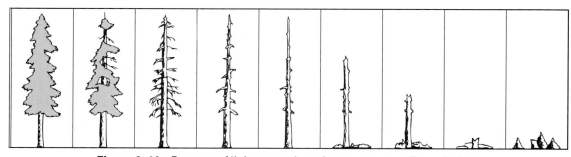

Figure 6–10 Process of living tree changing to a stump. (From Ref. 8.)

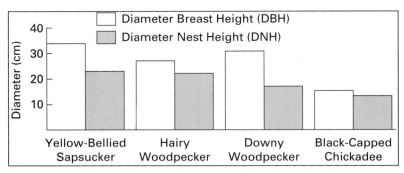

Figure 6–11 Mean diameter of trees used by selected cavity-nesting birds in Vermont. (Courtesy D. Runde.)

where P is the potential maximum number of pairs per 100 hectares and T is the size of the territory. Obviously, this formula must be considered in light of the number of species being managed. It is possible to take into consideration the proportion of each population that should be managed for, that is,

$$N = P \times M$$

where N is the number of pairs per 100 hectares and M is the percentage of the population managed. The snag requirement is based on the fact that there are approximately 16 snags without nest cavities for each one that has a nesting pair. Thus,

$$S = (C)(16 - 1)N$$

where S is the snag requirement per species and C is the number of cavities excavated per year. This formula must be related to the distribution of snags.[7]

Downed woody material is a component of the habitat structure used by many species of wildlife. Downed trees in a forest, an isolated log in the grassland, and dead shrub in a shrub community all provide variations in community structure that are attractive to wildlife. The natural process of timber fall and decay can add nutrients to the soil in each habitat. Logging operations often leave a great deal of wood debris. Biologists in Oregon recommend that 10 percent of the debris in a clear-cut area be left for wildlife cover. They feel that excessive slash may block the movement of many animals, particularly big game.[8]

Size.   The size of a habitat influences the presence of species in it. Ecologists find that with increasing habitat size, the number of species of wildlife increases. In eastern deciduous forests, neotropical migrant birds are declining because of a decrease in the size of their habitats. (See Chapter 20.) Wolves throughout the United States declined for the same reason. Sharp-tailed grouse decline when the open fields they require mature into shrub and trees. In effect, each community is an island. When a community is subdivided or has its configuration changed so that its size is reduced, some species will no longer use the area. On the other hand, new species move into the area—ones that are not always desirable, however.

Tropical forests that are reduced or fragmented lose some wildlife solely because of their decreased size. Tropical ecologists are working with private landowners to compensate for this decrease, encouraging those landowners with adjacent land to leave forests next to one another, thereby increasing the total size of the forest for wildlife.

Layers. A community can be divided into different layers. Most forests in the United States consist of three layers: a shrub layer, a midstory layer between shrubs and trees, and a canopy layer. Each of these layers supports different species of wildlife. The middle layer can be composed of saplings or small trees, or it can be an open area. It is important to herbivores and many birds. Snags or dead trees are part of the layering. The presence of all three layers enhances the diversity of wildlife, particularly in the eastern deciduous forests (Figure 6–12). Each layer affects species in the other layers. In some of the pine forests managed for pulp, all trees are of the same height, and the ground vegetation is often kept clear to prevent fire. The result is little variety in the wildlife species found there.

Layering that results from natural changes in any forest is one of the most important components in maintaining wildlife species. Grassland and shrub land also exhibit layering. This diversity of structure helps maintain the species' habitat.

Features. By reducing the time between seeding and harvest in forests, changes in the rotational cycle alter the habitat structure and thus the wildlife structure. Woodcutting activities also affect forest habitat structure. Many geomorphic features, such as rock piles, cliffs, caves, and ledges, can be important to wildlife. Prairie falcons, for example, prefer to nest in holes, which may extend 1.5 m (5 ft) into a sandstone ledge. Different species of wildlife use different features of cliffs[9] (Figure 6–13).

The Ecotone. When two communities meet, the area of junction is called the **ecotone.** This area can be obvious, as when a field meets a forest. Ecotones where different forests meet, in contrast, can be wide and difficult to discern. Biologists refer to these areas as **edges.** The ecotone is a combined community where two other communities meet. It can have a structure, pattern, and diversity different from that of the surrounding communities. The result is wildlife consisting of some species from the adjacent communities and some edge species. The phenomenon in the ecotone that creates a community different from either adjacent community is called the *edge effect.*

Factors such as fire, logging, roadway construction, transmission-line corridors, and pipeline connectors serve to increase the amount of edge. The fact that there are many species in ecotones has led to the idea that edge is good for wildlife. Edge,

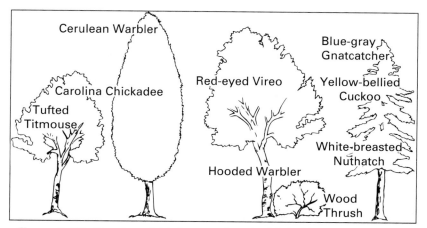

Figure 6–12 Species of birds associated with habitat layers in an eastern deciduous forest.

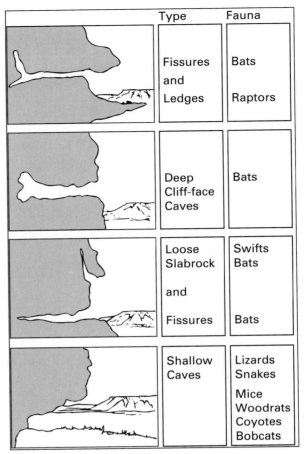

	Type	Fauna
	Fissures and Ledges	Bats Raptors
	Deep Cliff-face Caves	Bats
	Loose Slabrock and Fissures	Swifts Bats Bats
	Shallow Caves	Lizards Snakes Mice Woodrats Coyotes Bobcats

Figure 6–13 Use of cliffs by wildlife. (From Ref. 8.)

however, may attract species that become a nuisance. In some areas of Pennsylvania, increasing edge has led to an increase in deer that destroy cultivated crops. Coyotes also appear to increase with additional edge of forest and field. Nevertheless, some forms of management require that edge be increased. In grasslands, where most of the natural habitat has been used for farming, fencerows can create an edge important to the maintenance of wildlife. Managers, working with private landowners, have shown that pheasants and songbirds increase as natural vegetation is left around fences. Fencerows also provide areas for natural predators, which can help keep down pest species.

Diversity. Measurements of diversity are sometimes used to compare the composition of species in different communities. A low diversity generally means that a few species are common and a comparatively large number are rare. A high diversity generally means that all species present are common. Some use the terms *diversity* and *index of diversity* synonymously. Most comparisons are made between communities of related species. Thus, comparisons are made between the small-mammal communities of two fields, of bird communities in two forests, and of invertebrate species before and after a stream alteration project. A number of indices are used to compare diversities of wildlife. One in common use is the Shannon–Weiner index,

$$H' = \sum_{x=1}^{S} P_i \log_e P_i$$

where P_i is the proportion of the community belonging to the ith species, s is the number of species, and $\log_e$ is the natural logarithm.

The Shannon–Weiner index can be used to illustrate measurements of diversity. When a community has one species, the diversity is 0. When a community has 100 individuals, 99 in one species and 1 in another, the diversity calculation is

$$H' = (-0.99 \log_e 0.99) + (-0.01 \log_e 0.01)$$

$$= 0.010 + 0.046$$

$$= 0.056$$

A community with 100 individuals, 50 in each of two species, has a diversity of

$$H' = (-0.50 \log_e 0.050) + (0.50 \log_e 0.50)$$

$$= 0.347 + 0.347$$

$$= 0.694$$

A useful reference in diversity calculations is the maximum diversity possible with a given number of individuals and species. Thus, the value can be equal in all cases and be based on the total number of individuals present.

Another diversity index is the Simpson diversity index,

$$D = \sum \frac{n(n-1)}{N(N-1)}$$

where N is the number of individuals of all species and n is the number of individuals of a given species.

Diversity indices are used in three different kinds of comparisons. First, comparisons can be made within a community, as when small mammals on the north- and south-facing slopes, canyon bottoms, and ridge tops of a watershed are compared. Second, comparisons can be made between similar communities, as with bird species in deciduous forest sites in Pennsylvania. Third, comparisons can be made over a large heterogeneous area. For example, planners may be interested in the diversity of game species in one county compared with that in another county. Biologists call the results of diversity comparisons within communities *alpha* diversity, between similar communities *beta* diversity, and between large heterogeneous areas with several communities *gamma* diversity.[9]

Diversity measurements are a useful tool for management. Comparisons over time can indicate a community's stability or changing nature. The measurements are, however, only indicators; they should be used with other data and not thought of as absolute.

Biodiversity

Today, wildlife managers need to address biodiversity, which is the total diversity of living organisms in an area. Biodiversity can occur at different levels of biological organization. It can be considered either from the genetic or species point of view. For example, genetic diversity is the diversity of genetic material in a population or community. Community diversity is the diversity of all organisms in a community. In the case of the ecosystem, community diversity includes the organisms and the physical environment. As we decrease the diversity of species in an ecosystem, we reduce genetic and community diversity. This can occur during land-use changes such as urbanization. In urban areas, there are usually fewer species, many of which are well adapted to living around humans. In some cases, such as the introduction of the exotic zebra mussel, biodiversity is reduced as the mussel destroys habitat for other organisms. Legislation to preserve the biodiversity of our natural systems has been introduced in the U.S. Congress. Unfortunately, this is a difficult concept to present to politicians.

As managers, we must understand the short- and long-term impacts of our actions on biodiversity. We must consider the local and regional impacts of our actions on the ecosystems involved. We must also educate the public on the importance of biodiversity in maintaining a stable food web and thus the wildlife community.

Succession

The natural community is the result of a long evolution in which all the species have evolved to survive in an area with broader or narrower physical characteristics. Over time, the habitat has influenced the type of species, and the wildlife in turn has modified some of the physical characteristics (think of beaver damming streams, for instance). The tendency is to form a relatively coherent community and for organisms to become specialized in a stable environment. Where environments are particularly stable, the organisms may become very specialized, as is the case with the Everglade kite.

Disruptions are more pronounced in a stable environment, because evolution has increased specialization. It is thus more likely that organisms will be lost when a more stable environment is disrupted. Most endangered species are highly specialized, requiring a particular habitat, food, or other features found in their stable environment. They are also usually near the top of their food chain.

Bats and Succession

A study of bats in New Hampshire and Maine provided information on which components of their habitat were important to bats. Bats were found in higher numbers along trails and near the edges of bodies of water. Flight activity was highest along trails and still water, while feeding was concentrated along still and moving water. When the biologists examined forests in terms of their age, they found flight activity high in regenerated stands and overmature stands of hardwood trees. Feeding and flight activity was high in regenerated softwood stands and low in the intermediate stages of both hard- and softwood forests. Cutting removed some of the mature forests, thereby reducing the bats' habitat. Nonforested habitat along ponds and streams, but near forests, was also important to bats. Large dead trees were important for roosting. Beneficial management techniques involved cutting some trees and using controlled fires, which also left some overmature and dying trees.

Krusoc, R. A., M. Yamasaki, C. D. Neefus, and P. J. Pekins. 1996. Bat Habitat Use in White Mountain National Forests. *Journal of Wildlife Management* 60:625–31.

Generalist species are usually able to withstand major changes in the environment. Most harsh environments are inhabited by generalists, which can reproduce quickly or survive under adverse conditions. High-altitude grasslands (over 2,200 m, or 7,218 ft) often have bird species that can occupy a broad-range habitat. They nest for a short time after snowmelt and make several nesting attempts when snowfall disrupts the nesting cycle. Because generalists adapt quickly, some species, such as blackbirds and starlings, are considered pests.

Within a community, changes occur constantly. *Evolution* is a long-term change in which organisms evolve adaptations that lead to a stable community. *Succession* is a shorter term change in which the biological component of the environment changes the physical component in such a way that organisms formerly living in the community cannot survive there. When there is an open field, seeds are brought in by animal droppings or the wind, and grasses start to grow, followed by shrubs, seedlings, and, finally, forests. This type of succession, which can occur over a few hundred years, changes a community from a relatively unstable one to one of greater stability. The stages of succession, called *seres,* lead to an ultimate *climax community.* Climax communities vary, the differences depending on the type of physical environment present. Each sere is composed of a dynamic series of lower successional seres. Variations in successional stages lead to different wildlife associated with different seres.

Within communities, changes occur—a tree falls; there is a rock slide. The succession of the community as a whole may be at one stage, while individual parts may be at different stages. These small areas of disturbance are called *gap phases.* They are important to wildlife because they result in a more diverse habitat. Several examples of changes in wildlife as succession occurs illustrate this phenomenon (Figure 6–14).

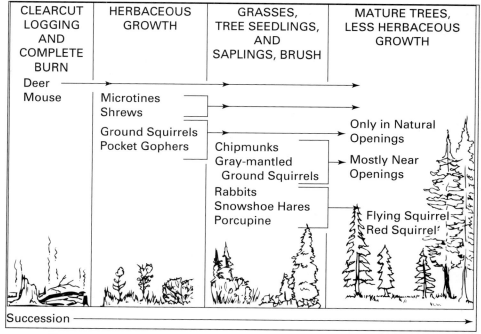

Figure 6–14 Changes in small-mammal species associated with community succession. (From *Impact of Emerging Agricultural Trends on Fish and Wildlife Habitat,* National Academy Press, Washington, DC.)

In an old-growth forest, trees have been shown to provide structural diversity with much specialized habitat. In addition, fallen trees provide sites that concentrate nutrients in long-term cycles. These relationships are very important in maintaining the climax forest.[11]

Lakes also progress through successional stages. Those with relatively few nutrients are called **oligotrophic.** When the process of **eutrophication** occurs, in which nutrients are added to the lake over time, the species there change. Lakes with many nutrients are called **eutrophic.** Some lakes, over hundreds of years, actually fill in and become, first, marshes and, eventually, terrestrial habitat.

One form of management involves sustaining particular successional seres to maintain the community desired. Deer, pheasant, grouse, and quail are all species associated with lower successional seres. If succession is allowed to proceed normally, these species will not remain in an area, and other species will move in as forest mature. All forms of habitat disturbance alter succession. In some cases, preventing disturbances such as fire can be an unnatural course for land managers.

THE ECOSYSTEM AND BIOSPHERE AS SUPPORT SYSTEMS

The ecosystem consists, as we have seen, of the physical and biological components of the community, which interact in such a way that any major changes are disruptive of the entire system. The interaction is most stable in a climax community, in which a natural balance appears to have been reached and in which there is little excess energy.

Energy

A concept basic to any form of ecological study or wildlife management is that of energy flow. Sunlight stored by green plants as chemical energy during the process of **photosynthesis** is available to wildlife. In this process, carbon dioxide and water serve as raw materials to produce sugar and oxygen. Animals that eat plants obtain their energy from the chemical breakdown of sugar. The process of **respiration,** by which sugar produced by green plants is broken down, by both plants and animals, into energy for growth, reproduction, and tissue repair, is quite complex. Most biology texts and plant physiology books have detailed descriptions of respiration. For our discussion, it is necessary only to recognize that energy, which becomes available to plants from the sun, flows from green plants to consumer organisms as each population eats and is eaten. The course that this kind of energy takes is called the *food chain.* Food chains involve energy movement from one population to another. Environmental contaminants can also pass from one organism to another along the food chain. Environmental contaminants can affect organisms if they are retained or built up in the food chain. Complex or interlinked food chains, in which one population feeds on a number of others, are called *food webs.*

Plants that convert solar energy to chemical energy, usable by life, are *autotrophic.* Most of the species that we deal with in wildlife work are *heterotrophic*—that is, they depend on green plants or autotrophs for converting energy from the sun to a usable form. Generally, *herbivores* are animals that feed only on plants. *Carnivores* are flesh-eating animals that devour the herbivores. A number of animals, such as bear and people, are *omnivores*—they feed on both plants and animals. Most carnivores also take in some plant products.

Ecologists use **trophic levels,** a term denoting the structure inherent in food chains, to describe the sequence of energy flow in ecosystems. All green plants (**producers**) are members of the first trophic level. Herbivores constitute the second trophic level. Animals that feed primarily on herbivores make up the third trophic level. The fourth and fifth levels are composed of animals that feed on consumers of the trophic level just below them. Some animals, including human beings, can occupy more than one trophic level.

Natural systems have two types of food webs: **grazing** and **detritus.** The terrestrial grazing food web moves energy and minerals from green plants to herbivores to carnivores. The detritus food web, a process of decomposition, becomes operative when organisms die. Millions of decomposer organisms break down dead bodies, using energy and releasing nutrients from plant and animal matter back into the mineral cycle. Organisms such as earthworms and beetles, called *macrodecomposers,* initiate the process by removing large pieces of the dead organism. *Microdecomposers,* such as bacteria and fungi, finish the process.

Phytoplankton are minute plants that form the base of the grazing food web in aquatic systems. They are eaten by small floating animals called *zooplankton,* which in turn are food for small fish and filter feeders. (Filter feeders obtain their food by straining plankton from the water.) The filter feeders are then eaten by other animals. Decomposers, including crabs, worms, and bacteria, tend to operate rapidly in the aquatic system, beginning to break down organic matter immediately after death, sometimes even before death.

Energy Conversion

The use of the sun's energy to form new biomass (the amount of living organisms), a process called **productivity,** varies in different types of ecosystems. Typically, it is expressed as the amount of usable energy produced per unit of area per unit of time. Examples of such units are kilocalories per square meter per day (kcal/m^2/day) and grams of food per square meter per year (g/m^2/yr). The *gross primary productivity* is the rate at which green plants convert solar energy (by means of photosynthesis) to chemical energy usable by life. Plants use much of this converted energy to maintain respiration. Thus, *net primary productivity*—energy available for growth and reproduction of plants or for consumption by animals—equals the gross primary productivity minus the rate of plant respiration:

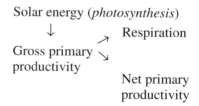

Communities such as estuaries, springs, marshes, and eutrophic lakes can have relatively high rates of productivity, but they constitute a relatively small proportion of the earth compared with deserts, deep oceans, or other areas with low rates of productivity (Figure 6–15).

Both grazing and detritus food webs are important energy-flow systems. While it takes energy to accumulate biomass, it also takes energy to break it down. Some decomposers in the detritus food web, such as algae or other plants, are able to convert energy absorbed from the sun. Thus, organic matter is an energy source for the algae at

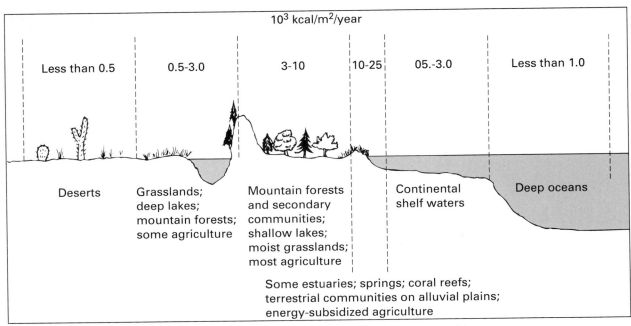

Figure 6–15 Productivity varies in different communities.

the same time that the algae are converting the sun's energy during primary productivity. Although it is convenient to separate energy processes in living systems, one should keep in mind that they are intimately linked in a homeostatic process that, when disrupted, alters the entire system.

The proportion of incoming solar energy converted to chemical or living energy is called *efficiency*. The efficiency of green plants in converting solar energy varies with location: about 0.3 percent on land and 0.13 percent in the ocean. Although some spring-fed ponds have recorded efficiencies of up to 7 percent, most ecosystems remain below 1 percent.[11] Ecological techniques for measuring efficiency vary considerably, but under any technique it is obvious that a very small proportion of incoming solar energy finds its way into food chains and webs.

The transfer of energy from one trophic level to the next is not 100 percent efficient, because the animals in each trophic level require energy for survival and reproduction. Energy is also lost as organisms consume one another. Not all the animals in each trophic level are eaten by others; some die and decay, transferring this energy to the detritus food chain. Producers use energy for respiration and lose energy as heat in the photosynthetic reaction. Energy uptake by herbivores represents the total amount of energy available not only to herbivores, but to all animals. The result is that there are very few fifth-trophic-level animals.

To summarize, three things can happen to energy assimilated at each trophic level:

1. It can be used for respiration by organisms at that trophic level and lost as heat.
2. It can become part of the detritus food chain, either when the organisms of that level die and decay or as it passes through the bodies of other animals without being assimilated.
3. It can be passed on to the next trophic level as animals are consumed and assimilated.

During the transfer of energy from one trophic level to the next, 80 to 90 percent of the energy is lost through respiration or decay, leaving only 10 to 20 percent.[12] Thus, energy available at each trophic level is represented by a decreasing pyramid. Energy, however, is basic in our examination of trophic levels, for it provides the power necessary to sustain life.[12]

Ecological Efficiency

Ecological efficiency, or the ratio of the energy received divided by the energy available at each trophic level, is important to many of our managed ecosystems, such as agriculture. The lower the trophic level from which people derive food, the greater is the available energy. Only a very small biomass of top-level consumers, such as lions, hawks, and sharks, can be supported in comparison with the total biomass produced. The total efficiency of the energy flow in living systems therefore dictates the quantity and type of available biomass in a given area. Smaller organisms use more energy per gram of body weight than do larger organisms. This phenomenon is partially accounted for by the fact that larger plants and animals have less surface area per gram of weight, so that less heat is lost.

Mineral Cycles

All biologically important minerals and compounds have some form of cycle through which they move. These cycles can include gaseous, liquid, and solid stages, depending on the mineral. Sometimes minerals may be bound with chemical elements or compounds, so that they are not readily available. Many minerals, including water,

---

Steelheads Require Several Habitats during Their Lives

Steelheads, relatives of trout and salmon, range from southern California to the Gulf of Alaska and from the coast to the interiors of British Columbia and Idaho. Some migrate from fresh to salt waters and return like salmon. Others, like trout, are resident and live their entire lives in fresh water. The migrating species require several habitats to complete their life cycles. They spawn in clear, cold streams where the eggs incubate for up to four months. Young steelheads spend from two to four years in fresh water, where they reach about 25 cm (10 in) in length. The young fish, or smolt, move downstream to salt water. On the way, they undergo physiological changes that allow them to live in salt water. Some smolt travel as far as 560 km (900 mi). Biologists are not sure where they go in the ocean, but oceanic nutrients and currents apparently affect them and their reproducibility. After two to three years, the fish return to the streams and undergo another physiological change whereby they can swim up the stream to spawn. The fish appear to return to the stream where they were born by using a chemical map in their brain. Unlike salmon, which die after spawning, steelhead can return to the ocean and then back to the stream to spawn again for more than one cycle. All of these fish require clear stream headwaters with rocks to spawn in and to allow smolt to grow. Dams present one of the biggest challenges to migrating fish. This illustration serves to show how some animals require different habitats in different stages of their life history. Some animals even require different habitats on a daily or seasonal basis.

Di Silvestro, R. 1997. Steelhead Trout: Factors in Protection. *BioScience* 47:409–14.

carbon, nitrogen, sulfur, and phosphorus, are required by living organisms. Minerals are often toxic when ingested by a living organism in excess or in a chemical state that should not be in the organism (Chapter 8). The ingestion of cadmium, normally bound to zinc, but released during coal mining, can be toxic to animals. Lead from paint has severe toxic effects. If we understand how these minerals normally cycle through the ecosystem, we can frequently prevent or correct problems.

SUMMARY

Wildlife habitats include both physical and biological factors. Among the physical factors are water, solar radiation, temperature, air currents, the length of the day, soils, and atmospheric pressure. Temperature is an important component in regulating the distribution of animals. Warm-blooded animals are able to survive in a broader range of temperatures than are cold-blooded animals, and aquatic animals are generally more temperature-sensitive than terrestrial ones.

Biological needs include food, a social structure, and a community, which are often closely interrelated. Thus, energy flow within a community is a basis for receiving food energy for survival. The community influences the energy availability through food chains. Attributes of the community include its pattern or distribution, structure, size, layering, edge, and diversity. Community structure is influenced by both living and nonliving components of the vegetation. Special features of the community, such as ledges, rock piles, and caves, are important to the diversity of wildlife.

Ecotones are areas where two communities meet, and that contain attributes of both communities. Changes that occur in one community, such as those of succession, influence the ecotone. Populations strike a balance in the community, with homeostatic mechanisms functioning between the biological and physical components. Disruption of these systems, such as alterations of mineral cycles or food chains, changes the components of the community or the type of community present in the ecosystem.

DISCUSSION QUESTIONS

1. How can we determine the needs of a population?
2. How can the manager best use community and ecosystem concepts in management?
3. What factors that occur in succession must be considered by managers?
4. Is it a good policy to allow community succession to proceed (as in the case of national parks) in wildlife management? Defend your answer.
5. How can edge improve or hinder wildlife management?
6. Describe the movement of energy through a food chain, and discuss factors that limit energy flow.
7. How does one population influence the biological needs of another?
8. What does *species diversity* mean? How is the concept used in management?
9. What components of structure should be considered in managing forests, grasslands, streams, and lakes?
10. How can you convince people that snags should be left for wildlife when they go to the woods on a weekend to gather firewood?

LITERATURE CITED

1. Wood, J. E., T. S. Bickle, W. Evans, J. C. Germany, and V. W. Howard. 1970. *The Fort Stanton Mule Deer Herd (Some Ecological and Life History Characteristics with Special Emphasis on the Use of Water).* Las Cruces, NM: Agriculture Station Bulletin 567.

2. Everhart, W. H., and W. D. Youngs. 1975. *Principles of Fishery Sciences.* Ithaca, NY: Comstock.

3. Fisher, K. C., and P. F. Elson. 1950. The Selected Temperature of Atlantic Salmon and Speckled Trout and the Effects of Temperature on Response of Selective Stimuli. *Physiological Zoology* 23:27–34.

4. Anderson, B. L., R. D. Peper, and V. W. Howard, Jr. 1974. Growth Response and Deer Utilization of Fertilized Browse. *Journal of Wildlife Management* 38:525–30.

5. Thomas, J. R., H. R. Cosper, and W. Bewer. 1964. Effects of Fertilizers on the Growth of Grassland and Its Use by Deer in the Black Hills of South Dakota. *Agronomy Journal* 56:223–26.

6. Runde, D. 1981. Trees Used by Primary Cavity-Nesting Birds in a Northern Hardwood Forest. Master's thesis, University of Vermont.

7. Thomas, J. W. (Ed.). 1979. *Wildlife Habitats in Managed Forests in the Blue Mountains of Oregon and Washington.* Agricultural Handbook 553. Washington, DC: USDA Forest Service.

8. Maser, C., J. M. Geist, D. M. Concannon, R. Anderson, and B. Lovell. 1979. *Wildlife Habitat in Managed Rangeland: The Great Basin of Southeastern Oregon.* General Report DNW 99. Washington, DC: USDA Forest Service.

9. Pielou, E. C. 1978. *Population and Community Ecology.* New York: Gordon and Breach.

10. Maser, C., and J. M. Trappe. 1984. The Fallen Tree: A Source of Diversity. In *New Forests for a Changing World,* pp. 335–39. Bethesda, MD: American Society of Foresters.

11. Boughey, A. S. 1973. *Ecology of Populations.* New York: Macmillan.

12. Odum, E. P. 1983. *Basic Ecology.* Philadelphia: W. B. Saunders.

7
HABITAT MANAGEMENT

Habitat management is the major method by which people control wildlife. Often, without considering wildlife at all, people change habitats as the human population expands. Much of our understanding of what happens to wildlife is based on tracking habitat changes.

Habitat management has many aspects. Sometimes managers manipulate the habitat through the use of fire, fertilizers, or water impoundments. Other times, they let the habitat alone. In this chapter we examine some of the methods used to manage habitats. These methods are closely related to habitat alteration, discussed in Chapter 8. Most of the techniques discussed are tools for manipulation. Each has different effects in different habitats.

To assist managers in predicting some of the results of habitat management for wildlife, there are a number of habitat-classification systems. These systems provide information about the size and location of various habitats, enabling managers to assess the impacts of manipulative activities.

HABITAT CLASSIFICATIONS

Some of the wildlife-classification systems are valuable for worldwide, some for regional, and a few for local habitat descriptions. Resource agencies sometimes develop their own systems of classification.

Biomes

Ecologists have divided the world into large, recognizable communities or associations of vegetation called **biomes** (Figure 7–1). Biomes are the biological expression of the interactions of organisms, mostly plants, with the physical elements in different regions. Climate and soil usually influence the type of vegetation found in a particular area. Vegetation and the physical factors, in turn, influence the type of wildlife present. Although normally named for its climax vegetation community, each biome is composed of life in all developmental stages. Thus, these areas provide a broad opportunity for viewing wildlife management, since within each biome a variety of different management units have been devised. In this section, we describe the major characteristics of worldwide biomes and how they have been subdivided.

Tundra. The far northern parts of the North American continent, Europe, and Asia are known as the tundra. The tundra has very short summers and growing seasons and very long, cold winters. The ground below the surface is frozen (and hence is called *permafrost*) all year. During the short summer months, grasses grow quickly and attract a number of species of breeding birds and mammals. Insects also thrive during this period. Very few sizable woody plants can live here because of the short growing season, so most of the vegetation is in the form of grasses, lichens, sedges, and dwarf willows (Figure 7–2).

Two categories of animals are present here: summer residents and those that remain all year. A number of migratory birds, including many waterfowl, are attracted to the area. Caribou, reindeer, and some small mammals, such as rabbits, are also found here. Most of the food chains are relatively simple. Thus, destruction of one organism in the predator–prey cycle may have a harmful effect on the population dependent on it. Because of the permafrost and the short growing season, disturbances such as tire ruts are very slow to disappear. Widespread mineral development projects could disrupt wildlife dependent on the area for food and could pollute Arctic waters, where an abundance of fish and mammal species are found. Some habitats are being reduced to the extent that species' needs cannot be met. One of the best forms of management for the Arctic would involve preplanning to avoid disturbance of the habitat. Although wildlife management has been relatively limited in the tundra region, effective management techniques could be introduced to avoid destruction of the habitat.

Alpine tundra occurs above the timberline in temperate and sometimes even in tropical areas. These areas have high levels of solar radiation. Snowfall is usually heavy, and wind stunts plant growth. As a result, the regions are relatively unproductive, although some ungulates use them for their summer range, and a number of birds and small mammals also use them.

Boreal Forests. South of the tundra region are a large number of coniferous forests, mostly spruce, fir, larch, and tamarack. This biotic region includes most of Canada, Scandinavia, and the northern part of Russia. The climate is slightly warmer than in the tundra, with much more precipitation—about 38 to 102 cm (15 to 40 in) a year. The soil thaws, allowing tree roots to penetrate more deeply and develop more fully. The boreal forests (taiga) in northern Europe and Asia have rather poor soil because decay there is slow. The acid produced with decay is carried into the soil by rain or melting snow, making the soil relatively infertile for most crops. Although the growing season is short, a large number of birds migrate here, and many mammals, such as moose, caribou, wolverine, and snowshoe hare, live in these forests (Figure 7–3).

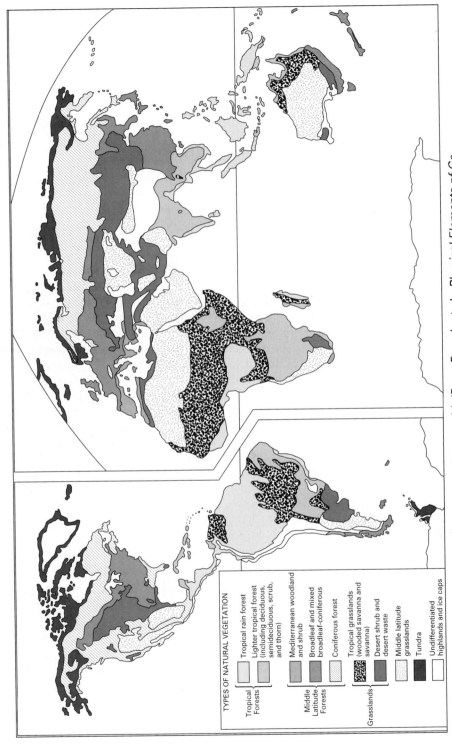

Figure 7–1 Biomes of the world. (From French et al., *Physical Elements of Geography*. New York: McGraw-Hill Publishing Company.)

TYPES OF NATURAL VEGETATION

Tropical Forests
- Tropical rain forest
- Lighter tropical forest (including deciduous, semideciduous, scrub, and thorn)

Middle Latitude Forests
- Mediterranean woodland and shrub
- Broadleaf and mixed broadleaf-coniferous
- Coniferous forest

Grasslands
- Tropical grasslands (wooded savanna and savanna)
- Desert shrub and desert waste
- Middle latitude grasslands

- Tundra
- Undifferentiated highlands and ice caps

Figure 7–2 White-tailed ptarmigan are found in the United States and Canada, in northern tundra as well as above the timberline. (Courtesy of the Wyoming Game and Fish Department; photo by L. R. Parker.)

The best management strategy for the area is to plan carefully to reduce disturbances. Any mineral development must be carefully thought out to avoid the destruction of large expanses of the habitat. Roadways, pipeline corridors, and transmission-line corridors should be consolidated or placed so as to avoid destroying migratory routes or reducing the habitat.

Studies show that warming trends in the boreal habitat due to logging and burning as well as acid deposition from sulfur oxide in the atmosphere are altering the plant and animal species composition.[1]

Deciduous Forests. Continuing south into the United States, Europe, and central Asia, we come to deciduous forests. In the United States, they extend throughout the east and down into parts of the south. The dominant trees are oak, maple, hickory, beech, and other hardwoods. Most become dormant as they shed their leaves in the late fall and early winter (Figure 7–4). Precipitation is relatively high and is distributed throughout the year. Rainfall averages 76 to 150 cm (30 to 60 in) a year, so there is abundant plant growth. The warm, humid summers and cool winters make for dense vegetation. Resident animal species characteristic of the area are the white-tailed deer, ruffed grouse, cottontail rabbit, red fox, raccoon, flying squirrel, and wild turkey. A

Figure 7–3 Caribou are found in both the tundra and boreal forest. (Courtesy of the U.S. Fish and Wildlife Service.)

Figure 7–4 Eastern deciduous forest in the winter.

number of migratory species are found here, including the wood thrush, red-eyed vireo, black-and-white warbler, worm-eating warbler, and scarlet tanager.

Numerous management techniques can be practiced in the deciduous forest region. Indeed, many of the forest techniques we thus far have discussed were developed and first practiced in this region. It is possible to manage for a single species or for a diversity of wildlife. The relatively long growing season in many areas contributes to management effectiveness.

Grassland or Prairie. Between the eastern deciduous forests and the western desert are the grasslands or prairie communities of the central United States. Most continents have similar biotic regions in their central areas. Grasslands, with their relatively low rainfall, are intermediate between forest and desert. Wet and dry cycles often alternate for periods of years. The summers are hot and the winters are cold.

There are several types of grasses in this region. A prairie community maintained by periodic fires is dominant next to the forest. Prairies have an ideal climate for crop production, and the soil is rich in organic matter, because litter deposited in the grass decays. Minerals are not leached out because the rainfall is light—under 76 cm (30 in) per year. Furthermore, the leaching processes are retarded during the winter. These are not only prime agricultural lands, but top grazing areas for many large ungulates such as bison and pronghorn antelope and nest sites for birds (Figure 7–5).

Management strategies here include maintaining native habitats for the wildlife species, avoiding land and water pollution, maintaining fencerows and shelter belts, keeping migratory routes open, and controlling hunting. Conflicts often occur between

Figure 7–5 Burrowing owls use abandoned prairie dog and badger burrows in the grasslands.

livestock owners and wildlife enthusiasts over land use. Some livestock growers argue that if wildlife species were removed, more land would be available for grazing livestock. This is not always true: Resources used by livestock are not necessarily the same as those used by wildlife.

Deserts. Deserts are usually found where mountain regions block the flow of water (from both rivers and rain) into an area where evaporation exceeds rainfall. Desert areas also exist in continental interiors and where prevailing winds sweep inland from cold ocean currents. Arid regions in the southwestern United States have less than 25 cm (10 in) of rainfall a year. The two main areas are the high desert or Great Basin, extending between the Rocky Mountains and the Sierra Nevada, and the low desert, including the Mohave in California and the Sonora in New Mexico, Arizona, and southern California. The vegetation of the Great Basin is chiefly sagebrush, low shrubs, small conifers, and juniper. The low desert has desert shrubs, creosote bush, and a wide variety of cacti. While most plants reproduce each year, taking their clues from the length of the day, desert plants are often triggered to flower by rainfall. A variety of wildlife is found in the desert, but most people do not see the large number of mule deer, rabbits, grouse, and nongame species, including amphibians and reptiles (Figure 7–6).

Like the tundra region, the desert is a fragile system. Because of the shortage of water, the growing season is brief, so that disturbances are slow to heal. Thus, off-road vehicle use, expanding human populations, and mineral extraction are greatly reducing the size of our deserts. Management techniques are aimed at retaining natural vegetation and controlling the harvest of animals. Attempts to increase the water supply and fertilization have succeeded in maintaining some desirable wildlife species.

Chaparral. The biotic region along the coastal area of southern California is part of the chaparral biome. Its relatively stable year-round climate is sometimes referred to as the "Mediterranean climate," since it resembles that around the Mediterranean

Figure 7–6 Southwest desert biome.

Sea. Many people consider this climate ideal and move there in great numbers. The chaparral is interspersed with trees and shrubs. Thoroughly dry in summer, it receives most of its rainfall in the short winter season. Thus, fires are common in the summertime. Many of the trees and shrubs in this area are fire-climax species. Deer, quail, and a variety of nongame species commonly live in the area. The large influx of people has resulted in a decrease in the number of wildlife species.

Tropics. Tropical biotic regions exist in southern Mexico, Central America, northern parts of South America, Africa, and Asia. The rainfall in these regions exceeds 229 cm (90 in) a year. Instead of the usual four seasons, most tropical forests have two: the wet season, when most of the rainfall occurs, and the dry season.

We once thought of the tropical forests as impenetrable, but we now know that they are quite susceptible to human impact. Wildlife management has been slow in coming to many of the tropical forests because the people who live in the areas are generally poor and must give all their attention to extracting a meager living from the land. Removal of the vegetation to raise crops has been disastrous: Most of the nutrients are stored in the vegetation itself, so that when plant life is removed, the nutrients are lost. As a result, crops in the area can be maintained for only two or three years.

Major efforts are now under way in several countries in the tropics to preserve the diversity of the forest by maintaining diverse botanical gardens, test plantations, and banks of seeds and plant tissue. Such efforts seem necessary because of the alarming rate of deforestation and pollution. These tasks focus on maintaining a diversity of plants in the wild.[2] There are areas of heavy rainfall along some seacoasts, as in the Olympic Peninsula of Washington state. Here the plants and animals resemble those of the dense tropical forests.

International organizations are attempting to maintain tropical forests and so to preserve wildlife. One of the more effective efforts has been to get landowners to leave part of their land—especially adjacent portions of different landholdings—in natural vegetation, thus retaining a relatively large clump of such vegetation. (See Chapter 6.) Maintenance of habitat size and minimization of disturbances are keys to managing the tropical forests.

Geographic Classifications of Vegetation

The biomes of the world are divided into broad categories, which are further broken down into subunits for outlining management strategies. Some of these strategies are quite useful for wildlife managers.

Kuchler's Potential Natural Vegetation. One system which has been used extensively is that of potential natural vegetation, developed by A. W. Kuchler. This system shows the types of natural vegetation throughout the United States. The major types of vegetation in the country are placed into easily recognizable categories (Table 7–1), which are in turn broken down to show the different types of forest within deciduous and coniferous forests. Redwood forests, mixed coniferous forests, red fir forests, lodgepole pines, and subalpine forests are listed on the map and described in some detail.[3]

The Kuchler system can be used to predict which wildlife species should be present in different regions. Since it is based on potential vegetation, the wildlife-species composition could change as a result of agricultural practices, logging, and development. Overlay maps based on satellite imagery or aerial photographs can be used to determine current land practices.

TABLE 7–1

Examples of Major Categories of Vegetation in Kuchler's System

Type of Land	Vegetation
Western forests	Needle leaf
	Broadleaf
	Broadleaf and needle leaf
Western shrub and grassland	Shrub
	Grassland
	Shrub and grassland
Central and eastern grassland	Grassland
	Grassland and forest
Eastern forests	Needle leaf
	Broadleaf
	Broadleaf and needle leaf

Ecoregion.   Another useful system is ecoregion classification, which was developed for forest management.[4] The ecoregion classification of North America has nine levels, from a broad national level to local habitats.[5] Three of the levels are shown in Figure 7–7. The levels are based on regional variations in climate, vegetation, and landform. The first level, the domain, is based on subcontinental areas of related climates. The next category is based on a finer subdivision of climate. Ecoregions, then, use provinces based on vegetation and soil, followed by sections using climate, vegetation, and districts based on geomorphology (Table 7–2). For example, the humid, temperate domain occupies latitudes of 30 to 60 degrees. These areas have strong annual climatic cycles. Winter frost determines six divisions: warm continental, hot continental, subtropical, marine, prairie, and Mediterranean.

Figure 7–7   Three levels of ecoregions in the United States.

TABLE 7-2

Hierarchy of Ecoregions

Name	Defined as Including:
1. Domain	Subcontinental area of related climates
2. Division	Single regional climate at the level of Koppen's types
3. Province	Broad vegetation region with the same type or types of zonal soils
4. Section	Climatic climax at the level of Kuchler's potential vegetation types
5. District	Part of a section having uniform geomorphology at the level of Hammond's land-surface-form regions
6. Landtype associations	Group of neighboring landtypes with recurring pattern of landforms, lithology, soils, and vegetation associations
7. Landtype	Group of neighboring phases with similar soil series of families and similar plant communities at the level of Daubenmire's habitat types
8. Landtype phase	Group of neighboring sites belonging to the same soil series with closely related habitat types
9. Site	Single soil type or phase and single habitat type or phase

In the eastern United States, warm continental dominates in the north, but merges into hot continental in the mideastern seaboard and Appalachian mountains and tropical in the south. Warm continental consists of northern hardwood spruce forests in northern New England. Northern hardwood forests become common in upstate New York and northern Pennsylvania. The eastern deciduous forest, with mixed deciduous trees, is common in the hot continental area. The subtropic includes beech, sweet gum, magnolia, pine, and oak forests and southern floodplain forests.

Integrating more than one system to identify homogeneous units of land creates what we call an *ecoclass*. Categories can be related to the major biotic divisions discussed earlier. Thus, the prairie division, one of the major classifications of the ecoregion boundaries, is a subcategory of the grassland biome. The prairie provinces are broken down into the ecoregion category according to prairie parkland, prairie brushland, and tallgrass prairie. Each of these has a specific descriptor and can be related to animals found in the area. Ecoregion classification can thus be an aid in planning the use of resources. Most effective at national rather than local levels, it can be used to organize information about resources for data systems. It can also be used for data interpretation and possibly for the selection of indicator species.

National Wetlands Habitat Classification. Following a national survey of wetlands in 1974, a classification of different types of wetlands, including a mapping scheme, was developed for the United States.[6] Wetlands, where the water table is usually near ground level or shallowly covers the land, are transitional between terrestrial and aquatic systems. Classification as wetland requires one or more of the following: (1) at least periodically, the land supports predominantly hydroplants, (2) the substrate is predominantly undrained hydrosoil, and (3) the substrate is nonsoil and is saturated with water or covered by shallow water at some time during the growing season of each year. Wetland classification systems identify five major areas: (1) those with hydrophytes (plants growing in water) and hydrosoil, such as marshes, swamps, bogs, and fens; (2) those without hydrophytes, but with hydrosoil flats, where drastic fluctuations in water level, wave action, or high concentrations of minerals may prevent the growth of some hydrophytes; (3) margins of impoundments, where hydrophytes have become established, but there is no hydrosoil; (4) areas without soil, but with hydrophytes, such as the seaweed-covered portions of rocky shores; and (5) wetlands without soils and without hydrophytes, such as gravel beaches or rocky shores without vegetation (Figure 7–8).

Since wetland habitats are rapidly being lost and those that remain are changing because of human impact, managers can effectively utilize the classifications to indicate which areas are critical to wildlife species. Knowledge of the habitat requirements of species makes it possible to manage wetlands for communities of wildlife. Because wetland communities have an unusually large concentration of wildlife species that utilize these areas for either part or all of their activities, wetlands are among our most important habitats for wildlife management.

Aviregions. The National Breeding Bird Survey (BBS) was used in classifying bird communities in North America (Figure 7–9). (See Chapter 4.) The idea behind stratification is that each species of bird has its own geographic limits. Within these limits are several zones of abundance, representing the availability of suitable habitats. In mountainous areas, there are zonal boundaries of vegetation types resulting from differences in temperature, precipitation, or wind speed. Typically, the abundance of many species of bird changes abruptly across such boundaries. In flat country, boundaries are more obscure and sometimes very irregular, often extending for miles along a stream valley, where differences in soil type or moisture produce habitats not found a short distance from either side of the stream.[7]

Because the distribution and abundance of birds are so strongly influenced by habitat, particularly in the breeding season, the use of ecological rather than political boundaries is most logical. Ecological boundaries are based largely on areas of abundance in North America.[8] Since the original classification, there have been many minor adjustments in strata boundaries. Such refinements in the United States have come largely from *Physiography of Eastern United States,*[9] *Natural Land-Use Areas of the United States,*[10] *Potential Natural Vegetation,*[3] and various publications for individual states.

The names of the strata, as defined by BBS data, are shown in Table 7–3. The strata are grouped into eight larger regions, which contain broadly similar types of habitat. In the eastern part of the United States, for example, the southeast mixed forest, eastern deciduous forest, and northern coniferous forest are broken down into a number of distinct BBS strata. Managers interested in working with birds commonly associated with the BBS will find this system useful.

Uses and Limitations of Habitat Classifications

Habitat classification systems provide opportunities for managers to evaluate changes that may occur in a habitat. By knowing the total number of hectares of specific types of habitat and the types of wildlife species associated with each of these broad ranges, a manager can make plans to preserve an area. Changes that occur in one component of a classified land can be assumed to occur throughout the region. Thus, important features of the habitat can be correlated with the wildlife species there.

Computers can be helpful in evaluating different types of habitat according to land classification systems. Comparisons can be made of different areas, using common terminology. Management can occur at several levels, regional as well as field.

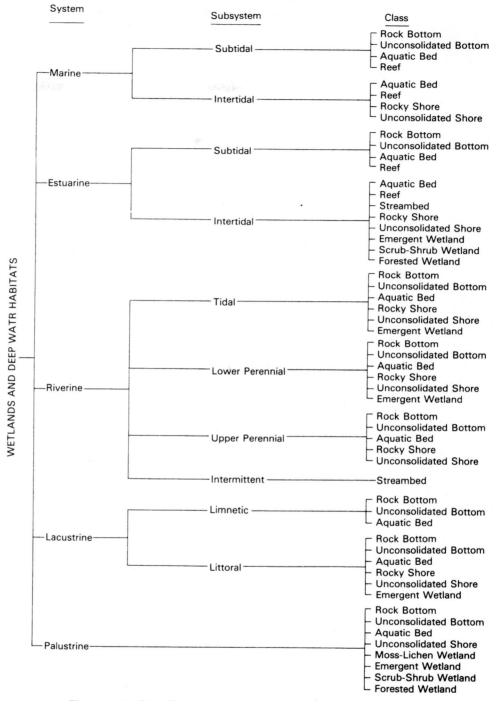

Figure 7–8 Classification of wetlands and deepwater habitats.

Figure 7–9 Aviregions based on physiography strata in the United States and Canada.

TABLE 7–3
Breeding Bird Survey Strata

Southeastern Mixed Forest

01 Subtropical
02 Floridian Section
03 Lower Coastal Plain
04 Upper Coastal Plain

05 Mississippi Alluvial Plain
06 West Gulf Coastal Plain
07 Nueces Plain
08 Glaciated Coastal Plain

Eastern Deciduous Forest

10 Northern Piedmont
11 Southern Piedmont
12 Southern New England
13 Ridge and Valley
14 Highland Rim
15 Lexington Plain

16 Great Lakes Plain
17 Wisconsin Driftless Area
18 St. Lawrence Plain
19 Ozark–Ouachita
20 Great Lakes Pine Belt

Northern Coniferous Forest

21 Cumberland Plateau
22 Kanawha Plateau
23 Blue Ridge Mountains
24 Allegheny Plateau
25 Open Boreal Forest

26 Adirondack Mountains
27 Northern Hardwoods
28 Spruce–Hardwood Forest
29 Closed Boreal Forest
30 Aspen Parklands

Prairie and Plains

31 Till Plains
32 Dissected Till Plains
33 Osage Plains
34 High Plains Border
35 Staked Plains–Pecos Valley
36 High Plains

37 Prairie Pothole Section
38 Missouri Plateau–Glaciated
39 Missouri Plateau–Unglaciated
40 Black Prairie
53 Edwards Plateau
54 Colorado Plateaus and Canyonland

TABLE 7–3

(continued)

Western Mountains	
61 Black Hills	65 Dissected Rockies
62 Southern Rocky Mountains	66 Sierra–Trinity Mountains
63 High Plateaus of Utah	67 Cascade Mountains
64 Central Rocky Mountains	68 Canadian Rockies
Arid Interior	
81 Mexican Highlands	85 Klamath–Pitt Plateau
82 Southern Sonoran Desert	86 Wyoming Basin
83 Northern Sonoran Desert	88 Great Basin
84 Piñon–Juniper Woodland	89 Columbia Plateau
Pacific Slope	
91 Central Valley	94 Northern Humid Coastal Belt
92 California Foothills	95 Southern California Mountains
93 Southern Humid Coastal Belt	

There are limitations, however, on the use of classification systems for wildlife. Most classifications were not developed specifically for wildlife management. Thus, the animals associated with each classification unit are based on general attributes and so are often difficult to relate to one species. In addition, habitat classification systems that have been developed for wildlife work, such as the aviregions based on the BBS, are usually applicable to just one group of animals. Managers and planners should be aware of these limitations.

Forecasting Ecosystem Changes

The U.S. Forest Service has developed a vegetation simulator model (FVS) to simulate changes in succession in the ecosystem. The model was based on many years of research on forest succession, from forest inventory information down to data on individual trees. Information on ecosystem components, including other plant and animal species and soils, was linked to the inventory model. To make predictions, the user needed a list of trees (species and size), a description of methods used to collect the data, and site and location information. FVS was used to predict tree and nontree vegetation growth and mortality in a forest. It was able to simulate regeneration and the effects of proposed management activities. While this model was primarily designed for predicting changes in vegetation, it has been used to forecast changes in which wildlife species would exist, based on management activities. For example, the model was used to predict changes in plant and animal species due to an outbreak of spruce budworm. These types of tools have provided managers with important means of making management decisions.

Teck, R. M. Mocur, and B. Eav. 1996. Forecasting Ecosystems with the Forest Vegetation Simulator. *Journal of Forestry* 94:7–10.

The greatest impact on wildlife—and the world—today has resulted from habitat modification, which results in changes in wildlife species. The presence of desirable wildlife species in a habitat is at once a result and a measure of ecologically sound uses of lands and waters. Wildlife fares well where management principles are sound. Of course, healthy wildlife is a necessary element in successful, long-term management of the land. How wildlife has actually fared and what factors can be used by managers to control wildlife are the subjects of this section.

Wildlife managers have developed many techniques to change habitats and thereby control or manage wildlife species. The land classification systems can provide one avenue for managing habitats. Some of the tools used for this purpose are discussed in this section.

Fire

Species such as the Kirkland's warbler require early stages of jackpine growth as nesting sites. Unless fire burns through the pine forest in which the warbler lives, reducing much of it to the early successional stage, the forest proceeds in succession until it becomes so dense that it is no longer inhabitable by the warbler. Fire can be managed in such a way as to renew the forest's understory, allowing game animals to browse. In the southeast, fire is used regularly to keep the understory down in southern pine forests. Fire therefore influences the type of wildlife found in those forests (Figure 7–10).

Fire has been a tool of mixed blessing in wildlife work. In the early to mid-20th century, forest and plant ecologists felt that fires should be strictly controlled. These feelings resulted from the massive control efforts needed during the destructive fires of the late 19th century. Some of those fires had burned large areas of the forest in the north-central region, destroying entire towns and occasionally killing hundreds of people.

Figure 7–10 Fire sets back succession, creating a diversity of wildlife in a habitat. (Courtesy of the USDA.)

But wildlife biologists observed changes in the composition of species in wildlife communities when forests formerly subjected to periodic fires were not burned for decades. Timber grew tall, but crowns grew closer together, resulting in the gradual suppression and sometimes elimination of understory food plants. Many areas lost traditional or desirable species.

In Michigan, changes in game-bird habitats occurred as the grassland habitat of the greater prairie chickens grew into brushland, thereby attracting sharp-tailed grouse. This habitat gradually grew into open woodlands, becoming a habitat for ruffed grouse. Finally, the vegetation in the habitat became too dense for ruffed grouse.

By the 1950s, biologists began to realize that fire increases the diversity of wildlife species in a habitat because the habitat becomes more diverse. The diversity of northern ecosystems, such as those found in Wisconsin, Minnesota, Michigan, and southern Canada, was maintained in part by fire.

A great volume of literature had appeared on fires and wildlife.[11] Studies showed that a number of bird species benefited from fire in jack pine forests. Breeding bird populations changed in a burned area of that type of forest in northern Minnesota. The diversity of biological organisms, as well as diversity in biological processes, can be maintained by including fire, which alters both the structure of a habitat and the succession of plants that exist there.

Beyond the immediate influences on animals, fire can cause major long-term modifications of a habitat. Obviously, food, cover, water, and structure are drastically modified by this disturbance. For a considerable number of years after a fire, plant succession will continue to change substantially. Some larger animals, such as moose, white-tailed and mule deer, elk, cougars, coyotes, and black bear, and some smaller ones, such as beaver, hare, turkeys, pheasant, bobwhites, sharp-tailed grouse, ruffed grouse, red and blue grouse, prairie chickens, willow ptarmigan, heath hens, and some waterfowl, benefit from fire. On the other hand, fire may temporarily displace or eliminate species, such as caribou, marten, red squirrel, grizzly bear, wolverine, fishers, and spruce grouse, depending on late stages of the development of the community of plants.

Among the smaller nongame animals, the effects of fire are frequently undetermined; on some species, there may be no observable effect. New species do move into an area after a fire. Some of the nongame birds that forage on trees and in the canopy disappear, while ground feeders increase. The shift in small-mammal population results from a loss of forest-dwelling species and gain in grassland and shrub species. Studies show that burned and fire-dependent forests support larger numbers of birds and mammals than do unburned or relatively fireproofed forests, but 80 percent of both mammals and birds remain about the same in both density and trend.

Some biologists assume that the beneficial aspects of fire relate to food production. More than that, burns create a mosaic of different types of vegetation, modify the water supply, and produce a variety of habitat niches. Following a fire, the concentration of plants near the ground causes an increase in mammals and seed-eating birds.

In a three-year study following a wildfire on the Seney National Wildlife Refuge in the upper peninsula of Michigan, investigators found that wildlife changes tended to follow the successional changes in plant communities. Characteristic species for each successional sere appeared in forest and marshland habitats that had been burned. As grasses began to grow on the burned ground, species such as sharp-tailed grouse and common snipe appeared. Woodcock were found in the grassy areas between forests two or three years after the fire. In the burned-over forested area, early plant colonizers, including blueberries and sedges, became very apparent the year after the fire. Black bears increased by the third year after the fire, because of the increase in blueberries and wild raisins. Black-backed three-toed woodpeckers became common in the burned forested area immediately after the fire, but declined as vegetation growth continued and reforestation began.[12]

Fire has been shown to benefit a number of big-game species. In an area north of Flagstaff, Arizona, a study of elk and deer following a lightning fire in a ponderosa pine forest in May 1977 showed that for the first two years, elk use shifted from an old, seeded clear-cut to the newly seeded burn. The third year revealed an equalizing trend between the two habitats. Deer did not use an area thinned by cutting, but increasingly used the wildlife area.[13]

In northern Minnesota, the density of moose increased from less than 1.3 per square kilometer (0.5 per square mile) to more than 5 per square kilometer (2 per square mile) in two growing seasons following a burn. This increase was related to immigration, especially of yearlings, rather than to increased production and survival of calves.[14]

In a study in the Rocky Mountain region, the use of winter range by bighorn sheep and mule deer was evaluated following a fire. Comparisons were made between burned and unburned areas of sagebrush–bluebunch grass–wheatgrass. Grass production decreased slightly the first year after burning, but returned to preburn levels two years later. Big sagebrush seedlings were noted on the burned site two years after burning. Burning was considered beneficial to bighorns; however, it did not seem to have an impact on mule deer, even though the deer utilize sagebrush to a large extent in their diet and sage was removed from the range for at least four years.[15] The nutritive quality of marsh plants was found to increase following burns.[16]

Fire, therefore, can be an effective tool in habitat management. One technique used by managers is to allow a fire to burn as long as public or adjacent lands are not jeopardized. A number of federal land-managing agencies utilize fire in their overall plans. The absence of fire carries the risk of an accumulation of fuel to such an extent that widespread, intense fires and consequent loss of floral and faunal diversity result.[17] This was part of the cause of the Yellowstone fires of 1988. Fire is thought by some biologists to assist in preventing the spread of disease in vegetation and wildlife, reducing parasitism in some species. Last, but not least, fire can be one of the wildlife manager's least expensive management techniques.

Fire can be a source of air pollution and therefore the source of complaints from both near and far. Pollution can be minimized when it is understood and when burns are managed in harmony with local natural conditions. Fire must be integrated into the total land-management practices and the interests of neighboring landowners. Foresters and wildlife managers differ on how burns should be controlled. Foresters tend to favor a hot burn, which usually eliminates much of the debris; wildlife managers, in contrast, prefer not to burn areas with quantities of downed debris, recommending instead a spotty fire and a cool burn for habitat manipulation. Cool burns destroy only some of the surface litter and are used early in the season, so that new grass has time to grow. Wildlife managers also like fires that produce erratic edge patterns, leaving much standing and downed debris for shelter and food for wildlife.

In the wildlife perspective, it is the total habitat that is important. Management of smaller areas (less than 8 hectares) by clear-cutting or fire will encourage some species of birds and small mammals. Big game that prefer edge will appear. Animals that utilize the interior of the forest will not be found around the opening. Thus, by using controlled burns in and around mature forests, managers can maintain a diversity of wildlife.

As a wildlife-management tool, fire requires careful study of the biotic community and its successional stages. For example, in the upper peninsula of Michigan, fire immediately following a logging operation has a greater effect on the succession of vegetation and thus on wildlife than does fire in later successional stages. An earlier fire tends to produce a persistent early stage in succession, whereas later fires tend to retard succession, slowing progress toward the climax. The time of the year can also be important in prescribed burning. Rangeland plant species may be encouraged to grow or decline, depending on spring, fall, or winter burns in relation to the time of seeding. Fire, then, can be a means of maintaining wildlife habitat.

Fertilizers

Fertilizers have already been discussed in relation to grassland habitats. Fertilizers are very successful in improving range habitats in some parts of the country, the success being related to amount of moisture and other soil conditions. Fertilizers are used in eastern hardwood forests to produce rapid growth, with the objective of getting young trees above levels where they can be damaged by deer. Unfortunately, fertilizers are expensive.

Vegetation Management

Manipulation of the vegetation is another tool commonly used by managers to aid in maintaining wildlife species. Vegetation management can take a variety of forms. The removal of understory vegetation in the forest can enhance accessibility for large animals. Increasing the edge can maintain areas for sage grouse in an otherwise densely packed sagebrush habitat.

A study of the use of piñon–juniper woodland by elk and deer in New Mexico revealed that the dense canopies developed in this type of woodland reduced the amount of midstory browse and understory herbage, which in turn reduced the deer and elk's use of the area. When small clearings within the woodlands were made, populations increased. However, when extensive clearings were made, the production of herbage increased, but the number of deer and elk did not.[18]

The size and shape of clear-cuts and their position in relation to uncut timber are important considerations in the improvement of wildlife habitat. One study revealed that deer and elk in New Mexico made greatest use of logged areas adjacent to uncut timber. Circular openings of approximately 8 hectares in spruce–subalpine fir and 18 hectares in ponderosa pine were the most beneficial.[19]

Thinning to improve tree growth can also be beneficial. Herbaceous vegetation generally decreases as a forest ages and the canopy closes. Usually, the production of forage is inversely related to the base area of the remaining trees. An increase in forage production resulting from forage-management practices encourages the use of an area by ungulates. Thinning practices in some parts of the eastern United States, together with other changes in land-use practices, have led in recent years to a major increase in the white-tailed deer population.

The management and improvement of rangeland for livestock offers major opportunities for modifying existing practices to benefit wildlife. In the past, most range-improvement plans have given priority to one or two species of domestic or wild grazing animals; however, evaluation of the effects of such improvement on other species is appropriate.

Millions of hectares of sagebrush and piñon–juniper have been plowed, burned, chained, or sprayed to allow increased growth of grasses. Such treatment appears to increase the production of forage for livestock, but the impact on the wild-ungulate winter range may be quite detrimental. Public lands producing abundant browse for wild ungulates in areas inaccessible to livestock should be excluded from such treatment. It has been suggested that when piñon–juniper stands in the southwest are cleared and seeded with grass in order to improve range for livestock, slopes steeper than 15 percent and some northern exposures should be left intact to provide cover and refuge for deer and elk.[18]

The application of herbicides on livestock range can also improve conditions for wild ungulates. However, treating large tracts of land can be detrimental to small mammals and raptors, among others, by reducing cover, desirable plant species, and food. In north-central Maine, herbicide application on clear-cuts resulted in a reduction of small-mammal herbivores.[20]

In Colorado, the U.S. Forest Service found that putting carbon black on snow causes the snow to melt faster, making deer forage more available on the animals' winter range. Managers found that with the carbon black, the winter range could support more deer than it could under normal conditions.[21]

Seeding

Particularly in shrub areas, grasses or other plants introduced by seeding can result in earlier greening following winter. Species that supply high-quality plant food help debilitated animals recover faster, which is critical after a stressful winter. In many areas, seeded species can supply significant amounts of green herbaceous vegetation. Sometimes seeding is done in and around sagebrush winter range, where native species are dormant or unavailable. Fall regrowth of crested wheat grass, Russian wild rye, and some other grasses provides browse, extending the foraging season into the fall and snow-free periods into the winter. Thus, seeding increases the often-limited amount of high-quality browse so necessary for big game when snow cover lowers the food supply.

Wildlife, particularly big game, can benefit from range seeding when livestock and cattle are present on the same range. The availability of forage for wildlife can decline, particularly in higher altitude sagebrush and grass zones. Consequently, the condition of these ranges declines, and competition for the palatable browse and the herbaceous forage becomes intense. Grass seeding at lower elevations has relieved the situation by attracting wildlife species to those areas in the late summer and early fall.

Grass seeding is also effective in areas where extensive destruction of grassland vegetation has occurred. Seedlings hold the soil, prevent erosion, and, at the same time, provide browse for animals that move into an area quickly. Crops are planted to attract wildlife on both public and private land. Corn is used to attract migratory geese in Michigan, Indiana, Ohio, and other parts of the country.

Water Impoundments

Water impoundments increase habitat utilization for a variety of animal species (Figure 7–11). Aquatic species and waterfowl are attracted to water impoundments. The number of big game, particularly when there is a lack of water in the area, can be increased considerably if water is made available all year. Water impoundments can be

Figure 7–11 Water impoundments provide habitat for both terrestrial and aquatic animals. (Courtesy of the U.S. Forest Service.)

constructed by digging and blasting or by building small dams made with rock, soil, or vegetation. Beavers often build water impoundments.

Water impoundments provide drinking water for some animals and nesting or wintering areas for others. Waterfowl use adjacent, drier areas when water impoundments are created. Water impoundments in arid habitats attract many species of wildlife. In the western states, these impoundments are frequently grazed extensively by cattle. Here, fencing can allow the vegetation to recover so that wildlife can use the area. Ranchers can pipe water from the impoundment to a drinking trough outside the fence for use by cattle.

Added Features

A number of habitat features, such as cliffs and rock piles, are important to some wildlife species. This fact gives a clue to management options. In areas where raptor nest sites appear to be the limiting factor, structures placed on the tops of poles or platforms on towers can encourage these birds to nest in the area. Perch sites in the form of rock piles and poles can be effective in increasing the population of raptors. Cover sites, constructed by leaving some form of natural habitat or making rock piles and mounds, can help maintain wildlife species. Placing rocks in streams can increase the trout population in them. Canada geese are quickly attracted when nesting structures are placed near water impoundments, which are quite easily constructed simply by putting old tires on top of platforms. In some areas, as a matter of fact, the geese have come in such numbers that they are now considered a nuisance. In the eastern United States, brush piles have been used to increase rabbit populations.

Placing nest boxes in forests has been tried in a number of areas, but the construction of nest boxes does not always increase the number of cavity-nesting species in forested areas. An increase in these species can occur, however, if food is available and nest sites have been the limiting factor.

A number of states under the sponsorship of the U.S. Fish and Wildlife Service now have a GAP analysis program. GAP is a broad-level approach for looking at landscape features. It helps managers determine areas of high biodiversity or large number of species of concern to them. These areas or species can then be monitored and set aside for special protection if appropriate.

GAP analysis involves preparing a series of maps that can be viewed together or in groups with the assistance of computer overlays. For example, vegetation, soil, and water maps can be overlaid with maps showing the location of groups of vertebrate species. The distribution of vegetation and the use of habitats by birds can be overlaid to show where there are rare bird species. All vertebrate species can be overlaid on a state or habitat map to show areas of high biodiversity. GAP analysis then helps us locate areas needing special consideration. We can use it to show potential impacts of proposed development on communities of wildlife.

The GAP process uses actual types of land cover mapped from satellite imagery, as well as existing surveys and species–habitat information to identify unprotected species, plant communities, and sites of high biodiversity. The end result of such analysis is to reduce the rate at which species require to be listed as threatened or endangered. GAP analysis is a complement to, and not a substitute for, the protection of individual rare species. It functions as a preliminary step to the more detailed studies needed to plan for biodiversity.

Scott, J. M., F. M. Jennings, R. G. Wright, and B. Csuti. 1996. Landscape Approaches to Mapping Biodiversity. *BioScience* 46:77–78.

PUBLIC LAND MANAGEMENT

There is a need to maintain natural vegetation in order to keep all wildlife species present. The management philosophy in a number of land-management agencies, as well as the National Park Service, is to allow areas to be left in their natural state. It is a philosophy that, unfortunately, does not always work, since successional changes force some species out and allow others to move in. This is particularly true in areas where fire prevention is practiced.

On many public lands, wildlife management may be a component of multiple use. Such management techniques can be employed to benefit wildlife. For example, forest management manipulates the forest environment to produce a combination of products that is desired by the public. These products change with time, economic conditions, public demand, legislation, and capability of the land. Managers of the federally owned forested lands have guidance in the form of laws passed by Congress as to what these products should be. A number of laws specify that wildlife should be considered in every management decision.[10]

One of the more extensive wildlife-maintenance systems in the United States is the National Wildlife Refuge System, an aggregation of national public lands set aside specifically to maintain species of wildlife. The National Wildlife Refuge System was established in 1903 by an executive order that set aside Pelican Island, off the east coast

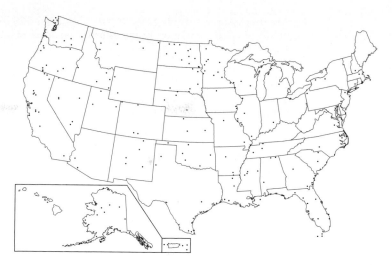

Figure 7–12 Locations of U.S. Fish and Wildlife refuges.

of Florida, as a refuge for herons and egrets, which at the time were hunted for plumes to be sold in the millinery business. Many of the initial refuges were created by taking public land and designating it specifically for wildlife management. This was done to protect threatened habitats, migratory birds, marine mammals, and resident species. Currently, there are more than 375 refuges (Figure 7–12).

Although most refuges have been created by withdrawing land from the public domain, some are on land purchased by citizen's groups. For example, members of the Boone and Crockett Club raised funds by a nationwide appeal to buy private land in northwestern Nevada that became the heart of what is now the Charles Sheldon National Antelope Range. In Jackson Hole, Wyoming, local concern for the elk herd that wintered in the Teton valley and farther south prompted successful campaigns to buy feed and land for the herd. This effort, spearheaded by the Issak Walton League, resulted in the creation of the National Elk Refuge, which, under the auspices of the U.S. Fish and Wildlife Service, National Park Service, U.S. Forest Service, and Wyoming Game and Fish Department, provides winter habitat for nearly 12,000 elk.

The Migratory Bird Treaty has been responsible for setting aside a large number of refuges in the central grassland region of the United States. Most of these refuges were purchased with funds from the sale of migratory bird hunting stamps to adult waterfowl hunters at the time they buy their licenses. The continuing income from the sale of these stamps provides funds for the purchase, management, and protection of waterfowl habitats. (See Chapter 17.)

PRIVATE LAND MANAGEMENT

In the United States, wildlife belongs to the state, while habitat belongs to, and is controlled by, the landowner. (See Chapter 10.) Efforts must be made to educate private landowners in how to maintain wildlife species that can be aesthetically pleasing and beneficial to them. This is sometimes a difficult task, especially since certain species of wildlife, such as prairie dogs, are thought by some cattle ranchers to be best dead. The truth is that poisoning prairie dogs can be detrimental to burrowing owls, black-footed ferrets, and other desirable species. Only through effective education and personal contact can wildlife managers persuade private landowners to develop and follow management plans. Most states have wildlife areas. Some are managed intensively by planting crops to attract desirable species. These management areas are usually attractive to sportsmen and sportswomen. Today, groups of people are buying private lands for fishing or hunting. These people must manage their lands to provide habitat to produce the desired species. Although such land is closed to the public, it does provide additional wildlife habitat.

Access to hunting and fishing is a big issue in many states. Private landowners sometimes charge fees to use their land. In other cases, they prevent access to public lands on the other side of private lands. Private landowners point out that the public often leaves gates open, throws out trash, drives off the road, and harasses cattle. Access is therefore a major issue of wildlife management on private lands that must be addressed.

The timber industry is in the unique position of being able to influence the diversity of wild ungulates and other species of wildlife because of its control over the direction of land-management programs. Logging forests, for example, often induces successional patterns similar to those resulting from fire. Timber-management practices can be manipulated to enhance game habitat without seriously hindering the economic objective. For example, it was found in northeastern and southeastern Washington that elk use during the five growing seasons after logging was highest in forests with small clear-cuts and lowest in forests with selective timber removal. Deer showed a similar, but less marked, preference for clear-cut areas.[22]

More range-improvement and timber-management practices should be implemented for a multiple-use concept. Multiple use need not mean that a range must always be shared by cattle and wildlife; rather, there may be *provisions* for both in the same area. On public land, the relatively high cost of many improvement projects may be more economically justified when multiple benefits accrue.

To ensure that wildlife-habitat values are coordinated with other uses, the wildlife manager must maintain close contact with other resource managers. Land-use planning on a local, statewide, or national scale should include wildlife preservation as an important consideration.

While public land is managed by an agency representing the public, private land is managed by an individual or a group of individuals. Most private land is managed for profit. In some cases, wildlife on private land comes into conflict with the landowner's objectives. When this happens, the landowner can file damage claims against the state or federal government. (See Chapter 22.)

Private landowners are demanding a say in how wildlife is managed on their land. They are indicating to the government that legislation, such as the Endangered Species Act, which affects how they can use their land, should also provide compensation for the loss of any income. These concerns are going to continue to involve the wildlife manager.

Wildlife managers must work with private landowners to help them with their activities and, at the same time, allow wildlife habitat to exist. The landowners must understand the value and benefits of wildlife and community organization so that they can conduct their operations more effectively. When management practices are implemented, the private landowner must be involved. In some cases, interested local groups can solve problems for an entire region. These working groups need to include landowners. Wildlife managers provide the link between public wildlife and the owners of private land that the wildlife uses.

Wildlife habitats can be managed in many ways. By leaving a habitat alone, managers allow natural changes to determine the type of wildlife present there. Manipulative techniques, however, are often used to maintain species or communities of wildlife. Vegetation and topography can be altered by removing vegetation through many processes. Seeding, water impoundments, fertilizers, and the construction of topographic features can bring new structure to habitats, thereby altering the composition of species present in them.

The impact that habitat management has on a particular area is in part dependent on the physical characteristics that are found in the region. These characteristics influence the type of vegetation that provides a component of the wildlife habitat. Habitat classification systems based in part on vegetation have been developed to aid managers.

Biomes of the world provide a broad classification of vegetation based on climate. Regional vegetation classifications, such as Kuchler's potential natural vegetation, ecoregions, and national wetlands classifications, have been developed. Each of these systems provides managers and planners with tools for evaluating habitat management techniques. Each has limitations that must be known to the user.

Management techniques often differ on public and private lands. Public lands are managed according to criteria prescribed to the agency controlling the lands. Private lands are managed by the owner. Normally, the laws that affect private lands are limited to the state's hunting and fishing quotas, unless endangered species are found on the lands.

DISCUSSION QUESTIONS

1. Discuss the uses of classification systems by wildlife managers at the field and regional levels.
2. How can a wildlife manager use the biome system?
3. What characteristics should be considered in developing a wildlife classification system?
4. What are the best uses for each of the classification systems discussed in this chapter?
5. Relate habitat management to classification systems.
6. Since fire sets back succession, is it always a useful tool for managers? When is it and when is it not?
7. What is the impact of new water impoundments on the natural food chain?
8. What changes occur in the composition of wildlife species as a result of fire?
9. How does the placement of nest boxes in a woods affect energy-flow patterns in the forest ecosystem?
10. What types of habitat are "easiest" to manage for wildlife? Why?

LITERATURE CITED

1. Schindler, D. W. 1998. A dim future for boreal waters and landscapes. *Bio Science* 48:1457–164.
2. Ledig, F. T. 1988. The Conservation of Diversity in Forest Trees. *Bio-Science* 38:471–78.
3. Kuchler, A. W. 1964. *Manual to Accompany the Map, Potential Natural Vegetation of the Contiguous United States.* Special Publication 36. Washington, DC: American Geography Society.

4. Ohmann, L. F. 1979. Northeastern and North Central Forest Types and Their Management. In R. M. DeGraaf and K. E. Evans (Eds.), *Workshop Proceedings, Management of North Central and Northeastern Forest for Nongame Birds*. General Technical Report 51. Washington, DC: USDA Forest Service.

5. Baley, R. G. 1980. *Description of the Ecoregion of the United States*. Miscellaneous Publication 1391. Washington, DC: USDA Forest Service.

6. Cowardin, L. M., V. Carter, F. C. Golet, and E. T. Laroe. 1979. *Classification of Wetlands and Deepwater Habitats of the United States*. FWS/OBS-79/31. Washington, DC: U.S. Fish and Wildlife Service.

7. Anderson, S. H. 1980. Habitat Selection, Succession, and Bird Community Organization. In R. M. DeGraaf and N. G. Tilghman (Eds.), *Workshop Proceedings, Management of Western Forests and Grasslands for Nongame Birds,* pp. 13–25. INT 86, 535. Washington, DC: USDA Forest Service.

8. Aldrich, J. W. 1963. Life Areas of North America. *Journal of Wildlife Management* 27:530–31.

9. Fenneman, N. M. 1938. *Physiography of Eastern United States*. New York: McGraw-Hill.

10. Marschner, F. S. 1933. *Natural Land-Use Areas of the United States*. Washington, DC: USDA Bureau of Agricultural Economics.

11. Kirby, R. E., S. J. Lewis, and T. N. Sexton. 1988. *Fire in North American Wetland: Ecosystems and Fire–Wildlife Relations: An Annotated Bibliography*. Biological Report 88(1). Washington, DC: U.S. Fish and Wildlife Service.

12. Anderson, S. H. (Compiler). 1982. *Effect of the 1976 Seney National Wildlife Refuge Wildfire on Wildlife and Wildlife Habitat*. Resource Publication 146. Washington, DC: U.S. Fish and Wildlife Service.

13. Kruse, W. H. 1972. *Effects of Wildfire on Elk and Deer Use of a Ponderosa Pine Forest*. Research Note RM-226. Washington, DC: USDA Forest Service.

14. Peek, J. M. 1974. Initial Response of Moose to a Forest Fire in Northeastern Minnesota. *American Midland Naturalist* 91:435–38.

15. Peek, J. M., R. A. Riggs, and L. L. Laner. 1979. Evaluation of Fall Burning of Bighorn Sheep Winter Range. *Journal of Range Management* 32:430–32.

16. Smith, L. M., J. A. Kodlee, and P. V. Fonnesbeck. 1984. Effects of Prescribed Burning on Nutritive Value of Marsh Plants in Utah. *Journal of Wildlife Management* 48:285–88.

17. Alexander, M. E., and D. E. Dube. 1983. Fire Management in Wilderness Areas, Parks and Other Natural Reserves. In R. W. Wein and D. A. MacLean (Eds.), *The Role of Fire in Northern Circumpolar Ecosystems*. New York: Wiley.

18. Short, H. L., W. Evans, and E. L. Bocher. 1977. The Use of Natural and Modified Piñon Pine–Juniper Woodlands by Deer and Elk. *Journal of Wildlife Management* 41:543–51.

19. Scotter, G. W. 1980. Management of Wild Ungulate Habitat in the Western United States and Canada: A Review. *Journal of Range Management* 33:16–27.

20. Santillo, D. J., D. M. Leslie, and P. W. Brown. 1989. Response of Small Mammals and Habitat to Glyphosate Application on Clearcuts. *Journal of Wildlife Management* 53:164–72.

21. Regelin, W. L., and O. C. Wallmo. 1975. *Carbon-Black Increases Snow Melt and Forage Availability on Deer Winter Range in Colorado*. Research Note RM-296. Washington, DC: USDA Forest Service.

22. Edgerton, P. J. 1972. Big Game Use and Habitat Change in a Recently Logged Mixed Conifer Forest in Northeastern Oregon. *Proceedings of the Annual Conference of the Western Association of State Game and Fish Commissioners* 52:239–46.

8

HABITAT ALTERATION

Storms, natural succession, and animals themselves are some natural causes of change in a habitat that make an area more or less usable by different species. Many forms of human activity also affect a habitat to varying degrees. Various forms of pollution, building and development, and even a walkway through the woods all create changes for some species.

A habitat can be partially or completely destroyed. The destruction of a habitat for one wildlife species can also mean an improvement in the habitat for another. For example, while logging causes woodland species to move away, it attracts species that prefer open areas. Frequently, alterations can be controlled by first studying an area and then planning around the wildlife that is present. By knowing the impact of such changes, managers can direct those that will produce the greatest benefit to wildlife. In this chapter, different types of habitat alterations and their effects on wildlife are discussed. The destruction of a habitat, the impact of air and water pollution, and the introduction of other animals, as well as foreign chemicals, are examined. In Chapter 9, we will look at some of the results of these alterations and methods of mitigating them.

PHYSICAL CHANGE

Many forms of alteration are obvious, such as those caused by fire, floods, logging, and heavy pollution. Many are less noticeable, but cause far-reaching changes. In any case, new habitat of some sort results, often of an earlier successional sere. Animal populations are changed or may be displaced. In some instances, less desirable forms of wildlife begin to thrive in the new habitat. This happens readily in aquatic habitats when a rise in temperature eliminates a sport fish, but not less acceptable species.

Destruction

Habitat destruction generally results in a new community with different forms of population interactions. When a hillside is logged and the slash burned, the bare soil is exposed and becomes susceptible to erosion. Logging in small sections can be less destructive (Figure 8–1). A community of microorganisms exists; and quickly, seeds carried by rodents, birds, and the wind form the earliest sere.

Mineral extraction is another human activity that frequently results in more devastating habitat destruction, since it disrupts the soil to a greater extent than logging in small sections does. More obviously, forests and grasslands are eliminated, along with food and shelter needed by wildlife. These areas take much longer to recover than logged land, unless restorative measures are instituted.

In the last 40 years, changes in American agriculture have usurped approximately 810,000 additional hectares (2,000,000 acres) annually.[1] Some 202,000 hectares (500,000 acres) a year of wetlands were lost between 1954 and 1977. Current estimates of the amount of wetland acreage remaining in the nation vary. The U.S. Fish and Wildlife Service judged that only 40 million hectares (99 million acres) of wetlands remained in the mid-1970s. Conservative estimates conclude that this has been reduced to approximately 38 million hectares (95 million acres) as of 1987. Some authorities believe that as few as 30 million hectares (80 million acres) of wetlands now remain in the 48 contiguous states, of which 12 million hectares (30 million acres; 37 percent) are so badly contaminated and degraded by toxic substances as to be useless.[2]

As a result, many species of upland game and shorebirds that used to breed throughout the grasslands of the central United States have declined in number. The long-billed curlew, mountain plover, willet, marbled godwit, upland sandpiper, and mourning dove declined as a result of tillage, grazing, and dragging (a process in which

Figure 8–1 Clear-cutting in a forest. (Courtesy of the U.S. Forest Service.)

ranchers drag trees, branches, or metal scraps across a field to break up piles of cattle manure). Ring-necked pheasants, which were harvested at a rate of 10 million in the 1940s, now number less than 3 million.[1]

Decrease

Farming and ditching projects reduce both the actual and the usable size of habitat for animals that need large, contiguous areas. Destructive events do not always eliminate habitat, but may only reduce it. When something happens that increases edge, deer and gallinaceous birds are likely to increase, but golden eagles and wolves, which need large areas, will decrease. At the same time, predator–prey interactions, competition, and food all change. The results of a reduction in habitat size cannot always be accurately predicted.

Subdivision

One of the more significant ways in which a habitat is altered is by subdivision, as we saw in Chapter 6. Numerous studies show that the size of a habitat is very important to wildlife species. This means that broken blocks of habitat—bits and pieces near one another—will not do. Roadway construction, power-line right-of-way construction, and such major activities as farming and mineral extraction decrease both the contiguity and total size of the available habitat.

Most people think that subdivisions do not take away much habitat and that roadways and transmission lines remove just small fractions of habitats. The fact is that roadways in the United States take up no less than 22 percent of the land and transmission-line corridors 7 percent. But the real problem is that some species require an undisturbed habitat.[3]

Corridors such as those used for transmission lines often attract wildlife species if there are no human activities nearby. Raptors perch and sometimes nest on power poles and towers. Many migratory birds have been found dead at the base of transmission lines, through either collision or electrocution. Raptors and waterfowl seem most vulnerable. Techniques have been developed to reduce the impact by changing the height of the pole and guide wires.[4] (See Chapter 9.)

Wolves, which once were found throughout a good part of the northern United States, have decreased dramatically because the home range they require no longer exists. Bighorn sheep, which once lived throughout much of the western United States, have now been relegated to a few small, isolated areas. Even in high-altitude meadows and grasslands with steep, rocky, mountain slopes, bighorn sheep are affected by the influx of humans, cattle, and dogs.

Studies of the petroleum industry's land-use activities in Canada showed that habitat alteration affected wildlife in various ways. Big game that was hunted avoided roadways more than did unhunted species. Animals adapted to stationary disturbances, such as a drilling rig, much more readily than to a moving disturbance, such as traffic on a road.[3] Animals with smaller home ranges were more tolerant of disturbances, and migratory or nomadic species were less tolerant than resident species.

Fencing of habitat and highway construction can reduce available habitat. Migratory pathways, access to water, and paths that lead to protected habitat can be lost.

Structural Change

Some kinds of habitat disturbances change the structure of the habitat. By *structure,* we mean the layering of vegetation and the spatial arrangement and form of the physical components of the habitat, such as rocks and slope. All changes in the structure of a terrestrial or aquatic system affect the wildlife species that inhabit the system. When

Figure 8–2 Ring-necked pheasants require grassy areas and edge to feed, as well as shrubs and tall cover for nesting. (Courtesy of the Wyoming Game and Fish Department; photo by L. R. Parker.)

a quantity of dead, dying, and down vegetation is removed, the structure is changed, and food sources and shelter sites are removed for some species. Selective cutting—removing trees of a certain size—changes the area structurally because it removes one layer of the forest system. This can mean the exodus of the entire array of wildlife species dependent on that layer, as well as other species dependent on the food sources in the layer. Changes made from farming operations encourage many generalist species and eliminate specialized ones. This disrupts predator–prey interactions and generally creates a nuisance with which farmers must contend (Figure 8–2).

Cover for Rabbits

A study in central Pennsylvania provided information on behavioral responses of eastern cottontails to a terrestrial habitat scheduled to be used for domestic wastewater disposal. The rabbits used old fields and shrub land during the day, switching to oat and corn fields when those crops matured. The rabbits needed dense cover for aboveground bedding sites. In the cold weather, they used underground burrows. The vegetation structure seemed to be most important to cottontails. Shrub habitat was especially important during late winter and early spring, because of an increase in cover associated with early-leafing shrubs. If wastewater operations were to alter the dense, woody, herbaceous cover near the ground, the quality of the cottontails' habitat would decline.

Althoff, D. P., G. L. Storm, and D. R. Dewalle. 1997. Daytime Habitat Selection by Cottontails in Central Pennsylvania. *Journal of Wildlife Management* 61:450–59.

Competition

Competition among wildlife can come about in several ways, of which grazing is one of the most noticeable. Grazing decreases the amount of food available for wildlife and destroys the structure of some species' habitats. As cattle and sheep are introduced into grasslands, the amount of forage available for wildlife is reduced. Domestic livestock, wild ungulates, and small herbivores all compete for the grass. When fences are erected and cattle cannot move freely, overgrazing frequently occurs as the number of cattle increases beyond the ability of the land to support them.

Human beings also compete with wildlife when they intrude on undisturbed habitat. This can occur through affording hunters access to the habitat, constructing roadways, or developing the land. Not only do roadways bring disturbance in the form of traffic, but they open up once-isolated areas. This intrusion, a form of competition for space, is one of the major forms of impact on wildlife. Animals are no longer free to give birth and raise their young, a fact borne out by studies of oil and gas development in isolated areas.[3]

Access is sometimes very sudden and voluminous. For example, construction of the Alaskan pipeline brought large numbers of people to the affected area for a short period, after which a minimal number remained. This type of impact is very difficult to assess, because wildlife responses have not been clearly documented. Therefore, any statement about the effects of human intrusion must be carefully evaluated. Are there fewer nesting birds? Do wild ungulates leave the area? Are there fewer fish in streams? All these questions require comparison with previous populations before conclusions can be drawn.

Introduction of Exotics

Another form of habitat loss comes from the introduction of new organisms into an area. This has resulted in competition for habitat and, often, loss of native organisms. On the other hand, there have been successes in the introduction of exotic organisms. Some of these fall into the category of **biological control,** which is the use of parasites, predators, or pathogens to contain a pest population at a lower average density than it would otherwise have. The object is to keep the pest species at a population level below which economic damage will result. In most cases, biological control programs are initiated when an exotic pest species is introduced into an area and gets out of control, causing damage to crops or livestock. An extensive search is then made for another organism that can inhibit the pest population.

The success stories in biological control are relatively few, but one of the most notable concerns the cotton cushion scale, a small, scalelike insect that looks like a bark or leaf coloration on citrus fruits and, occasionally, other plants. Introduced into California during the last quarter of the 19th century, it began to cause economic losses in the citrus industry. A search of agricultural areas in New Zealand turned up a number of potential parasites and predators, including the ladybird beetle, which multiplied rapidly and fed voraciously on the scale (Figure 8–3). Within a year, the beetle controlled the scale, containing its density well below the economic threshold. This balance was disrupted in the early 1950s, when DDT destroyed the ladybird beetle population. The scale increased rapidly, making it necessary to reintroduce the beetle, which resumed its winning ways.

A number of introductions of foreign plants and animals, however, have had quite disastrous results. In an attempt to reduce the mosquito population, a tree called

Figure 8–3 Ladybird beetle feeding on cotton cushion scale. (Courtesy of the University of California, Riverside, Department of Entomology, Division of Biological Control.)

Melaleuca was brought into Florida. Since this is a dense-growth tree that flourishes around swampland, the idea was that it would reclaim the area from being a breeding ground for mosquitoes. In 1936, the owner of a nursery in Florida spread seeds of this exotic tree from an aircraft over central and western Broward County. The Melaleuca is a rapidly growing tree, flowering from two to five times each year. One-year-old trees can bear seeds; an individual seed pod contains between 200 and 350 seeds, so that one 40-ft tree can produce 3 million seeds. Fire appears to assist the dispersal of the seeds, which are also carried by wind. Quickly, the exotic import spread over vast areas. Now, where hummocks of natural cyprus and pine formerly stood, the Melaleuca reigns. Parts of water conservation areas have become covered from dike to dike with the trees. Stands are so dense that deer will not enter them to find shelter or food, and in any event, most food plants have disappeared. So the Melaleuca tree is doing its job, and doing it so well that the habitat for many of the native species is gone. There is concern that in time Melaleuca trees may alter vast amounts of the Everglades habitat.[5]

During the latter half of the 19th century and early part of the 20th, two species of exotic fish were introduced into the United States. One, the carp, can be considered a mistake and the other, the brown trout, a success. The carp, native to Asia, has been widely cultured in Europe and Asia for centuries as a food fish. Selective breeding in captivity that has produced carp with few bones and scales has made it a popular food on European dinner tables. It was not realized at the time the fish was imported that the cultural variety would revert in the wild to its bonier condition, with coarse flesh and an increased number of scales. Not only was this variety generally rejected by the American public, but it caused substantial destruction to the habitat of native fish.

The carp is now established in 46 of the 50 states and in most Canadian provinces. The fish grew rapidly when first introduced, uprooted vegetation, created muddy waters, and displaced many native fish. Carp damaged wildlife habitat by destroying aquatic vegetation, the loss of which frequently caused massive plankton growth, adding to the turbidity resulting from the carp's feeding habits. While the carp constitutes only a small fraction of the commercially harvested freshwater fish in America, its detrimental habits have resulted in the expenditure of millions of dollars on largely ineffective control and removal programs.

The brown trout, which has been successfully introduced into many streams in this country, appears not to compete with native trout species or to harm the habitat. Generally, it inhabits waters not suited for native trout (Figure 8–4).

Two species of introduced bird species have become major pests. The house sparrow, introduced for use by shooting clubs, has competed with native birds and become a pest in many cities. The starling, with its great reproductive capacity, was introduced because some people wanted to have all the animals named in Shakespeare's writings represented in the United States. It was introduced despite awareness that the bird had destroyed fruit crops after its introduction into New Zealand.

Ring-necked pheasants have been successfully introduced in the United States as a game bird. They came from China and were introduced into Oregon in 1880. Since then, they have become established in 18 states.[6]

One of the animals accidentally introduced into this country that few managers can escape having to deal with at some time is the nutria. In the early part of this century, some entrepreneurs, concocting a "get rich" scheme, brought in this large, rodentlike creature, which some people had compared to the beaver and muskrat. Touted as a superior fur animal, nutrias were sold and shipped to breeders in many states. Some of the animals dug their way out of the enclosures in which they were put to breed; others were turned loose by breeders when they found that the animals' pelts did not bring the prices promoters had claimed they would. Some nutrias were moved from state to state in an attempt to cure certain aquatic plant problems. By 1956, the nutria was established in at least 18 states. The animal destroys crops, particularly rice, and damages leaves. In waterways, nutrias feed on emergent vegetation, thus modifying aquatic habitats.[7] Because of all these detrimental effects, wildlife managers have tried, with relatively little success, to remove nutrias from impoundments. Because they are so prolific, even poisoning and trapping are unsuccessful in reducing the nutria population.

The reintroduction of species and introduction of new species have increased dramatically since the 19th century. But some new species not only compete and reduce habitat, but also change the energy transfer system. There may be direct competition for trophic and spatial resources or indirect competition that results in subtle changes in native populations. The degree of change that can occur in native populations must be evaluated before exotic species are introduced.

Figure 8–4 Brown trout have been introduced into U.S. waters.

Zebra Mussel

The zebra mussel is a striped mollusk about the size of a person's thumbnail. It is native to the Caspian and Black Seas and, as of 1988, is also found in North America. Biologists think that this exotic species arrived by transatlantic ship and was discharged as larvae into Lake St. Clair between Lakes Huron and Erie. The prolific creature has spread rapidly throughout the lakes and waterways of eastern United States and Canada. Apparently, no predators or parasites have affected the mussel.

The zebra mussel is disrupting the food web in many of our waterways by outcompeting many of the native microorganisms for food. It is outcompeting native mussel species that play an important part in the detritus food web and in the reduction of sediment. The U.S. Fish and Wildlife Service, which is very concerned about this problem, estimates that this exotic will cause more than $5 billion in damages by 2002.

Obviously, because of the complex interactions of natural systems, managers should be cautious when someone wants to introduce a nonnative species. Many years of work—commonly more than 20—can be needed for the discovery or development of a successful biological control agent. We should not be fooled into thinking that importing species is always a quick cure for what ails a habitat.

AIR POLLUTION

The impact of air pollution (for example, SO_2, NO_2, and particulates) on plants and animals can be both short and long term. Long-term impacts are difficult, if not impossible, to evaluate because of lack of research. In particular, the effects of exposure of ecosystems to low, but persistent, concentrations of trace elements from power plants, mostly from emissions, can only be surmised. Acid precipitation, a secondary pollutant associated with coal power plants, is known to have direct and indirect effects on terrestrial and aquatic ecosystems. Such effects include injury to foliage, leaching of nutrients from plants and soils, and the elimination of certain plant and animal species from lakes and streams. Habitats can be altered so that new communities develop.

Today, *acid rain* is a major worldwide pollutant. While some politicians want to debate the amount of damage done by acid rain, many scientists believe that it will have a greater impact than the chlorinated hydrocarbons discussed in Rachel Carson's *Silent Spring*. Acid rain results from atmospheric pollution by chemicals of various industries that burn fossil fuels, particularly those with a high sulfur and nitrogen oxide content.

The effects of acid rain were first noted in Scandinavia, where it was found that the rain, which removes sulfuric materials from the atmosphere, was lowering the pH in freshwater lakes to such levels that they were made uninhabitable for some fish populations. In southern Norway, salmon and trout populations were lost at a number of popular fishing sites. Calcium was removed from the forest floor when it was chemically bound to some of the sulfuric material ion the rain, causing a reduction in forest growth. In western Sweden, the pH of many lakes has decreased. Not only are the fish species lost, but phytoplankton and many other food sources are no longer found in these lakes.[8]

Lichen, the primary winter food of reindeer, is very sensitive to sulfur dioxide. In Scandinavia, it appears that atmospheric pollution generated farther south in Europe is brought to earth in the form of acid rain, which inhibits the growth of lichen.[9] Fish populations are disappearing in many parts of northeastern United States and Canada, apparently because of acid rain.

In North America, the acid rain problem is under investigation in many quarters, but little coordinated effort has been made to look at its impact on wildlife and fish habitats. A generalized map (Figure 8–5) indicates the acidity of rainfall. The normal pH of rainfall is 5.6. The lower the pH, the higher is the acid content. Because the pH scale is logarithmic, there is a tenfold difference between adjacent numbers. Thus, water of pH 4 is 10 times more acidic than that of pH 5 and 100 times more acidic than that of pH 6 (Figure 8–6). The north-central and eastern parts of the United States have rainfall below this pH level. The pH of George Lake in central Ontario, Canada, 65 km (40.4 mi) downwind from mining and smelting activity, is decreasing at an estimated annual rate of 0.13 pH unit per year. A number of species of fish, including lake trout and smallmouth bass, are no longer found there. Waterfowl populations are declining because of a loss of their food base.[9]

Similarly, spotted salamanders in ponds near Cornell University are no longer able to breed there; apparently, because of the low pH in winter snows, the embryonic development of the salamander is arrested, and malformation occurs.[10]

Laboratory studies have shown that artificial mists with acidities ranging from pH 5 to pH 2.3 leach nutrients from the leaves of plants. Other direct and indirect effects on plants include reduction of forest growth, erosion of the cuticles of leaves, and leaching of nutrient salts from the soil, all of which have a long-term effect on wildlife. One subtle effect could be an increased input of cadmium into food webs, as studies show that the absorption of cadmium in soil is dependent on pH. Low pH releases cadmium, making it available to plants in soils that contain a toxic level of the element. Plants and animals can tolerate cadmium in small amounts—in fact, some require it. However, a larger amount can be toxic.

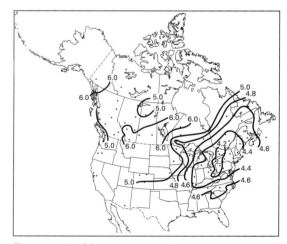

Figure 8–5 Mean annual pH of precipitation in the United States and Canada. Normal rainfall has a pH of 5.6. (Courtesy of the National Research Council.)

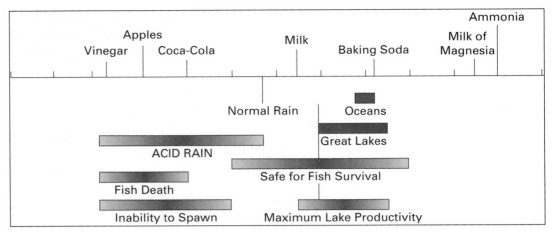

Figure 8–6 The pH scale.

There appears to be little question that acid rain is going to have a major impact on wildlife and wildlife management in this country. More than that, human health is involved. Presumably, with the removal of sulfuric chemicals from the atmosphere, the pH of waters can return to normal. The solution is simple; its implementation, given human greed and political shenanigans, is complicated and expensive.

WATER POLLUTION

Water resources have been affected by primary contamination from such practices as dumping improperly treated waste into waterways and by secondary contamination as occurs with surface runoff from overfertilized farmland. At present, public and private groups spend about $275 million annually to collect data on water quality. As a result, there has been some improvement in water resources in the past decade, but much remains to be done.[1]

Changes in water quality resulting from the introduction of toxic material can affect both aquatic and terrestrial wildlife. Organic wastes use dissolved oxygen, so that fish and invertebrate organisms cannot survive in contaminated waters. Timber harvests and grazing result in greater runoff, increasing the number of suspended solids in streams. Coal mining and acid mine runoff are particularly troublesome in many parts of Appalachia. Some minerals form complex interactions, with prolonged exposure for wildlife.

The Federal Water Pollution Control Act of 1972 has been the major stimulus to improving water resources. The act and its later amendments brought about a major change in water-resource management. Before 1972, when the act was passed, cleanup efforts consisted of work on the receiving bodies of water, rather than prevention of the pollution. Treatment of waste was the order of the day. This strategy, unfortunately, was generally unenforceable because of political interests, technical weaknesses, and loopholes in the legislation. Even since the passage of the Federal Water Pollution Control Act, it has remained difficult to control nonpoint sources of water pollution, and it appears that such sources now account for more than half of the pollution entering the nation's waterways. For example, siltation from tillage on farmland causes a sizable amount of water pollution.

Groundwater

About half of the U.S. population gets its drinking water from wells. These wells are becoming contaminated and disappearing at an alarming rate. In addition to the contamination, groundwater levels are falling. In Arizona, the water table is dropping at a rate of 2 to 3 m (6.5 to 10 ft) per year.[11]

Nationwide, groundwater for agriculture will be largely depleted by the year 2000. Even in the high-rainfall area of the southeast, expanded irrigation programs are lowering the water table. The Department of Agriculture uses the word "mining" to describe the rate at which groundwater is being depleted. These changes affect vegetation, the availability of water, and topography (Figure 8–7). Still, there is no rational policy designed to alleviate the situation.

Sources of Water Pollution

There are a large number of aquatic pollutants. In considering threats to the water system, we can look at point sources, nonpoint sources, toxicants, heated wastewater, suspended solids, organic material, and rapid changes in the habitat itself as a result of dams, channelization, and dilution.

In the United States, public attention has centered largely on point-source discharges. When material entering a body of water makes the water look bad, sooner or later there is a public outcry. Effluents leaving a pipe have effects that are often readily identified. Regulatory agencies can more easily see where pollutants come from and collect a sample of water to determine its chemical and physical qualities. Municipal sewage systems, electrical power units, the chemical and pulp industries, petroleum and paper-producing factories, food-processing plants, and more can be monitored.

Ground- and surface-water contamination is occurring due to leaks in underground storage tanks, particularly gas tanks. Estimates indicate that 350,000 of the nation's 1.2 million service station tanks may have leaks.[12]

Nonpoint sources are more difficult to identify, but major pollution can be traced to agriculture, forestry, mining, ranching, and construction activities. Simply constructing

Figure 8–7 Loss of groundwater results in changes in topography.

streets in a city and covering areas with asphalt creates a tremendous runoff into streams and rivers. In draining city streets by storm sewers and irrigation processes in which we collect large quantities of runoff, we provide a large influx of water into lakes or streams at certain times of a year. Farming results in runoff consisting of topsoil, as well as fertilizers. Cities are beginning to ask people not to "curb" their dogs because of the tremendous increase in pollution from storm sewers. All these activities create physical changes in the aquatic system that alter or destroy habitats.

The chemical makeup of most toxicants can be identified, but it is often difficult to identify the sources. In areas in the midwest where roadways are heavily salted, the wash water can be toxic for some species of fish and other aquatic life. Some forms of toxicants accumulate in aquatic organisms; others can cause immediate death. Planning ahead in areas where there is runoff, particularly in urban and industrial developments, should involve some method of extracting toxic material from wastewater before it is returned to the natural system.

A number of power plants utilize water in their cooling systems. The heated water is then returned to lakes or streams. As indicated earlier, aquatic organisms have specific zones of tolerance, so that raising the temperatures of a stream or lake will change the composition of species living there. Suspended solids are a real problem in areas where logging, farming, or construction activity is going on. Soil runoff increases the turbidity of the water, reduces the amount of light and oxygen available to aquatic life, and destroys the habitat for many species.

Organic pollution is caused by people. Human sewage, treated and untreated tannery wastes, and runoff from feedlots constitute nutrients for lakes or streams. The resultant increase in bacterial activity lowers the dissolved oxygen, which makes the water uninhabitable for a number of organisms. The entire process of aquatic succession or eutrophication of lakes is accelerated by the disposal of organic wastes in excessive quantities. Lake Erie and Lake Washington (which forms the eastern boundary of the city Seattle, Washington) are two of the major aquatic systems in the United States that have been affected by the discharge of organic material.

An aquatic habitat can change as a result of increased runoff from paved areas or changes in nearby vegetation. Once the velocity changes, in effect a new area is created for different forms of life. Dams also cause extensive habitat changes. As water accumulates behind the dam, the river's temperature, movement, and other physical components change. The dam itself blocks the movement of many species of fish. The entire process of building a dam brings about major changes in the aquatic system.

Channelization

Channelization, the process of deepening or straightening streams, is common in urban areas (Figure 8–8). If we view a river as merely flowing water with a collection of organisms inhabiting it, we will have missed a very important aspect of an aquatic system: the relationship of the bed of the river to the aquatic ecosystem as a whole. The bottom of the river provides a home for organisms, is part of the chemical and physical equilibrium, and is a resource of suspended solids and other important materials. When rivers are straightened, many unfortunate things happen to the structure and function of the aquatic community. A naturally meandering river has a complicated pattern of currents and a wide array of velocities. Straightening a river is like creating a rain gutter, which substantially changes the diversity in velocities, ranges, and characteristics of the system. It also shortens the river, so that organic materials reach the ocean and estuaries much more rapidly and have less time to degrade. In channeliza-

Figure 8–8 Concrete-lined stream channel in an industrial area. (Courtesy of the U.S. Soil Conservation Service.)

tion in which cement is used for the bottom, the bed of the river system is removed. Through channelization, the relationship of the river to the floodplain may also be changed, and the energy-absorbing and energy-dispersing capabilities of the river may be decreased, making the region less suitable for both aquatic and terrestrial organisms.

Loss of Wetland

Another various aspect of water pollution and channelization is the loss of wetland habitat for wildlife. Changes in vegetation that can occur in all wetland habitats in which development is taking place affect the structure of the habitats, as well as the food supply for animals. The flowering and growth cycles of plants can be altered by roadway or construction dust.

Studies by the U.S. Department of Agriculture indicate that wetlands in the United States are disappearing at an alarming rate. The losses have varied by region and with time as "civilization" moved west and urbanization increased and agriculture expanded. The eastern states were the first to lose wetlands through drainage. The prairie pothole region extends from Wisconsin and western Minnesota through the Dakotas, northeastern Montana, and central Canada. This region, which contains some of the most desired waterfowl breeding areas in the country, is deteriorating rapidly.

The most dramatic loss of wetland habitat has been from the conversion of river bottoms, swamps, and overflow wetlands to croplands. Numerous studies show that the decline in the diversity and abundance of wildlife has been the result of drainage and channelization. For example, in Missouri, the diversity and abundance of birds decreased, while those of small mammals increased as the result of the loss of bottom-land forest.

Today, 92 percent of California's wetlands are gone. As a result, wintering habitat for nearly 60 percent of the Pacific flyways of ducks, geese, and other waterbirds no longer exists. The remaining habitat is heavily used. Pollution and the spread of disease among the birds create further loss of populations.[13]

Insecticides and Herbicides

Insecticides and herbicides affect wildlife directly through chemical interaction and indirectly by destroying food and shelter. Some pesticides, such as the *organochlorines,* which include DDT, dieldrin, and aldrin, are used to destroy insect pests. These insecticides tend to accumulate in the food chain, so that a top-level carnivore retains a great deal of the DDT from organisms that it eats. Aquatic food chains, which often lead to waterfowl or fish that are consumed by people, can pass organochlorines from one organism to another. When an organism higher in the food chain has higher levels of pesticides than those beneath it, we call the process **biological amplification.** Most organochlorines tend to remain in the fatty tissue of animals. Fortunately, we do not eat much of the fatty tissue of waterfowl or fish.

Another major problem with organochlorine insecticides is that they do not break down quickly in the environment. Twelve years after application, 10 percent of the DDT is still active in the soil. There are other pesticides that break down much faster. Beside accumulating in the food chain, some insecticides are known to cause disruptions of animals' physiological systems. DDT and other organochlorines cause thinning of eggshells in birds, particularly upper-level consumers such as raptors and pelicans (Figure 8–9). As a result, many raptor populations declined or even became extinct in the 1950s and 1960s. The ban on DDT has helped restore the bald eagle population in the United States.

The chemical makeup of genetic material has been changed in many natural populations. Insects such as mosquitoes developed resistance to DDT through genetic selection. Mosquitoes, fish, and frogs are known to have become resistant to pesticides used in fields bordering the creeks in which they live. Experiments on laboratory fish have shown that a number of pesticides cause cancer in wildlife, although the exact impact on the total population has not been determined. Wild animals may succumb for other reasons before cancer can cause death.

Perhaps the major way in which herbicides affect wildlife is through habitat alteration. All the components of a habitat—structure, pattern, and density—can be

Figure 8–9 Pesticides in the adult bird's food chain cause eggshells to become fragile.

altered by herbicides. Since these are communitywide changes, the impact can be complex. The treatment of perennial forb and shrub-grass ranges with 2, 4-D, a herbicide, in western Colorado usually produced an increase in grass cover and a decrease in the cover of most herbs and shrubs. The recovery time of herbicide-sensitive species varied. Dandelion reestablished dominance within six years following treatment of a perennial forb range, but big sagebrush showed little signs of recovery five years after treatment. The density and litter size of the deer mouse was little affected by the 2, 4-D treatment, but densities of northern pocket gophers and chipmunks were reduced. The number of captured montane voles was correlated with the amount of herbage (forbs and shrubs). After treatment, the number of voles increased, but did not change on untreated areas. With the reestablishment of herb dominance, pocket gopher and vole populations returned to pretreatment levels.[14] Changes in the density of pocket gophers on treated ranges were due primarily to altered food availability, those of chipmunks to both food and cover, and those of voles to change in cover.

Heavy Metals

Industrial processes and mineral extraction now use and release a number of heavy metals that adversely affect wildlife and wildlife habitats. Because some metals undergo chemical change in water, aquatic organisms are particularly affected by heavy metals, either directly in the water or through atmospheric pollution. Several heavy-metal cycles and problems are discussed in this subsection.

In its natural cycle, a mineral often goes through reservoir stages, in which a large amount of the mineral is present. Reservoir stages can be gaseous, liquid, or sedimentary. Minerals such as oxygen, carbon dioxide, and nitrogen move freely between gaseous and liquid states. Many of the common soil chemicals, such as silica, are sedimentary reservoirs.

Mercury. Between 1953 and 1960, industries around Minamata Bay, Japan, emptied organic mercury compounds into the bay. Deaths or serious injuries to people who ate contaminated fish resulted some years later. In 1970, mercury was the focus of much public attention when toxicologists found abnormally high concentrations of its compounds in freshwater fish from Lake Erie and in saltwater tuna and swordfish. Mercury has many uses. When released into the water in small amounts, some mercury compounds, such as the chloromethyl mercury used by chemical companies, can accumulate in the fatty tissues of animals. This is similar to the process that occurs in organochloride pesticides. Thus, animals higher in the food chain consume more concentrated amounts.

Once in animal tissue, mercury interacts with body chemicals. Some of the mercury compounds formed in the body pass readily through membranes, and neurological disorders can result when the compounds reach the brain. Most organisms—fish, birds, and mammals—receiving excessive doses of mercury show symptoms of neurological disorder. Mercury is then taken up by people who eat contaminated wildlife. Prolonged exposure to some organic mercury compounds creates a variety of neurological disorders, including temporary insanity. The phrase "mad hatter" was derived from hatmakers who used to chew hatbands soaked in mercury compounds to shape men's hats and so became temporarily intoxicated.

Lead. Lead occurs naturally, primarily as lead sulfide ore, and is mined in many parts of the world. It is a useful metal because it is very dense, has a low melting point, and can be worked with easily and bent into different shapes. It is used for many metal objects. Lead is normally found only in the earth's crust, but with our use of leaded gasoline, there has been a great increase in the concentration of atmospheric lead: Ninety-eight percent of the lead in the atmosphere comes from the combustion of gasoline products. The lead settles on the earth's surface and concentrates near the source of the pollution: Urban lead levels are up to 100 times higher than those in rural areas. As a result of distribution by atmospheric wind currents, beginning in 1950, higher-than-normal levels have been recorded in the Greenland ice cap. Industrial nations add lead to the air in the combustion process; the lead moves to the earth's poles on wind currents and settles as air descends to the earth's surface.

Plants obtain lead from soil, water, and air, animals from food, water, and air. A small portion of the ingested lead (5 to 10 percent) is absorbed by the digestive tract, and 40 percent is absorbed by the lungs. Lead then enters the bloodstream, where it can be deposited in tissue to disrupt enzyme systems and neurological processes or excreted in the urine. In the blood, lead interferes with the synthesis of hemoglobin, causing anemia from impaired oxygen transport. Lead deposits also interfere with normal kidney functions. Behavioral changes accompany lead poisoning. Some people become listless and tired and seem to lose mental capacity. Others exhibit symptoms associated with drunkenness. Some historians suggest that the Roman Empire declined partly because lead poisoning affected the mental capacity of the Romans. Traces of lead have been found in the bones of ancient Romans, probably the result of drinking wine stored in lead-glazed pottery vessels.

Because of the high levels of atmospheric lead and its presence in some products, lead poisoning occurs in humans more often than it should. It can be an occupational hazard: Workers near lead smelters and police on traffic duty in tunnels often display symptoms of lead poisoning. Some cooking utensils are still made with leaded solder, and pottery that is incorrectly glazed releases lead into the food or liquid it contains. Because of the high atmospheric levels of lead, the element is taken up by wildlife in food. The levels passed on to people are not yet well documented.

Selenium. Selenium is a naturally occurring elemental metal that is both an essential nutrient and a potential poison. It is required for metabolism of vitamin E. In the past, selenium has generated commercial interest. The electrical conductivity of the metal varies with the intensity of light. Its photoelectric properties enable selenium to be used in electronic eyes, light meters, and electrical rectifiers. The steel industry uses the metal to produce stronger, corrosion-resistant products. Selenium is used as a rubber vulcanizer, in dandruff shampoo, and as an alloy. Red, pink, and ruby glass obtain their tint from selenium, which is also used in insecticides and paints applied to the hulls of ships to prevent the growth of barnacles.

Metallic selenium is odorless and tasteless. The gaseous, methylated forms have an odor similar to that of garlic. Selenium is naturally found in volcanic dust, coal deposits, and some shales. Selenium weathered from rocks and found in soils is taken up by plants, which may then be ingested by animals. Thus, much of the interest in selenium in the west has centered around the metal's known or potential toxcity to animals (Figure 8–10).

The concentration at which selenium becomes toxic to animals depends on the form of selenium; the presence of oxalates, alkaloids, or other natural products; and the quantity of vegetation consumed. Mildly toxic effects can generally be observed if a quantity of plants containing 5 parts per million of selenium is ingested. Livestock deaths have been associated with selenium poisoning.

Irrigation drain water has recently become the focus of studies to determine the extent of adverse affects of fish and wildlife resources in habitats receiving irrigation drainage in the western United States. Concern for the effects of the quality of drain water on natural resources was prompted by U.S. Fish and Wildlife Service studies in 1983 at the Kesterson National Wildlife Refuge in the Central Valley of California. High concentrations of selenium and other trace elements in impounded drain water at Kesterson were associated with reproductive failure, deformities in embryos, and mortality in several species of waterfowl. This discovery prompted more detailed investigations of the impact of contaminated drain water on fish and wildlife species. Drain water from federal land to irrigate croplands in the western San Joaquin Valley was

Figure 8–10 Some species of waterfowl, such as this blue-winged teal, are adversely affected by selenium and other heavy metals. (Courtesy of the U.S. Fish and Wildlife Service, photo by D. McLauchlan.)

subsequently terminated because of the high concentrations of selenium in soils and groundwater. This action was required in order to comply with provisions of the Migratory Bird Treaty Act.

EFFECTS OF HABITAT CHANGE ON WILDLIFE

The composition of wildlife changes when food or shelter in the habitat is altered. Behavioral and biological changes can also occur.

Loss of Food

One of the principal ways in which habitat loss affects wildlife is through the loss of food. For instance, as fields are turned into developments, rodent and rabbit populations diminish, so that raptors are no longer able to find food. Removal of marshlike areas along streams reduces feeding areas for rails and waterfowl. The influx of human wastes into streams results in an increase in bacteria, which utilize the oxygen in the streams, destroying much of the food used by fish.

Movement Patterns

Many big-game animals perish when highways block traditional movement routes. Dams prevent salmon from returning to their breeding grounds. When not properly placed, transmission lines interfere with migratory birds. Whooping cranes have died as a result of colliding with such lines.

Stress

Stress in wildlife often is the result of habitat alteration. Stress causes loss of body weight, a lower reproductive rate, and susceptibility to disease. This much we know, but these results of stress are very difficult to measure, so we have little precise information in the literature. We do know from studies of areas where development has taken place that the addition of heavily used roadways reduces nearby wildlife, particularly big game. This may be flight from the stress caused by moving vehicles or noise. But some animals increase their numbers, particularly edge species when vegetation is present.

In Scandinavia, snowmobiles are replacing reindeer-drawn sleighs in reindeer herding. While the machines are more maneuverable and easier for the herder, they disturb the reindeer. The approach of a snowmobile can cause these animals to panic and become unmanageable.[15] Animals change their activity patterns when snowmobiles are in the area. In Minnesota, home range and movement patterns of white-tailed deer increased as snowmobile trails decreased. When snowmobile traffic stopped, deer began to use the area again.[15]

Stress appears to cause chemical changes in animals. Hormone levels and other biological characteristics of the population may be altered. Apparently, stress can also affect reproductive potential.[17,18] One study of sheep provides some insight into the effects of stress. The telemetered heart rates of unrestrained female bighorn sheep were recorded under various behavioral and environmental circumstances. In all ewes, the heart rate varied positively with the surrounding activity level and inversely with the distance to a road traversing the study area. The heart rates of animals moving at night or through timber by day were higher than those moving during daytime.

Responses to transient stimuli varied greatly. The appearance of free-ranging dogs or coyotes brought maximal increases in the heart rates of all ewes. Vehicular traffic and aircraft caused heart-rate responses only at close range (less than 200 m, or 656 ft). Most (78.1 percent) responses to disturbing stimuli preceded or occurred in the absence of visible behavioral reactions. The rate usually peaked within 60 seconds of the onset of the response and recovered to the predisturbance baseline in less than 200 seconds. The appearance and continued presence (1 to 10 minutes) of a human within 50 m (164 ft) of the sheep resulted in a 20-percent rise in mean heart rate.[19,20]

The Mechanics of Stress. What does stress do? That is a difficult question to answer, but an example may help. On the average, caribou or sheep take in about 145 kilocalories (kcal) of metabolizable energy, or about 80 g (2.8 oz) of fodder per $kg^{0.73}$ of body weight each 24 hours. Mild excitation, during which the animal does not run, uses about 25.30 $kcal/kg^{0.73}$ per hour above and beyond normal daily needs. On a winter day with little wind and temperatures above $-34°C$ ($-20°F$), the normal cost of existence is about 5 $kcal/kg^{0.73}$ per hour. Exertion such as fast, sustained running costs about 1,100 $kcal/kg^{0.73}$ per 24 hours, or nearly 1 $kcal/kg^{0.73}$ per minute. Steady, fast walking over rough terrain costs about 0.4 $kcal/kg^{0.73}$ per minute[21] (Figure 8–11).

With the preceding information, one can calculate the minimum cost of energy, forage, or body fat to a 90-kg (199-lb) caribou that is chased for 10 minutes, then walks for an hour, and remains excited for another hour. Such exercise costs the caribou about 665 kcal above and beyond the 3,200 kcal it uses during the day just to stay alive. This is a jump in the cost of living of almost 21 percent and about 3 percent more than the animal's total possible forage consumption. Since, this additional cost of 665 kcal will be drawn from the energy stores of the animal, it represents about 74 g (2.6 oz) of body fat oxidized. It takes between 1,300 and 1,600 kcal, or 2 lb of good forage, to synthesize 74 g of fat. Thus, a running herd of 100 caribou requires a minimum of 91 kg (200 lb) of forage. The situation is not serious until it becomes repetitious. Ten such episodes would require an extra ton of dry forage.

Figure 8–11 Increased or sudden noise can cause stress in bighorn sheep as they increase their energy use. (Courtesy of Wyoming Game and Fish Department; photo by L. R. Parker.)

Noise. Noise appears to be a component of stress that can produce changes in animal activity. As with stress in general, measuring the impact of noise is difficult. For example, the U.S. Air Force found that domestic chickens and mink were not adversely affected by noise.[22] Field studies in areas of seismic activity, snowmobiles, and aircraft, however, indicate that noise as a component of a moving stimulus causes animals to leave. For example, an increase in traffic on rural roads reduces the number of animals seen or heard on the roads and in the surrounding areas. Sometimes animals acclimate themselves to continuous noise; however, the animals' energy intake may increase.

Reduction of Stress and Noise. One way to reduce stress is to develop parks and wilderness areas. Parks, however, are generally edge communities when established in urban areas.

When large regions are to be disturbed, as with mining in the west, plans can be made to extract minerals from one area before moving to another. The wildlife manager must demonstrate the value of wildlife to the private owners or users. These people, in turn, can develop plans such as decreasing their use of roads to reduce stress to wildlife.

The manager can also do some things to reduce stress. Managers often have an easier time of reducing stress to wildlife on public lands. Noise can be reduced by concentrating activity in one area. Preventing seismic activity or blasting near calving ranges or during raptor nesting can be very helpful to those species. Developers should be encouraged to leave raptor nests and other unique wildlife regions isolated. Rerouting roads and reminding people who fly small aircraft and helicopters that they should not frighten the animals will help.

One of the most effective tools managers have is an education program. All public wildlife agencies should be active in showing how developers, hunters, fishermen, and landowners can maximize the benefits from wildlife. The agencies can also assist landowners in getting compensation or tax incentives for providing benefits to wildlife.

SUMMARY

Both natural and human-made habitat alterations change wildlife communities. Habitat alteration can be caused by a minor modification to the habitat. Alteration occurs when an entire community is destroyed and a new community of organisms becomes established. Decreasing the size of a habitat, changing the habitat's structural components, and introducing or removing species alter the habitat for wildlife.

Many human activities also affect habitats. Air pollutants destroy vegetation, while acid rain destroys vegetation, wildlife food, and sometimes wildlife itself. Water pollution activities have been a major source of habitat change. Runoff from adjacent land activity or disposal of toxicants changes the food base and chemistry of the habitat. Channelization is a major source of aquatic habitat alteration. Terrestrial and aquatic habitats are both affected by the introduction of pesticides, herbicides, and heavy metals. These chemicals often cause subtle changes that cannot be detected immediately.

Habitat alteration usually takes its toll on wildlife, through loss of food or by changing behavioral traits or causing stress. When managers can meet with people planning activity in a habitat, they can often prevent major impacts on wildlife.

DISCUSSION QUESTIONS

1. Describe different types of habitat alteration, and tell what managers can do about them.
2. When a habitat is destroyed, does it mean that wildlife can never live in the area again? Why or why not?
3. How would you as a manager argue for wildlife habitat when national food and energy needs of people are pressing?
4. How can wildlife managers deal with the problem of acid rain?
5. What is channelization, and how does it affect fish and wildlife?
6. Why are wetlands so important in wildlife management?
7. How can aquatic pollution occur? What means can be taken to reduce its impact on wildlife habitats?
8. How can exotic organisms be successfully introduced into an area? What precautions are needed?
9. How should a manager measure alterations to big-game habitats caused by mining?
10. Why must stress be considered a component of habitat change?

LITERATURE CITED

1. Karr, J. R. 1981. An Integrated Approach to Management of Land Resources. In R. T. Dinke, G. V. Viger, and J. R. March (Eds.), *Wildlife Management on Private Lands,* pp. 164–92. Madison, WI: Wisconsin Chapter of the Wildlife Society.

2. Feierabend, J. S., and J. M. Zelazny. 1987. *Status Report on Our Nation's Wetlands.* Washington, DC: National Wildlife Federation.

3. Prism. 1982. *A Review of Petroleum Industry Operations and Other Land Use Activities Affecting Wildlife.* Edmonton, Alberta, Canada: Prism Environmental Management Consultants.

4. Howard, R. P., and J. F. Gore (Eds.). 1980. *Workshop on Raptors and Energy Development.* Boise, ID: Idaho Chapter of the Wildlife Society, U.S. Fish and Wildlife Service, Bonneville River Administration, and the Idaho Power Company.

5. Courtenay, W. R. 1978. The Introduction of Exotic Organisms. In H. P. Brokaw (Ed.), *Wildlife in America,* pp. 237–52. Washington, DC: Council on Environmental Quality.

6. Burks, C. D. 1985. Habitat Selection by the Ring-necked Pheasant during Breeding Season. Master's thesis, University of Wyoming. Laramie, WY, p. 155.

7. Laycock, G. 1966. *The Alien Animal.* Garden City, NY: Natural History Press.

8. Begley, S. 1987. On the Trail of Acid Rain. *National Wildlife* 25:6–13.

9. Luoma, J. R. 1987. Black Duck Decline: An Acid Rain Link. *Audubon* 89:18–27.

10. Anon. 1982. *Impact of Emerging Agricultural Trends on Fish and Wildlife Habitat.* Washington, DC: National Academy Press.

11. Powledge, F. 1987. The Poisoned Well. *Wilderness* 51:40–41.

12. Edwards, J. W. 1985. Trouble in Oiled Waters. *National Wildlife* 23:28.

13. Steinhart, P. 1987. Empty the Sky. *Audubon* 89:70–97.

14. Johnson, D. R., and R. M. Hansen. 1969. Effects of Range Treatment with 2,4-D on Rodent Populations. *Journal of Wildlife Management* 33:125–32.

15. Klein, D. R. 1971. Reaction of Reindeer to Obstructions and Disturbances. *Science* 173:393–98.

16. Dorrane, M. J., P. J. Savage, and D. E. Huff. 1975. Effects of Snowmobiles on White-Tailed Deer. *Journal of Wildlife Management* 39:563–69.

17. Gibson, P. S. 1970. Abortion and Consumption of Fetuses by Coyotes Is Following Abnormal Stress. *Southwest Naturalist* 21:558–59.

18. Arvay, A. 1967. Effect of Exteroceptive Stimuli on Fertility and Their Role in Genesis of Malformation. *GIBA FDNO Study Group 5* 26:20–24.

19. MacArthur, R. A., R. H. Johnston, and V. Geist. 1979. Factors Influencing Heart Rate in Free-Ranging Big Horn Sheep: A Physiological Approach to the Study of Wildlife Harassment. *Canadian Journal of Zoology* 57:2010–21.

20. MacArthur, R. A., V. Geist, and R. H. Johnston. 1982. Cardiac and Behavioral Responses of Mountain Sheep to Human Disturbance. *Journal of Wildlife Management* 46:351–58.

21. Geist, V. 1971. Is Big Game Harassment Harmful? *Oilweek,* pp. 12–13.

22. Brewer, W. E. 1974. Effect of Noise Pollution on Animal Behavior. *Clinical Toxicology* 7:179–89.

Part 4
SETTING GOALS

Now we can see how populations might regulate themselves in the absence of people. Add people, however, and the dynamics of wildlife and habitats change dramatically. With this in mind, we realize that effective wildlife managers must also be effective people managers.

In Part 4, we consider human-related impacts occurring to the habitat and how they can be mitigated. We look at legislation with which managers must be familiar in order to manage wildlife. We also discuss the duties and approaches used by wildlife managers and wildlife administrators. Ways to achieve goals that are important components of the planning process are discussed in two chapters in this section.

As the student will see, a good grasp of the concepts presented is important for effective management. At the same time, managers must understand the political system in order to communicate and negotiate effectively for the welfare of both habitats and wildlife.

Wildlife management can involve many facts. Here a great Horned Owl is being banded to show how birds are traced.

9
IMPACTS AND MITIGATION

Impact may be defined as the change in a population's natality, growth, and/or survival caused by some disturbance. It is an alteration of the homeostatic relationship within the animal population or between animals in a community and their environment. *Impact,* in other words, is the result of the impingement of *A* on *B*. Say that *A* is a tornado and *B* is a wildlife population. The impact of *A* on *B* is that some or all members of the population may be killed or forced to use different habitats. Of course, an impact can be positive, causing a population to increase in size or distribution. It can also be so subtle that it is difficult to measure. Impact on wildlife can come from natural or human sources, the presence of other species, or additional members of the same species. *Impact* is also used to indicate the act of impingement itself. Thus, we say that *A* has an impact on *B*.

Analyzing impacts, actual or speculated, on wildlife has become a major function of the wildlife profession since the advent of the National Environment Policy Act (NEPA) in 1969. (See Chapter 10.) Wildlife must now be considered in the evaluation of federal projects. State laws related to private lands vary considerably and are administered through different agencies. It is necessary to determine what legislation and which agencies are responsible for performing impact analyses in each state.

Descriptions of the impact of various sources on wildlife are now abundant. This is true particularly in areas where logging, fire, mining, or water pollution have occurred. Alterations of migration patterns, for instance, are related to the placement of roadways, transmission-line corridors, and fences, as well as to development in general. But few studies document measures that can mitigate or reduce impacts. Because of this lack of data, planners cannot always predict the changes from impacts on wildlife. Wildlife managers generally are quick to point out adverse impacts, but slow to suggest alternatives.

In this chapter we discuss different forms of impact, including those of some human activities. We look at some methods of moderating or preventing adverse impacts.

FORMS OF IMPACT

Many natural and human-made changes affect wildlife. These changes, which vary from obvious to subtle, affect either the population or the habitat. The impact can be *primary,* resulting from the disturbance or change itself (e.g., removal of habitat to extract minerals), or *secondary,* hinging on, but following, the change (e.g., increased human use of areas opened through oil and gas activity).

Population

Both deliberate and unintended population diminution or destruction has been documented many times. Passenger pigeons, eagles, and other predators have been shot and poisoned. People have altered marine populations; some whale populations are so small that individuals in them cannot find mates, and sardines have been so overharvested that their schooling behavior has been disrupted. Wild-ungulate populations have contracted diseases from domestic cattle. The presence of people in remote areas has caused stress in isolated populations.

Habitat

Habitat alteration was examined in detail in Chapter 8. Alteration can take the forms of complete destruction, subdivision, structural change, or introduction of exotics, making the habitat unavailable. Alteration of a habitat, negative for one group of animals, can be positive for another.

The Jordan River as a Migratory Corridor

Riparian ecosystems are very important to bird communities, particularly in areas where there is limited water. In the west, riparian habitats make up less than 1 percent of the landscape, yet they have more than 50 percent of the bird species. In Utah, 75 percent of the neotropical migrants depend on riparian habitat for nesting and migrating. The Jordan River flows from Utah Lake to the Great Salt Lake through the heavily developed Salt Lake Basin. Each spring, hundreds of thousands of birds are seen along this river habitat. To passing migrants, the Jordan River offers a travel corridor and stopover area with food, water, and roosting sites between Utah's western desert and the Wasatch Range.

The extensive degradation and loss of wetlands have increased the importance of the Jordan River as stopover habitat for migrant birds. Despite the fact that the Jordan is much altered from being drained, filled, and rerouted, it still serves as a focal point for migratory birds. If such habitat is lost, major changes can be expected in the migratory bird population. Some changes are being made in the habitat as people realize the importance of "green areas" to enhance their own living style. Hopefully, this trend will continue, as it provides important areas for migratory birds.

Norvell, R. E. 1997. Avian Use of Riparian Habitats within a Gradient of Urban to Rural Matrix of Urbanization, Salt Lake Valley, Utah. Master's thesis, University of Wyoming.

Urbanization

The development of cities and suburban areas has led to one of the major impacts on wildlife habitat over the last hundred years in the United States and many other parts of the world. We only have to look at the types of wildlife habitat and activities in Europe today to see what happens when a large number of people live in a confined area. In the United States, cities and agricultural activities have changed millions of hectares (Figure 9–1). The entire east coast is now virtually one urban center (Figure 9–2). The result has been major changes in the composition of species living there. In Massachusetts, where comparisons of bird communities were made between urban and suburban areas, more breeding species were found in the suburban areas, although the total diversity of birds was higher in the city. Migrant birds that feed on insects and ground-nesting birds were not found in cities. Rather, introduced exotics, such as the house sparrow in the breeding season and starling in the winter, were common.[1]

In Wyoming, one of the least populated states in the union, the land-use change that has had the potentially most damaging impact is human expansion, including expansion of cities, subdivisions, mobile homes, and commercial areas. Between 1967 and 1977, when approximately 388 km^2 (150 sq mi) was altered, big-game movement patterns were disrupted and critical winter range was lost, thereby reducing populations of such species as elk and mule deer.[2]

Urban areas reduce both food and habitats available to native species. Since most urban wildlife species are generalists, they can usually adapt to a variety of habitats. These are the species we often see colonizing disturbed areas or inhabiting ecotones. By contrast, many species with specialized requirements do not appear to be able to live near people.

Energy Development

Impact can occur at all stages of oil and gas field exploration and development (Figure 9–3). Usually, careful planning can reduce the adverse effects on wildlife.

Seismic Activity. Seismic activity is one process of exploring for oil or gas. Crews often use large vibrator trucks (Figure 9–4), which pound on the ground and record the vibrations that pass through different materials in the earth. Another form of

Figure 9–1 Housing development near citrus grove in Arizona. (Courtesy of the U.S. Soil Conservation Service.)

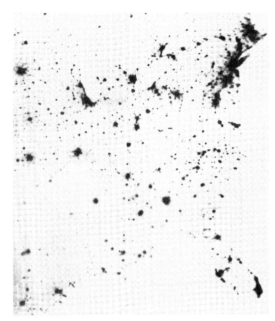

Figure 9–2 Satellite view of the east coast at night. (Courtesy of NOAA.)

seismic exploration involves the detonation of explosive charges below or above the surface of the earth. Geologists are able to predict the possibility of oil and gas concentrations on the basis of the records of the vibrations. Seismic crews also use helicopters and bulldozers to gain access to some areas. Temporary road construction can change vegetation patterns and encourage public access into wildlife habitats.

Seismic activity causes animals to move away. Normally, this movement would not be serious if animals were not forced out of their habitat or subjected to unusual stress. But during the fall rut, bull elk must establish and maintain areas against

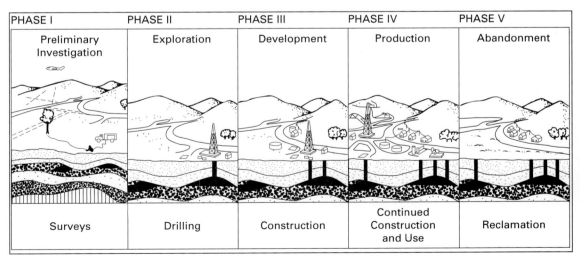

Figure 9–3 Sequence of operations in an oil and gas field.

Figure 9–4 Vibrator truck used to explore for oil and gas.

challenges to their dominance by other bulls, and during the calving period, cows favor warmer exposures with gentle terrain. Seismic activity that drives the bulls away can change dominance patterns, while the calves of cows driven to suboptimal areas may not survive.[3]

To avert an adverse impact on big-game offspring, it is recommended that seismic exploration be seasonally foregone in crucial big-game fawning, lambing, or calving areas. By the same token, seismic exploration should be avoided near waterfowl nesting areas in the spring.

Since human disturbance is a component of the overall effect of seismic exploration on wildlife, many of the avoidance measures depend on controlling people. Seismic camps should be located at least 100 m (328 ft) away from important meadows used by various wildlife. Camp personnel should not use firearms or all-terrain vehicles, such as motorcycles, four-wheel-drive trucks, dune buggies, and snowmobiles. Personnel behavior extends to the operation of helicopters. Helicopter travel lanes should be at least 250 m (820 ft) above wildlife, especially during critical reproductive periods.[4] Seismic lines should not be closer than 0.5 km (1.6 ft) to other lines, and in all cases they should be kept from looking like roads, which might encourage off-road traffic. Activity within 4 km (13 ft) of critical big-game winter ranges should be restricted from the time that game concentrations occur until the spring migration. Flexibility in this restriction, which has been incorporated into some leasing agreements, may be exercised when wildlife biologists observe that game is not concentrated in a particular winter or when some animals leave the winter ranges earlier than usual. Observing these caveats would help reduce the impact of seismic exploration on wildlife.

Construction. The construction phase of energy-development activities can be stressful to wildlife. This is a time when various activities are carried out by a large number of people, and animals are not always able to adjust to the changes. Take the coal-fired energy plant, for example. Among the features and activities of a coal-fired power plant that can have an adverse impact on fish and wildlife during both construction and operation are coal slurry pipelines, coal cleaning and storage, limestone preparation (when flue-gas desulfurization is employed), particulate and gaseous emissions, ash, and desulfurization sludge. The impact of coal slurry pipelines results from right-of-way construction and accidental spills. Coal cleaning and storage cause noise, dust, loss of habitat, and runoff of material similar to acid mine drainage into surface waters. Limestone dust and runoff from limestone storage piles may have a measurable effect on soils and vegetation and can increase surface-water hardness and turbidity.

Disposal of combustion-waste products (collected ash and flue-gas desulfurization sludge) requires sizable land areas and has the potential for adverse effects, including seepage of trace elements into groundwater, solids, and surface waters.[5] The amount of seepage will vary from region to region (Table 9–1).

Transmission-Line Corridors. Transmission-line corridors that cut through forests must be maintained so that maintenance crews have access to them and vegetation does not interfere with the electric wires (Figure 9–5). This is not an easy task, especially in areas such as the eastern deciduous forests, where vegetation grows rapidly. In some areas, vegetation can return within five years after a corridor is cut.

A number of studies undertaken to determine the impact of corridors on wildlife show that corridors create an edge in the forest. Corridors wider than 20 m (66 ft) actually create a shrub or grassland community. White-tailed deer, woodcock, indigo buntings, cardinals, and other animals are found in edge habitats.[6] Electric utilities use herbicides, in addition to removal by mechanical means, to keep the corridors clear. Managers have expressed concern about the impact of broadcast spraying (spraying all the vegetation, usually from aircraft) on the corridors. Field studies have revealed mixed results. Vegetation is set back in the succession sequence, so that different animals populate the area. Less stability is found in population numbers when broadcast spraying is used frequently (every five years or less).

Studies in several areas of the east, including Maryland and Pennsylvania, that use selective spraying have shown good results when rapid-growth shrubs and saplings are sprayed, but other shrubs and much of the ground cover are not. This form of spraying encourages a combination of plants and animals that does not fluctuate greatly. Unfortunately, the cost of selective spraying is high.

Blueberry bushes 10 to 15 cm (6 to 24 in) high are food for raccoons, bear, and birds. Goldenrod and fern covers harbor an insect population needed by nestling grouse and turkey chicks. The taller shrubs, such as the various dogwoods, viburnums, and hollies, retain their berries until late in the winter, providing food for resident birds. All these, as well as the colorful azaleas, give ample browse for deer. Rhododendrons, laurels, and junipers provide the necessary protective cover for some mammals and birds.

The use of mechanical devices to clear vegetation is becoming more common. If the operators of these devices leave vegetation on the ground, it provides cover for

TABLE 9–1

Elements That Can Appear in Aquatic Systems Following Onset of Mining Operations

Arsenic	Mercury
Barium	Molybdenum
Beryllium	Nickel
Boron	Selenium
Cadmium	Tellurium
Chromium	Thallium
Cobalt	Tin
Copper	Uranium
Fluorine	Vanadium
Lead	Zinc
Lithium	Zirconium
Manganese	

Figure 9–5 Transmission-line corridor in the eastern deciduous forest.

small mammals and birds, as well as serving as a source of food for other animals. The environmental contamination from herbicides is also eliminated.[7]

Oil Spills. Oil spills adversely affect aquatic organisms. For example, 50 percent of cutthroat trout exposed to 2.4 mg of crude oil per liter of water died in 96 hours,[8] and this acute toxicity value was less than the 10-mg-per-liter effluent limitation for oil and gas in Wyoming, Colorado, and Montana. A threshold concentration (the amount above which harm occurs) of approximately 100 mg per liter for 90-day exposures, based on reduced growth of cutthroat trout, was determined. Thus, a safe concentration for long-term exposure may be as much as 100 times lower than current effluent limitations for oil and gas. It is an important consideration for streams and rivers that continually receive oil spills or pipeline seeps.

Some zooplankton and aquatic macroinvertebrate populations decrease when crude oil is spilled into aquatic systems, while others increase in size. For example, oil perturbation caused a decrease in zooplankton populations and a reduction of primary production in only three to four days in arctic tundra thaw ponds. At the same experimental spills, oil killed macroinvertebrates only in peripheral areas, where the animals contacted the oil on plant surfaces, and at the water surface; and long-term exposure interfered with metamorphosis and mating.[9] Finally, it appears that the composition of the macroinvertebrate community changes and its diversity decreases when oiled substrates are recolonized.[10] A few tolerant species of microinvertebrates (Ephemeroptera and Chironomidae) dominate these surfaces, sometimes showing a positive response to oiled substrates. This could result from a competitive advantage, since other organisms cannot live there under the altered circumstances.

Reserve Pit Fluids. Reserve pit fluids are complex chemical mixtures produced during oil and gas exploration. Because of the variety of drilling muds and chemical additives used in the industry, probably no two reserve pit fluids are alike, a fact that makes it difficult to determine acceptable water quality standards and predict effects. Although several chemicals in these fluids are known to be toxic, little is known of the specific toxicity of many of the components or of the whole fluids. For example, in a recent test of a reserve pit fluid, chemical analyses (of pH, conductivity, and major ions) of the conventional pollutants and two heavy metals (zinc and chromium) indicated acceptable concentrations for all parameters. Nevertheless, cutthroat trout died after 12 minutes in this water.[8] Apparently, the combined (synergistic) effect of all the pollutants was greater than the sum of the individual effects. In some pits, birds and mammals become stuck in the oily substance and die (Figure 9–6).

In short, too little is known about the effects of reserve pit fluids and reclaimed reserve pits to predict their impact on aquatic and terrestrial animals. Once again, complex chemical mixtures associated with oil and gas development should be

Figure 9–6 Oil reserve pits can attract and kill many wildlife species, especially migratory birds. (Courtesy of the Wyoming Game and Fish Department.)

evaluated as whole effluents, rather than for the effects of individual chemicals on water quality.

Overall Impact

Among many studies of how wildlife reacts to energy development was one designed to determine the impact of hydrocarbon development activities on elk, bear, and bobcats in the north-central portion of lower Michigan. Track data for elk, deer, and bobcats collected twice monthly from designated routes in each subregion, served as an index of animal use over time.[11]

Levels of elk activity throughout the area were recorded. Hardwood forests, which contained the greatest concentrations of elk, especially in winter, were also used for ongoing hydrocarbon development activities. Only limited elk activity was found in the southern portion of the Pigeon River County State Forest, a scene of hydrocarbon activity before a court order banned further development. Elk activity, however, had been historically low in this area.

Bear and bobcats were widespread throughout the regional study area, although both species seemed to favor lowland habitats. Analyses of the collected data indicated that hydrocarbon development activity during the study period had only short-term, localized impact on the species examined. Elk, bear, and bobcat activity decreased within $\frac{1}{4}$ mi of active drilling operations, but returned to predrilling levels within two to four weeks after the site was abandoned—that is, after drilling was completed. A density of one well per 520 hectares (2.0 sq mi) appears to have had little adverse effect on the species studied.

The effect of oil drilling on wildlife on the Mackenzie River delta in Canada was studied during June, July, and August 1971.[12] The results from aerial surveys of wildlife populations within 30 km (19 mi) of the rig site were compared with those from a survey taken the year before. Numbers and species of birds found in eight selected plots within 2.5 km (1.5 mi) of the site and in eight plots of comparable habitat in a control area of similar size 8 km (5 mi) distant were compared. Observations of drilling activities at the rig and of resupply operations were made in an attempt to isolate disturbing influences.

Of the more abundant bird species, 43 percent were found to be noticeably less abundant than normal within 2.5 km (1.5 mi) of the oil rig during the summer drilling operations, 52 percent were not affected, and 5 percent (two species) were more abundant. Geese and swans, when molting or in family-group flocks with downy young, moved from or strayed more than 2.5 km (1.5 mi) from the drill rig. Other species apparently became accustomed to activity associated with the rig. Helicopters at low levels were apparently the most disturbing factor, directly affecting wildlife within a 2.5-km (1.5-mi) radius.

Other forms of human intrusion into wilderness areas can result in direct encounters. In Alaska, people–bear interactions, particularly those involving grizzly bear, have increased.[13] These matches involve not only oil drillers, but hunters and fishing enthusiasts who now use the area. Human intrusion, rather than the oil activity itself, seems to have the major impact on black bear in Alberta, Canada.[14]

Logging and Grazing

Logging often creates greater access to an area. In wilderness areas, harvests usually displace big-game animals, at least temporarily. But elk are found in and near logged-over areas after the activity is completed.[15] Food supplies change. Logging slash increases cover for some animals, but obstructs the movement of others (Figure 9–7). Elk select bedding sites that will increase control of their body temperature. As summer temperatures increase, elk often move to north slopes in the day and south slopes at night. Logging operations appear to limit access to choice sites and to disrupt the movement of these animals.[16]

The composition of small-mammal and songbird species reacts to logging and grazing. In both cases, a change in the successional stage causes a change in the composition of wildlife species.

Figure 9–7 Logging slash. (Courtesy of the U.S. Forest Service.)

How Does Grazing Affect Small Mammals of the Riparian Community?

Cattle often graze in concentrated numbers along riparian areas in arid regions of the world. In New Mexico, a study of small-mammal populations in a desert wetland indicated that the abundance and diversity of such mammals was greater in riparian communities in which cattle were excluded. Over a 10-year period, some populations of small mammals decreased 50 percent or more in grazed plots over ungrazed plots. These results were significant because small mammals provided an important food base for many raptors and larger mammals in the desert community. One management solution to this problem was to limit access to wetlands, which would allow cattle to find water but not spend extensive time grazing.

Hayward, B., E. J. Heske, and C. W. Painter. 1997. Effect of Livestock Grazing on Small Mammals at a Desert Cienaga. *Journal of Wildlife Management* 61:123–29.

Agriculture

Agricultural activities on cropland, rangeland, pastures, and forests have been altering wildlife habitats in both positive and negative ways since the beginning of American history. Agriculture and forestry production has increased to meet an expanding and affluent population's demand for food, fiber, and forest products. During the early periods, clearing forests for small, scattered farms attracted species such as robins, woodchucks, bobwhites, and rabbits. At the same time, it lessened habitats for others, such as wild turkeys, black bear, and moose. The massive westward movement brought drastic changes in prairie habitats, to the detriment of bison, elk, and pronghorn and the benefit of rodents and rabbits.

These early changes were gradual and minor in comparison with those brought about by the mechanization of agriculture. Modern agricultural technology and economic considerations have favored large, contiguous fields planted in single crops. The increasingly efficient drainage of lowlands has increased the number of crops grown on marginal soil. More efficient use of fertilizers and pesticides and the development of irrigation have expanded cropland carved out of natural ecosystems. Thus, the variety of habitats essential to wildlife has decreased. Species that responded positively to early patterns of agriculture have now declined. The introduced ring-necked pheasant is an excellent example. The large pheasant population of the midwestern and northern plains of 30 years ago has dwindled rapidly with the intensification of agricultural operations.[17]

Many agricultural and forestry land-use practices affect the potential of an area to support wildlife. The unit of land planted to grow crops or grains and plowed immediately after the fall harvest results in a poor habitat for most wildlife. If the same unit of land is planted using conservation tilling practices, such as leaving buffer zones or rows of natural vegetation, or is used for pasture, it becomes more attractive to some wildlife species. Plant diversity, the presence of weeds, field size, and the distribution of fields in relation to one another and to neighboring natural habitats all help determine the attractiveness of an area to wildlife. Often, small changes in land-use practices can result in major differences in the amount of habitat available to wildlife.

Silt from erosion and runoff from organic and inorganic wastes, fertilizers, chemicals, irrigation, channelization, and cutting deep into slopes all affect aquatic systems negatively. Species diversity decreases in these habitats. Diversifying crop and tree species can increase yields and change the texture and fertility of the soil, while at the same time creating wildlife habitats superior to those without such diversification. Erosion control, both direct and indirect, improves terrestrial habitats and protects water quality, thus increasing agricultural production. Fall plowing, double-cropping, and increased use of herbicides may aid production, but usually are detrimental to wildlife habitat (Figure 9–8).

The adoption of no-till farming may prove beneficial to game birds, as well as to farms in the prairie pothole region. Spring-planted wheat is grown in the region that is the prime duck-nesting area of the United States: North Dakota, South Dakota, eastern Montana, and part of Minnesota. The wheat is planted during the peak of the mallard nesting season, when lack of cover and the disturbances caused by tillage render the land unsuitable for nesting mallards. The no-till technique enables farmers to switch from spring wheat directly to winter wheat, with seeds being planted into full stubble. The stubble serves to trap an insulating blanket of snow, which protects the wheat from extreme winter temperatures. Then, in the spring, ducks, pheasant, and other birds nest in the stubble and emerging green wheat. Except for minor disturbances by weed sprayers, the nests are not exposed to dangers such as spring tillage. The stubble and growing wheat provide the cover needed for successful nesting, and the nesting activities do not damage the wheat (Figure 9–9).

Whenever a change in agricultural or forestry practices provides food or cover for wildlife, a number of species quickly adapt to the new resource. Canada geese winter by the thousands in areas where there is little water or natural food, but where corn has been spilled during the harvest. Bluebirds and small rodents quickly invade abandoned orchards.[17]

The primary economic trends in agricultural production are related to the large increase in capital required to engage in modern, technologically based agriculture. These reflect (1) a greater reliance on purchased input, such as chemicals and machinery; (2) rising land costs; and (3) increases in energy prices. Today as never before, American agriculture is influenced by economic forces. It is also influenced by continued population growth, both domestic and foreign; income growth; and government

Figure 9–8 Soil erosion can mean habitat loss for many species of wildlife.

Figure 9–9 Some ducks benefit from no-till farming.

regulations.[4] All these factors have had a significant impact on agricultural production in the last 30 years. So one very important component in agriculture–wildlife interaction is the value of farm real estate. The massive increase in agricultural land costs since 1960 explains in part why more land is being removed from agricultural use. In addition, costs associated with all aspects of farming have increased. Capital outlay has gone up greatly, land use has increased only slightly, and labor has decreased. With these costs in mind, it is clear why ranchers and farmers feel that they cannot spend their money to benefit wildlife.

While fuel costs have increased for everybody since the mid-1970s, the increase has been more than 50 percent for farmers. Some experts foresee a decline in the use of fossil fuel, and there is evidence that land preparation practices (plowing and cultivating) are being modified to reduce fuel consumption. Although interest in gasohol is developing, the use of gasohol will be limited, probably to less than 10 percent of total fuel consumption. The degree to which substitutes for fossil fuel will come from agricultural products is not clear. The current federal policy of subsidizing the production of gasohol could lead to competition for grain between exporters and domestic-use buyers. Such a development would increase the demand for land to grow corn. But if deregulation of domestic oil and natural gas increases supplies and stabilizes prices, it seems unlikely that gasohol production will increase. Even with continued moderate increases in fossil fuel prices, it appears improbable that the production of fuels from agricultural products will increase significantly, although conversion of some grain, crop residue, or biomass for fuels will probably continue. However, if major increases in fuel costs return, it is likely that significantly more of our agricultural land will be devoted to raising crops that can be used in fuel. Attempts may be made to put less optimal land into crop production. All this has implications, not only for the nation's food supply, but for the use of habitats and so for wildlife and wildlife management.

MEASURING IMPACT

To measure the impact of a practice or activity, it is necessary to have some form of documentation—ideally, before-and-after studies. Control and experimental studies are also very effective. When such studies are not possible, other measures must be used. Biologists measuring the impact of humans on wildlife should use standardized scientific techniques, so that comparisons of sites can be made. Methodology should always be clearly stated, so that a study can be replicated. Value judgments such as "ex-

Part 4 / Setting Goals

citing results," "interesting," and "excellent value" should be avoided. The distinction between facts and conclusions should be observed. That a new roadway does not cause animals to move is a fact and should be so stated. That it does not cause stress in the animals is a conclusion based on facts derived from studies of physiology, energy, and reproduction. The validity of the fact depends on the accuracy of the observation. The validity of the conclusion depends on the extent and accuracy of the studies. Although we sometimes state conclusions as if they were facts, we must remember that their validity is in proportion to the evidence on which they are based.

When preliminary data show the predisturbance species that are present, comparisons can indicate changes in number, in composition, in reproductive status, and in movement patterns. Opponents of such studies argue that we cannot prove one-to-one relationships between habitat alteration and a change in the wildlife population. For example, there is much controversy concerning the effect of acid rain. Some people hold that no cause-and-effect relationship has been shown between a change in the pH of lakes in the northeast and declines in wildlife populations. There are conflicting opinions, too, regarding the significance of the effect of construction activity on the winter range of elk: Some maintain that because elk should be able to move to new sites, construction should not necessarily be considered detrimental. Unfortunately, new sites may have varying plant and animal interactions. Elk originally moved and utilized a winter range because they found conditions there conducive to their survival. Whether a new site would supply their needs is uncertain.

IMPACT ANALYSIS

There are two kinds of impact situations that wildlife managers have to handle. First, for cases in which some form of habitat change could occur, managers need to develop plans to reduce the impact on wildlife. Second, when some form of impact has already occurred, they need to develop plans either to restore wildlife to the affected area or to reduce any further impact on the wildlife.

Planning for Impacts

When some form of change is intended, it is common to develop plans to initiate that change. Thus, in roadway construction, topographic maps of the area are obtained and routes are planned. If logging operations are planned, maps assist in evaluating the classes of timber and marking routes through which equipment can move. In large-scale operations, such as the development of power plants, mining operations, or sewage treatment facilities, extensive plans are developed involving the use of land, water, and air resources. All this material is documented, and in many cases, blueprints are developed. These planning stages should include an evaluation of the impact of the operation on wildlife and a consideration of alternatives with wildlife in mind. Ideally, the planners should confer with wildlife managers before any changes are made.

Approaches during the Planning Phase

Wildlife managers must be familiar with plant and animal species in areas of proposed change and should have good inventory data available. They must evaluate the data and understand the interactions that are likely to be affected. For example, a new gas-sweetening plant in one area might create air pollution that would affect distant wildlife populations because of changes in water quality or the composition of plant species.

Placing a power plant along a river may change the temperature of the water until it becomes lethal to fish and their food supply.

Most studies start with an evaluation of the area that is expected to be affected. Data evaluation covers the wildlife present, soils, water, and the potential for improving the habitat. For this evaluation, a search of the literature and the many data banks currently in use often provides information. These sources can often answer questions about the types of wildlife species and plant communities present in a habitat and about which field evaluations are needed. But very seldom can accurate conclusions be drawn from the literature and available data, unless previous surveys of the area have been made. Often, references show that some endangered species could live in the area, but not whether they do indeed live there. Thus, survey-type field studies involving population and habitat sampling need to be made. These need not be extensive, time-consuming studies, but should be designed to give an adequate representation of existing wildlife populations and the habitats they require on a seasonal basis. Such initial surveys should help identify particularly important habitats, as well as habitats that might be used in trade-offs. Trade-offs involve the creation or enhancement of wildlife habitat for desired species, as when disruption of the golden eagle's nesting habitat requires finding another suitable habitat for the nest platforms.

Once the population and habitat survey information is available, a reasonable evaluation of wildlife needs can be made by synthesizing field data with information from the literature and a data bank. Areas that are particularly important to wildlife, such as migratory corridors, raptor nesting areas, and winter or summer ranges, should be delineated and marked as critical to wildlife.

Wildlife biologists need to consider the following in the planning process:

Effects of the proposed activity on wildlife. For example, the effects of mineral operations on wildlife are determined by the type of exploration, type of extraction process, characteristics of the site, and wildlife present on the site.

Different effects on species. The biologist must identify which species are expected to be positively affected, which adversely affected, and which unaffected by the proposed activity.

Duration of effects. The effects of some projects are short term or temporary; others have long-term or permanent consequences. Knowing how long the expected effect will probably last helps the biologist determine the significance of the activity for wildlife.

Scope of effects. Activities differ in their intensity of effects and extent of area affected.

Season of activity. When the activities are to occur is obviously important, especially if they will conflict with nesting or calving seasons.

Adaptability of wildlife species present in the area. Information concerning the adaptability of the species that are present is essential for sensible planning.

Sensitivity of the area regarding wildlife. Some areas are extremely sensitive to alteration because they contain restricted habitats of specific wildlife species. An example of a restricted habitat is a riparian zone (Figure 9–10).

Resiliency and tolerance of vegetation. If the vegetation in a habitat is to be altered or exposed to stress, such as air pollution, the degree of various plant species' resiliency to, and tolerance of, the disturbance may need to be analyzed. Where there is high precipitation, for example, disturbed areas will normally be revegetated more readily than arid regions. The type of vegetation and its ability to recover from disturbance is related, of course, to the duration of the effect.

Figure 9–10 Riparian community.

Habitat diversity—minimum habitat requirements and viable populations of wildlife species. The quantity, quality, and distribution of components of a habitat required by the wildlife species in the area must be taken into account. The biologist must determine whether the requisites for maintaining viable populations in the area will be present when the project is in process and after it is completed.

Potential for the area's rehabilitation. The potential for habitat rehabilitation and/or opportunities for improvement should be determined for projects expected to affect wildlife habitat.

Contributory effects. Some changes occur naturally in a community. Succession and water runoff are examples. These changes need to be considered when evaluating human-induced impacts.

Consequences and risks associated with unplanned events. Some projects involve risks for wildlife because of unplanned, but possible, events, such as sedimentation or the chemical pollution of streams.

Human population growth. Whether or not a project is expected to bring about human population growth is an important consideration in the biologist's assessment. The size of the human population affects the use of the wildlife resource, the degree of harassment, and such secondary matters as changes in water quality and quantity.

Accessibility of the area. Human accessibility to previously remote areas may significantly affect some wildlife species because of the greater number of encounters with people.

Cumulative effects. The relation of other forest activities and use of resources to the proposed activities is an important part of the wildlife analysis. The cumulative effects of projects that are sequential and occur in a chronological order—for example, oil and gas exploration leading to development and production—should also be considered. Cumulative effects can be spatial. For example, each new mine site or set of drilling pads in an oil field requires a series of access roads. Land becomes more and more fragmented. Fences are often associated with roads, and fencing, although it benefits edge species, has a negative effect on species that require large habitats or that are migrating through the area.[18]

One big problem with our ability to deal with impacts is our inability to measure cumulative impacts. Generally, the combination of impacts such as urbanization, energy development, contaminants, logging, grazing, and so on must be considered. Different species of wildlife may respond differently to each combination of impacts. Thus, survivorship curves may vary with impacts. Armed with all the information just indicated, the wildlife manager can meet with planners and developers to discuss techniques for reducing the impact of humans on wildlife.

MITIGATION

Most of the techniques just discussed fall into the category of **mitigation.** The purpose of mitigation is to avoid a change in, or reduce the impact of a change on, wildlife living in an area. There are several types of mitigation: (1) avoidance—that is making no change in the habitat; (2) trade-off, or creating a new habitat to replace the one that will be disturbed by the change; (3) special techniques to reduce the impact on the habitat during the development phase and the phases following development of the habitat; and (4) reclamation or restoration of the habitat following impact.

Avoidance

When highway engineers plan interstate highways, they try to spare areas of historical interest, which are marked on a map for their guidance. In the same manner, critical wildlife habitats, such as some riparian habitats, important cliffs or structures, drainage systems, habitats of endangered species, and critical winter habitats, can be avoided when development is planned. Sometimes avoidance is expensive. An oil well placed some distance from its optimum location requires a longer shaft. Rerouting roadways around a streambed where runoff may create water pollution is a costly business.

Actually, avoidance is customarily approved only when the impact on a population is likely to be severe. Since this type of mitigation is extreme, the wildlife manager can seldom insist on complete avoidance of any considerable area. In any event, to argue the case, the manager must have accurate field information as to which areas are critical for wildlife.

A good example of avoidance is the preservation of old-growth forests for wildlife species. In the west, a number of species—the pine marten, spotted owl, and some others—are associated with old-growth habitats. The U.S. Forest Service has drawn up a list indicating the amount of old growth in each of the forest systems. It has been decided to retain a certain percentage of old-growth forest, so each forest-management plan must provide for the preservation of an area where timber harvest operations will be allowed. To develop these plans, Forest Service biologists calculated the size of old-growth habitats necessary to support each desired species associated with those habitats.

Trade-Offs

Trade-offs involve creating suitable habitats for the wildlife that is to be affected by a project. This necessarily involves value judgments on the part of managers: Habitat changes made for the species to be moved will affect the resident species. Trade-offs are likely to be time consuming and are often expensive. It is important to recognize that improving a habitat for one species is made at the expense of habitat for other species.

There are many examples of trade-offs. The woodcock, a highly sought-after game bird in the east, is declining because of urbanization and agriculture. The bird prefers forest edge, with some openings. Its habitat has been increased in areas of development by selective burning of forests. Dragging (removal of vegetation by tractors) of piñon–juniper areas for use by mule deer is another example of a trade-off. In this case, the dragged area substitutes for areas taken over by agriculture and ranching interests.

The U.S. Fish and Wildlife Service has proposed a methodology called *habitat evaluation procedures* (HEP) as a means of evaluating habitat trade-offs. HEP evaluates habitats for a particular species in the light of data collected where the species is found. Other areas are then given classification ratings of good to poor for that species. When the habitat for a desirable species is altered, less desirable habitats can be made more suitable by incorporating attributes of the habitat required by the species.

To determine the suitability of a habitat for a particular species, evaluators develop a habitat suitability index (HSI), or model, for the species. Most models developed thus far have been based on literature reviews and lack extensive field verification. Literature data on habitat and life history are combined to formulate the model. Often a model is presented in a graphic, verbal, and mathematical format.

The model for the southern kingfish can serve as an example. This fish—also called the channel mullet, ground mullet, black mullet, and king whiting—has been collected along coasts from Long Island Sound to Port Isabel, Texas. It is taken commercially and as a sport fish in some regions. It uses different habitats during different life stages. The adults are found in areas of high salinity—20 parts per thousand (ppt)—near shoreline or strong countercurrents. They are usually found where white or pork shrimp are abundant.

Southern kingfish eggs are found floating offshore in waters with a salinity of 20 ppt and a temperature of approximately 15°C. The larvae lie in shallow estuarine nursery grounds with lower salinity. The juvenile fish move around a great deal and are found in waters with a range of salinities. They also move up tidal rivers with bottom currents. Juvenile kingfish are bottom feeders; like the adults, they are found in areas where shrimp are available.

The HSI model for this fish separates the marine and estuarine environments. Based on food and water quality, the two HSI models provide information about the southern kingfish's needs (Figure 9–11). A graphic display of each variable is made, with a value scale from 0 to 1. A suitability index of 1 indicates what is best for a particular variable. For example, the preferred salinity (index 1) for juveniles and adults ranges from 10 to 27 ppt.[19] (Figure 9–12).

The habitat variables (eight in number) can be combined to determine whether an area is suitable for a particular life stage of the fish. There are management options, particularly in the estuarine environment, to improve a habitat if pollution is present. The model can also be used to predict the impact of such habitat changes as channel dredging, shoreline construction, or wastewater disposal.

Techniques to Reduce Impact

When plans are made that will mean changes in a wildlife habitat, it is possible to mitigate the effects of these changes if one understands the wildlife to be affected. In many parts of the eastern United States, cavity-nesting species have declined as dead trees have been removed for firewood or to prevent forest fires. Nest boxes, particularly for eastern bluebirds, have helped maintain the affected species. (It is important to construct the boxes so that undesirable species, such as starlings, cannot use them.) In

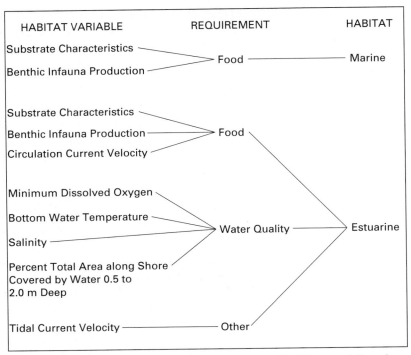

Figure 9–11 Habitat variables associated with life requisites for southern kingfish model.

many marsh areas where there are no longer snags, wood ducks, too, are maintained by nesting boxes placed in appropriate spots.

In a technique used to prevent electrocution of raptors by transmission lines, poles and towers are extended above the lines, which are strung below normal perching level. This lessens the problem, but does not solve it completely, even though in some areas the raptor population has actually increased with the added perch sites. The distance between wires and towers has been widened, so that the birds' wings will not catch there. The technique is promising but expensive.

An effective but also expensive project was undertaken in Wyoming along Interstate 80. During and shortly after the construction phase, many mule deer and antelope

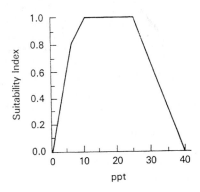

Figure 9–12 Average summer salinity requirements for southern kingfish.

Part 4 / Setting Goals

were hit by automobiles. Damage to cars and wildlife was considerable. A study of the area indicated that this was a traditional migratory route for both animals—in fact, the interstate subdivided some antelope herds. On the advice of biologists, highway engineers built higher-than-usual fences, which funneled the deer and antelope into specially constructed underpasses. At first the animals were reluctant to use the underpasses, but after mirrors were placed to give the illusion that other animals were in the tunnels, they began to move freely back and forth (Figure 9–13).

Minimizing Disturbances

Planning ahead to log timber can mean avoiding wholesale destruction of a forest. If timber is removed first in one area and then in another, wildlife species will move and sustain themselves. If some slash is left on the ground, soil erosion is likely to be less, and wildlife can move back into the area as grass begins to grow. Reseeding will help restore the site quickly to predisturbance levels.

Reduction of disturbances to aquatic habitats can be initiated by keeping pollutants out of lakes and streams. Projects designed to purify sewage systems, cool hot water, and leave natural vegetation around streams and lakes have been effective in reducing pollution levels in many parts of the country.

Reducing human activities is sometimes important. When development is planned for an area, an education program teaching people to avoid disturbing wildlife is in order. It is always good to have as few roadway systems as possible and to minimize the time in which the disturbed areas are used. Avoiding excessive habitat subdivisions and respecting animals' activity periods will lessen any deleterious effects of humans.

Reclamation

When there seems to be no way to reduce the impact of humans on a habitat, reclamation work after the fact can help the environment. This is common procedure for mine sites, pipelines, and areas disturbed by fire. One form of reclamation is to let nature take its course, a method effective in areas where the vegetation grows rapidly. Wherever there is a lack of moisture, however, active reclamation is needed. Often, this involves planting vegetation or spraying grass seed to hold the soil. Plants to be used for permanent revegetation are sometimes selected by people who have no knowledge of

Figure 9–13 Deer fence and underpass prevent deer from entering highway. (Courtesy of L. Ward.)

wildlife problems. Fortunately, there is a growing tendency to favor native species, which are usually better suited to site conditions than are some of the more rapidly growing exotics. The native species are also more likely to be beneficial for wildlife. Wildlife managers can advise which plant and grass species are attractive to wildlife when this type of reclamation is planned.

Proper management during and after initial reclamation work is extremely important, particularly for wildlife. If the land has been drastically disturbed, many years of work are often required. Usually, some effort should be made to salvage and stockpile topsoil for redeposition before development begins. This is advisable even though topsoil resources, especially in dry areas, are often of mediocre or poor quality. Unfortunately, this is the case where much of our oil and gas are found. After regrading, furrows should be cut in the topsoil to relieve compacting, but the practice is by no means universal (Figure 9–14).

Some method should be used to stabilize or conserve moisture and topsoil. Water-diversion bars are frequently constructed on access roads and in pipeline corridors to reduce water erosion. Various organic mulching methods have been tried, but their use is certainly not universal and their benefits not well documented.

In any event, the reclamation process involves some form of manipulating soils to regrade the area. It may mean leaving banks or piles of soils to attract wildlife. It may mean grading the area to prevent runoff. Whatever is used should be evaluated in the light of conditions in the particular area. Plantings follow, and it is possible for managers to plan them so that particular successional seres can occur, thereby attracting desirable wildlife species.

Even with proper mitigation and reclamation, follow-up management is necessary. If, for example, grasses are planted, water catchment may be used to attract some wildlife species. Animals themselves may be introduced into the area to ensure desired neighbors. Undue human intrusion should be guarded against, but wildlife managers should play a key role in all this. Follow-up management, unfortunately, has often been ignored.

Figure 9–14 Planting trees can help reclaim mined sites and attract wildlife.

Much reclamation for wildlife has been undertaken in the Appalachians, where deciduous forests often grow rapidly. One project is to grade and contour the land for vegetation attractive to wildlife in such a way that wildlife species can move around and so that drainage from acid mines, which degrades the quality of water in the receiving streams, will be prevented.

Heavy equipment can be used to prepare a desired wildlife habitat. In relatively flat areas, the land can be contoured into valleys and mounds. In the more mountainous areas of Appalachia, the overburden can be contoured and sloped so that wildlife can use the slope. Topsoil saved in the original mining effort can be replaced to ensure fast recovery of vegetation.

A diversity of vegetation should be planted to attract and keep wildlife. Large tracts of uniform forest or large areas seeded with fescue, corn, or soybeans are not conducive to wildlife productivity. Habitats are best created by using patches of vegetation. Rocks and logs can be placed in piles for cover for rabbits and small rodents, which will provide food for predators and assist in revegetation by bringing in plant nutrients. Again in the Appalachian region, a few years after the beginning of reclamation, surface-mined areas can support populations of deer mice, white-footed mice, cottontail rabbits, voles, and shrews. These animals, which feed on vegetation, fungi, and invertebrates, provide food for foxes and birds of prey. Woodchucks, opossums, raccoons, deer, and bats are also soon to be found in these reclaimed areas. Muskrat, mink, and beaver will inhabit areas with streams or water impoundments (Figure 9–15). A variety of birds, including ruffed grouse, bobwhite quail, mourning doves, and many nongame birds, utilize these areas. Fish species suited to the type of habitat created can be established in the lakes and impoundments.

In relatively flat arid environments, reclamation practices can increase topographic heterogeneity, creating additional wildlife habitat. By leaving terraces and cliff nesting sites for birds, as well as areas that mammals can use, wildlife diversity is increased. For instance, these sites afford antelope protection from the wind and winter elements.[20]

Sometimes the proper management after an impact on a habitat is to let the area alone, so that nature can take its course. Succession will proceed, and changes that occur naturally will come about. This is often the proper course in burned-out or logged

Figure 9–15 Beaver often impound water in reclaimed habitats.

areas, where the wildlife species change as each successional sere appears. In some instances, foresters plant in these areas to change the successional sequence and to produce desirable wood-producing species. This is often the case in clear-cut areas from which Douglas firs have been removed; Douglas fir is not the first tree that moves into an area following clear-cutting, but it is one of the more desirable timber species. Wildlife associated with this successional sere will then appear.

Research

Research in the general area of impacts is needed. Whereas many papers document the results of some forms of impact, techniques for modifying impacts and thereby maintaining wildlife are few. Research is needed to discover better methods of reducing impacts through roadway construction and creating time frames so that wildlife will not be destroyed while an impact is occurring.

One of the major weaknesses in the development of environmental stipulations is a lack of background information. It is significant that pipeline construction could not proceed until technical studies had provided the information required by design engineers, yet similar delays were not experienced because of a lack of biological information. The actual impact on wildlife, the ability to use alternative sites during an activity, and limits of tolerance can be determined for some populations by research biologists.

Education

One of the most important roles of a wildlife manager is as teacher of the public. When habitat-improvement techniques are used, it is important to let people know why and to indicate how the public can participate in the effort. Some agencies put signs up indicating what has been done and how it will help wildlife in affected areas. Newspaper and magazine articles help. Often the cost of mitigating efforts following disturbances is not great, and this should be made clear to the public. If it is to participate in the decision-making process, the public must be informed as to the environmental compromises that will have to be made when major development projects are undertaken.

The trans-Alaskan pipeline project was unlike any previous major development in that, because of tremendous public concern, a detailed set of stipulations was designed to minimize the environmental impact.[13] A team of state and federal biologists worked jointly to assure compliance with the environmental stipulations and to address problems relating to fish and wildlife that occurred as the project progressed.

SUMMARY

Most forms of human activity affect wildlife and wildlife habitats. Whether through urbanization, mineral extraction, or the use of resources, people affect the environment and therefore change the composition of wildlife species. Impacts on wildlife can best be measured by collecting preactivity data on the interactions of wildlife and habitat and correlating these data with comparable postactivity data.

Plans to reduce the impact (mitigation) should be an important part of the planning phase of any development. Mitigation techniques include the avoidance of important wildlife areas, the creation of new habitats to replace destroyed ones (trade-offs), the elimination or reduction of impacts to minimize disturbances, and reclamation. Reclamation can also be practiced to make previously disturbed land more suitable for wildlife. Reclamation practices should be developed as a result of research and field evaluation studies.

Perhaps the best method of minimizing impacts on wildlife is public education. Public support can generally be effective in maintaining our wildlife resources and in solving problems relating to fish and wildlife that arise as human projects are planned and implemented.

DISCUSSION QUESTIONS

1. Discuss why the key to reducing the impact of humans on wildlife is habitat management.
2. How can the impact of noise on wildlife be documented?
3. Discuss the different approaches to use in working with proposed impacts and those used where impacts have already occurred.
4. What is the value of an environmental impact statement?
5. How can education reduce adverse impact on wildlife? Discuss some forms of education.
6. What wildlife species are likely to be affected by vibriosis? How?
7. Why does mineral development have a major impact on wildlife?
8. Discuss the different approaches to impact analysis in desert, grassland, coniferous forest, deciduous forest, wetlands, and aquatic communities. What factors would you consider in each community?
9. What wildlife data should be available to conduct an impact evaluation prior to development?
10. What techniques should be used in conducting an impact evaluation?

LITERATURE CITED

1. DeGraaf, R. M. 1978. Avian Community and Habitat Associations in Cities and Suburbs. In *Johns S. Wright Conference Proceedings, Wildlife and People,* pp. 1–24. West Lafayette, IN: Purdue University.
2. Rippe, D. J., and R. L. Rayburn. 1981. *Land Use and Big Game Population Trends in Wyoming.* W/CRAM-81-W22. Washington, DC: U.S. Fish and Wildlife Service.
3. Stubbs, C. W. B., and B. J. Markham. 1979. Wildlife Mitigation Measures Oil and Gas Activity in Alberta. In G. A. Swanson (Ed.), *Mitigation Symposium.* General Technical Report RM-65. Washongton, DC: USDA Forest Service.
4. Johnson, B. K., R. Kominsk, R. Nelson, J. Reynolds, and R. Woodward. 1982. *Coordinating Geophysical Exploration with Seasonal Distribution of Wildlife.* Boise, ID: Western Association of Fish and Wildlife Agencies.
5. Dvorak, A. J. 1978. Impact of Coal-Fired Power Plants on Fish, Wildlife, and Their Habitats. FWS/OBS-78/79. Washington, DC: U.S. Fish and Wildlife Service.
6. Anderson, S. H., K. Mann, and H. H. Shugart, Jr. 1977. The Effect of Transmission Line Corridors on Bird Populations. *American Midland Naturalist* 97:216–21.
7. Doucet, C. J., and J. R. Bider. 1984. Changes in Animal Activity Immediately Following the Experimental Cleaning of Forested Right-of-Way. In *Proceedings, 3rd Symposium on Environmental Concerns in Rights-of-Way Management,* pp. 592–601. Palo Alto: Electric Power Institute.

8. Woodward, D. F., R. G. Riley, and C. E. Smith. 1983. Accumulation, Sublethal Effects, and Safe Concentrations of a Refined Oil as Evaluated with Cutthroat Trout. *Archives of Environmental Contamination and Toxicology* 12:455–64.

9. Mozley, S. C. 1978. Effects of Experimental Oil Spills on Chironomidae in Alaska Tundra Ponds. *Verhandlungen International Vereinigung für Theoretische und Angewandte Lemnologie* 20:1941–45.

10. Rosenberg, D. M., A. P. Wiens, and J. F. Hannagan. 1980. Effects of Crude Oil Contamination on *Ephemeropterg,* in Trail River, Northwest Territories, Canada. In J. F. Hannagan and K. E. Marshall (Eds.), *Advances in Ephemeroptera Biology.* New York: Plenum Press, pp. 443–55.

11. Bennigton, J. P., R. L. Dressler, and J. M. Bridges. 1982. The Effect of Hydrocarbon Development on Elk and Other Wildlife in Northern Lower Michigan. In R. J. Rand (Ed.), *Land and Water Issues Related to Energy Development.* Ann Arbor, MI: Ann Arbor Science.

12. Barry, T. W., and R. Spencer. 1976. *Wildlife Response to Oil Well Drilling.* Progress Note 67. Ottawa, Ontario: Canadian Wildlife Service.

13. Hinman, R. 1974. The Impact of Oil Development on Wildlife Populations in Northern Alaska. *Proceedings, Annual Conference of the Western Association of State Game and Fish Commissioners,* pp. 156–64. Juneau, AK: Western States Game and Fish.

14. Tiete, W. D., and R. L. Ruff. 1983. Response of Black Bear to Oil Development in Alberta. *Wildlife Society Bulletin* 11:99–112.

15. Lyon, J. L., and J. V. Basile. 1980. Influences of Timber Harvest and Residue Management in Big Game. In *Environmental Consequences of Timber Harvest in Rocky Mountain Coniferous Forests,* pp. 441–54. GTR IM-90. Washington, DC: USDA Forest Service.

16. Beall, R. C. 1976. Elk Habitat Selection in Relation to Thermal Radiation. In *Proceedings, Elk Logging-Roads Symposium,* pp. 97–100. Moscow, ID: Forest, Wildlife, and Range Experimental Station.

17. Anon. 1982. *Impact of Emerging Agriculture Trends on Fish and Wildlife Habitat.* Washington, DC: National Academy Press.

18. Anon. 1982. *Wildlife User's Guide for Mining and Reclamation.* General Technical Report INT-126. Ogden, UT: USDA Forest Service.

19. Sikora, W. B., and S. P. Sikora. 1982. *Habitat Suitability Index Models: Southern King Fish.* FWS 10BS-82-10.31. Washington, DC: U.S. Fish and Wildlife Service.

20. Ward, J. P., and S. H. Anderson. 1988. Influences of Cliffs on Wildlife Communities in South-central Wyoming. *Journal of Wildlife Management* 52:673–78.

10

LEGISLATION AND WILDLIFE MANAGEMENT

Like any other form of resource management, wildlife management is based on legal documents and procedures. Federal statutes and regulations, executive orders, and treaties and other international agreements govern the action of federal agencies, while state laws, administrative orders, and court decisions provide the authorization for action at the state level. In this chapter we discuss some of the ideas important in the evolution of our national wildlife laws, types of current regulations, and legislative acts that have an important bearing on wildlife maintenance.

HISTORICAL BACKGROUND

In examining the legal basis for managing wildlife, one gets bewildered by the many interrelated, overlapping, and frequently ambiguous regulations. The situation has not been helped by the fact that the word *wildlife* has been difficult for the legal profession to define. European hunters originally associated the word only with animals taken for food and sport. Today, however, *wildlife* includes nongame vertebrates and invertebrates as well.

Furthermore, wildlife does not observe human-drawn boundaries, although when an aquatic species comes near the shore of a country, that country may try to make regulations governing it. Nor, of course, do terrestrial animals stop at state boundaries. So, within our country, many wildlife regulatory measures have been the subject of states' rights debates. Jurisdictional disputes between federal and state agencies have sometimes increased the difficulty of managing our wildlife resources.

Wildlife law can be traced to various decisions and proclamations from the Roman Empire, through feudal European history, to the beginning of the United States as a sovereign nation. In England before the signing of the Magna Carta (1215), wildlife was the property of the king, who granted hunting and fishing rights to the nobility.[1] Later, Parliament assumed the right to control the harvest of wildlife.

This concept of a legislative body's controlling wildlife was carried over to the United States in 1842, when Roger Taney, chief justice of the Supreme Court, wrote that the states (people) were the successors to Parliament and the crown in managing wildlife (*Martin* v. *Wadell*). The case involved a New Jersey landowner's right to prevent others from taking oysters from his riparian mudflats on the Raritan River. Title of the land had been given to the Duke of York by King Charles II in 1664. Chief Justice Taney ruled that the people of New Jersey were successors to the king and Parliament and therefore that they, not a private individual, controlled the use of wildlife. Implied in the decision was the notion that ownership rights were subject to conditions of the Constitution. Some, in using the Taney decision as a justification for states' ownership of wildlife, have ignored the implication.

Most early wildlife laws passed by states were to regulate hunting of game birds, waterfowl, and deer—obviously, in favor of special interests. Thus, the dense waterfowl wintering populations along the Chesapeake Bay and Currituck Sound along the east coast were protected as a result of pressure from sportsmen. In 1872, a Maryland law prohibited the use of vessels for taking wildlife within a half mile of shore, prohibited the use of some weapons, and restricted hunting to the daylight hours of Mondays, Wednesdays, and Fridays.[2]

The real dispute between federal and state wildlife laws did not surface before the 20th century, because federal wildlife legislation up to that time was limited and relatively insignificant.[1] Many states and territories did pass a variety of wildlife regulations, however, and the Supreme Court generally upheld the states' rights in cases that were challenged.

In 1896, the Supreme Court had ruled in favor of state ownership of wildlife and interstate transportation of wildlife (*Geer* v. *Connecticut*). Geer had appealed a conviction for possessing game birds taken in Connecticut with the intent of taking them out of the state. The court ruled that the state owned the game species and had the right to preserve them as a food supply for its people. This decision is regarded as the cornerstone of state ownership of wildlife. The decision did indicate, however, that the state powers could not be incompatible with rights conveyed to the federal government in the Constitution.

ASPECTS OF REGULATIONS IN THE UNITED STATES

The Constitution is the ultimate source of authority for governmental actions in the United States. Thus, state and federal governments both look to the Constitution in establishing wildlife law. States generally have been given authority over wildlife that resides within their boundaries. States enforce hunting regulations, but they must abide by treaties on migratory species made by the federal government. The federal government can exercise control over fish and wildlife by virtue of the powers conferred on it in the Constitution.[3] These powers have been expressed in laws passed by Congress and interpretations of the courts.

The authority for the conservation and protection of wildlife resides primarily in three legal sources. The first is *statutory laws*. These are laws enacted by Congress either for the protection of specific wildlife or for the protection of resources, including wildlife. Some of the latter are the Clean Air Act, Water Pollution Control Act, National Wild and Scenic River Act, Solid Waste Disposal Act, Environmental Noise Control Act, Resource Conservation and Recovery Act, and National Environmental Policy Act. Some of the legislative acts protecting specific wildlife, such as the Bald Eagle Protection Act and Wild-Free-Roaming Horses and Burro Act, are discussed later in the chapter.

Common law is the second major authority for wildlife regulations. This is the body of court decisions deriving from custom and traditional practices. Common laws have affected wildlife in the areas of negligence, nuisance, and trespass. Thus, the right of a landowner to prevent access for hunting or fishing on his or her private land falls under common law (Figure 10–1).

The third major area from which wildlife regulations derive authority is *case law.* Legislative acts are often written in general language, allowing a number of interpretations. Common law, too, is often susceptible to different interpretations. Conflicts in the interpretation of common law are resolved in the courts, and the decisions of the courts become case law. Case law, which often reflects changes in people's attitudes, has constituted much of the authority for the federal government to control commerce in wildlife and to manage wildlife on federal lands.[4] Courts resolve federal–state wildlife conflicts.

Other legislative acts and decisions influence wildlife management. Zoning laws and permits that control or direct land development are critical in wildlife management. Leasing rights on federal and state lands, as well as access rights, impinge on wildlife. The regulatory mechanisms for land development and reclamation practices can have a profound influence on wildlife. When one works with wildlife, no one document furnishes all the answers; many regulations and documents must be considered. Thus, a good understanding of the many legal ramifications makes life easier for the wildlife manager.

Figure 10–1 Prevention of access to wildlife under common law. (Courtesy of the USDA.)

Appropriations

The legislative branches of state and federal governments have a great deal to say about wildlife activities through the control of appropriations. When legislation such as the Federal Endangered Species Act is passed, funds must be voted to carry out the provisions of the act. But funds are not always there. For example, nongame legislation was passed in 1980, but no funds were appropriated.

Treaties and Conventions

Treaties have been the basis of much federal involvement in wildlife actions. On August 16, 1916, the Convention for the Protection of Migratory Birds was signed between the United States and Great Britain (signing for Canada). A group of migratory birds listed with the convention was specifically protected. The convention allowed for the establishment of open hunting seasons on game birds and provided protection for nongame birds. It prohibited taking nests or eggs, except for scientific or propagation purposes.

Subsequent treaties pertaining to the conservation of migratory birds were closely patterned after the 1916 convention. Such treaties were signed with Mexico in 1936 and Japan in 1972.[1] In 1977, the U.S. Fish and Wildlife Service, through publication in the *Federal Register,* clarified the conventions by publishing a list of species covered by the Migratory Bird Treaties.[5] In 1978, a convention with the Union of Soviet Socialist Republics on the Conservation of Migratory Birds and Their Environment was concluded. A convention on the nature, protection, and preservation of wildlife in the western hemisphere was signed by the United States and 11 other American republics in 1940. This treaty expressed the wish of governments to "protect and conserve their natural habitats for wildlife and to preserve representatives of all species in general of their native flora and fauna including migratory birds" and to protect regions and natural areas of scientific value. The signatory nations agreed to take certain actions to achieve these objectives, including "appropriate measures for the protection of migratory birds of economic or aesthetic value or to prevent the threatened extinction of any given species."

The government's treaty-making power as a means of federal involvement in wildlife management has been challenged in court. When federal agents moved to enforce the 1918 Migratory Bird Treaty Act (which mandated enforcement of the 1916 convention) in Missouri, the state initiated legal action to prevent enforcement (*Missouri v. Holland*). Chief Justice Oliver Wendell Holmes wrote the Supreme Court's majority decision: The federal government can control a human food supply, in this case waterfowl, that does not remain within state boundaries. The Court denied the contention that state ownership of wildlife precluded federal wildlife regulation. The decision established federal treaty-making powers as authority for such regulation (Figure 10–2).

There are other federal regulations stemming from international agreements. For example, treaties have been developed with regard to fish, polar bears, and antarctic seals. There are also treaties that protect wildlife indirectly, such as the 1920 Convention on Nature Protection and Wildlife Preservation in the Western Hemisphere, which established special areas for wildlife preservation and control of international trade.[6] This treaty has been the basis for legislation establishing special wildlife areas, such as wilderness areas and national monuments, and, later, protection of our endangered species. While treaties form the basis for some of our wildlife laws, Congress has also

Figure 10–2 Sandhill cranes are protected under the migratory bird treaties.

passed both specific and general legislation relating to wildlife protection, particularly with reference to removal and commerce.

Fur Seals. In 1911, the Soviet Union, Japan, Great Britain, and the United States concluded the Treaty for the Preservation and Protection of Fur Seals. The treaty banned harvests at sea and required that harvests on each of the island rookeries be conducted under the supervision of the nation controlling the land. Limits were placed on the importation of sealskins. The treaty was superseded by the Interim Convention on Conservation of North Pacific Fur Seals (1957), still in force.

The fur seal treaty was concluded by the United States in part because it had been found that individual states were unable to protect the seal against exploitation by commercial interests, including those of other nations. Northern fur seals had been heavily exploited in the latter part of the 19th century and were harvested on breeding grounds, in rookeries, and along migratory routes. Some of the lands and waters were under the jurisdiction of the Soviet Union and Japan (Figure 10–3).

Whales. In 1931, years of effort by conservationists resulted in a Convention for the Regulation of Whaling. The first of several international whaling agreements, the convention prohibited killing calves, suckling whales, immature whales, and females accompanied by calves. A 1946 convention established an International Whaling Commission to designate certain species of whales as protected species, to fix open and closed seasons and areas, and to specify size limits, overall catch limits, and methods of whaling.[6]

Commerce

The Constitution gives the federal government power to regulate interstate commerce. Just four years after the Geer case, the federal government stepped into the wildlife regulation business with the passage of the Lacey Act (1900), which prohibited the interstate transportation of "any wild animal or birds" killed in violation of state law. The

Figure 10–3 Fur seals are protected against overharvesting by international treaty. (Courtesy of the U.S. Fish and Wildlife Service.)

act upheld the authority of a state to prohibit the export of game lawfully killed in the state and allowed states to prohibit the importation of game. It also authorized the secretary of agriculture to adopt measures necessary for the "preservation, distribution, introduction, and restoration of game birds and other wild birds," subject to laws of the various states and territories. The importation of some species, such as the mongoose, fruit bat, starling, and English sparrow, was prohibited. Permits were required for importing other species. The Lacey Act has been amended several times to include various regulations governing the importation of wildlife and to control interstate commerce in wildlife.[6]

The Lacey Act was passed for two reasons: to supplement and strengthen state wildlife law and to promote the interests of agriculture and horticulture by prohibiting the importation of certain types of wildlife determined to be injurious to those interests.[1] The immediate effect of the act was to strengthen the states' authority to regulating wildlife. In the long run, however, states' powers would be limited, because, with the act, the federal government entered the wildlife regulation business.[2]

The Black Bass Act was passed in 1926 and was later amended to regulate the importation and transportation of black bass and other fish.[1] The Black Bass Act was passed because the Lacey Act applied only to terrestrial wildlife.

Ownership

The use of more than one-third of the nation's land is controlled by agencies of the federal government (Figure 10–4). Under the property clause of the Constitution, the federal government has broad powers over this land. Those powers include the management of wildlife.[3,7] While the National Wildlife Refuge System is the only extensive federally owned land system managed exclusively for wildlife, many legislative acts empower the federal government to manage wildlife on other federal lands. As we have noted, the U.S. Forest Service, the U.S. Bureau of Land Management, and other land-management agencies must consider preservation of fish and wildlife in researching land-use decisions.[3] Individual states have legislation relating to the management of wildlife on state-owned property.

Acquisition of Wildlife Habitat

Land Acquisition. The Migratory Bird Treaty Act was a stimulus for the establishment of a systematic program of acquisition of wildlife refuges. The original act

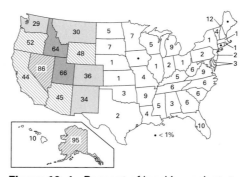

Figure 10–4 Percent of land in each state controlled by the federal government. (Courtesy of the Council on Environmental Quality.)

did not provide for the acquisition of habitat, a deficiency remedied by a 1929 amendment. The Migratory Bird Hunting Stamp Act (1934) provided funding for the refuges. The result was the establishment of a series of wildlife refuges along major migratory bird routes. Originally designed primarily for the protection of migratory waterfowl, the refuges came to serve many species of animals.

Several other laws allow the federal government to acquire land for wildlife—for example, the Fish and Wildlife Coordination Act, the Land and Water Conservation Fund Act, and the Endangered Species Acts. The Conservation Act has been the major act providing for land acquisition.

Acquisition of land by the federal government has become difficult because of political pressure. Private conservation agencies, such as the Nature Conservancy and Ducks Unlimited, have bought or received donations of land that can be used for wildlife. The Nature Conservancy has owned more than 250,000 hectares (617,750 acres) and often acts as a middleman, buying and holding land until a public agency can complete the purchase or assume management of the land. Acquisition by a conservation agency or organization remains the best and perhaps the most favored method of maintaining wildlife habitat, especially wetlands. Nonprofit private conservation organizations are taking an increasingly important part in advising which wildlife habitats to purchase. Converting private holdings to public ownership has been effective in the midwest and east, where few public lands were reserved.

Easement.   It is not always necessary or even desirable to take full ownership of land and water to preserve wildlife: Acquiring easements or development rights may often get the desired results. Easements involve a greatly reduced initial outlay and lower management expense. The U.S. Fish and Wildlife Service has approximately 942,000 hectares (2,327,685 acres) under lease or easement, and more than half the acreage in waterfowl protection areas has been acquired through easement.[8]

Zoning.   Land-use zoning, the control of privately owned real estate by public law, came into being in 1916. Zoning is an exercise of police power. First used to prevent such nuisances as slaughtering horses in residential neighborhoods, zoning has been expanded to control land for many public benefits. Zoning has been effectively used to maintain wildlife habitats in a number of states; Alaska, for instance, has developed a coastal management program with land-use control, and California's coastal management program allows the zoning of special areas, providing significant wildlife habitats, such as forests, wetlands, estuaries, and streams.

A precedent-setting approach to the use of zoning laws involves the cooperation of federal, state, and local governments with private conservation groups in the New Jersey pinelands, a million-acre forest expanse in the midst of the country's most densely populated region (Figure 10–5). This region, shaped by both natural and human-related factors over the past 300 years, contains a wide variety of fish and wildlife species.

In 1978, Congress designated the pinelands a national reserve. The objective of a national reserve is to combine the capabilities and resources of local, state, and federal governments with those from the private sector for better land management. The plan developed by the Pinelands Planning Commission designates areas that should be acquired by governments for protection of the ecology and prescribes methods for the formulation of local land-use ordinances that will conform to the master plan for the region.[9] Thus, while dealing with all forms of environment management, the New Jersey Pinelands Management Plan is in effect a method of using zoning ordinances to protect wildlife habitats.

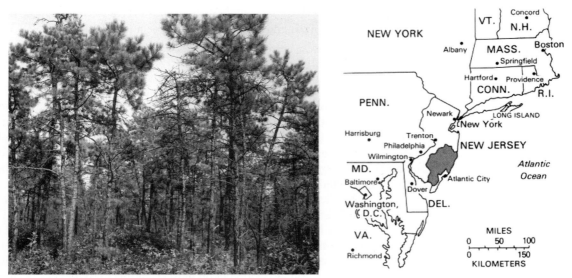

Figure 10–5 New Jersey pinelands. (Photo courtesy of the U.S. Forest Service, by L. J. Prater; map courtesy of the New Jersey Pinelands Commission.)

Despite the many acts relating to wildlife, gray areas remain. Most of the regulations apply only to federal and state lands, and more than 60 percent of the land is privately owned.[10] On private land, the owner is the manager. Furthermore, jurisdictional disputes arise among local, state, and federal governments. Even when the wildlife manager is thoroughly familiar with the pertinent legislation, he or she must develop skill in applying it, sometimes in ambiguous areas.

Wildlife and Fish Removal

Regulation of the removal of migratory birds is tied closely to the treaty-making and commerce powers of the federal government. Most states have specific removal quotas, but these must be within the limits prescribed in federal treaties and commerce legislation. Federal wildlife-law enforcement is necessary principally to enforce regulations governing the possession of some wildlife species and interstate commerce in wildlife. Federal fish and wildlife agents issue permits for the transport and possession of migratory birds, including raptors. Since the age-old sport of falconry is still alive, permits are issued to take and keep birds of prey. Possession without a permit is punishable by heavy fines and even jail sentences.

State game wardens are responsible for enforcing state wildlife laws, many of which are related to hunting. State hunting and fishing regulations are normally recommended by a state fish and game commission and approved by the state legislature. Almost all of our hunting and fishing regulations are of this nature. In addition to enforcing these regulations, some state game wardens enforce boating and gun regulations.

When the state of Montana was challenged for charging higher fees for out-of-state hunters (*Montana Outfitters Action Group* v. *Fish and Game Commission of the State of Montana,* 1976), the U.S. Supreme Court ruled that wildlife within a state's boundaries is entrusted to the care of the people of the state and that charging higher fees for recreational hunting by nonresidents does not violate the Constitution's guarantee to a person of equal rights under the law.[1,3,7] The state must still observe the limits imposed by treaties, endangered-species legislation, and federal land policies.

Federal–State Conflicts

Federal–state wildlife conflicts have arisen over the enforcement of state regulations on federally controlled land. In 1894, all hunting was prohibited in Yellowstone National Park; now the federal government controls the taking and possession of wildlife in that park. No fishing license is required, because the park was established before Wyoming became a state. But Grand Teton National Park, just south of Yellowstone, operates under Wyoming state laws: State game wardens routinely patrol the park, and Wyoming fishing licenses are required.

In 1928, the Secretary of Agriculture directed the removal of excess deer in the Kaibab National Forest in Arizona, and state officials arrested people carrying out the federal directive in the 1920s. In that case, the Supreme Court ruled that the federal government has the right to protect its lands (*Hunt* v. *United States,* 1928). The decision presumably established the federal government's right to carry out its management policies on federal land, even though on most federal land, state laws are enforced. But some federal lands, particularly national parks and Department of Defense installations, have more restrictive rules, even to the extent of forbidding trespassing.

Another dispute over federal–state land regulations occurred in New Mexico when National Park officials at Carlsbad Caverns, contrary to state law, allowed deer to be taken for research purposes (*New Mexico State Game Commission* v. *Udall,* 1969). Although the courts ruled that the secretary of the interior had the right to carry out the research, the clash between federal property rights and state enforcement activities continued. The state removed some wild burros from federal land, in violation of the federal Wild-Free-Roaming Horses and Burro Act of 1971. (See next section.) The federal government demanded the burros back, but instead of returning the animals, New Mexico sued the secretary of the interior to have the act rescinded. A lower court ruled in favor of the state, but the U.S. Supreme Court unanimously reversed the lower court's decision. The Supreme Court's ruling thus reinforced the federal government's right to manage wildlife on federally owned land.

FISH AND WILDLIFE LEGISLATION

Fish and wildlife legislation has been passed by the federal government in response to different pressure groups. Conservation organizations have been most influential in getting through legislation on bald eagles, marine mammals, and endangered species. Fish and wildlife have been components of the Fish and Wildlife Coordination Act and the National Environmental Policy Act, which controls development. Some of the wildlife legislation acts are reviewed in this section.

Fish and Wildlife Coordination Act

In 1934, Congress passed one of the first federal wildlife statutes to consider the impact of people on wildlife. The original Fish and Wildlife Coordination Act authorized research to determine the effects of sewage, trade, wastes, and other polluting substances on wildlife. It encouraged state and federal cooperation in developing national programs, with the goal of having an "adequate supply" of wildlife on public lands.

Amendments to the act, however, changed the thrust away from the idea of a unified, nationwide program.

The 1934 act directed that waters and lands utilized by important groups of wildlife be administered by the appropriate state wildlife agency or by the secretary of the interior if the area had value for migratory bird management. Water development projects were specifically emphasized in the act. There were two major results: (1) When federal water development projects, such as dams, reclamation, and channelization, were planned, the U.S. Fish and Wildlife Service and state wildlife agencies would evaluate the expected impact on wildlife; and (2) wildlife agencies would have a say in issuing and denying permits in any federal water development project.

Unfortunately, the vague language of the coordination act apparently contributed to its ineffectiveness. In 1974, a General Accounting Office report indicated that the policies of the act have been carried out in fewer than 28 water development projects.[11]

National Environmental Policy Act

Originally proposed as an amendment to the coordination act, the National Environmental Policy Act (NEPA) became an independent directive to all federal agencies in 1969. The agencies were to evaluate the impact of all actions that had a significant effect on the quality of the human environment.

NEPA figures prominently in litigation arising under the coordination act, but it has a broader scope than that of the act itself. Some feel, indeed, that NEPA is the most comprehensive federal environmental statute. It is generally considered the most important statute for wildlife, yet it never uses the word *wildlife*. NEPA's declared policy is to promote efforts that will prevent or eliminate damage to the environment and biosphere, to create and maintain conditions under which human beings and nature can exist in productive harmony, to fulfill the responsibility of each generation as trustee of the environment, to preserve important natural aspects of our national heritage wherever possible, and to enhance an environment that supports diversity and the quality of renewable resources.

The act directs that an environmental impact statement covering the following points be prepared for federal projects: (1) a description of the proposed action and its environmental effect; (2) the relationship of the proposed action to land-use plans, policies, and controls for the affected area; (3) alternatives to the proposed action, including, when relevant, any not within the existing authority of the responsible agency; (4) any probable adverse effects that cannot be avoided; (5) the relationship between local short-term uses of the environment and maintenance and enhancement of long-term productivity; (6) any irreversible and irretrievable commitments of resources that would be involved; and (7) an indication of what other interests and considerations of federal policy are thought to offset the adverse environmental effects of the proposed action (Figure 10–6).

The development of environmental impact statements was a significant innovation in the field of wildlife management. Managers had previously been trained simply to make field observations and recommend management options based on their knowledge from field experiences. With the advent of NEPA, the manager must also research the potential impact on wildlife that a change would have, and the manager's statement must stand the scrutiny of the legal profession, which could now call for legal hearings, with additional expert opinions.

More than anything else, however, NEPA has illuminated the point that wildlife management is not an exact science. Predictions can be made, but much leeway must be allowed for verifying options. NEPA has made it clear to the many levels of wildlife

Figure 10–6 Alteration of wetlands requires an environmental impact statement if the federal government owns the land or is involved in the project.

managers that where we lack information, basic surveys are needed before we an even discuss the impact of a project.

The wildlife profession grew tremendously after the passage of NEPA. Federal and state agencies in charge of managing land more than doubled their wildlife staffs. Organizations such as the Oak Ridge National Laboratory added people who did nothing but write environmental impact statements, and private consultant groups, most of which considered all areas of impact, including that on wildlife, flourished.

Many of the early environmental impact statements were based on brief site visits and massive references to literature. Others ended up being just lists of animals and plants that were *probably* present in the areas. There was little discussion of an actual impact, for a very good reason: Little was known. The statements did prevent some problems, nevertheless. The Kings Point nuclear reactor on the Hudson River had waste intake flues in bass spawning grounds, and the discharge ponds had elevated temperatures that could affect the fish. Because of the impact statement and after lengthy litigation, elaborate measures were taken to reduce the impact of the reactor system.

Getting NEPA procedures into the routine of the established federal wildlife agencies has not been without difficulty. For several years after NEPA's passage, most of the environmental impact statements prepared by federal agencies concerned individual projects. For example, the U.S. Fish and Wildlife Service prepared individual statements for specific areas in the National Wildlife Refuges or National Wilderness Preservation Systems, or in connection with special regulatory actions, such as the proposed injurious wildlife regulation under the Lacey Act or the migratory waterfowl lead-shot regulations. But for many actions that were part of long-established, ongoing wildlife programs, the proper scope of the required statement was not clear. Thus, in 1974, certain plaintiffs challenged the 1974–1975 migratory-bird-hunting regulations because the Department of the Interior had not prepared an accompanying environmental impact statement (*Fund for Animals, Inc.* v. *Morton*). The case was settled out of court on the Fish and Wildlife Service's agreement to prepare a "programmatic" statement applicable to the annual regulation-setting process for migratory birds. When the service proposed in the following year to permit the hunting of certain species that had not been hunted in immediately preceding years, the same plaintiffs went back to court, claiming that a new impact statement was required. That claim was rejected, however, partly because the programmatic statement had discussed, in general terms, the effects of regulated bird hunting.

Both the amended coordination act and NEPA can be used to benefit fish and wildlife resources. The act is a means for making wildlife-resource conservation a part of water-resource planning and development, and NEPA provides a broad mandate that all federal agencies formulate policies calculated to meet national environmental goals.[12] (Table 10–1).

Endangered Species Acts

In 1966, the federal government began to list endangered species and designate habitats that should be preserved for them. The Endangered Species Act of that year was extensively revised in 1969 and 1973 and was renewed in 1982. When the Endangered Species Act was considered for renewal in the late 1990s, some members of Congress used stalling techniques to have special-interest items added. Agency personnel proceeded with their work as though the act were intact.

The Endangered Species Act requires that the Secretaries of the Interior and Commerce departments determine which species will be listed and indicate courses of

TABLE 10–1

Summary Comparison of NEPA and the Fish and Wildlife Coordination Act

Criterion	Coordination Act	NEPA
Coverage	Only federal water resources projects and federally granted permits and licenses	All federal actions significantly affecting the environment
Relationship to water resources planning process	Built into the planning process on a continuing basis	Overall check on the planning process at widely spaced intervals
Requirements on lead agency to respond to comments	Direct responses to fish and wildlife agency with attempts at resolution required	Direct responses to fish and wildlife agency not required
Involvement of the public	Limited involvement, but increasing due to influence of NEPA	Heavy emphasis on public involvement
Effect on major decisions—consideration of alternatives	Scope often limited on project framework	Emphasis on consideration of major alternatives
Effect on secondary decisions—project alterations	Mechanism geared to negotiating secondary or incremental decisions	Mechanism not primarily geared to negotiating incremental decisions

Source: Ref. 12.

action to help the species recover. (Recovery plans and other parts of the legislation are discussed in Chapter 21.) Although much concern has been raised about the restrictive nature of the act, most cases are resolved by compromise.[13]

Bald Eagle Protection Act

Originally passed in 1940, the Bald Eagle Protection Act made it illegal to take or possess any bald eagle or any part, egg, or nest thereof (Figure 10–7). Specific penalties were imposed and have since been increased. A 1962 amendment extended the protection to golden eagles, because the young of bald and golden eagles are very difficult to tell apart. The name was changed to the Eagle Protection Act.

A number of conflicts with mineral development companies and with ranchers have occurred because the golden eagle is included in the regulation. Golden eagles nest on cliffs and isolated trees in many western areas where mines or oil wells are proposed. Sheep ranchers, for their part, point out that golden eagles prey on lambs, sometimes causing considerable economic loss. (Commonly, however, golden eagles catch jackrabbits and rodents.)

Wild-Free-Roaming Horses and Burro Act

In 1971, Congress passed the Wild-Free-Roaming Horses and Burro Act to protect the wild horses and burros on western land, particularly public land (Figure 10–8). In addition to protecting a type of animal thought to be on the verge of extinction, the act involved national pride: Wild horses and burros were symbols of the historic and pioneer spirit of the west. Controversy between those with grazing rights and land management agencies has resulted.

The Wild-Free-Roaming Horses and Burro Act is an example of how a special-interest group can bring about federal legislation by exerting pressure. The act was the result of many years of effort. It requires that wild horses and burros be treated as an

Figure 10–7 Bald eagle. (Courtesy of the U.S. Fish and Wildlife Service; photo by M. Lockart.)

Figure 10–8 Wild horses. (Courtesy of the Wyoming Game and Fish Department.)

integral part of the natural system of public lands. It also provides specific restrictions on federal land-management agencies in the management of wild horses and burros. Among these restrictions is a limitation on the removal of these animals, which means that the horses end up competing with other wild ungulates and domestic livestock for limited resources and rangeland. This provision is particularly controversial; when landowners find that their cattle must compete for food, they naturally want the protected animals removed. Federal agencies find their hands tied and thus have been unable to solve the problem. The Bureau of Land Management has instituted an "adopt a horse" program, which has had limited success in some parts of the country.

Marine Mammal Protection Act

When the Endangered Species Act was passed in 1969, eight marine mammals were considered endangered, but little effort was made to protect them. States seemed to have neither the authority nor the ability to deal with the problem. So in 1972, Congress passed the Marine Mammal Protection Act, which established broad general policies. The act was a drastic departure from the general scheme of regulation, for it deprived the states of all authority over marine mammals, substituting a single federal program for the many state programs. The central feature of the federal program was a moratorium of indefinite duration during which, with limited exceptions for scientific investigation and public display, no marine mammal could be taken by any person within the United States.

The act has had many ramifications. It served as a springboard for developing treaties and the International Marine Mammal Protection Act, to which the United States is a party. But it causes a certain amount of friction between the federal government and some states, such as Alaska, since it is closely tied in with conservation of the fur seal. The act became a center of controversy, too, when it was discovered that tuna fishing nets kill marine mammals.

Tuna and Porpoises, an Example of How a Wildlife Law Functions

Porpoises are protected under the Marine Mammal Protection Act. To enforce the protection, observers from the National Marine Fisheries Service have often accompanied fishing vessels to see that porpoises tangled in tuna fishing nets were released. Regulations originally allowed incidental taking of porpoises killed in the nets. However, there was no clear definition of "incidental." Furthermore, people reported that boats without observers often captured many porpoises without trying to save them. In this case, there was a law to protect porpoises, but agents responsible for enforcing the law could not always do so because of the difficulties involved, including a lack of funds.

Research may eventually develop methods to ward off the porpoises or nets that will not harm the animals. Meanwhile, the problem continues, with no clear definition and solution in sight.

Fisheries Resource Act

Early in this century, Congress, recognizing the importance of the offshore fisheries resource, passed the Fisheries Resource Act. The early versions protected both Great Lakes fisheries and offshore fisheries. Besides establishing cooperative agreements with states and other federal groups, the act encouraged surveys and research activities to preserve oceanic fishes.

The Fisheries Conservation and Management Act of 1976, which became operational in 1977, attempts to establish a comprehensive program of fishery conservation. It designates a 197-mile zone, in addition to the 3-mile zone that at the time was considered to be territorial seas of the United States. In that 200-mile zone, the federal government was to manage wildlife and control harvests. The act introduced ideas not found elsewhere in the body of wildlife law. For instance, in the 200-mile zone, the United States claims authority over all forms of marine animals and plant life other than marine mammals, birds, and migratory species, which are covered under other acts. Eight regional councils were established to develop management plans for fisheries resources in the zone. This rather complex act has been the subject of several international incidents, as well as being the stimulus for other nations to extend their control limits to 200 miles from their coasts.

Swampbuster Legislation

Although not specifically termed wildlife legislation, the "swampbuster" provision of the Food Security Act of 1985, reauthorized in the late 1990s, provides significant support for wildlife and fish. Set up as part of the U.S. Department of Agriculture farm support program, the swampbuster discourages the conversion of wetlands for agricultural purposes. If farmers use wetlands to plant crops, they lose eligibility for support programs. Later provisions were added to include not only wetlands, but also highly erodible land.[14]

Provisions of the act, sometimes called the Farm Bill, include the Conservation Reserve Program, which provides assistance in improving water quality and in increasing wildlife habitat for native and migratory species. Swampbuster legislation under the Food Security Act is a very significant step in legislation that affects wildlife.

It specifically addresses the question of maintaining habitat for wildlife. Thus, a move is made from focusing on a species to focusing on the entire community of wildlife (Figure 10–9).

Some of the provisions of the Farm Bill not mentioned earlier include the establishment of conservation priority areas where significant water and natural resource problems exist; the establishment of 3- to 10-year contracts to provide technical assistance and pay up to 75 percent of the costs of conservation practices; the requirement of a conservation plan; the prohibition of large-scale livestock operations in order to be eligible for other than technical assistance; a limit of annual payments to an individual; and the provision of cost sharing with landowners for developing wildlife habitat. In 1996, legislation allowed landowners to rest their cropped wetlands for an indefinite period of time and then convert them back to agricultural production.

Figure 10–9 Sharp-tailed grouse increase in numbers on Conservation Reserve Lands. (Courtesy of the Wyoming Game and Fish Department.)

SUMMARY

Wildlife legislation can be traced from British parliamentary control to state control in the United States. In the United States, the Constitution became the authority for government actions. States control resources within their boundaries, making their own hunting and fishing laws. The federal government controls wildlife through its appropriation, treaty-making, commerce, and property charges in the Constitution.

Contemporary wildlife protection involves statutory, common, and case laws, together with zoning laws and permits. Some legislative acts and zoning laws do not specify wildlife in their language, but include it as part of our natural resources. Thus, the National Environmental Policy Act requires environmental impact statements for federal projects, and wildlife has been interpreted to be part of our natural resources that can be affected. Major legislative acts, such as the Eagle Protection Act, the Fish and Wildlife Coordination Act, and the Endangered Species Act, have influenced the direction of various developments in the country. Some legislation, such as the Wild-Free-Roaming Horses and Burro Act, has been the result of special interests.

Currently, many different legislative actions affect the management of wildlife. The exact actions that apply in local areas are sometimes difficult to interpret. Comprehensive plans, such as those for the California coastal zone or the New Jersey pinelands, are helping to reduce this ambiguity.

DISCUSSION QUESTIONS

1. What is the basis of wildlife law in the United States?
2. How do federal and state jurisdictions differ?
3. Distinguish between the Fish and Wildlife Coordination Act and the National Environmental Policy Act.
4. How do multiple-use laws protect wildlife?
5. Why is there so much confusion about wildlife law?
6. What major points need to be covered in an environmental impact statement?
7. When is an environmental impact statement required?
8. Can any federal laws prevent construction activity? Illustrate.
9. What political bodies and legislative acts would have to be considered in a development project on the shores of the San Francisco Bay marsh habitat?
10. How do federal treaties relate to wildlife management?

LITERATURE CITED

1. Bean, M. J. 1983. *The Evolution of National Wildlife Law.* New York: Praeger.
2. Tober, J. A. 1981. *Who Owns Wildlife?* Westport, CT: Greenwood Press.
3. Hair, J. D. 1980. National Fish and Wildlife Policy: State–Federal Relationships. A Question of Jurisdiction. *Proceedings of the Annual Conference of the Western Associations of Fish and Wildlife Agencies* 60:10–19.
4. Heer, J. E., and D. J. Hagerty. 1977. *Environmental Assessments and Statements.* New York: Van Nostrand Reinhold.
5. Greenwalt, L. A. 1977. Migratory Birds: Revised List and Definition. *Federal Register* 42(221):59, 858–69, 362.
6. Committee on Merchant Marine and Fishery Laws, U.S. House of Representatives. 1977. *A Compilation of Federal Laws Relating to Conservation and Development of Our Nation's Fish and Wildlife Resources.* Washington, DC: U.S. Government Printing Office.
7. Lund, T. A. 1980. *American Wildlife Law.* Berkeley, CA: University of California Press.
8. U.S. Department of the Interior. 1979. *Annual Report of Land under Control of the U.S. Fish and Wildlife Service as of September 30, 1979.* Washington, DC: U.S. Fish and Wildlife Service.
9. Pinelands Commission, 1980. *New Jersey Pinelands Comprehensive Management Plan.* New Lisbon, NJ: Pinelands Commission.
10. Walton, M. T. 1981. Wildlife Habitat Preservation Programs. In R. T. Dumke, G. V. Buger, and J. R. March (Eds.), *Wildlife Management on Public Lands,* pp. 193–208. La Crosse, WI: Wisconsin Chapter of the Wildlife Society.

11. Bean, M. J. 1978. Federal Wildlife Law. In H. P. Brokaw (Ed.), *Wildlife in America, pp. 279–90. Washington, DC:* Council on Environmental Quality.

12. Schueler, R. L. 1973. The Matched Pair: National Environmental Policy Act and the Fish and Wildlife Coordination Act. *Proceedings, 27th Annual Conference of the Southeastern Association of Game and Fish Commissioners,* pp. 548–60.

13. Harrington, W. 1981. The Endangered Species Act and the Search for Balance. *National Resources Journal* 21:71–92.

14. Avon. 1987. Highly Erodible Land and Wetland Conservation. *Federal Register* 52(180): 35,194–208.

11

THE WILDLIFE MANAGER

Who are these people called wildlife managers? What is it they do? To answer these questions, we need to look more closely at the skills and background needed by a manager and at the specific tasks within the wildlife profession. We distinguish between managing and administering. In this chapter, we speak of the wildlife manager as *manager.* We discuss the characteristics of the manager as administrator in Chapter 12.

SKILLS AND TRAINING

Interaction with People

One of the most important skills of managers is the ability to interact positively with people (Figure 11–1). Managers should be able to make others feel at ease and inspire confidence in their knowledge and abilities (Figure 11–2). People who prefer working by themselves or who are constantly aggravated by others do not make good managers. Managers must feel comfortable with landowners, legislators, tourists, and public-interest groups—people of highly varied ability, education, and background. Unfortunately, this skill is not taught in most curricula, nor is it always easy to learn—but it can be learned.

Communication

Another important skill needed by wildlife managers is the ability to communicate, both orally and in writing. Managers must develop the art of writing clearly and concisely. Long, wordy reports or statements that get to the point only somewhere on page

Figure 11-1 Managers interact with all sorts of people. (Courtesy of the Wyoming Game and Fish Department.)

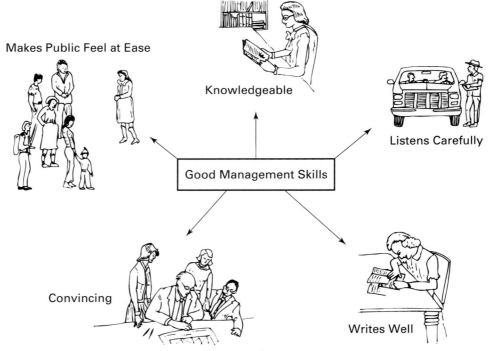

Figure 11-2 Characteristics of a manager.

20 can hardly be effective. Since administrators are usually buried under paperwork, lengthy reports will be set aside and concise ones acted on first. Clear communication among the field manager, supervisor, and administrator is equally important. Nobody likes to spend time trying to figure out what someone else has tried to say.

The arts of negotiation and persuasion are also needed in this field because of the many diverse, often conflicting ideas that need to be moderated in implementing programs. The manager must understand the planning and integrating process: Compromises and trade-offs are important in wildlife management. However, the manager must understand the limits of compromise and not antagonize others within those limits. These skills are effective, of course, only when the person using them has a sound knowledge of the wildlife principles involved.

Most wildlife managers spend less than half their time in the field. Paperwork—plans, reports, and data analysis—take most of their time. Many people new on the job are not aware of how much office time is involved. Some, who associated wildlife management only with animals, lakes, and nature's greenery, drop out during the first few years of work.

Innovativeness

Even though managers may know a great deal about the biology of animals, they are often confronted with situations that have no precedent. Thus, the ability and willingness to innovate is important. Managers deal with living organisms, and the one constant is that they will all be different. Modifications of the radio collar, for instance, will be needed for different species and perhaps for individuals within a species.

Rupert

Biologists have been called for a wide variety of reasons. The following is one of the more humorous examples involving people management. It was late at night when the biologist received a call from a distraught woman. One year before, she had found a gosling apparently abandoned by its mother at a pond near town. She took the gosling home and cared for it in her house. She and the gosling, Rupert, quickly became attached. In all likelihood, the baby became imprinted, which occurs when hatchlings identify themselves as a member of the same species as their mother or, in this case, foster mother. Imprinting is a form of learning. Youngsters have been known to imprint on humans or other animals during their short, impressionable youth. In this case, Rupert sat on the couch with his foster mother, ate with her, and slept on her bed. But now, a year later, Rupert was getting too big for the house. The lady was desperate and insisted that the biologist find a home for Rupert where it could be returned to the wild and yet she could still visit it. The biologist tried to explain that the bird would never be able to survive in the wild, but still the lady persisted. Finally, the biologist decided to call a friend who had some domestic geese in the country. Sure enough, Rupert could go and live there.

This story had a happy ending, which was unusual. People have often tried to deal with cute baby animals, which, if they survive, often become unmanageable adults. Here the biologist came up with a good solution to a problem that should not have occurred. Had the lady left the gosling alone, the mother would have returned and taken care of it.

Education

What type of education is needed by managers? Obviously, people entering wildlife work must have a good background in animal biology: courses in biology, physiology, genetics, and wildlife (Table 11–1). They must understand how to use and analyze data, which requires education in statistics. Chemistry and mathematics are also important science courses that the prospective manager should take.

Because communication and public relations are so important, wildlife managers should have good training in the humanities, including courses in oral and written communication, government, economics, psychology, and sociology. Students who do well in these courses are ready to gain experience as managers. Of course, nobody can carry all the information that a manager will need. A manager must be able to locate information in published and unpublished sources. Every college student, including prospective wildlife managers, should learn how to use the library. This means taking very seriously the preparation of research (library) papers in English and other courses. Changes occurring in the wildlife-management field today make other courses important: courses in botany, computer programming, mapping, remote sensing, geology, soils, and, in some parts of the country, Spanish.

Most of all, students need to learn the techniques used in wildlife work: sample design, census techniques, and a variety of data-analysis techniques. While students can expect techniques to be modified in field situations, theoretical knowledge can make these techniques more useful in the field. Educators must make sure that students receive an extensive background in basic skills. Students must try to apply techniques while still in school through internships and summer jobs.

Managers who are critical of their formal education, saying that nothing is the same in the classroom as in the field, either did not plan their careers very well or did not do a good job of putting the building blocks in place. The field is, and should be, different from the classroom; but it is knowledge gained in the classroom, coupled with field experience, that makes for good management decisions.

TABLE 11–1

Education Requirements for Certification as a Wildlife Biologist by the Wildlife Society

---

36 semester hours in biological sciences, including:
  6 semester hours of wildlife management
  6 semester hours of wildlife biology
  3 semester hours of ecology
  9 semester hours of basic zoology
  9 semester hours of basic botany
  3 additional semester hours in one of the foregoing five categories
  9 semester hours in physical sciences, such as chemistry, physics, geology, or soils,
    with at least two disciplines represented
  9 semester hours in quantitative sciences, including:
  3 semester hours in calculus
  3 semester hours in statistics
  3 semester hours in sampling, computer science, or applied statistics
  9 semester hours in humanities and social sciences
12 semester hours in communications, such as English composition, technical
    writing, journalism, public speaking, or the use of mass media
  6 semester hours in policy, administration, or law

Note: One semester course hour at a university usually equals 0.67 quarter hour.

In the late 1970s, information was gathered on the academic training that managers were receiving. Comparisons were made of the opinions in universities and those in wildlife management agencies.[1] Most agencies felt that new graduates were adequately trained in science, but needed more courses in public relations, wildlife law, business management, administration, social science, and the humanities. They also suggested that colleges and universities offer programs in fisheries and/or wildlife management that could be accredited according to guidelines furnished by the American Fisheries Society or the Wildlife Society. In addition, the survey asked whether there was a substantial communication gap between management agencies and academic institutions and, if so, what could be done about it. An overwhelming majority of the agency people felt that a liaison with the universities was good, but that there was a need to improve relationships through formal agreements. Eighty percent of the academic heads reported good relations. In many instances, the presence of the Cooperative Wildlife and Fisheries Research Units made liaisons formal and routine. A variety of other formal and informal links, such as contracts and intern programs, were cited by agencies and academic departments. Both agency and academic leaders strongly supported practical work experience during graduate and undergraduate years.[1]

Data Systems. Competence in computer science and statistics is becoming essential for students entering wildlife management today. One reason is that statistical analyses and models are being used more and more. Managers need to understand how to use models and, indeed, must be able to do some modeling. If not actually capable of developing models, managers must understand how models are developed and their limitations. It is important to remember that models are only one tool and that they will provide answers only as good as the data supplied.

Wildlife managers must also be able to analyze and interpret data. A whole array of data files, or banks, is available. (Examples are discussed in Chapter 13.) Some are simply retrieval systems, in which filed data can be retrieved quickly. Knowledge of how data systems are constructed and of the types of systems available will enable the manager to use such systems (Figure 11–3).

Scientific Literature. Some people use scientific literature only in their educational training, but it can be used by managers in many situations. Field managers often complain, with reason, that they do not have time to keep up with the scientific literature. That is why abstracting services and data-retrieval systems are valuable. Together with such publications as *Wildlife Review,* they make the search for data much less time consuming.

A wealth of information is also available from unpublished resources. Efficient managers use card files and other indexing systems to indicate where such information is available. Many set up data systems based on field observations. Sometimes new developments in an area, such as mining, mean that data are needed very quickly. Sometimes people or companies in the local area have such information. Wildlife managers need to be aware of these resources.

Mapping. A variety of maps can be made, including cover, distribution, soil-type, and potential-vegetation maps. Aerial photos are useful for developing a vegetation cover map. How accurate are they, however? What limitations do they have? How, specifically, can they be used for the job at hand? The manager needs to answer these questions about any maps that are to be used.

Figure 11–3 Data systems are becoming an important asset for wildlife managers.

Remote sensing can be effective. Often, satellite imagery or aerial photographs are used to produce maps, some of which may be effective at the planning level, but not in the field. The manager should know the uses and limitations of all these systems.

Experience

Seldom is a person equipped to move directly into management from college. Academic learning provides the background for understanding wildlife populations and habitats, but this information must be adjusted to specific management situations under a variety of conditions. Theory is general information; it is experience that teaches us to identify and locate the specific within the general.

Like other professionals, most managers say that their greatest learning occurs during the first few years on the job. That is the way it should be. They are acquiring experience, a necessary component of any good manager. Those with a good academic background have a good base on which experience can build. Those who barely got through the academic training will be left behind.

Most agencies put new managers through a probationary or apprenticeship period. Agencies that do the best job during this time produce the best managers, people often sought by other employers. Agencies that simply give the new manager time to acquire experience (sometimes by moving from one job to another) often end up with people who are very superficial in their abilities.

The best experience record is broad, but not too general. Employers are wary of applicants who jump from one job to another at short intervals, because they do not have time to profit from any one job experience. On the other hand, applicants who stay in one job all their working lives lack breadth of experience.

Let us summarize: Well-qualified wildlife managers have taken the appropriately broad college curriculum, have a sound knowledge of wildlife populations and their habitats, and have acquired the necessary experience to apply this knowledge in new situations. Such managers can express themselves well orally and in writing, know how to present a case persuasively, know how to manage, and like people.

Certification

Certification of wildlife biologists is made by the Wildlife Society, which grants two kinds of certificates. The first, based on education and experience, is certified wildlife biologist. The second, associate wildlife biologist, is granted to a person who has completed the education requirements, but lacks the necessary experience. The education requirement is at least a bachelor's degree, with an appropriate curriculum, from an accredited college or university (Table 11–1). The experience requirement for a certified wildlife biologist is at least five years of wildlife work after the baccalaureate.

Those wanting certification submit their applications with a fee to the certification board of the society. Few professional wildlife managers have applied for certification, since most federal and state agencies do not require that their employees be certified. The employment standards of these agencies are similar to the requirements for certification. A comparable program exists in the American Fisheries Society, which certifies fisheries biologists and associate fisheries biologists.

NEW DEVELOPMENTS

Universities

Universities can help wildlife managers develop and evaluate population and habitat techniques and design field surveys. Many game and fish laboratories are stationed at universities, where faculty members are available for short-term research projects lasting one or two years. Student help is there, too, with mutual advantages: It is relatively inexpensive and students get valuable experience.

The Cooperative Research Unit Program has integrated the research of agencies and universities. The units have educated many current resource managers and university faculty and have produced a wealth of research data. Cooperative Park Service Units at several locations operate on a smaller scale, but in a similar fashion (Figure 11–4). Some federally funded programs for fish and wildlife research are located in universities. Usually, federal funds go to the state wildlife agencies, which write contracts with university faculty members.

Few independent or government research laboratories can duplicate the work of research facilities at major universities. Among the most important of these facilities are (1) libraries; (2) statistical services; (3) computer facilities that allow storage retrieval; (4) museums, including specimens; (5) sophisticated equipment feasible for small research jobs, such as shops to build specialty items; (6) audiovisual services, including drafting, photography, and display material; and (7) specialized research laboratories, aided by advisory services on the use of special material, such as radioisotopes and X-ray analysis.[2] Obviously, the university connection is a valuable resource for the wildlife manager. In addition to the advantages we have discussed, managers can take or audit university classes or short courses and participate in workshops. Continuing education is a part of maintaining proficiency.

A survey of universities and wildlife management in the 1970s revealed a positive response to the suggestion of establishing an in-service educational program for wildlife management agencies, including short courses in wildlife techniques and current problems. A number of universities indicated that, with proper funding, they would

Figure 11-4 Cooperative Research Unit personnel working on an aquatic project.

be willing to offer such a program. The feeling about programs in existence was generally good.

Professional Societies

Many wildlife managers join professional societies to keep abreast of developments in their field. Most professional societies hold national and local meetings, where managers can exchange ideas with professionals from other regions. (See the appendix for a partial listing of professional societies and their publications.)

THE WILDLIFE PROFESSION

The term *wildlife manager* covers a wide variety of people who perform many different jobs. A small landowner and a refuge manager can both be wildlife managers. A wildlife manager can specialize in endangered species, animal damage control, migratory birds, big game, waterfowl, nongame, or habitat management. Of course, many wildlife managers—wildlife refuge managers, for instance—must operate in all these areas. Both the manager with specialized responsibility and one with general duties must understand the principles of management through population regulation, habitat manipulation, and legal restraints. We commonly associate the term *wildlife manager* with the person responsible for managing an area of land for wildlife. Some of the more specialized wildlife professions are discussed in this section.

Law Enforcement

Most wildlife law-enforcement officers are wildlife managers with special law-enforcement training. Federal law-enforcement officers are employed by the U.S. Fish and Wildlife Service and the National Marine Fisheries Service. Enforcement responsibilities for federal wildlife laws also fall to the Coast Guard, Customs Service, Forest Service, National Park Service, Bureau of Land Management, and Bureau of Indian Affairs. Each state has wildlife-enforcement specialists, whose titles vary in different states, to enforce both state and federal wildlife laws. In 1987, state wildlife agencies had some 9,000 law-enforcement employees and spent more than $300 million on law enforcement. Approximately 32 percent of the employees in state fish and game agencies are listed as law-enforcement specialists.[3]

Wildlife law-enforcement officers have responsibilities very similar to those of other law-enforcement people. Unfortunately, wildlife laws are often quite general and so require interpretation by the individual officers. This problem makes it hard to get convictions and anything but minimal sentences or fines. For example, there are laws against the harassment of wildlife, particularly raptors, but what constitutes "harassment"? Too, the enforcement staff is thin. Laws prohibit the possession of migratory birds without permits, yet many people find injured birds and, in violation of the laws, nurse them back to health. Since these are federal laws, notification of the state game and fish office is not in order. But there are fewer than 300 federal enforcement officers, and it may not be easy to find one.

Wildlife law enforcement therefore requires not only training in the law, but tact in dealing with people. In addition, wildlife law-enforcement people need to be trained in wildlife biology. They must be able to recognize the various animal species and must know the habitats of the animals, their activity patterns, and something of population dynamics (Figure 11–5).

Law-enforcement duties are frequently different from the public images of them.[4] The agents often have quite varied duties. They check hunters bag limits and creel limits. Endangered species legislation, harassment of wildlife, and the issuance of permits all fall within their province. They may be called to development sites to solve problems, such as what to do about raptors nesting near a mine, or to handle a nuisance wildlife situation. (See Table 11–2.) They also census wildlife, recommend

Figure 11–5 Checking the success of hunters is part of wildlife work.

TABLE 11–2

Examples of Calls Answered by a
Wildlife-Enforcement Specialist

Time	Call
0810	Dead animal on highway
0915	Squirrel in chimney
0933	Dog harassing wildlife
1009	Possible poaching violation
1055	Fishing without license
1125	Truck–deer collision
1235	Assist state police
1403	Time out for school presentation
1530	Cattle on public land
1547	Wild goose in swimming pool
1618	Wildlife harassment
1654	Injured animal
1718	Dogs chasing deer
1850	Raccoon in yard
1920	Gunshots heard in woods
2019	Concern about bats
2049	Vehicle–animal collision

hunting quotas, operate check stations, assist in trapping, perform hunter safety activities, investigate hunting accidents, write reports, and otherwise maintain an office.

First and foremost, though, law-enforcement officers in wildlife are educators. Through public discussions, news articles, display posters, and pamphlets, they make the public aware of laws that exist to protect wildlife.

Public involvement in wildlife law enforcement has increased in recent years as more people have developed an interest in our wildlife resources. A nationwide antipoaching campaign, with people calling an 800 number to report violators, has been successful. In some states, rewards are given to those who report violators (Figure 11–6).

Has wildlife law enforcement had any real impact on the preservation of our resources? This is, of course, a very difficult question, but some cases point to success. The American alligator was declining in the southeast, primarily because of poaching. Efforts by state and federal law-enforcement offices to bring violators to court were

STOP POACHING

**CALL TOLL-FREE TO REPORT
ANY GAME & FISH VIOLATION**

1-800-442-4331

Figure 11–6 A nationwide campaign has been launched against poachers.

instrumental in the recovery of this species. In Arkansas, the law-enforcement division is given major credit for the restoration of wild turkeys.[5]

Planners

Many wildlife organizations employ professional planners, often with wildlife training, to develop, implement, and coordinate management plans, as well as to integrate these plans into the budgeting process. In the federal government, new wildlife-related programs must be packaged and presented to Congress for funding, which can take up to three years. Planners in such private corporations as oil and gas companies must develop strategies to avoid or reduce the impact on wildlife of their operations. We discuss the planning process and the role of planners in Chapter 13.

Game Classification

Wildlife managers must have a wide range of tools at their disposal to implement management strategies effectively. Ideally, they would like to know the exact number of animals in the population being examined. If the population is small or consists of highly visible animals, managers might be able to get a reasonably accurate estimate by doing a visual count. If they have enough time and money, they can use statistical techniques, such as line transects or mark-recapture data to obtain an estimate. If they have a good idea of how many animals are in the population, they can make recommendations to assist their agency in achieving its goals. Knowledge of a population's size helps managers to determine the number of game animals to harvest, to decide whether there are excess animals available to be transplanted to another area, or to assess the stability of an endangered population. Unfortunately, most wildlife populations are not easy to count, and estimation techniques are often too expensive or time consuming to use on a regular basis. Without a population estimate, how can managers make appropriate decisions?

Commonly, managers use classification data to help them make recommendations. Such data are collected at different times of the year to give managers a good idea of the reproductive and mortality status of a population, as well as the proportion of desirable animals—for example, mature bucks in a deer population. Although it is nearly impossible to count all the animals in a population, it is always possible to count some of them. If animals are tallied in specific categories during counts, figures such as juvenile–adult female ratios can be obtained. If a large enough sample of animals is counted, the numbers usually approach the true ratio in the population. Typically, the sample required is much lower than the total number of animals in the population. The number of animals needed to accurately estimate a population ratio can be computed by several different statistical formulas described in biometry texts.

As a practical example, let us say that you are asked to manage a deer population in a region. The deer in this region are fairly well dispersed throughout the year, so an accurate count of each individual is not possible. The agency you work for wants to maintain a high proportion of bucks in this population to provide good hunting opportunities. You do not have the time or money available to get mark-recapture population estimates each year. What tools can you use to set

---

quotas for hunting seasons from year to year in order to reach the agency's objective? This is a good time to use classification data. For this example, let us say you want to achieve the ratio of 35 bucks per 100 does. During your first year of classification, you find 23 mature bucks per 100 does and 15 yearling bucks per 100 does. Let us also say that 2,000 deer hunting licenses have been issued in each of the past 10 years. Since your buck–doe ratio is high, it appears that the population can handle the pressure of 2,000 deer hunters, so you decide to issue that many licenses again. That year, however, there happens to be an extremely hard winter, and you think many of the deer might have died. When you do your classifications before hunting season, you find 22 mature bucks per 100 does and 5 yearling bucks per 100 does. In addition, you classify 30 fawns per 100 does, whereas past classifications show around 70 fawns per 100 does. These classification ratios can be effectively used to explain to the public about the hard winter for deer. You find that the yearling ratio is down considerably, indicating the death of many of last year's fawns. Reproduction in the population is also down about 60 percent. As things stand, if you issue the typical 2,000 deer tags, the same number of bucks can be taken as in past seasons, but there will not be very many yearlings to move into the adult age class next year. In addition, there are going to be very few fawns moving into the adult age class in two years. Although hunters might experience a typical hunting season this year, unless the buck harvest is reduced, the following two years are going to provide extremely poor hunting. If you reduce the number of licenses this year, the decline in the mature-buck ratio is going to be less severe in the next two hunting seasons.

The preceding example is just one of many illustrating how classification ratios can directly affect a manager's recommendations. Even without a good idea of the overall deer population numbers and ratios, the manager could still strive towards the agency's objectives by using the information gathered from the classifications.

Supervisors and Administrators

Typically, promotions move field managers to supervisory and then to administrative positions. Supervisors and administrators formulate policy, develop budgets, approve plans, initiate major actions, and deal with personnel matters. Managers who relate well with people and have broad experience often make good administrators. (See Chapter 12.) Unfortunately, our political system often results in the appointment in federal and state agencies of top-level administrators with no experience in wildlife management. Their decisions frequently impede good management and cause severe morale problems.

Research Biologists

Most federal and state wildlife-management agencies, and some private groups, have research divisions. The principal research biologist here generally has more academic training than others in the wildlife profession. Principal research biologists usually have a doctorate, although technicians with bachelor's or master's degrees are commonly employed as assistants. Research biologists use standard scientific methods to investigate wildlife questions: The problem is identified and a null hypothesis is established. For example, the problem may be to ascertain the difference in the movement patterns of two deer herds between their summer and winter ranges. The biologist will set up the null hypothesis: There is no difference between herds *A* and *B* in this respect. The study plan the biologist develops will include a description of the methods used in investigating the problem, a timetable, and a budget. Once the plan is approved, fieldwork can begin. Data on movement patterns for a specified time are gathered and analyzed, conclusions are reached, and a report is written. The manager uses the report as a guide to handling the deer as a single herd or as two herds.

Research biologists employed by wildlife agencies or organizations not only provide data to wildlife managers, but keep current on scientific developments in their field, so they spend a good deal of time in professional meetings and reading professional journals. Research biologists must be guided by the needs of managers. Some research biologists have a tendency to want policy to be based solely on their research, failing to recognize that management decisions must be made in the light of political, as well as biological, considerations. The researcher is there to aid the manager, not to set policy. There are many forms of specialty research in wildlife work. Disease, genetics, pesticides, damage control, migration, and endangered species all involve questions to be answered by experts (Figure 11–7).

Educators

Most educators who teach courses on wildlife management are members of university faculties. Typically, they hold the doctoral degree and have done field research. More and more universities are hiring educators who have also had management experience.

Figure 11–7 Research biologists work on wildlife problems in the field and laboratory.

There are, of course, very few courses in wildlife at the secondary-school level, but some schools do include a bit of wildlife management in their science curricula. The wildlife manager can do missionary work with the teachers in these schools toward their common goal of developing among students a positive attitude toward wildlife.

Museums, zoos, and national parks hire people with wildlife training to present wildlife courses to the general public. Millions of people enjoy wildlife through these organizations, and many take the courses.

Writers and Photographers

Some people make a profession of writing about and/or photographing wildlife. The best writers and photographers are often those who have had extensive experience working with wildlife populations in the field.

Impact Analyzers

NEPA has directed that the natural resources be evaluated when changes are proposed in the natural system. Many biologists earn an income by writing environmental impact statements for such projects. Private consultant groups do so almost exclusively. This type of work requires extensive knowledge of wildlife biology and habitat requirements, a sound academic background, writing skill, and field experience.

Hobbyists

Millions of nonprofessionals enjoy wildlife. Some, indeed, have become better known in the field of wildlife management than many professionals. Wildlife provides a recreational outlet for almost everyone. People who enjoy wildlife as a hobby usually have some education or have done extensive reading in the subject. The more people know, the greater is their pleasure.

JOB PROSPECTS

One question people always ask is "Can I get a job?" There is no simple answer to this question, because it depends on the person and the times. Most wildlife jobs are with state and federal agencies. When their budgets are tight, jobs are scarce. More people interested in a career in wildlife are getting jobs with private consulting or other private companies, such as land and timber management companies and oil or mineral companies. Conservation groups hire wildlife biologists, as do some health care companies.

There may never again be such plentiful jobs as in the period following the passage of NEPA, with its requirement for environmental impact statements. In all likelihood, fluctuations in the number of jobs available are to be expected. Right now, there are a few opportunities for people with bachelor's degrees and a larger number of jobs for those with master's degrees. The market for people with doctorates is very tight and will probably remain so for the foreseeable future.

Students wanting to get wildlife-related jobs should start planning early. Volunteer positions and internships help build credentials to compete in the market. Students need to realize that most groups hiring wildlife people want them to have more than a college education. They look for good writing and speaking skills, as well as experience. More nontraditional jobs are appearing. Many leisure groups, such as ski companies and cruise lines, are looking for people who want to work in the business and who also have a wildlife background. In looking for jobs, students should carefully consider what they enjoy doing and find jobs that suit those interests.

ORGANIZATIONS USING WILDLIFE MANAGERS

As we have noted from time to time, a variety of organizations employ wildlife biologists.

Public Organizations

Most people trained in wildlife are employed by public organizations. The federal government hires wildlife biologists for many different departments, prominent among them the Department of the Interior (which hires the most), Energy, Agriculture, Defense, Commerce, and State. Candidates for employment by the federal government must have the qualification specified in a federal register for the position sought. The fisheries and wildlife biology registers, for example, require specific academic course work. (See Table 11–3.) There are also biology, zoology, and ecology registers which wildlife agencies use in hiring.

The applicant sends credentials to the Office of Personnel Management (OPM) for the register. (As a rule, OPM accepts applications only at specified times during the year.) OPM then lists people who are qualified on each register. For some positions, it is necessary to take a test and receive a rating. Federal agencies with openings then use the registers to select people. Besides being on the register, it is important for the applicant to make contact with people doing the hiring. Currently, changes are under way so that each agency will create a register for each job. Applicants apply directly for the job. They must, however, meet all qualifications of the register under which the position is listed.

State agencies generally have examinations for such positions as conservation officer, biologist, and warden. People interested in getting a state job should consult the state employment office to find out what they should study to pass the tests, most of which are both written and oral. Public utilities, land boards, counties, and others may employ wildlife professionals. Individual contact must be made.

Private Conservation Organizations

Some private conservation organizations that employ wildlife biologists manage land, some have research programs, and others carry on public information as well as research activities. They all hire their employees through applications sent to their personnel offices. People interested in working for private organizations should contact them to find out what employment opportunities are available. (See the appendix.)

Private Businesses

Many private corporations that produce paper, extract minerals, or construct large facilities employ wildlife biologists to guide them in carrying out their responsibilities under the law. These corporations look for people experienced in wildlife management and with know-how in legal matters and public relations. Consulting firms employ people with wildlife training to evaluate the potential impact of projects on wildlife, to prepare documents relating to wildlife, or to make wildlife-management recommendations for public and private organizations. Most of a consulting firm's business comes

TABLE 11-3

Courses and Experience Required to Be Qualifed on a Federal Register

General Qualifications

GS-5   A bachelor's degree from an accredited university or four years of experience in approved work.

GS-7   One year of approved professional experience or one year of approved academic work beyond the bachelor's level.

GS-9   Two years of approved professional experience, a master's degree, or two years of approved academic work beyond the bachelor's level.

GS-11 Three years of approved professional experience, or three years of approved academic work beyond the bachelor's level, or a Ph.D.

Course Requirements

Fishery Biologist—30 semester hours, or equivalent, in biological sciences, including (1) 6 semester hours in aquatic subjects such as limnology, ichthyology, fishery biology, aquatic botany, aquatic fauna, oceanography, fish culture, etc.; (2) 12 semester hours in the animal sciences in such subjects as general zoology, vertebrate zoology, comparative anatomy, physiology, entomology, parasitology, ecology, cellular biology, genetics, or research in those fields. Extra course work in aquatic subjects may be used to meet this requirement when appropriate.

For research positions—major in biology, zoology, or biological oceanography, including 30 semester hours, or equivalent, in biological and aquatic sciences and 15 semester hours in physical and mathematical sciences, including (1) 15 semester hours of preparatory training in zoology beyond that provided in introductory courses in zoology and biology, in such courses as invertebrate zoology, comparative anatomy, histology, physiology, embryology, advanced vertebrate zoology, genetics, entomology, parasitology; (2) 6 semester hours of training applicable to fishery biology in such subjects as fishery biology, ichthyology, limnology, oceanography, algology, planktonology, marine or freshwater ecology, invertebrate ecology, principles of fishery population dynamics, etc.; (3) 15 semester hours of training in any combination of two or more of the following: chemistry, physics, mathematics, and/or or statistics.

Wildlife Biologist—30 semester hours, or equivalent, in biological sciences, including (1) 9 semester hours in such wildlife subjects as mammalogy, ornithology, animal ecology, wildlife management, or research courses in wildlife biology; (2) 12 semester hours in zoology in such subjects as general zoology, invertebrate zoology, vertebrate zoology, comparative anatomy, physiology, genetics, ecology, cellular biology, parasitology, entomology (excess wildlife biology courses may be used to meet this requirement where appropriate); (3) 9 semester hours in botany or related plant sciences.

For research positions—courses must include (1) 12 semester hours in such zoological subjects as invertebrate zoology, vertebrate zoology, comparative anatomy of vertebrates, embryology, animal ecology, entomology, herpetology, parasitology, genetics; (2) 9 semester hours in training applicable to wildlife biology, animal ecology, wildlife management, principles of population dynamics, etc.; (3) 9 semester hours in botany and related plant sciences; (4) 15 semester hours in any combination of two or more of the following: chemistry, physics, mathematics, statistics, soils, and/or geology.

Wildlife Refuge Manager—9 semester hours, or equivalent, in zoology; 6 semester hours in such wildlife courses as mammalogy, ornithology, animal ecology, or wildlife management, or equivalent studies in the subject-matter field; 9 semester hours in botany.

Figure 11–8 In some agencies, wildlife must compete with other, multiuse activities on the land. (Courtesy Fred Lindzey, U.S. Geological Survey.)

through bidding or contracts. Newspapers, magazines, tour groups, and outfitters sometimes employ people with wildlife training for their public relations programs.

Multiuse Agencies

Like private businesses, multiuse federal agencies, such as the U.S. Forest Service and U.S. Bureau of Land Management, need wildlife biologists who understand not only wildlife, but also the legal aspects of management. It is the wildlife specialist who must determine exactly how wildlife legislation and wildlife concerns impinge on the major objectives and activities of these agencies. For example, wildlife must be considered in any timber operation planned by the U.S. Forest Service, and it is the function of the service's wildlife specialist to make such a determination (Figure 11–8).

SUMMARY

A wildlife manager must be able to interact and communicate effectively with all sorts of people. The better managers generally have a strong background in biology and formal training in management. They are innovative people with field experience. Wildlife managers need to keep informed of developments in the management field. They can do so by continuing to take university courses, participating in workshops, and being active in professional organizations.

There are many people in the wildlife profession other than managers. Law-enforcement agents, planners, researchers, writers, photographers, and administrators are also important members of the cast. They are employed by both public and private agencies.

DISCUSSION QUESTIONS

1. Are law-enforcement officers wildlife managers? Explain your answer.
2. What courses do you think a prospective wildlife manager should take in college? Could a liberal arts curriculum include a major in wildlife management?
3. In what way is wildlife management an art? A science?
4. Comment on the following: "A good college curriculum should produce a wildlife manager who is able to be immediately effective in the field."
5. What skills do managers need? Can they all be learned?
6. How can a university contribute to the wildlife-management skills of its students? Of practicing managers?
7. Discuss how communication skills are used by managers.
8. For what different jobs can a wildlife degree prepare people?
9. How can the political climate affect the job of the manager? What can the manager do about the political climate?
10. How are the jobs of research biologist and wildlife manager related?

LITERATURE CITED

1. Cookingham, R. A., P. T. Bromley, and K. H. Beattie. 1980. Academic Education Needed by Resource Managers. *Proceedings, 45th North American Wildlife Resource Conference,* pp. 45–49.
2. Weller, M. W., N. R. Kevern, T. J. Peterle, D. R. Progulske, J. G. Teer, and R. A. Tubb. 1979. Role of Universities in Fish and Wildlife Research. *Proceedings, 44th North American Wildlife Conference,* pp. 209–16.
3. Morse, W. B. 1987. Conservation Law Enforcement: A New Profession Is Forming. *Transactions, 52nd North American Wildlife and Natural Resource Conference,* pp. 169–75.
4. Psikla, E. J. 1979. Citizens' View of Law Enforcement. *Proceedings, 44th North American Wildlife Resource Conference,* pp. 97–101.
5. Hunter, C. 1980. A Game and Fish Commission Administrator's View of Wildlife Law Enforcement. *Proceedings of the Annual Conference of the Southeast Association of Fish and Wildlife Agencies* 34:644–46.
6. Hodgdon, H. E. 1988. Employment of 1986 Wildlife Graduates. *Wildlife Society Bulletin* 16:333–38.

12

WILDLIFE ADMINISTRATION

The administration of wildlife management agencies is big business. Budgets range from a few thousand dollars for small units to more than half a billion for larger agencies. Obviously, there will be many people and groups trying to influence the decisions made by a top-level administrator in a large conservation organization. In this chapter, we examine the function of administration in a wildlife agency and describe some types of wildlife administrators.

PROFILE OF A GOOD ORGANIZATION

A good organization has many characteristics, but perhaps the most important is that it has a good administration and administrator.

Personnel

The selection of effective personnel is a responsibility of the administrator. When square pegs are pushed into round holes, morale problems usually develop. Resulting mismatches cause misery for the people hired and for those with whom they work. Finding the right people for jobs is difficult under the best conditions, but when politics interfere, headaches ensue. Weak administrators who succumb to policy dictated by political pressure find that the morale among employees in the agency drops.

Chain of Command

Wildlife management organizations that function effectively have a clearly stated chain of command. Employees know who their immediate supervisor is and where supervisors get their guidance. With a clear chain of command, an agency can speak with one voice. This is perhaps the most important characteristic of an efficient wildlife-management agency. To be effective, an administrator must have a clear chain of command and control the release of information. When the individual field person, field supervisor, or manager publicly expresses opinions in a policy matter, confusion results. Controlling the release of information bothers some people, who feel that they should be able to speak out on any issue that they choose. An effective administrator certainly allows people to express their ideas within the organization and to show why those ideas should be incorporated into agency positions. But once policy is set, there is no surer way to undermine confidence in an agency than to have its employees disagreeing with policy in public.

Public Relations

Another important component of an effective organization is its public relations program. Some organizations have public relations offices that function to educate the public about the mission of the organizations and seek public assistance in reaching their objectives. (See Chapter 14.) In one sense, all employees are public relations people. In any wildlife activity, the employees meet the public. Employees are usually identified with their organization, especially if it is a government agency, and people have a tendency to judge organizations by the image their employees present. Public relations, then, involves every person in an organization (Figure 12–1).

SETTING POLICIES

Public Agencies

The overall policies of public agencies are generally set by elected legislative bodies, by governors, or by the President of the United States. The executive or legislative branch tells wildlife-management agencies what general policies it wants carried out. The agencies are then responsible for developing structures and procedures (including specific internal policies) that will implement the general policies. Sometimes, the general instructions contained in legislative acts have broad latitude for agency interpretations of policy. Of course, even then, the agencies are responsible to the governor, President, or legislature. Thus, if the interpretations are unpopular, changes can be

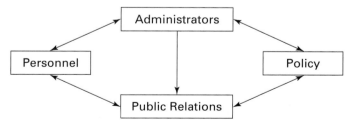

Figure 12–1 Components of an effective wildlife organization.

made. The Endangered Species Act, adopted by Congress in 1973, is a good illustration of how a federal law can undergo change.

Endangered Species Act.   The Endangered Species Act gave broad powers to the Secretary of the Interior, who could halt major construction projects and designate appropriate areas of critical habitat for endangered species. This caused concern on the part of some members of Congress, particularly with reference to the snail darter controversy in Tennessee. (See Chapter 21.) As a result, an amendment in 1978 required formal congressional approval to implement portions of the act. The U.S. Fish and Wildlife Service, which administered the act, now had a secure authorization from Congress to designate critical habitats by having them published in the *Federal Register.*

Because of its concerns, Congress set up a process by which a project can be exempted from the act. A seven-member Endangered Species Committee was formed, consisting of the Secretaries of Agriculture, the Interior, and the Army, the chairman of the Council of Economic Advisors, the administrators of the Environmental Protection Agency and National Oceanic and Atmospheric Administration, and a representative of the state affected by the project.

The exemption process works like this: If the U.S. Fish and Wildlife Service has concluded that a proposed federal project might harm the continued existence of an endangered or threatened species by destroying or adversely modifying its critical habitat, and the agency proposing the project or the governor of the state in which the project is to occur does not agree, the agency or governor can apply to the Secretary of the Interior for an exemption to the Endangered Species Act. A three-member review board is then appointed to consider the application. If this board determines that a conflict does indeed exist between the proposed project and the mandates of the Endangered Species Act, and compromise does not appear possible, the review board prepares a recommendation for the Endangered Species Committee. Although not required to do so, the review board may conduct one or more hearings. Upon receiving the review board's report, the committee will grant or deny the requested exemption (with granting possibly conditioned on the performance of certain enhancement measures) if five or more members so vote. Any outcome of the exemption process may be reviewed in the Court of Appeals.[1]

The exemption process is perhaps less important than it seems. In the first place, Congress apparently envisioned that few exemptions would be requested and fewer granted: The Endangered Species Committee is composed of seven highranking officials—busy people with little time to be involved often with the committee. Moreover, even though Congress did place time limits on each step, almost two years must pass before an irreversible commitment of resources can be made. This means that little work can be done on a project until after the endangered species problem is resolved.

The delay can be even greater. First, of course, is the possibility of judicial review. Second, there is no penalty for missing a deadline: Many deadlines in other federal legislation are routinely missed. Third, the applicant is financially responsible for any mitigation or enhancement measures the committee may require. Before the exemption can be granted, the committee must have assurance that the funds are available and that such measures will be completed. For federal agencies with development responsibility, this means that the funds for mitigation and enhancement measures must be appropriated by Congress.[1] (Figure 12–2).

State Policies.   State legislatures set the policies of state game and fish agencies. Some states also require that the legislature establish hunting seasons, a measure that fosters a great deal of political haggling and pressure. Most states turn the matter of hunting seasons over to a commission appointed by the governor.

Figure 12–2 Endangered species such as this peregrine falcon occupy a great deal of administrators' time. (Courtesy of the U.S. Fish and Wildlife Service.)

Commissions review the recommendations made by the administrators of the U.S. Fish and Wildlife Service. Commissions that concern themselves with the daily activities of these agencies—in effect, looking over the shoulder of the chief administrator—tend to confuse the operation of the agencies by weakening the director's position.[2] No director can do a good job or keep the confidence of the staff unless the commission leaves administrative decisions to his or her judgment. This does not mean, of course, that the director is always right or that a decision should never be challenged. But it is the better part of wisdom for the commission to replace an administrator in whom it cannot have confidence.

Some commissions are provincial—that is, the members feel that they represent local rather than state interests. Obviously, with such a commission, it is doubtful that a game and fish agency can be effective.[2] Good commissions, in contrast, mediate between citizens and the agency, bringing public input into the system and "selling" to the people the program developed by the agency.

Because political factors have an impact on the efficiency of wildlife management in states, a positive relationship between state legislatures, commissions, and wildlife administrators is important. States with the most beneficial arrangement of organizational and political factors are more effective in carrying out policies. Such states generally have larger proportions of hunters and wildlife observers.[3]

Private Organizations

Private conservation organizations are usually established by a group of people whose goals dictate the policy of the organization. A number of national wildlife organizations lobby legislatures on behalf of wildlife. Some publish magazines, which serve the dual purpose of raising funds and educating people about wildlife and wildlife policies.

Private corporations that in some way affect wildlife must have a policy toward wildlife, even if it is one of total disregard. Utilities, oil, gasoline, and mineral companies, construction companies, paper producers, and agricultural interests—all must formulate policies with regard to wildlife. The policies of most corporations have been shaped largely in the light of legislation affecting their operations. Some companies have found public relations value in doing more for the conservation of wildlife than is required by law. Others have turned public-resources management, including wildlife management, into profit—for example, by setting aside areas where people can go to enjoy wildlife, developing campgrounds, and so on.

Since corporations operate for profit, their administrators must see some value in undertaking wildlife-management projects. It is commonly thought that corporations are opposed to wildlife conservation. This is partly because wildlife managers and conservation organizations have failed to convince corporations that supporting wildlife projects can be good for business.

THE ADMINISTRATOR

In a wildlife refuge with only two employees, one will hold the post of refuge manager. Part of that person's time will be spent planning to acquire and expend funds for operation of the refuge. The refuge manager will also prepare reports, on everything from gasoline used to travel accounts and expenses. He or she makes sure that the refuge follows the policies and strategies set up in the agency's plan, answers correspondence, attends meetings with local interest groups, and attends conferences in the supervisor's office. In a small operation, the administrator spends only part of the time in administration and the other part carrying out the responsibilities of the organization. Each level, from the refuge manager up, has administrative needs. Department and agency heads devote most of their time to administrative duties. The duties of administrators can be divided roughly into two types: (1) carrying out the policies of the organization and (2) managing people (Figure 12–3).

Policies

Organizations with highly related objectives are easier to administer than are those with a wide variety of objectives, some of them not obviously related to the mission of the organization. Most major corporations in the nation have as their goal producing a product or rendering a service to make a profit. A utility company sells gas or electricity to make a profit for its shareholders, an automobile company makes cars to turn a profit for its stockholders, and so on. Thus, administrators of corporations must be able to anticipate and adjust to their clients' demands, as expressed in purchases or the absence of purchases. An auto manufacturer must design and build cars that will sell—in

Figure 12–3 Providing fish for sports fisheries is an important part of most state agencies.

other words, must make decisions about quality. When there is a downturn in the economy, the manufacturer may need to close plants, laying off thousands of workers. All these decisions are based on the profit motive.

Wildlife-management agencies are generally not operated for profit, although wildlife administrators can use most of the procedures employed by corporate administrators. The goals of most wildlife-management agencies are stated in such terms as "conservation of our nation's wildlife resources," "providing a wildlife experience for the public," or "maintaining our wildlife heritage." These goals are translated into policy, which is implemented by the agency personnel. As we shall note later, policies may be formulated by the public, by politicians, or by administrators.

Since most wildlife-management agencies are a part of government, there are broader, governmentwide policies they must also observe. Some of these are time consuming: Energy-conservation policies can be implemented only through reports on such things as use of vehicles, travel, and temperatures of buildings. Equal-opportunity policies usually lengthen the hiring process. Desirable as these policies may be, they can seem to detract from the main business of wildlife-management agencies. Weak administrators are inclined to complain about them as "red tape" and use them as an excuse for not doing a better job. Strong administrators see them for what they are, needed supports for a healthy total system, and find ways to use their contribution to reaching the agencies' primary objectives.

People Management

Managing people is perhaps the most important task of the administrator. A successful administrator must be effective in working with people inside and outside the organization. "Must be effective in working with people"—the words are easy to say, but how does one go about becoming effective? The matter is complex—so complex that there are full-semester college courses (called "personnel management" or something similar) in the subject. Every prospective administrator should take such a course.

Here we have space to list only a few rules of thumb—a foreshortened thumb—that good administrators use within their organizations:

1. Be fair—in evaluation reports, in recommendations for promotion, and in hiring. (Don't always hire your own children or a friend's children for summer work, etc.) Do not play favorites, even though you may be tempted to.
2. Like all the employees. Will Rogers said he'd never met a man (today he'd add "or woman") he didn't like. Don't fake it—with proper effort you can find something in everyone that you can like. Find it! In general, we like people who like us.
3. Respect all the employees. Again, you can find something to respect. To paraphrase Rogers, you will never meet a person who doesn't know more about something than you do. Make employees aware that you respect them for what they are and what they know. It does not have to be job-related knowledge, either. Letting a baseball fan know that you admire her memory for batting averages can be an "in" to a complete change of climate.
4. Communicate. Write and speak in such a way that there is never any doubt about what you mean.
5. Listen. Listening is the surest signal of respect. It also has several advantages: It establishes wholesome equality, makes employees part of management, makes employees feel responsible for succeeding in projects they helped formulate, and lets you pick up good ideas—for which you should scrupulously give credit to those to whom it is due.
6. Command respect—for the job and for yourself. Do not be unreasonably demanding, but neither should you tolerate sloppy, lazy work. If you will observe this rule and the preceding rules, respect for you will follow.
7. If you have not already done so, take a course in personnel management!

Becoming an Administrator

How do you become an administrator? By *planning* to become an administrator! Successful applicants for administrative positions have generally prepared themselves by reading, meeting with people, and learning about an agency's operations. Of course, if you are fortunate enough to work for an organization that believes in promoting people from within, your best preparation is doing excellent work in whatever nonadministrative jobs you are assigned. In addition, take further college course work—perhaps a more advanced degree—especially work in management, psychology, and the humanities. Attend seminars and professional meetings. Show interest in the whole work of the agency. Work on your writing and speaking skills—to be an administrator without liking to communicate must be one of the world's most uncomfortable positions! Actually, many organizations have a path through which people can gain administrative experience.

Of course, not everyone should try to become an administrator. "Know thyself" is always good advice. Anyone who does not like the types of duties involved in administration should stay away from it. That, too, is easier said than done. Administrators are usually better paid than other employees, and money has assumed unseemly importance in today's economy. But many an administrator, in all fields, longs for the day when he or she can afford to go back to a job paying less, but "inspiring" fewer headaches.

In any event, young people entering the wildlife profession should not seek an administrative job without having had experience in the field of wildlife management.

Conflict

How do you deal with conflict? Wildlife administrators are often in a position in which they need to deal with conflict. For example, when people become irritated with an issue, they complain to the agency and write letters to the editor of the local paper. Sometimes public meetings turn into screaming matches, with some members of the public trying to get their way or just get attention in a public forum. It is always a difficult situation. Keeping cool is a major key to the problem. You need to realize that in a public meeting, very few people want to hear one person voice all sorts of complaints. In this case, try to get the person to direct his or her concerns elsewhere, or ask the person to get in touch with the agency later, when, hopefully, he or she will have calmed down. If a public meeting is being held with many hostile people, let them vent their anger, and give them few, if any, comments. If a public meeting has two or more points of view, and it looks like a major conflict is in the making, it may be necessary to change the subject.

Getting experience in dealing with the public and being able to think quickly are important attributes of the wildlife administrator. A sense of humor can help, as well as the ability to explain things in layman's terms. Go to meetings well prepared, especially on major issues the organization is facing. Try to help the people with the problem understand the issue. Recognize that sometimes the individual has other problems and is just using the wildlife administrator as a substitute target for venting anger.

There are many young men and women without such experience who have been lured by the money, title, and prestige to accept administrative posts in wildlife management organizations and have been initially successful. But lack of experience invariably catches up with them, they are stuck when they try to get new positions, and often they lose the respect of their colleagues. Too late, they realize the necessity of wildlife management experience, and because of the salary scale or agency promotion rules, they are unable to move back to gain the basic experience they should have gotten earlier. Often, they lack the people management skills required of an administrator.

TYPES OF ADMINISTRATORS

There are a number of different types of administrators within wildlife-management agencies.

Public Agencies

Operational. Operational management is the day-to-day direction of an agency. At the national level, most agencies have a chief administrator, who is assisted by various officers—for budget and planning, personnel, procurement, property, safety, and so on. These operational administrators, who may or may not have had training in fish and wildlife management, handle the mechanical procedures necessary to keep an operation going. An effective operational management system frees the chief administrator to do the kind of creative thinking that any organization needs to be successful. This is not to denigrate operational personnel. Without good operational management, any organization, regardless of how inspired, will fall flat on its face.

Management. The management administrator, such as the chief of the game or fish section, is the key person in most state and federal agencies. Most fisheries and wildlife management administrators have had formal training in management and have worked in the field. These are the people who, although not specifically trained in administration, have the responsibility of carrying out the policies of the organization.

Management is often broken down into a number of levels. For example, in the U.S. Fish and Wildlife Service, endangered species, environmental contaminants, and federal aid fall under one associate director. Refuges and wildlife, as well as fisheries, also have associate directors, each of whom has subordinate managers for specific tasks (Figure 12–4). In most state agencies, management is the work of the chief administrator and a number of assistants who report to the commission and are responsible, depending on how the particular agency is organized, for setting hunting and fishing regulations, as well as indicating which populations can be taken during hunting and fishing seasons. The administrators are involved in policy and decision making in all areas of wildlife, including environmental protection, land acquisition, research, construction, public information and education, endangered species, damage claims, and landowner relationships.

Other. Federal and state wildlife agencies also employ administrators in other capacities. Law enforcement generally has administrators trained in the law. Wildlife research has administrators, experienced as research biologists, who supervise biologists working on practical wildlife problems.

Multiuse Public Agencies

The administration of wildlife concerns in multiuse agencies, such as those that manage land or timber resources, generally falls to people without wildlife training. Because of this, agencies hire wildlife biologists, who are often supervised by people with training in one of the other applied science disciplines. This can cause friction when wildlife managers present ideas for wildlife management, and effective wildlife managers in those organizations have had to learn the art of communicating with adminis-

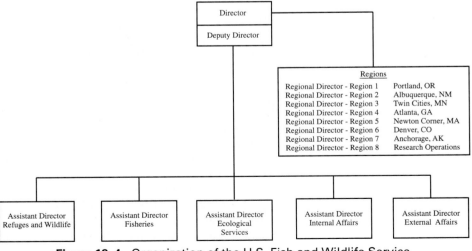

Figure 12–4 Organization of the U.S. Fish and Wildlife Service.

trators in other disciplines. In the past, it has been difficult for wildlife biologists in those agencies to move into higher paying administrative jobs. Fortunately, this situation is changing.

Private Conservation Organizations

Administration in private conservation organizations generally includes the task of raising funds to support the organization, in addition to carrying out policy. Fundraising is an art (or service) in itself, one usually learned in an apprenticeship. When private organizations hire an administrator from "outside," they usually look for someone who, in addition to possessing wildlife know-how and fieldwork, has some fundraising experience.

Administrators in organizations that lobby should be experienced in communicating with the public as well as with legislators. Of course, to be convincing, they must also have a strong background in wildlife and must know the experts in various aspects of wildlife well enough to be able to call on them as witnesses when needed.

More and more, private conservation organizations are becoming the voice of wildlife in the nation (Figure 12–5). Unencumbered by political pressures, volumes of governmental red tape, or bureaucracy, these organizations are assuming a leadership role in wildlife preservation and are a valued support for the wildlife programs of state and federal governments. This is the result of carefully defined goals, effective administration, and growing public support (Figure 12–6).

Private Corporations

The administration of wildlife policies in private corporations, like that in multiuse resource agencies, is usually handled by people with little or no training in wildlife. Typically, they hire wildlife managers who report to administrators trained in business, engineering, or public relations. As in the multiresource agency, effective wildlife managers are good communicators who can persuade the corporation of the benefits and importance of wildlife programs. Like multiresource agencies, private corporations do not give their wildlife managers much chance to move into administrative positions. These people do, however, serve a very important function for both wildlife and the corporation.

The Legislature

Many wildlife administrators have to deal with elected officials. Sometimes these officials are supportive, and sometimes they are hostile. You must know what the organization's policies are in dealing with elected officials. When you deal with such officials, be sure to have all the facts correct. Still, remember that elected officials may not be interested in the facts; instead, they may having a burning issue, or maybe they are being pressured by a special-interest group. Still, the more information that you as an administrator have, the better off you are. A pleasant personality and good speaking skills are important. Let elected officials know that they can come to you for information. Try to learn as much as you can about a person before any planned meeting. The more information you have, the better you will be able to present your point of view.

Figure 12–5 Private conservation organizations have an important role in managing our wildlife. For example, funding for the recovery of the black-footed ferret was received from many private groups.

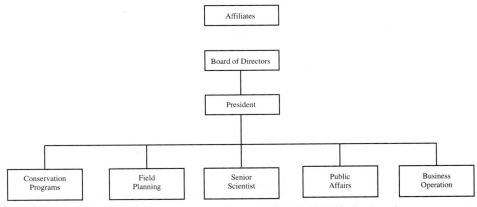

Figure 12–6 Organization of the National Wildlife Federation.

SUMMARY

Although their voices carry more or less weight in the formulation of agency or organizational policies, the prime function of wildlife administrators is seeing that the wildlife mission is fulfilled. Wildlife conservation is the primary objective of government and private wildlife agencies and a secondary objective of some government multipurpose agencies, private conservation organizations, and many companies. In wildlife agencies, the wildlife administrator is the boss, although the policies the administrator implements are set by legislatures or (in private agencies) usually determined by charter or a governing board. In multipurpose agencies, private organizations, and companies, the wildlife administrator is responsible to a general administrator.

In any case, if the wildlife component is large, the administrator, in carrying out the mission, hires managerial personnel and coordinates their work. Ultimately, it is the administrator who is responsible for the smooth operation of a wildlife agency or division within an agency. The administrator achieves this goal primarily by hiring the best qualified people, knowing how to use their talents, and fitting actions to policies.

DISCUSSION QUESTIONS

1. How do the duties of a wildlife administrator differ between public and private agencies?
2. What are the characteristics of a good administration? A good organization?
3. What is the role of the administration in obtaining funding?
4. Should everyone aspire to move up the administrative ladder? Why or why not?
5. Is it good experience to move between jobs in federal, state, and private wildlife management agencies?
6. Should all administrators have had field experience? Why?
7. In state agencies, what is the role of a commission?
8. What is the role of administrators in setting policy in federal agencies?
9. Why is "the chain of command" important in an organization?
10. Can an organization be effective when people are not in the proper positions? Discuss.

LITERATURE CITED

1. Harrington, W. 1982. The Endangered Species Act and the Search for Balance. *Natural Resource Journal* 21:71–92.
2. Towell, W. E. 1971. Role of Policy Making Boards and Commissions. In R. D. Teague (Ed.), *A Manual of Wildlife Conservation*. Washington, DC: The Wildlife Society.
3. Langenan, E. E., and C. W. Ostrom, Jr. 1984. Organizational and Political Factors Affecting State Wildlife Management. *Wildlife Society Bulletin* 12:107–16.

13

PLANNING

A road map gives people options for reaching a destination. Some may want to go by the fastest route, others by scenic routes, still others to visit friends. By looking at the map, people can plan their trips to best suit their needs. Organizations, be they private corporations or public agencies, must also have routes to their goals. They must, so to speak, develop maps. These maps are called *plans* and making them is called *planning*.

For efficient planning, it is essential that managers know the consequences of their actions. Planning for resource management therefore requires knowledge of functional input–output relationships. The inputs in fish and wildlife work include budgets, expenditures, the consent level of access, fish and wildlife regulations, and agency regulations and laws. The output should be expressed as consumer satisfaction.[1] In this chapter, we discuss the planning process and the resources used by planners.

THE NATURE OF PLANNING

We can identify four major components of planning: objectives, time, substance, and contingencies. Let us use a simple illustration. Suppose your objective is to spend two months traveling in Europe next summer. You want to visit Paris, Berlin, Vienna, Florence, Rome, and Madrid, and you also wish to make a number of side trips. This objective obviously indicates a short-range plan. The term is relative, of course. When does a plan become long range? We can be arbitrary and say when more than a year is involved—or two years—or five years. It depends on the context, too. For the individual, five years is certainly long range; for a company, it may be the outer limit of short range.

As for substance, suppose you have saved about $3,000 for this holiday. Now you can make your plans. You will want to make at least three different plans, perhaps more. In one, you will fly first class and stay in first-class hotels. This will give you less time in Paris, which is more expensive than the other cities. The only side trips the first plan will accommodate will be two days in Switzerland and two days in Frankfurt. Another plan calls for flying coach and staying in pensions. This plan permits additional side trips to Munich, Naples, and Toledo. Obviously, thinking about substance can lead to modifying your objective.

So, too, can allowing for contingencies. Suppose that a medical bill or increase in tuition for next year reduced your substance to $2,500. Should you cut a week from the vacation? Cut out the side trips to Munich and Naples? Or the trip to Toledo? You will plan for such a contingency.

Meanwhile, you make out your itinerary—or have your travel agent do it for you. This means another plan—a plan within a plan, as it were. It is a closer plan, specifying days, hours, and even minutes: you need to know exactly when your trains leave, for example.

Now, suppose that you are the manager of a wildlife area. You are asked to make a five-year plan for protection of species, improvement of habitat, and handling of possible intrusion of an automobile-assembly plant in the area. You have your objectives and the time frame. Probably you will not know the substance you will have for the entire five-year period, so you will have to include contingencies in your budgeting. At any rate, the budgets you make will depend on the strategies you plan to use to meet the objectives. This plan will be much more complex than the holiday plan, of course, but it will have the same components: objectives, time, substance, and contingencies.

THE PLANNING PROCESS

The mission statement of most agencies that manage wildlife as their primary or secondary objective is expressed as broad, general goals. In their planning process, agencies subdivide the mission into categories that can be used as planning units or programs.[2] (Figure 13–1).

Figure 13–1 The planning process.

A *program* is a series of related activities that results in a product. Most wildlife agencies organize their programs around species or groups of species, such as an elk or waterfowl program. The product could be to control the number of individuals in a wildlife population, to establish a habitat to support a specific population, to provide recreation activities associated with hunting or sport fisheries, or to foster a diversity of wildlife species. These products are called the program *objectives*. *Strategies* are prepared for reaching the objectives, and an *evaluation* is made to see whether the strategies have been successful.

Objectives

Often, goals are established within which management objectives can be pursued. Broad goals are frequently set, such as the goal of creating the opportunity to have an enjoyable wildlife experience. These broad goals must be translated into output for a program. Objectives need to be established in measurable terms such as actual population numbers. For example, in the U.S. Fish and Wildlife Service Endangered Species Program, meeting the objective of supporting 550 bears in the Yellowstone ecosystem might reach the goal of maintaining a viable population of grizzly bears in the United States.

In the U.S. Fish and Wildlife Service, the Migratory Bird Program includes a list of bird species of high federal interest—a list based on knowledge of the status of the species (is it declining?) and public interest. The program uses this list in establishing its objectives. State programs for big game often set maintaining an optimum herd size as an objective. This may require increasing the size of the population if more recreation days are desired or decreasing the size if the herd is causing damage to, or is otherwise adversely affecting, the community.

Strategies

Once program objectives are established, problems can be identified that prevent us from reaching them. Strategies can then be developed to address these problems and move us toward the attainment of the objectives. Strategies can include inventory, management, law enforcement, and research efforts. For example, the initial strategy in a program for supporting an endangered species, migratory birds, or big-game species probably will be to estimate how many animals there are. Sometimes an early strategy is collecting age–sex distribution data. Next, the general objectives can be made more specific: For instance, instead of the general objective of increasing the number of deer in an area, the objective may be to add 200 deer to the area. Then specific plans are made to achieve the numerical objective. Sometimes the strategies must be sequential. If, for example, the objective is to increase the distribution of the black-footed ferret, one of the first strategies must be to determine the animal's physical and biological habitat requirements. Next, biologists find areas that meet these requirements. During this period, a captive-breeding program should be under way, unless ferrets located in the wild can be transplanted. At this point, a program introducing the ferrets to the area can be instituted.

Sometimes, an objective is to decrease a population that is becoming a nuisance. If it is a game species, the strategies may include changes in hunter's fees, raising quotas, surveying for age, and changing the sex ratio of animals taken. For nongame species, habitat-management programs may be instituted, with biological control or eradication efforts.

In all cases, program objectives and strategies recommended by field personnel are evaluated by administrators. Generally, the objectives and individual strategies are given priorities. Funding levels may prevent working on all segments of each proposed

Figure 13–2 Lesser scaup. Waterfowl regulations are set jointly by the federal government and the states.

objective and strategy. Thus, work on a newly listed endangered species may not be funded for several years, until it is high on the priority list (Figure 13–2).

Evaluation

An important part of any planning is provision for the evaluation of results. Evaluation is the measure of progress toward obtaining objectives. Objectives that are quantitative and measurable can most easily be evaluated. Often, this involves a population or habitat inventory. If the objective is an increase or decrease in the population of a wildlife species, the techniques described in Chapter 4 can be used. Sometimes, populations must be censused or sampled for a number of years, or continuously, to determine whether objectives are being reached. Sometimes, an evaluation will indicate that new strategies should be tried.

Should Wolves Be Reintroduced to the Olympic Peninsula?

The Washington Department of Fish and Wildlife wanted to know what the general public thought about reintroducing wolves to the Olympic Peninsula in the state. This area represented a potential site for reintroducing gray wolves because it contained a large area (about 60,000 hectares) in national park- and forestlands with a large ungulate prey population. The department had received many requests from individuals favoring the reintroduction of wolves into this region. A public survey on the subject found that 48 percent of the respondents wanted wolves reintroduced and 40 percent did not. A total of 12 percent had no opinion. Of those who wanted the wolves reintroduced, 50 percent realized that the proposed habitat was historically a part of the wolves' territory. Of those opposing the reintroduction. 10 percent felt that livestock losses would be too high, 37 percent felt that the cost of reintroduction would be too high, 8 percent felt wolves would have an impact on big game, and 45 percent had other reasons for their opinions. The survey results indicated that there was considerable support for reintroducing wolves and that the public had a variety of thoughts on the costs and benefits of doing so. The surveys felt that a public education program would be beneficial.

Surveys are one way to get information about potential wildlife management decisions. In this case, the information about the need for public education could be most valuable. Managers need to give much consideration to the questions asked in a survey, as well as to how it can best be distributed so that it accurately reflects public opinion.

PUTTING IT ALL TOGETHER

The planning process is really a continuum (Figure 13–3). We can ask ourselves a series of questions[1] and then answer them:

1. Where are we? We are at the x deer (program) population data level of 1,000 animals (inventory).
2. Where do we want to be? We want to increase this population (goal) to 1,100 animals (objective), but the winter range will support only 1,000 animals (problem).

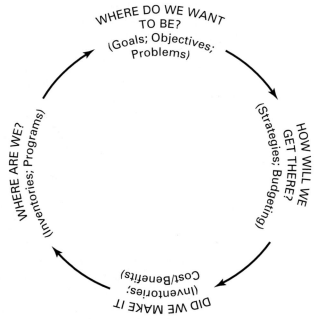

Figure 13–3 The planning process can be viewed as a continuum.

3. How do we get there? We would have to improve the winter range (strategy) at a cost of $10,000 (cost).
4. Did we make it? The new population estimate of 1,100 deer (inventory) has been reached at a cost of $100 for each additional animal (cost–benefit).

THE USE OF PLANS

Some people feel that it is not necessary to indicate whether goals (objectives) are long term or short term,[3] although most wildlife management agencies find the distinction an aid in planning. As we indicated earlier, time is one of the components of a plan. Keeping the time limits in mind can often help planners determine the proper pace and sequence in a program.

Wildlife and other natural-resource management agencies have not always developed adequate methods for coping with projected situations. Such lack of planning has often resulted in the use of emergency measures. It is not uncommon to hear employees complain about having little time to do the work of the agency because they are always fighting "brush fires."

Agencies use the approach of environmental planning to look ahead and identify problems. By examining trends, managers can develop plans to anticipate conditions. Thus, agencies must consider not only future problems, but the future of present decisions. Plans give direction to an agency, making it responsive and pursuing an identified course of action rather than more organization charts and brushfire fights.

Direction

Plans supply direction for employees. With documented objectives and strategies, both administrators and employees know where the agency is moving. If plans are adhered to and modified as necessary, employees are neither surprised nor left ill informed. They know what the agency is all about. Naturally, both overall and departmental plans are vital to coordination within an agency.

Plans are useful not only for employees, but for the public as well. In wildlife management agencies, public input is important. For example, if habitat acquisition is a high priority, but local interest groups oppose it, the best way to win them over is often to produce a clear, well-documented plan. This allows conservation organizations to counter landowners' concerns. The plan is out in front of the public; the statement of the agency's goals is clear, and people can react to it and not feel left in the dark.

Plans are especially important for supervisory people above the agency level and for legislatures. In state systems, the commissioners need clear, coherent plans, written in language as nontechnical as possible, to back up their appropriation requests. It is sometimes difficult to get adequate funding for wildlife programs in multipurpose agencies, so here carefully made plans have special relevance.

Urgent Responses

As noted previously, identifying long-term trends is a major consideration in planning. However, plans also need to include strategies to respond to emergencies.

For example, the U.S. Fish and Wildlife Service has a planned strategy for handling problems of migratory birds associated with oil spills. A team of people with different backgrounds, who together can recommend procedures for avoiding further damage to the birds, is on call in the event of an oil spill as well as advising on cleanup

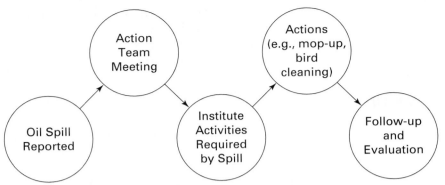

Figure 13–4 Plan of action following an oil spill.

procedures (Figure 13–4). Field visits may be necessary. Experts from other areas may be called in to assist.

If an agency manages land, plans must include actions to be taken in case of wildfires. Although each fire has its own characteristics, the plan should include general guidelines on whether to extinguish fires or let them burn, in the light of the affected area's state of succession, wildlife needs, and danger to human life and property. There is simply no way to improvise basic principles in these kinds of situations—they must be thought through and recorded in advance. Cleanup strategies, although not subject to such time pressure, should be in the plan—matters such as preventing runoff into waterways, reseeding, and leaving cover and shelter, for example, snags or brush piles. It is amazing how many slipups can occur if these things are not included in *written* plans.

Alternatives

Planning must be flexible enough to allow for periodic reevaluation of objectives and the development of new or alternative strategies. For example, it is often desirable to compare the benefits of developmental projects and wildlife recreation days. By having information available, wildlife planners can point out the number of recreation days that would be lost as a result of a development that would change wildlife habitat. Alternatives also come into play when a habitat is threatened by an influx of people. After it has been determined what wildlife species are important to people, habitat-improvement techniques can be used in other areas to attract the dislodged species. It is also possible to set up programs to attract wildlife species to areas where people already live. Alternatives can be designed in the light of possible changes in population size or license fees.

TYPES OF PLANS

As we have noted, each division within an agency will have strategies for reaching its own objectives within the agency's overall program. Thus, refuges, regions, and the central office all have assigned strategies or components of strategies. One important function of the central office is to coordinate these plans, which will include procedures for handling everything from fires, accidents, and personnel matters to visits of VIPs. Well-made, coordinated plans show how the plans for each programmatic unit or field installation contribute to the agency's objectives.

Operations

It is obvious that the plans for the day-to-day running of an agency or organization should reflect its overall purpose or mission. The type of physical plant needed, its size and maintenance, the equipment (how many cars? are trucks needed? etc.), how many people, what type of people (what academic training? what experience? etc.), where people's offices should be, how many people will spend time in the field, and how much time—all this and more may go into the operational plan. How detailed and closely structured the plan is will depend largely on the style of the manager, even when, as is sometimes the case, the form of the plan is prescribed by the head of the agency.

We have said that operational plans reflect the organization's mission. Thus, in wildlife management, plans for different seasonal activities are very important. Plans for reducing the water level in lakes on refuges in order to provide nesting habitat must allow for the possibility of flooding private property. Trails must be open in time for the tourist season. Temporary personnel must be recruited and hired within the proper time frame. These activities are part of the operational strategies.

Regional managers need to schedule the collection of data for monthly, quarterly, and annual reports so that there will be time for input from field people and adequate time for completion. Annual wildlife surveys must be coordinated so that data are collected at the proper time and are compatible with data from other areas.

Budget

Making a budget is perhaps the clearest example of what is meant by planning. Actually, a budget functions in two ways in the overall planning. In one, it is a reflection of all plans. This is its preliminary stage. Then reality enters, and the budget disciplines the planners, forcing them to set priorities in the light of available funds. A budget, then, is at once a reflection of, and a disciplinary constraint in, plans.

Ordinarily, every level in an organization makes its own budget—actually, its request for funds. Each level presents the tale of its financial needs to the next structural level for approval. This means that the lower on the organizational ladder you are, the farther ahead you must plan. On the lowest levels of federal government organizations, years may be involved. Since most wildlife-management agencies are public organizations, we will discuss the budget process in a public agency.

In the federal government, budget formulation begins roughly 36 months before the beginning of the fiscal year in which the budget will be operative—that is, before October 1. Thus, plans for the budget of fiscal year 2003–2004 would begin in the fall of 1999. The initial planning and strategy meetings occur in the central offices of the various agencies, which then request information from the different levels. Before this information gets to the central office, it goes through a series of reviews, each of which recommends a priority ranking for the requests.

Data are used in the formulation of a proposed budget in the Washington office at least 24 months before the beginning of the fiscal year. Activities on the budget during these two years may be categorized as (1) planning, (2) formulation, (3) negotiation, and (4) justification. (See Table 13–1).

TABLE 13–1

Budget Sequence for U.S. Fish and Wildlife Service

Phase	Period	Months Prior to Fiscal Year
Planning	October–December	21–24
Formulation	January–May	16–21
Negotiation	June–November	10–16
Justification	December–August	1–10

Planning.   During the planning phase, data from the field personnel are assembled by the program managers in the U.S. Fish and Wildlife Service. These people decide, on the basis of information supplied by the agencies, which objectives should be supported and how they should be supported. Thus, they may decide to seek funds to construct new facilities, to acquire more land, to increase research, or, on the other hand, to abandon current projects. Statements are normally required to justify new objectives or programs, showing how they are related to other objectives and programs of the agency. Once the top-level administrators have agreed on which major budget issues will be pursued, they set up a tentative schedule.

Formulation.   Sixteen to 21 months before the beginning of the budget year, the formulation phase begins. During this period, field and regional personnel are required to justify their request. Special justification papers are prepared for new major requests. In some agencies, the process involves developing a minimal standard of performance for each program, on the basis of information obtained from each organizational level.

Negotiation.   In the negotiation stage, conferences are held within the agency and the federal department; for example, the Fish and Wildlife Service will meet with officials from the Department of the Interior. This takes place from 10 to 16 months before the fiscal year. Budgets from all agencies are combined into a departmental budget. During the process, priorities are set for the entire department, and some good proposals disappear. Finally, the budget moves up to the Office of Management and Budget (OMB), which reviews the budget on behalf of the President of the United States. OMB often alters priorities and makes further changes.

Justification.   In the justification phase, the planning office of each agency writes justifications for budget requests recommended by OMB for funding. The proposed budget is then presented to Congress. Congressional hearings are held, and committees may ask individual agencies questions about the budget. Congress, which may increase or reduce items in the budget, notifies OMB of the budget it finally approves. The budget then goes to the president, who must approve or veto. Congress can override a presidential veto by a two-thirds vote. When the budget is passed and approved, program managers give field people the go-ahead to plan for or continue programs. With luck, all of this will be completed by August of each year.

Obviously, there are many participants in the federal budget-making process. Various congressional committees, executive review committees, the congressional budget office, and the president all have roles. (See Table 13–2.) As a consequence, there are many opportunities for things to go wrong or be delayed. Fairly often, interim budgets have had to be passed because Congress or the president has not acted within the prescribed time limits. The usual result is wasted effort and poor use of funds, because

TABLE 13-2

Duties and Functions of Participants in the Federal Budget Process

President	Authorizing Committees	Appropriations Committees	Revenue Committees	Budget Committees	Congressional Budget Office
Submits executive budget and current estimates of services Updates budget estimates in April and July Signs or vetoes revenue, appropriations, and other budget-related legislation May propose the deferral or rescission of appropriated funds	Prepare views and estimates on programs within their jurisdiction Report authorizing legislation for the next fiscal year Include CBO goal estimates in reports accompanying their legislation *Limitations:* 1. Legislation providing contract or borrowing authority is effective only as provided in appropriations 2. Entitlements cannot become effective before next fiscal year	Report regular and supplemental appropriations bills After adoption of a budget resolution, allocate budget authority and outlays among their subcommittees Provide five-year projections of outlays in reports accompanying appropriations and compare budget authority with amounts provided in latest budget resolution Can be directed by second budget resolution to report reconciliation bill repealing new or existing budget authority Review rescission and deferral proposals of the president *Limitation:* After second resolution is adopted, spending cannot exceed amount set by Congress	Submit views and estimates on budget matters in their jurisdiction Can be directed by second resolution to report legislation changing tax laws *Limitation:* Legislation cannot cause revenues to fall below level set in second resolution	Report two or more concurrent resolutions on the budget each year Allocate new budget authority and outlays among House and Senate committees Monitor congressional actions affecting the budget Advise Congress on the status of the budget	Issues reports on annual budget Estimates cost of bills reported by House and Senate committees Issues periodic reports on status of the congressional budget Assists the budget, revenue, appropriations, and other committees Issues five-year budget projections

the plans submitted cannot be followed. Failure to have a budget at the beginning of the fiscal year makes it virtually impossible for agencies to balance their expenses during the year. Because many wildlife activities are seasonal, delays in the budget process are particularly harmful to wildlife-management programs.

States

Each state wildlife agency has a budget-planning procedure that goes through a four-stage process similar to that of the federal government. In most cases, the process is condensed into one year. States that rely to a large extent on funds from hunting and fishing permits must plan ahead so that increases in fees will occur when needed. States generally have their budgets reviewed or approved by a commission. In a few cases, the commission retains final authority; however, in most cases, the state legislature must review and appropriate funds.

TECHNIQUES USED IN PLANNING

Agencies use two major methods of establishing budget priorities: management by objectives and zero-based budgeting. Most agencies use a combination of these techniques.

Management by Objectives

In management by objectives, program objectives are listed in order of priority. Thus, if the Endangered Species Program has as its primary objective maintaining a viable population of the California condor, strategies needed to reach that objective will be funded before funds for a second objective become available. As a result, objectives with a low priority may receive no funding. This has been the case with nongame objectives in some agencies.

Zero-Based Budgeting

In zero-based budgeting, an agency or company starts with the assumption that each budget expenditure must be justified anew each year or cycle. This method of budgeting forces a review of program priorities and a close examination of changes in costs in programs as these affect the priority or even continuance of the programs. Items that are justified one year may not be funded another year. Annual reviews prevent delays in the elimination of low-priority programs. Zero-based budgeting is a way to get rid of deadwood.

EXAMPLES OF PLANNING

Organizations that are most effective in planning usually have full-time planning units. In the most conducive atmosphere, people doing the planning are allowed to carry out their responsibilities without having conflicting responsibilities. They have the freedom to consider all feasible alternatives. This does not mean that field personnel should be excluded from the planning effort, but it does mean that planners must feel free to develop their own analyses.[4] There must be adequate support for the planners, so that they can find information that is required—for example, data systems and recommendations based on the data they are able to collect.

It is very important that planners utilize the talents of everyone in their organization. Planning units that operate in a vacuum often drift off and do not develop plans to effectively carry out the responsibility of the agency. If wildlife managers do not have the opportunity to participate in the planning process, they easily become dissatisfied.

Some organizations use committees and top-level administrators instead of a special planning unit. This can be effective in relatively small organizations, but in larger private organizations and federal and state agencies the job is better handled by specially trained personnel. Even in smaller organizations, the committee process sometimes results in a popularity contest. When top-level administrators are given an equal voice in voting on individual items, their biases come into play, and important ideas may be neglected or undervalued. The top-level administrator must then have extraordinary skill in developing comprehensive programs that fulfill the goals of the organization. Some examples of planning in wildlife agencies follow.

U.S. Forest Service

In the multiple-use area, each resource management agency has some form of directive to work on wildlife. For example, under the directive given the Forest Service,[5] the Department of Agriculture and the Forest Service are to develop, implement, and authorize program policies and actions that will support the economic, aesthetic, ecological, recreational, and scientific values of fish and wildlife, improve their habitat, and ensure a viable diversity of naturally occurring wildlife population—all the while considering fully the other missions of the Department of Agriculture and its resources and services. The directive further indicates that this goal is to be reached by (1) management action on departmental land; (2) educational, technical, and financial-assistance programs; (3) programs to improve the status of threatened and endangered species; (4) the alleviation of economic loss to agricultural, livestock, forest, and range resources caused by pest species; (5) support and encouragement of biological control to regulate insect-caused disease and pest vegetation; and (6) research needed to implement the programs.[6]

The National Forest Management Act, passed in 1976, became an important planning tool in an organization dedicated to the management of U.S. timber resources. The act provided for multiple-use planning and management and made timber management a mission of the Forest Service. Public participation in decision making became an objective.[7] As with wildlife, the act prescribes general guidelines consistent with the Multiple-Use Sustained-Yield Act. In 1981, recommendations of a presidential task force resulted in an amendment of the National Forest Management Act to the effect that an interdisciplinary approach must be used in land and resource planning and that both fish and wildlife must be considered in any activity planned by the Forest Service.[8]

The net result of the original act and its amendment was, on both national and regional levels, that the Forest Service must develop and implement plans to maintain all resources. Plans for fish and wildlife must be developed so that these resources will be protected when timber harvest, mineral extraction, or recreation development occurs. Each regional office within the Forest Service therefore establishes regional goals related to conservation of species, protection of habitat, and provisions for hunting and fishing.

Generally, for each forest within a region, a plan is developed indicating which species will receive special management. Indicator species may be listed, as well as specialized habitats at particular successional stages. Coordination with state game and fish agencies is stressed.

Private Corporations

Private corporations that utilize our natural resources are becoming aware of the importance of developing multiple-use plans that include wildlife. In the late 1970s and early 1980s, there was extensive oil and gas development in western Wyoming, eastern Utah, and Idaho—an area known as the Overthrust Belt (Figure 13–5). The development included not only exploration for, and drilling of, oil and gas wells, but construction of roadways, pipelines, worker camps, and treatment plants. The change in the landscape was immense.

There has been criticism throughout the nation that desirable wildlife populations, including big-game species and sport fish, were being reduced because of habitat loss. A number of conservation groups and some of the oil and gas companies formed a consortium calling itself the Overthrust Industrial Association (OIA). The association set up a committee of industrial, state, and federal wildlife managers, together with private conservation representatives, to do wildlife planning for the area.

The committee reviewed available data and undertook studies in species, habitats, and problem areas where the data were not adequate. The goal was to make enough data on wildlife available to the industrial planners to avoid adverse impacts on wildlife. Much of the data compiled by the OIA will be useful to other developers and public management agencies. The OIA also sponsors an educational program to foster wise use of wildlife resources. The scientific and educational facets of this group's project will greatly benefit the entire area of interest (Figure 13–6).

Figure 13–5 Areas of oil and gas activity in the Overthrust Belt.

Figure 13–6 Antelope are affected by oil and gas development.

THE ROLES OF PEOPLE

Internal

Suggestions from people within an organization are absolutely indispensable to the agency's planning efforts. For planning to be accepted and effective, plans must be built from the bottom up and administered from the top down.

The Public

The public needs to be involved in planning. Public meetings are held so that people can have an opportunity to express their ideas. Unfortunately, except where their own backyards, so to speak, are involved, members of the public are generally indifferent. Unfortunately, also, because many of the people who come to public meetings represent special interests,[4] the point of view that gets projected from such meetings is hardly representative. (See Chapter 14.)

Some agency personnel throw up their hands at the public's indifference. Game and fish agencies partially solve this problem by scattering meetings around the state, holding them often in small communities, where they attract more people. Other agencies contact individuals through mail or phone surveys, using such categories as applications for hunting and fishing licenses, occupation, and land ownership. For example, hunters' favorite type of hunting experience and desired number of hunting days can be learned by using hunting license files. General surveys can be made of landowners and city dwellers to determine what types of wildlife experience they prefer. Even so, mail surveys seldom elicit a representative response: People who feel strongly are likely to return the survey, whereas others do not bother. Also, mail surveys are not always clear to the reader. Telephone surveys do not suffer from this lack of clarity, but they are very time consuming.

Difficulties notwithstanding, wildlife resources belong to the public, so the public's input must be solicited by all means available. In today's environment, an agency is courting disaster if public input is not earnestly solicited and considered seriously during the planning process.

Data systems are of great assistance to planners.[9] By being able to recall information on the distribution of a species, its habitat requirements, and special management needs, planners can easily locate gaps in data. National surveys and local inventories also provide information useful in making and evaluating plans.

One type of data system that has been useful to decision makers in wildlife planning, management, and research is a geographic information system (GIS). Such a system is a collection of data, often on a computer, available to assist in decision making. These systems are often in the form of computerized maps. Data are put into the system and updated so that they can be retrieved on several levels. For example, data on vegetation, weather patterns, soils, topography, and disturbance of a habitat can be placed into a system. Information on the distribution of white-tailed deer can be collected on a daily or seasonal basis and placed into the system. Overlays can then be prepared to show how deer activity and movement relate to the variables in the system.

GIS systems may be prepared for many land units, such as rivers, lakes, states, or very local areas. In each case, the usefulness of the system is dependent on the accuracy of the information placed in it. The more up to date the system, the more useful it will be to the planner or manager.

Models that have been developed on the computer can be of assistance in the planning process. For example, by supplying information on a variety of environmental conditions, population factors, seasonal changes, and public desires, a simulation model developed by the Michigan Department of Natural Resources helped managers and planners develop alternative courses of action. The model indicated the number of hunting days that could be allowed in an area, on the basis of access to the area by hunters. It specified the total number of days of deer hunting that could be projected over a constant period of time, given a constant budget and an equal emphasis on management activities (Figure 13–7). This model was then varied to show what would happen if habitat improvement and law enforcement were emphasized at the expense of land acquisition and promotion. Changes were also made in the length

Figure 13–7 White-tailed deer.

of the hunting season and fees for licenses. The model shows the number of days of deer hunting that was projected as possible, given a declining budget, but emphasizing law enforcement and habitat improvement. A decline in the number of hunting days reflects lower numbers of deer if development is likely, so that fewer animals could be harvested.[1]

Two problems that planners have to solve concern the sufficiency and accuracy of data. Collecting data costs money, of course, but going ahead without enough good data can often be more expensive in the long run. How much data is enough? There is no rule of thumb, and what is enough in one situation may be inadequate in another. Experience is the best guide here—your experience and the experience of others. Again, we see the value of keeping up with the literature and attending seminars and professional meetings. The latter are especially important: Talk to others in the field—some of them are bound to have had the kind of experience you are having.

Some data become obsolete fast; other data need updating less often. An insurance company that did not continuously update its data would not last long. In contrast, given no marked changes in habitat or hunting practices, data on the deer population in an area would be good for a number of years. But danger lies in the temptation to use obsolete data. If there is any doubt, it is wise to collect new figures. Often, the quality of planning depends on the accuracy and sufficiency of the data used.

Managing Beaver with Data from a Survey

In many parts of the country, beaver get mixed reviews. Beaver dams located upstream create water impoundments that provide good fishing and keep up the water table. The areas also serve as watering holes in arid climates. In the dam-building process, beavers use trees that some people want left alone. Sometimes impoundments flood areas that landowners do not want flooded, and sometimes drainage ditches from beaver dams block water runoff.

In an effort to manage beavers better, biologists conducted a survey in the western United States. A select group of landowners previewed the survey, which was designed to evaluate the advantages and disadvantages of beaver. The survey was then mailed to 5,265 private landowners and 124 public land managers. Fifty-five percent of the former and 72 percent of the latter responded. Private landowners raised concerns about damage from beaver, including blocked irrigation ditches, girdled timber, blocked culverts, and flooding of pastures, roads, crops, and timber. Among the benefits they perceived to derive from beaver were elevated water tables, increased riparian vegetation, and increased stock-watering opportunities. Over 45 percent of the private landowners and all the public land managers expressed an interest in introducing beaver into their lands.

The survey came with a map that respondents were asked to use to indicate where there were problem beaver they wanted removed. They were also asked to indicate where they wanted beaver to be introduced. The biologists took the information and instituted a program of removing beaver from problem areas and placing them in areas where people wanted them.

SUMMARY

Plans are the road maps an agency uses—the guides to its day-to-day and long-term operations. The planning process involves translating the mission or missions of an agency into specific programs. Objectives are established for each program, and strategies are devised that, in the best judgment of planners, will enable an agency to realize its objectives. These strategies generally require teamwork by all employees, including managers, researchers, educators, enforcement officers, secretaries, and maintenance personnel.

An integral and vital part of the planning process is that of developing budget requests on a priority basis. The budget is, in effect, a concrete summary of how the agency sees itself—of what it considers its functions to be. A coherent budget usually bespeaks a coherent agency. In the federal government, the budget process begins three years before funding becomes available.

The public, agency people, and professional planners are all important to the planning process. Agencies must use people's ideas, along with a multitude of other tools, such as massive data systems, to formulate the most effective plans.

DISCUSSION QUESTIONS

1. How can field wildlife managers input their ideas into the planning process? How can the process help them do their jobs?
2. Why should all plans relate to the mission(s) of the organization? What happens when they don't?
3. How can wildlife enthusiasts convince others of the importance of multiple-use planning for wildlife?
4. What information should be available to the professional planner?
5. What background and experience should the professional planner have?
6. How does politics interfere with the budget-planning process?
7. Describe zero-based budgeting and management by objectives.
8. How can managers avoid spending a large proportion of their time planning?
9. What are the advantages and disadvantages of committees in planning?
10. What is the role of research people in planning?

LITERATURE CITED

1. Bell, E. F., and E. F. Thompson. 1973. Planning Resource Allocations in State Fish and Game Agencies. *Proceedings, 38th North American Wildlife and Natural Resources Conference,* pp. 369–77. Washington, DC: Wildlife Management Institute.
2. Crowe, D. M. 1983. *Comprehensive Planning for Wildlife Resources.* Cheyenne, WY: Wyoming Game and Fish Department.
3. Riecks, C. A. 1971. Program Planning: A Manager's Attitude. In R. D. Teague (Ed.), *A Manual of Wildlife Conservation,* pp. 14–17. Washington, DC: The Wildlife Society.
4. King, D. A. 1972. Towards More Effective Natural Resource Planning. *Proceedings, 37th North American Wildlife and Natural Resources Conference,* pp. 260–67.
5. Bergland, B. 1980. *Policy on Fish and Wildlife.* Secretary's Memorandum 2019. Washington, DC: U.S. Department of Agriculture.

6. Mealey, S. P., and J. R. Horn. 1981. Integrating Wildlife Habitat Objectives into the Forest Plan. *Proceedings, 46th North American Wildlife Conference,* pp. 488–500. Washington, DC: Wildlife and Natural Resources Conference.

7. Anon. 1976. *The National Forest Management Act of 1976.* Current Information Report 16. Washington, DC: USDA Forest Service.

8. Anon. 1982. National Forest System Land and Resource Management Planning. *Federal Register* 47:43026–52.

9. Bown, J. L., and D. J. Decker. 1982. Identifying and Relating Organized Publics to Wildlife Management Issues: A Planning Study. *Proceedings, 47th North American Wildlife Conference,* pp. 686–92. Washington, DC: Wildlife and Natural Resources Conference.

14

GOALS AND DESIRES OF THE PUBLIC

While wildlife management involves animals and their habitat, it also involves people and their interests (Chapters 1 and 11). Because most wildlife-management efforts are made to satisfy public desires, it is necessary to have some method of identifying those desires. Astute wildlife managers, like astute businesspeople, know what their clients want. In this case, the client is the public. Unfortunately, as we indicated in Chapter 1, the public's wishes are not always easy to determine. In addition, the great diversity of public interests makes setting wildlife goals difficult. By doing the best job of getting information about people's interests, the wildlife manager can do a better job of managing wildlife for all concerned. Data collected from the public directs management actions and shows the manager or the agency which educational efforts are needed to make the public more aware of wildlife issues. Because public desires are not always what is best for the natural habitat, managers must strive to maintain the most favorable habitat for wildlife in light of, but not necessarily in accordance with, public desires. It is important to convey to the public that wildlife is a key indicator of a healthy habitat. The loss of wildlife is one indication of habitat decay. Improper wildlife management can therefore result in a disruption or loss of natural resources for future generations.

Funds dictate objectives. Clearly, most funds go into the management of a few select species. Indeed, 97 of every 100 federal wildlife dollars go for game management.[1] This is so partly because hunters and fishing enthusiasts pay license fees. At the same time, landowners are expected to provide wildlife species for public use, although they usually get little or nothing from license fees. There have been conflicts between landowners and wildlife-management agencies when the landowners wanted to alter a habitat or reduce the number of species of wildlife living there. Further conflicts occur when, for a minimal fee, allotments are given to graze livestock on public land. Allotment holders have requested at times that wildlife species be ignored or reduced so that more domestic cattle or sheep could be grazed.

The role of government in setting goals for wildlife management remains somewhat unclear, despite government's control of funds. As we have noted, wildlife is considered to be owned by the states, with the federal government having major responsibility for migratory and endangered species.

In this chapter, we focus on several major questions, such as "What are the public's goals and desires?" "How can we determine what these goals and desires are?" "Can they be formed or changed?" "How do you translate goals and desires into management activities?" and "Who pays for implementing the public's goals and desires?" The latter question is explored from the user's, landowner's, biologist's, conservationist's, and politician's points of view. Current methods of financing are examined.

DETERMINING THE PUBLIC'S GOALS AND DESIRES

Public Meetings

Public meetings are common methods for providing information to the public and determining public opinion and goals. Many state agencies are required to hold public meetings on issues such as setting the dates of the hunting or fishing season. Public meetings are usually distributed throughout the state and are often held by state or federal agencies on issues such as damage to private property, the reintroduction of endangered species, and new forms of legislation. The whole idea is to involve the public. When announced properly, such meetings do get the public involved. There are, however, limitations to the value of public meetings. When a particularly contentious issue is raised, supporters and opponents of the issue sometimes bring a large number of individuals to the meetings. As a result, a truly representative public opinion may never be heard. Some people are particularly vocal at public meetings, making it difficult for an agency or the general public to deal with the real issue. If you want to see the distortion that sometimes comes about in a public meeting, attend one and then read the press report later. You may wonder if the press release is about the same meeting. All in all, public meetings have some value, but they must be taken in context with the whole picture. Agencies and managers must evaluate what is said on the basis of the individual and not simply by counting the number of people speaking for or against an issue. Now some agencies are hiring or contracting with individuals trained to facilitate public meetings in order to get the most representative information possible from the attendees.

Surveys

Resource agencies have used surveys to determine public opinion so that they can use that opinion as a guide. In the early part of the century, little was done to obtain public opinion except by stopping and talking to people on the road. Since then, we have developed a sophisticated science to sample public opinion. This has been very helpful to wildlife agencies, although the surveys continue to have their limitations. Wildlife surveys have often been stratified by sex, region of the country, age class, urban versus rural communities, occupation, education level, income, and age. Specialists have been able to design surveys to get the most helpful information from the questions, as well as to get the correct sample size and makeup to yield the most valuable results.

All forms of surveying have weaknesses. Mail questionnaires require time on the part of the recipient and so get a low percentage of response. Telephone surveys are skewed by the time of the day the survey is conducted. Personal interviews are best, but require work on nights and weekends and thus are expensive. Still, the survey is one of the best ways to sample public opinion.

In recent years, the public has shown a great deal of interest in wildlife issues. Data collected by nationally known pollsters have indicated that people have been asking the government to pay more attention to the environment and wildlife. People generally care about the quality of the human experience. The public is asking elected officials to do more to maintain a healthy environment.

National Survey of Fishing, Hunting, and Wildlife Associated Recreation.
This survey, conducted approximately every five years since 1955 by the U.S. Fish and Wildlife Service, includes questions about hunting and fishing, as well as about nonconsumptive uses of fish and wildlife in general. Most of the survey involves either mail or phone questionnaires, with a relatively small number of personal interviews.[2]

In 1996, 77 million Americans, or about 40 percent of the U.S. population 16 years of age or older, enjoyed some recreational activity related to fish or wildlife. This group's expenditures on fish and wildlife was $104 billion, which was about 1.4 percent of the gross domestic product of the United States. These expenses were primarily for equipment, but also for food, lodging, transportation, and licenses. There were over 39 million people 16 years or older who fished and hunted in 1996. They spent over $60 billion on transportation, food, equipment, and other items such as licenses, fees, and magazines (Figure 14–1).

Fishing was the favorite sport of people in the United States aged 16 years or older. In 1996, 18 percent of the population, or 35 million people, spent approximately 18 days fishing. Freshwater fishing was the most popular, with 29 million people par-

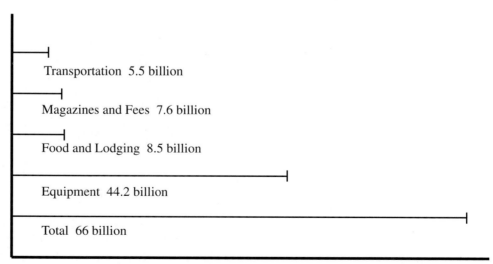

Figure 14–1 Expenditures by people hunting and fishing totaled $72 billion in 1996.

ticipating. The survey showed that 2 million fished the Great Lakes, while 9.4 million anglers fished in salt water.

The U.S. Fish and Wildlife Service found that the number of anglers remained constant since the last survey in 1991; however, the amount of money these people spent increased from $28 billion (adjusted for inflation) in 1991 to $38 billion in 1996. Expenditures for special equipment such as boats and four-wheel-drive vehicles increased 124 percent! Expenses included not only equipment, but also travel, food, and lodging (Figure 14–2).

In 1996, 14 million people 16 years and older enjoyed hunting. Each spent an average of 18 days hunting. Eighty-one percent hunted big game. A total of 59 percent hunted small game, including squirrel, rabbit, quail, and pheasant. Twenty-two percent indicated that they hunted migratory birds such as waterfowl, doves, and woodcock. (The percentages did not add up to 100 because some people hunted more than one group of animals.) Expenditures for hunters also increased after 1991. In 1991, hunters spent $14 billion (adjusted for inflation), and in 1996 they spent $21 billion (Figure 14–3). As with anglers, most of the increase occurred in big-equipment purchases.[2]

Nonconsumptive users added to number of people participating in wildlife-related activities. Since many people hunted and fished as well as participating in nonconsumptive activities, there were overlaps in the numbers of participants. In 1996, 63 million people aged 16 years and older were involved in nonconsumptive activities, including watching wildlife, photographing wildlife, feeding birds, visiting wildlife areas, and planting for wildlife (Figure 14–4). This group of people spent $27 billion in 1996, up from $21 billion in 1991 (Figure 14–5).

The type of survey represented by the National Survey of Fishing, Hunting, and Wildlife Associated Recreation is very useful to wildlife professionals in both state and federal governments. The data are used to point out the value of wildlife to the economy. Nonconsumptive users, as well as hunters and anglers, spent a great deal of money on wildlife. Managers can provide these data to people who want to capitalize on wildlife through both consumptive and nonconsumptive activities. In some places,

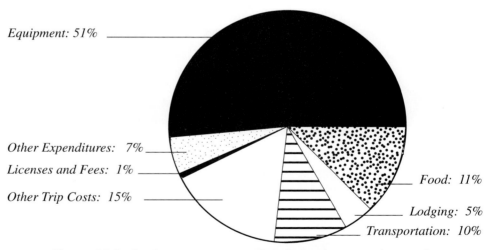

Figure 14–2 Anglers spent most of their funds on equipment in 1996.

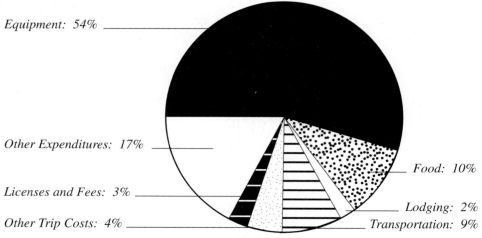

Equipment: 54%

Other Expenditures: 17%

Licenses and Fees: 3%

Other Trip Costs: 4%

Food: 10%

Lodging: 2%

Transportation: 9%

Figure 14–3 The $21 billion spent by hunters in 1996, divided into categories of expenses.

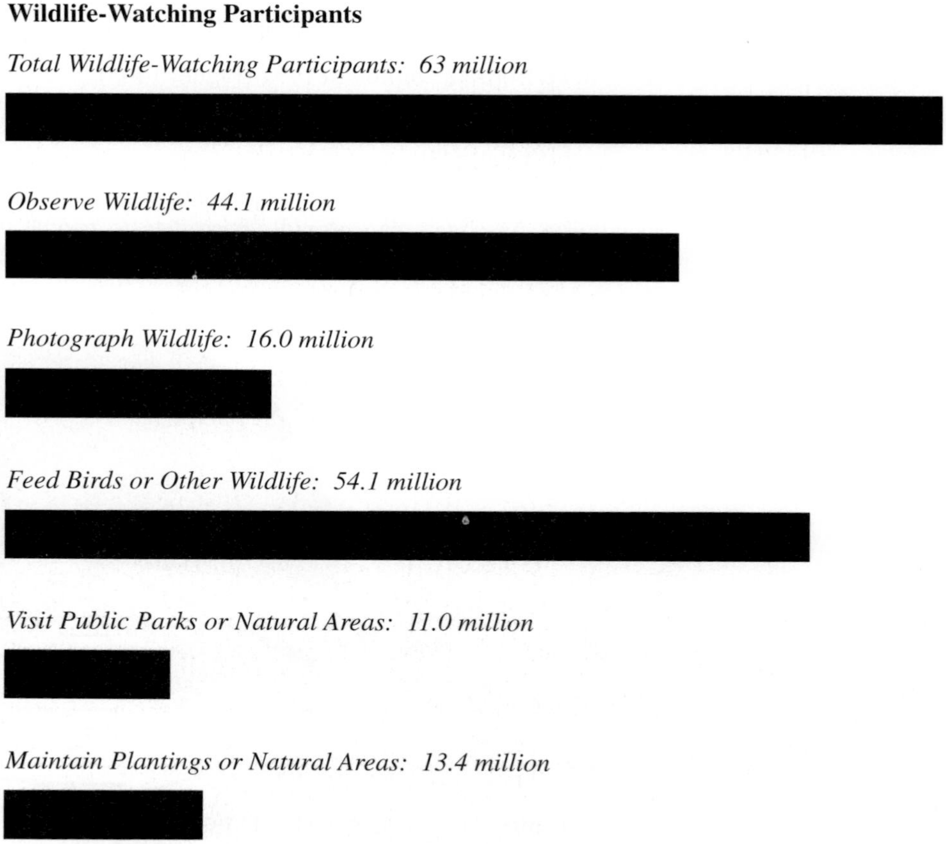

Wildlife-Watching Participants

Total Wildlife-Watching Participants: 63 million

Observe Wildlife: 44.1 million

Photograph Wildlife: 16.0 million

Feed Birds or Other Wildlife: 54.1 million

Visit Public Parks or Natural Areas: 11.0 million

Maintain Plantings or Natural Areas: 13.4 million

Figure 14–4 Major activities of nonconsumptive users of wildlife.

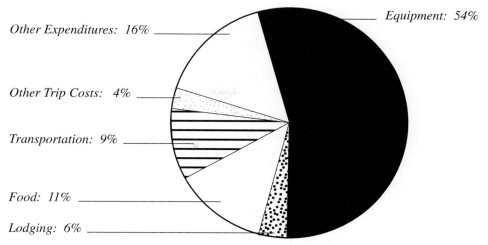

Equipment: 54%

Other Expenditures: 16%

Other Trip Costs: 4%

Transportation: 9%

Food: 11%

Lodging: 6%

Figure 14–5 Nonconsumptive users of wildlife contributed a great deal to the economy in 1996.

guides and outfitters use the data to lead hunters in the fall and provide nonconsumptive activities at other times of the year. The data also indicate how far people travel for activities, which helps agencies provide hunting, fishing, and nonconsumptive opportunities at the appropriate places.

Kellert Report. A survey conducted in the late 1970s by Stephen Kellert of Yale University under contract with the U.S. Fish and Wildlife Service is probably the most comprehensive wildlife survey ever undertaken.[3–6] In this survey, designed to provide data on public attitudes toward wildlife and wildlife-related activities, respondents were divided into consumptive and nonconsumptive groups. The consumptive group included hunters, trappers, fishing enthusiasts, falconers, ranchers, and those affected by property damage caused by wildlife. The nonconsumptive users included bird-watchers, photographers, backpackers, and those who visited zoos and museums, enjoyed wildlife photography, owned pets, or liked to ride horseback.

The Kellert Report showed an overwhelming public support for all wildlife issues, both game and nongame. It indicated that a high proportion of the public wanted funds to be provided for wildlife. For example, 89 percent of the public wanted to support and protect the bald eagle. Sixty-four percent of the public also voiced strong support for the protection of butterfly species.

The public as a whole seemed concerned about wildlife. Most of the people interviewed indicated an interest in maintaining wildlife even at the expense of development or higher taxes. Private-interest groups were sometimes exceptions—landowners favoring predator control, for instance. The general public, however—including the hunter and fishing enthusiast—was not as aware of major wildlife issues as were many of the conservation groups.

Responsive Management Report. In 1997, nearly 20 years after the Kellert Report, the private organization Responsive Management was operating to assist wildlife and fisheries agencies in collecting data on fish and wildlife issues and to publish information on public attitudes toward wildlife. Responsive Management conducted telephone surveys throughout the country. The company found that 8 out of 10 Americans wanted to use tax dollars to save endangered species. The company's results indicated that 43 percent strongly supported using tax dollars, 36 percent moderately supported the idea, 7 percent moderately opposed the idea, and 9 percent strongly opposed using tax money to save endangered species. Five percent of the public had no opinion. The data showed that 75 percent of the American population supported hunting, while nearly 95 percent supported fishing, as recreational sport.[7]

The data collected showed a big difference in public attitudes between animal welfare and animal rights. Animal welfare people advocated the humane treatment and responsible care of animals, ensuring freedom from unnecessary pain and suffering. Animal rights advocates generally did not want animals used for any purpose. They should not be eaten, hunted, used for pets, or kept in zoos. According to the survey, 79 percent of the American public felt that it was permissible to use animals to benefit humans if the animal did not suffer undue pain (Figure 14–6). About 15 percent of the public held an animal rights point of view.[7]

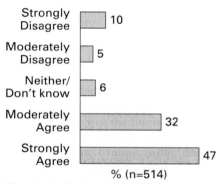

Do you agree or disagree that animals can be used by humans as long as the animal does not experience undue pain?

Figure 14–6 People feel that animals can be used by humans, as long as the animals do not experience undue pain (*Responsive Management Report*).

Responsive Management provided information on the public's perception of the Florida panther, an endangered cougar (Figure 14–7). The Florida panther survey showed that 90 percent of the public knew of the animal and 58 percent know that the population was endangered. Only 14 percent knew that the population numbered less than 50. Forty-four percent of the public's perception was gained from newspapers and 36 percent from television. When respondents were asked about the effort to support the panther and save the population from extinction, 91 percent of the public supported the idea, while 2 percent opposed it. With this showing of strong public support, Florida agencies were able to transform public opinion into actions to conserve the panther. Agency personnel realized that they needed to continue to provide information to the public via the news media in order to maintain the high level of support.[8]

Figure 14–7 Florida panther (Courtesy of the U.S. Fish and Wildlife Service.)

THE PRESS AND PUBLIC OPINION

Public goals are determined, to some extent, by individual experiences and background. Press coverage can also be influential. A public relations office that supplies the press with good wildlife articles educates the public and helps formulate public opinion. It is desirable, therefore, to have a good writer or two on the agency staff. Wildlife managers have not always realized how important good press coverage can be. Many managers have been inclined to leave the writing to journalists, but it is the informed agency person who can write authoritative copy.

Major news stories have been more difficult to control. Whenever an issue concerning endangered species has arisen, newspapers have rewritten the story so many times that the real issue often has been lost. In many cases, the Endangered Species Act is ignored. The snail darter/Tellico Dam issue in Tennessee was an example. While the dam was being constructed, this small endangered fish was found in one of the streams that would be flooded. News stories discussed people, money, and politician's idea's; they did not refer either to the Endangered Species Act or the life history or biology of the fish. Reauthorization of the Endangered Species Act took the same approach. Politicians who knew nothing about biology or the act itself were interviewed. The information that was obtained was quite distorted.

Press coverage of any story varies, based on the approach of the news company and reporters. Coverage also differs from the concerns expressed in letters to the editor and articles in the outdoor or editorial pages. Now, with more than five television channels broadcasting news on a 24-hour basis nationally, reporters are looking for information anywhere and everywhere. Stories must be made appealing to viewers, which sometimes means that the public is getting an inaccurate story (Figure 14–8).

Figure 14–8 Spotted owls are a major topic of discussion with the public. They have been declared threatened in the Pacific northwest and are the cause of a controversy between those wanting to log on public land and those who want to preserve the owls' habitat. Spotted owls require old-growth forests for nesting and foraging. These forests take many hundreds of years to regenerate. (Courtesy of the U.S. Fish and Wildlife Service.)

INFERRING PUBLIC ATTITUDES FROM WILDLIFE-RESOURCE USE

Attitudes toward wildlife in the United States have been influenced by lifestyle, national leadership, and wildlife use. Early settlers and explorers were attracted to regions with an abundance of wildlife, since they had come to America from countries where the human population was increasing, meat was scarce, grain was the principal food, firewood was no longer plentiful, landownership was only for the privileged, and wildlife was the property of the crown. The first arrivals in America could scarcely believe the abundance of food, wood, water, and space. Deer, turkeys, fish, and shellfish were there for the taking, as were rabbit, opossums, raccoons, squirrel, pigeons, and even bullfrogs. Sage grouse could be knocked down with a stick. Bison and elk roamed the eastern woodland edges, and out on the prairies were pronghorn to provide everything people needed.[1]

Early attitudes resulted, in part, from seeing the dependency of native Americans on wildlife for their food, clothing, ornaments, weapons, and other necessities. The early arrivals were strongly influenced by Indian attitudes toward native flora and fauna as means of survival. but the settlers also saw an opportunity to earn cash as the fur-trading companies moved in. What with the normal desire for gain and the seemingly unending abundance of wildlife around them, it is small wonder that people did not always recognize the need to preserve natural resources. Attitudes were also influenced by predatory animal attacks on livestock and competition between domestic and wild animals.

Nevertheless, many of our early colonial and national leaders helped form an awareness of the need for conservation as they provided the public with direction in, and opportunities for, wildlife experiences. The rapid decline of some species, such as the passenger pigeon, added to the impetus toward wildlife conservation. Thus, leadership in formulating wildlife goals was not entirely lacking.

As mentioned in Chapter 1, attitudes toward wildlife fluctuated during our country's history. Public concern about wildlife has, however, burgeoned, particularly in the last half of this century. In the spring of 1970, colleges and conservation organizations celebrated the first Earth Day, which focused on pollution, population, and conservation. It is no exaggeration to say that celebrations of this sort have grown into a movement that has had a major impact on the nation's attitude toward wildlife and even on legislation. In addition, people who use wildlife generate more interest in it. Thus, we find somewhat unexpected bedfellows, such as sports groups and private conservation agencies, pulling together many people with an interest in wildlife preservation, however diverse their motivation. Many factors, then, shape people's attitudes toward wildlife. Managers who are aware of this fact can use it to develop methods of promoting beneficial wildlife programs.

THE VALUE OF WILDLIFE

The most crass value of wildlife is its *consumptive* value—the food it puts on our table or the dollars in our pocket, as in the fur trade and commercial fisheries. Harvesting wildlife is still a good-sized business, and there is reason to believe that a wider range of ocean products will become an even more important part of the world's diet—as they are in Japan today.

Human beings once hunted and fished from pure necessity, and although it must have seemed like hard work then—as everything we *have* to do seems—there must have been fun in making the catch, and that fun must have entered into the cultural heritage. We call it sport. When John and Amy Smith climb out of bed at 4:00 A.M. and head for a trout stream, it is not because the refrigerator is empty. They may eat the catch or give it to their neighbors, but that is not why they went after it in the first place.

There is also an *aesthetic* value in wildlife. For some people, the *emotional* content of their reaction to wildlife is very real. They value wildlife resources, even though they do not use them directly (Figure 14–9). To these people, just knowing that wildlife exists is important. This attitude is quite common among people in urban areas. Surveys show that many people in large cities are very interested in wildlife issues and are concerned about the maintenance of wildlife species they never see in the species' natural habitat (Figure 14–10). Since the existence value of wildlife is independent of its current use, values must be assigned on the basis of a consideration of future generations or concern for the ecosystem.[9]

However great our benefits from wildlife, it is difficult to put a dollar value on the most important of them. Wildlife is a part of the natural system, and what figure should we put beside that item on the balance sheet? What do we debit when reduction or destruction of a species causes unanticipated damage to that system?

Figure 14–9 Some people hunt rabbit. Others just enjoy seeing the animals around. (Courtesy of the Wyoming Game and Fish Department; photo by L. R. Parker.)

Public's response to statement, "More tax money should be spent on programs to increase wildlife in urban areas."

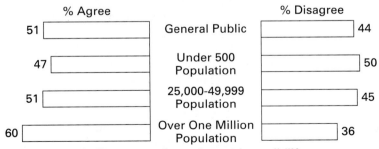

Figure 14–10 Response of people to urban wildlife programs.

How Much Does an Animal Cost?

This question poses a real dilemma for wildlife biologists when they are asked to compare the dollar value of any animal, such as a turtle, mouse, butterfly, or fish, with some form of activity that provides income or jobs for people. Not only are biologists asked to reduce wild animals to a dollar-and-cents value, but they must also use human criteria for the evaluation. By insisting that owls, rabbits, and snakes must generate money, people are really saying animals are valueless if they do not provide income for some people. Is this really a wise way to decide whether wildlife has a value to the planet? Maybe it is a warped way of looking at life in general, since all life, including the human race, is dependent on other life in one way or another.

Putting dollar values on animals and human experiences with them seems shortsighted. It is hard for most people, particularly politicians, to understand how an ecosystem functions and that there is survival value in have a functioning system with wildlife. If we try to put values on wild animals, maybe we also need to put a value on climbing a mountain or viewing a lake. It could be that some things have intrinsic value and cannot have a human dollar value assigned.

Methods of Evaluating Wildlife

We have just noted the impossibility of assigning a quantitative value to wildlife. Still, a number of methods have been designed to measure that value, including the *direct-expenditure method, market value of harvest, cost, unit day value,* and *willingness to pay.* These methods can be of some use to the wildlife manager, not for their accuracy, but for their thrust.

Direct Expenditures. The direct-expenditure method evaluates wildlife activities according to the total amount spent by participants in wildlife-related activities. The information is obtained by questionnaires concerning how much money was spent on food, lodging, travel, equipment, and license fees. The method is based on the assumption that the value of wildlife-related activity is indicated by the amount spent on enjoyment. When an entire trip is planned to hunt, fish, or bird-watch, the total amount of direct expenses can be used. But when wildlife activities are only part of a trip, it is difficult to measure the wildlife dollars. What part of the cost of a motel room on a family's vacation is an expenditure for wildlife?

Market Value. Computation according to market value measures the dollar profit from wildlife. For example, the salmon industry speaks of its net profit, and big-game or bird tours compile a profit (or loss) statement for each year. There are shortcomings in this method. For instance, the profits do not take into account the costs to others of maintaining the habitat, as when outfitters do not include the cost of land-management agencies or road maintenance in computing their profits. Similarly, fishing, evaluated by the market value of the fish crop, does not include the wide range of benefits associated with the activity, but assumes that the value of the harvest is the sole benefit. For many species harvested for sport, a commercial market may not exist.[1]

Cost: Unit Day Value. In the cost equation, the value of wildlife resources is equal to the cost of developing and maintaining it. One cost analysis, used by the Water Resource Council, is the unit day method, which combines direct expenditures and market value. Experts' opinions and judgments are used to estimate the value of a day's activities. The value is typically based on the commercial figures assigned to various wildlife-related activities. Thus, the value of a commercial catch of salmon is equated with that of recreational salmon or deep-sea fishing. Among the problems with this method are that (1) uniform values are not necessarily appropriate for both commercial and public wildlife resource use; (2) the values assigned are quite arbitrary; and (3) even though subject to adjustment, the values frequently do not adequately reflect variations in the quality of the recreational experience.

Willingness to Pay. A survey of how much people say they are willing to pay before they stop participating in a wildlife activity should give a monetary evaluation. The trouble with this technique is that respondents may say one thing but do another.[10] So more indirect methods are used to reach the willing-to-pay valuation. For example, the number of animals taken during the hunting season can be related to travel and material expenses. Or the number of people using guides and trappers can be an indication. A number of biases come up when these types of studies are made. People return to their favorite sites not only because they have found wildlife there, but because they enjoy the area. There is some evidence that people who enjoy particular areas are apt to make annual reservations or even buy land there and build their own places to stay.[11] Thus, subjective considerations reduce the monetary reliability of the technique. In short, there appears to be no clear-cut monetary method of evaluating wildlife. This complicates the problem of setting public goals in relation to the use of habitats by wildlife.

Kellert Results. In a 1980 survey to determine what people might be willing to pay for wildlife resources Stephen Kellert found that most people felt that users should be taxed.[4] For example, 82 percent of those questioned felt that a special tax should be levied on fur clothing made from wildlife animals. Entrance fees to wildlife refuges and public wildlife areas should be charged, according to 75 percent of the people, although only 54 percent felt that a sales tax on bird-watching equipment was appropriate. Kellert found that overall, support was very high for greater tax revenues for wildlife management.

FUNDING FOR WILDLIFE

Current funding for wildlife protection comes from a number of sources, the most productive of which are recreational hunters and fishing enthusiasts. A number of state agencies receive virtually all of their funding from hunting and fishing license fees. Since NEPA became effective, there has been a considerable increase in the amount of work done by state game and fish agencies, including reviewing impact statements. Yet the funding base for most of these agencies has not increased as a result of NEPA.

At the federal level, the concept of taxing select groups of wildlife users originated in the 1920s, when a waterfowl-refuge system was partially financed by fees that each waterfowl hunter had to pay for a stamp, sold at the local post office, to be affixed to the hunting license. This stamp is now quite popular.

Major forms of funding for wildlife-related work developed in the 1930s. In 1937, the Pittman–Robertson Act for wildlife restoration levied a 10-percent (later, 11-percent) manufacturers' excise tax on sporting arms and ammunition to support wildlife restoration work by state wildlife agencies. The funds are collected by the federal government and turned over to each state in proportion to its size and population. States can use the funds for appropriate projects involving wildlife, land acquisition, development, and research on a 75-percent federal–25-percent state cost-sharing basis. The funds are apportioned on the basis of states' area (50 percent) and number of license holders (50 percent). Each state is required to guarantee that the revenues will be used for wildlife conservation.

There have been several changes to the original Pittman–Robertson Act. In 1970, an amendment directed that the existing federal excise tax on pistols and revolvers be paid into the Wildlife Restoration Fund and that half the revenues from this source be apportioned to the states for hunter-safety programs, including the construction, operation, and maintenance of outdoor target ranges. Since fiscal year 1975, this special "fund within a fund" has included half the federal tax on bows and arrows. The money is apportioned among states on the basis of population and, at the discretion of the states, may be used for traditional wildlife restoration projects rather than for hunter-safety programs.[10]

The Dingell–Johnson Act, passed in 1950, is structurally similar to the Pittman–Robertson Act, but taxes fishing equipment to provide funds for fish restoration. The funds from federal excise taxes on fishing rods, creels, reels, artificial lures, baits, and flies are apportioned annually among the states on the basis of their area and number of paid fishing licenses. The states can use these funds for fish restoration and management projects. Since 1970, the money has been used for comprehensive fish- and wildlife-management programs. In 1984, the Dingell–Johnson Act was amended with the Wallop–Breaux Act. This act expanded the tax base to include a tax on additional fishing gear, the sale of small boats, and fuel for motors in boats.

Unlike Pittman–Robertson and Dingell–Johnson, the Land and Water Conservation Fund Act of 1964 generates funds from a wide variety of sources and applies them to a broad range of both state and federal recreational programs. The fund was originally supplied by user fees, proceeds from the disposal of surplus federal property, and the federal motorboat fuel tax. In 1968, an amendment provided that enough unappropriated Treasury funds and miscellaneous receipts from the Outer Continental Shelf Lands Act be added to guarantee an annual income of at least $200 million ($300 million as of fiscal year 1971).[12] Sixty percent of the annual appropriation from the fund is made available to the states for planning, acquisition, and development of needed land and water areas and facilities; the remaining 40 percent remains with the federal government for land acquisition and development. Two-fifths of the funds available for the states is divided equally among them, and the rest is apportioned on the basis of need, as determined by the secretary of the interior.

The Land and Water Conservation Fund Act generates and distributes revenues for outdoor recreational purposes, many of which substantially benefit wildlife, directly or indirectly. Together, the Pittman–Robertson Act, Dingell–Johnson Act, and Land and Water Conservation Fund Act provide for what are known as federal aid funds.

In 1983, federal aid funds were used by Ohio to study yellow perch in Lake Erie. New Mexico undertook a study designed to develop harvest recommendations for big game, such as Barbary sheep, javelinas, and wild turkeys. Connecticut used federal aid monies to acquire 421 acres of upland wildlife habitat. Virginia built boating access areas for fishermen, and Alaska studied salmon and trout distribution and migration patterns. Both the Pittman–Robertson and Dingell–Johnson funds are handled by the federal aid office in the U.S. Fish and Wildlife Service. This office receives money, evaluates state lands, and distributes the money to each state.

A change introduced by the 1970 amendment to federal aid programs gave the states the option of submitting a comprehensive fish and wildlife resource management

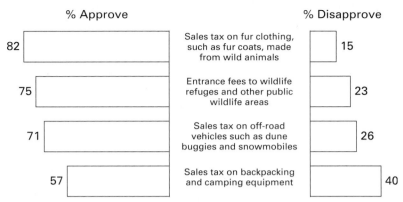

Figure 14–11 It is particularly difficult to obtain funds to manage nongame species.

plan or traditional individual restoration projects. The basic requirements are stated in very general terms. The plans must cover a period of at least 5 years, must be based on the projected needs of the people for at least 15 years, and must include provisions for updating every 3 years. Beyond that, the only substantial standard imposed by this change was that comprehensive fish and wildlife plans must ensure the perpetuation of these resources for the economic, scientific, and recreational use of the people. No procedural requirement to involve the public in developing the plans was imposed.

In 1979, Congress passed an act similar to Pittman–Robertson and Dingell–Johnson to provide funds for the management of nongame species. The act was to set up trial programs in several states. After debates concerning taxes on such items as binoculars and bird seed, Congress decided that the funds should be appropriated from the general revenues. Approximately $5 million was to be set aside in 1980 (Figure 14–11). Then budget cuts eliminated the funds. Still, the plans are to implement the program if and when funding becomes available. Other means of nongame funding are discussed in Chapter 20.

Teaming with Wildlife

Funding for nongame and endangered species has been a dilemma in wildlife management. In the late 1990s, a bill was introduced into Congress to improve funding for these species. Called Teaming with Wildlife, the bill would create a trust fund for state-level fish and wildlife conservation, recreation, and educational programs for nonconsumptive species. This funding proposal would be similar to the Pittman–Robertson and Dingell–Johnson/Wallop–Breaux federal aid for wildlife and sport fish programs. A user excise tax fee on a sliding scale of between 1/4 and 5 percent on the manufacturer's price of outdoor-related products such as binoculars, field guides, tents, and other camping gear would be imposed. Money would be returned to the states using an allocation formula based on land area and population. The bill would require a 25-percent, 75-percent state–federal match.

Teaming with Wildlife is an example of how more funding might be generated to support nonconsumptive wildlife programs. It is an indication that the public is interested in providing support for the wildlife management of all species.

MANAGEMENT OF PUBLIC AND PRIVATE LANDS

There is a tendency for people to feel that they should have access to land where wildlife is found, but landowners, not surprisingly, disagree. Many landowners have had bad experience with public invasion of their property, including people harming domestic animals, habitats, and fences; leaving gates open; and littering. The result is conflict between the people and landowners. There is no easy resolution of the conflict, although the economic system may provide some answers.

One nagging issue is how to compensate private landowners for such wildlife activities as hunting, fishing, and bird-watching on their land. Should the tax system supply funds to pay the owners? Should the owners be required to maintain the wildlife as part of the natural system and let the public come onto their land without paying for the privilege? The problem is particularly complex in many states with limited public land. For example, in virtually all states east of the Great Plains, less than 20 percent of the land is public property. Several states, including Maine and Iowa, have almost no federal landholdings. This contrasts with the western states, which, with the exception of Hawaii, have at least 30 percent federal land. In any event, there probably will be no solution to the problem of funding wildlife until a satisfactory formula for compensating private landowners is devised.[1]

Generally, private landowners have little direct incentive to manage wildlife or to preserve habitat, although there are some indirect incentives. For example, along the Chesapeake Bay, the abundance of waterfowl serves to increase property values.[10] Preservation, however, can also be an incentive in itself: When oil spills occur or wetlands are lost, property values decline. Some states allow private landowners to receive a portion of the state license fees by sending in license stubs from people who hunt on their land, and user fees for hunting and fishing may be charged. In some states, landowners receive compensation for damage caused to their land or livestock by game animals. In Texas, where there is very little public land, some landowners have charged leasing fees for the right to hunt on rangelands, the primary game being white-tailed deer. The owners have used a good portion of the fees to enhance deer habitat, thereby improving hunting for the lessee.[13]

These returns are hardly enough to advance any substantial wildlife-support efforts. How, then, can landowners be encouraged to maintain a diversity of habitat for wildlife species? There are some national programs run through the U.S. Department of Agriculture; for example, funds are available to private landowners to maintain woodlots for wildlife, for which the owners can also receive property tax reductions. But these programs are subject to political pressure and, often, budget cutting. With the vast private landholdings in our country, our wildlife-management programs must find better ways to create wildlife habitat on those lands. Tax reductions, economic benefits, and other incentives must be greatly increased, so that owners will find it feasible to maintain wildlife habitats and thereby contribute to a diverse ecosystem.

SUMMARY

Wildlife management is simply management of our wildlife resources in accordance with the desires of the public. These desires are often difficult to determine, because the public is such a diverse group. Public meetings and surveys are two methods used to determine what the public wants.

National and local surveys have been helpful in finding out how people view wildlife and what they expect wildlife managers to do in managing resources. Surveys have also been useful in assessing the public's attitude on funding wildlife management.

Currently, many state wildlife-management agencies receive funds from license fees. Some get funds from general revenues. Federal aid programs have assisted states in obtaining funds for fish and wildlife programs.

There has been much interest in pinpointing "values" of wildlife, partly for assistance in getting finances. Most methods involve an assessment of how much people pay to participate in wildlife activities or how much commercial operations pay to harvest wildlife.

Management of wildlife on private lands is generally at the owners' option. Few methods exist to compensate landowners adequately for maintaining wildlife habitats. As a result, conflicts between agency wildlife managers and landowners occur.

DISCUSSION QUESTIONS

1. Discuss the pros and cons of using a telephone survey to formulate wildlife goals.
2. What must be considered in developing a good survey?
3. How can wildlife values be compared with energy values?
4. Should wildlife have a price tag? Explain.
5. Should wildlife management be dictated by state and federal agencies for both private and public land? Give both sides of the argument.
6. Can you provide more equitable wildlife user charges than current state funding based mostly on hunting and fishing licenses?
7. Should wildlife management goals be based on public opinion?
8. Should the federal government manage all wildlife in the country? Why or why not?
9. How can more money be put into nongame management?
10. Do you feel that recent surveys of public opinion indicate that wildlife management will change course in the next 10 years? If so, how?

LITERATURE CITED

1. Brokaw, H. P. (Ed.). 1978. *Wildlife in America.* Washington, DC: Council on Environmental Quality.
2. Anon. 1997. *1996 National Survey of Fishing, Hunting, and Wildlife Associated Recreation.* Washington, DC: U.S. Fish and Wildlife Service.
3. Kellert, S. R. 1979. *Public Attitudes toward Critical Wildlife and Natural Habitat Issues.* Washington, DC: U.S. Fish and Wildlife Service.
4. Kellert, S. R. 1980. *Activities of the American Public Relating to Animals.* Washington, DC: U.S. Fish and Wildlife Service.
5. Kellert, S. R., and J. K. Berry. 1980. *Knowledge, Affection, and Basic Attitudes toward Animals in American Society.* Washington, DC: U.S. Fish and Wildlife Service.
6. Kellert, S. R. 1981. *Trends in Animal Use and Perception in 20th Century America.* Washington, DC: U.S. Fish and Wildlife Service.
7. Duda, M. D., and K. C. Young. 1997. Americans' Attitude toward Animal Rights, Animal Welfare, and use of Animals. *Responsive Management Report.* Harrisonburg, VA: Responsive Management Corp.

8. Duda, M. D., and K. C. Young. 1997. *Floridians Knowledge, Opinion and Attitudes toward Panther Habitat and Panther Reintroduction.* Report prepared for Florida Advisory Council on Environmental Education. Harrisonburg, VA: Responsive Management Corp.

9. Randall, A., and G. L. Peterson. 1984. The Valuation of Wildland Benefits: An Overview. In G. L. Peterson and A. Randall (Eds.), *Valuation of Wildland Resource Benefits,* pp. 1–52. Boulder, CO: Westview Press.

10. Kellert, S. R. 1981. Wildlife and the Private Landowner. In R. T. Dunke, G. V. Burger, and J. R. March (Eds.), *Wildlife Management on Private Lands.* La Crosse, WI: Wisconsin Chapter of the Wildlife Society.

11. Charbonneau, J. J., and M. J. Jay. 1978. Determinants and Economic Values of Hunting and Fishing. *Proceedings, 43rd North American Wildlife Conference,* pp. 391–403. Washington, DC: Wildlife Management Institute.

12. Bean, M. J. 1983. *The Evolution of National Wildlife Law.* Washington, DC: Praeger.

13. Applegate, J. E. 1981. Landowner's Behavior in Dealing with Wildlife Values. In R. T. Dunke, G. V. Burger, and J. R. March (Eds.), *Wildlife Management on Private Lands.* La Crosse, WI: Wisconsin Chapter of the Wildlife Society.

Part 5
MANAGEMENT APPLICATIONS

Deer have been the focus of many management activities, both for people who enjoy sports and for viewers of wildlife (Courtesy of the Wyoming Game and Fish Department; photo by L. R. Parker.)

In this part of the text, groups of animals and their needs are discussed in each chapter, along with management options. Special groups like endangered species or animals that are thought by people to cause damage make up individual chapters. We end the section with a chapter on managing wildlife nationally and internationally, using the information given in previous chapters. Keep in mind that the information in this part is based on information from Parts 2–4. Also, it is impossible to apply all the principles of management without being aware of some of the history and background of wildlife management and conservation given in Part 1.

Most of the chapters in Part 5 are based on people's interests, such as big game, waterfowl, and nongame species. We continue to stress that people are the driving force in wildlife management. Your educational background must be related to people, because the public needs to be educated and heard in order to have an effective wildlife management program.

15
BIG GAME

Many people associate wildlife management with big game. Perhaps this is so because, historically, changes in big-game populations have dictated changes in wildlife-management policies. Big game are affected to a tremendous extent by people: In both the United States and Canada, the movement of people westward so devastated big game that now they must be maintained by selective management techniques.

The enormous changes in big-game populations can be illustrated by the history of the bighorn sheep between 1800 and 1970, during which time population fluctuations on the eastern slope of the Canadian Rockies and the western slope in Kootenay National Park, British Columbia, were recorded. Early in the period, bighorn sheep underwent only sporadic fluctuations from severe winters, disease, and changes in habitat caused by weather, fire, and intraspecific competition. But beginning about 1860, thousands of railway builders, miners, traders, settlers, and resident Indians using firearms reduced the sheep from an original population of more than 10,000 to 2,600, by heavy, indiscriminate hunting.

Between 1910 and 1915, as part of an extensive preservation program, a 19,425-km² (7,500-sq-mi) area was closed to hunting. That was the beginning of Canadian national parks. During those years, hunting was also restricted in other areas. These restrictions, together with improved range conditions, resulted in a tripling of the bighorn population over the next 20 years, to an estimated 8,500 by 1936. Then, between 1937 and 1949, a series of die-offs cut the population back to 2,500. The heavy mortality was attributed to pneumonia, lungworm disease, a deteriorated range, heavy competition from elk and livestock, decreased grassland caused by forest succession, and three severe winters. A planned reduction in the number of elk and sheep then allowed the range to improve, so that by the summer of 1960, the bighorn population had gradually increased to 10,100. During the fall and winter of 1966–1967, it again declined by 75 percent in the national parks, because of a deterioration of the sheep's winter range due to grazing, lungworm disease, and a severe winter. However, populations east of Kootenay National Park continued to increase, despite internal parasites and a deterioration of their winter range.[1] Future population levels are expected to be determined by lungworm disease, harvests, and range-carrying capacities, which in turn will be affected by forest succession and human-made habitat modifications.

This account of the bighorn indicates how changes can occur in a big-game population. While early records show that the population was far from stable, the influx of human beings created even greater fluctuations. Although conditions such as climate cannot be controlled, management practices can alter succession, competition, and, to some extent, disease to reduce extreme fluctuations.

Today, most big-game animals in North America are managed for hunting or as a part of wildlife aesthetics. Some people in the United States hunt for outdoor recreation, food, and, sometimes, trophies. Others travel to national parks just to see the animals. Big-game management in the United States contrasts with that in Europe, where animals are managed largely on private lands and in public and private preserves. In Europe, the animals are generally smaller and caught less frequently. The hunt is more like a day in the country than an extensive trip.

In this chapter, we discuss some biological principles and describe the life patterns of various species. We examine census techniques and different habitat-management criteria. We look at hunting as a management tool in relation to big game and big-game diseases.

PEOPLE AND GAME ANIMALS

Big game can be divided into three major groups. The first consists of *ungulates,* also called herbivores or grazing animals, such as moose, elk, antelope, deer, sheep, and goats (Figure 15–1). Most of these animals are found in areas where they can forage on

Figure 15–1 Mountain goats are ungulates. (Courtesy of the Wyoming Game and Fish Department.)

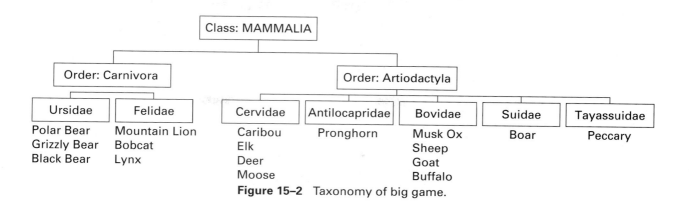

Figure 15–2 Taxonomy of big game.

low tree branches and shrubs or grass. The second group is the *carnivores:* mountain lions, bobcats, and some bear. Most of these animals live in forests or savannalike areas, where they can utilize the scattered trees for shelter, but also move freely. The third category, *omnivores,* consists primarily of bear, which eat both plants and animals and are frequent visitors to garbage dumps. Taxonomically, big game fall chiefly into two orders (Figure 15–2).

A big-game experience means different things to different people. Hunting for big game satisfies a number of desires. Some hunters seek big game for meat, others for

---

How Do Elk Respond to Roads?

Managers are always in need of information on wildlife species to determine how they are going to be affected by human impacts on the habitat. Elk require patches of forest cover for hiding and resting during the day. Some mountainous areas are subject to timber cutting which requires building roads.

In the Big Horn Mountains of Wyoming, biologists assessed how additional roads affected elk. By placing radio collars on 107 elk, the researchers were able to follow the movement of the animals and the habitat they used throughout the year. Generally, elk responded to roads if there was traffic, so no distinction was made between major and secondary roads. Based on the locations of the elk, the biologists developed a linear model comparing the number of miles of road with the quality of the elk habitat. The more elk locations found, the better was the quality of the habitat. Habitat quality was ranked from 1, which meant that there were many elk locations, so that the habitat was very good, to 0, which meant that they had few locations with elk, and was therefore very poor. The results showed that use of the habitat by elk was high when there were no roads. Use decreased very rapidly as roads increased. The quality of the habitat decreased to 0.5 as the density of roads increased to 0.3 per square mile. The quality went down to 0.2 as the road density reached 0.5 per square mile. These results showed how roads drastically reduced the quality of the elk habitat. Managers have used this information to improve habitats by closing roads in areas once timber removal is completed.

Sawyer, H. H. 1997. Evaluation of a Summer Elk Model and Sexual Segregation of Elk in the Bighorn Mountains, Wyoming. Master's thesis, University of Wyoming.

---

trophies. Some people simply like to hunt; still others want to be outdoors with friends or enjoy the hiking and wilderness and seeing the animals. These experiences may be combined, of course, in many different ways. Some people spend the day hunting within 100 miles of their homes; others spend a week or two vacationing in public or private hunting areas; still others make a major expedition of hunting, hiring outfitters and guides to get into remote areas. Many people travel long distances with their families primarily to see and photograph large animals. They look on big game as an exciting experience, just as the hunter does.

The wildlife manager is the person largely responsible for providing the big game for harvest or aesthetics for all these people. Managers may at times wish they had to deal only with the animals and their habitat, but they must remember that wildlife funds come from a wide variety of people, all of whom have their own ideas of what big-game hunting is and of their right to engage in it.

People, for their part, must understand that the behavior of many big-game animals changes as a result of contact with human beings. Normally, animals will adopt a flight response when danger, including that posed by people, appears. Since big-game species are tourist attractions, people stop to look at and photograph them. Many animals have thus become somewhat accustomed to humans and will tolerate photographers and sightseers at a distance. Unfortunately, most people, particularly in national parks, do not respect the space required by the animals. They find out too late, when buffalo kick, moose gore, or bear bite, that the animals are indeed wild. Managers have the job of trying to educate the public with respect to wildlife (Figure 15–3).

Figure 15–3 Wildlife managers educate the public to respect the space of wild animals such as bison. The map shows the original distribution of bison in North America. (From E. R. Hall, *The Mammals of North America.* New York: John Wiley & Sons.)

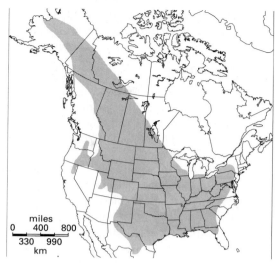

Figure 15–3 (continued)

ASPECTS OF ANIMAL BIOLOGY

Biology and life history patterns differ to some extent between artiodactyla (grazers and browsers) and carnivores.[2] (Figure 15–2). Some of the reproductive biology of these two orders is discussed under the descriptions of species that follow. Life-history data provide a great deal of information relevant to management: Decisions about hunting times, places, and numbers that can be removed all depend on life-history data. The type of herd desired (for example, for maximum reproduction or for trophies) is attainable only with this background information. Life history is pertinent in management determinations to increase or decrease the size of a herd. Field identification of these animals and their remains rests on a variety of characteristics, including skull and dental configurations. Knowing such characteristics helps determine the animals' movement patterns and habitat use. In the dental formulas shown in Table 15–1, numbers of incisors, canines, premolars, and molars on the upper and lower jaws are listed

TABLE 15–1

Dental Formulas for Selected Big Game

Type of Game	Dental Formula
Black bear	3142
	3143
Mountain lion	3131
	3121
White-tailed deer	0033
	3133
Bighorn sheep	0033
	3133

according to dental pattern. Since big game are normally bilaterally symmetrical, the number of teeth on only one side of the jaw is shown.

Movement

Big game move about for a number of reasons. Young animals move to set up their own territories, either displacing other animals or settling in unoccupied areas. Some species are nomadic by nature—that is, they move from area to area in their range in response to habitat conditions. Others move in cyclic patterns in response to a change in available food.

   Dispersal.   The young of the big-game species, particularly ungulates, disperse in two major ways. First, species that are solitary or territorial, such as bear and mountain lions, force their young out to colonize new areas, recolonize old abandoned areas, or displace other individuals. In the process, the young are vulnerable to hunting, predation, and shortage of food, water, and cover. Thus, the young often have the highest mortality rate in the population. They move varying distances, depending on habitat, human activity, and species. For mountain lions, dispersal appears to be a major population-regulating mechanism: The young are forced out of the limited habitat unless they can displace adults already present. In Idaho, young mountain lions were killed more than 160 km (100 mi) from the area in which they were observed with their mothers.[3] Normally, young lions are not able to carve out and defend a territory in an area where older males exist, unless one of the older males is killed or dies from disease (Figure 15–4). Territorial boundaries are then redefined by the new occupants. A disturbance of the mountain lions' habitat through human activity will usually result in changes in social structure and subsequent dispersal. Mountain lions whose territories are disrupted sometimes become predators of livestock.

   The second type of dispersal is found in more social animals. Here, the young are brought into the structure of the group. Young female antelope join the herd to feed and

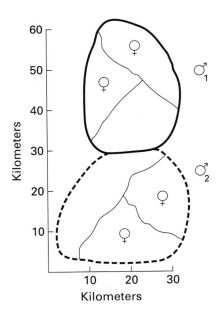

Figure 15–4 Area defended by a mountain lion.

migrate. Young males form isolated groups during the rutting season and become increasingly isolated during this period as they get older.

In populations in which females are forced into a harem and protected by a dominant male, other males form isolated groups until they can displace one of the dominant males. Some animals protect territories to keep their harems. When management decides to change the size of such populations, the nature of the territories must be understood.[4]

Managers can use this behavioral information in developing plans for maintaining species. Species whose young disperse can more easily colonize appropriate new habitats. Animals with strong herd behavior are likely to be slow to colonize new areas.

Migration. Many species of big game migrate. In North America, sheep, elk, moose, antelope, and some deer migrate between summer and winter ranges. An area is occupied each year by the same herd of animals, and disruption of either the area or the migration can cause major changes in the social structure and population dynamics of the group.

Caribou and reindeer are also migratory; however, they do not appear as attached to one winter range. This may be, in part, because of climatic differences. In Africa, some big-game species migrate in response to moisture. The zebra, wildebeest, and Thomson's gazelle all exhibit this migratory behavior. They tend to move toward areas where moisture is higher during the drier part of the year.

Big-game migration may involve movement to other specialized habitats. For example, some animals move from one water source to another. Special birthing areas are used by some species, such as the caribou and elk.[4] Rutting areas, which may be specialized or a component of the winter range, are important to moose and bighorn sheep.

Although many ungulates are migratory, there are exceptions within a migratory species. For example, migratory elk herds are generally found in mountainous regions, where they are able to move vertically in response to seasonal changes. Ranges in lower elevations often have less snow during the winter, while ranges at higher elevations, up to the timberline, are often snow covered, so that the elk cannot winter there. Thus, the distance traveled by different elk herds between winter and summer ranges varies considerably in different geographic areas.[5]

There is also a great difference in individual movement patterns. Elk that winter in the Jackson Hole area are from separate herds.[6] Five of those herds migrate, while a portion of the sixth does not. This herd is able to remain at an altitude of approximately 7,000 ft because of thermal springs and warm water that reduce snow cover in the critical area. Some big game in Arizona and New Mexico are able to remain in the area because of the relatively mild winters there.[7] Accordingly, weather conditions and food supplies appear to be major factors influencing the migratory behavior or movement of many big-game species.

Introduction to New Habitat. When a big-game species enters a new habitat or recolonizes a former habitat, its numbers generally increase rapidly. It appears that moose arrived on Isle Royale National Park, Michigan, early in the 20th century, probably by swimming during the summer or walking on ice in the winter.[8] The population increased to approximately 5,000 by 1930 (Figure 15–5). Habitat destruction between 1930 and 1934 caused the population to decrease to about 500. In 1949, wolves appeared probably by walking across ice from neighboring land. They have become the principal predator of the moose, keeping the population within the limits of the habitat's carrying capacity.

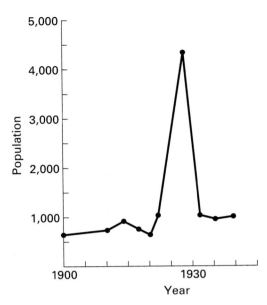

Figure 15–5 Population fluctuation of moose on Isle Royale.

The introduction of reindeer on St. Matthew Island, where 29 animals multiplied to 6000 between 1944 and 1963, is another example of the eruptive nature of big-game populations when they are introduced into areas without established population control mechanisms.[9] In the case of the reindeer, food finally acted as the limiting factor that caused the population to crash.

A study of the thar, a goatlike ungulate of northern India liberated in New Zealand, indicated that food was responsible for changes in the population density.[10] Biologists found that the population erupted and destroyed forage. Here, a balance was achieved with the available food, and the population fluctuated with changes in the food supply.

These results indicate the type of growth that can occur when a big-game species appears in a new environment. Without natural checks, the population generally increases to such an extent that the habitat is severely affected. (See Chapter 3.) Many years may pass before a balance is established between animal and habitat. Equally severe effects can occur if entire populations are reduced to very low levels or removed from a habitat.

Habitat

Big-game populations maintain their size and minimize their impact on a habitat through migration. Different types of habitat support different groups of big game. For example, wooded areas with edge appeal to bear, ungulates are found in the edge and grassland, mountain lions in open forests, and sheep on mountain slopes.

Habitats provide not only food and cover, but space as well. A minimum area is needed for each group of animals. (See Chapter 8.) Unfortunately, there are no definite figures for each big-game species, since miniimum habitat sizes vary with region, use of the adjacent habitat, and climate.

Often winter and summer habitats are different for big game, necessitating migration between these areas. Sometimes refuges are necessary for the species' survival,

Are Mountain Lions Becoming Less Wary of People?

A number of press releases have been issued in the recent past regarding conflicts between people and mountain lions. In the western United States, encounters or sightings of mountain lions, result in injury or death, are becoming increasingly common. Why is this happening? Typically, young mountain lions disperse and move to new sites to establish territories. Males in particular need to find an area not occupied by another male. They wander across increasingly larger expanse of territory as food becomes scarce. People who build homes in the foothills and into the wilderness are likely to encounter some of these animals as they disperse. Considering the wilderness areas that many people now occupy, it should not surprise us that these encounters occur. "The astonishing thing about our American lion," according to *Audubon*, "is not that it has been known to have killed a dozen humans since 1890, but that it is known to only have killed a dozen human" (p. 34).

Williams, T. 1994. The Lion's Silent Return. *Audubon* 96:28–35.

particularly if harassment is high. In a study in Illinois, investigators found that large blocks of forest were used as refuges in the winter so that successive generations of white-tailed deer could live long enough to use the sites of intensively farmed, dispersed woodlands. Deer used forest tracts of less than 100 hectares (247 acres) when more than 50 percent of the forest offered refuge protection. Forests greater than 400 hectares (1,790 acres) provided sufficient escape cover from harassment by hunters to shelter deer in the winter.[11]

Big game are an important component of the food chain; they convert plants to flesh that can be used for food by human beings and other animals. Mountain lions, although often criticized for their attacks on domestic livestock, are still important, in that they keep ungulate population in check. People also like to know that mountain lions are out there.

Habitat management requires a knowledge of all habitat needed to sustain the animals present. Habitats must be maintained if they are to provide food and shelter; when exploited, they can cause animals to starve or fall victim to disease.

FIELD TECHNIQUES

Counting

Wildlife managers frequently need to know how many big-game species and individuals are present. There are a number of ways to count animals. (See Chapter 4.) Obviously, the best is to make a total count. This is possible when animals are confined to a specific location, where a count can be made on the spot or from a photograph. When the animals move past a particular location at a certain time, such as when they are crossing a road or transmission corridor, total counts are possible. Usually, however, an estimate is made. With big game on open range, aerial counts are possible. Moose counts on Isle Royale and antelope counts in the west are commonly made with aerial surveys. Each area or species requires a different approach.[12]

Determining Movement

Biologists are able to record big-game movement by using various forms of radiotelemetry and satellite. Telemetry units attached to several animals help identify important sites for big game at different times of the year (Figure 15–6). Telemetry enables the researchers to obtain data concerning an animal's movement, dispersal, and range at a distance from the subject, thereby reducing stress to the animal. Radio waves usually carry this information, although sound and light are sometimes used.

By placing radio-signal-emitting devices on animals, biologists can locate the animals by means of a receiver (Figure 15–7). Movement patterns, territorial behavior, and habitat use can be determined through the use of telemetry data (Figure 15–8). Often, the most accurate location of an animal can be obtained by triangulation, using two receivers. Radio signals can be received by handheld receivers and receivers on trucks and in aircraft.

Other uses of telemetry are now being evaluated. Sound waves can be used to detect pregnancy in mule deer in the field without removing the does from the population.[13] The returning waves detect changes in tissue density. The most effective period (with 100-percent accuracy) for testing is from about 60 to 120 days' gestation. From 100 to 117 days' gestation, the number of fetuses can be counted. Temperature-sensitive transmitters can be implanted to monitor body temperatures of free-roaming grizzly bear.[14] The monitored heart and pulse rates of deer allow biologists to draw conclusions about the physiological condition and behavior of the animals.[15] Big game that have died in inaccessible places have been located by telemetry units.

The effects of radio transmitters on behavior, reproduction success, survivability, and movement have been studied in numerous big-game species. Early investigators generally found negative effects, but as techniques for attachment improved and units became lighter, these negative effects were no longer detected. Most researchers reported that animals required a three- to seven-day period of adjustment to collars before their behavior returned to normal. Mixed results are reported on the use of

Figure 15–6 A culvert trap baited with garbage is used to capture bear. (Courtesy R. Grogan.)

Figure 15–7 Placing a transmitter on an antelope. (Courtesy D. Inkley.)

telemetry on the young. Radio-collared white-tailed fawns were immediately accepted by their mothers.[16] However, biologists found female mule deer reluctant to accept their fawns for several days after radio collars were attached.[17]

In comparing the effectiveness of visual and telemetry monitoring, biologists found that nighttime habitats could be determined better with radiotelemetry.[18] They also found that deer selected significantly different habitats at night than during the day. In a study of radio-collared deer, biologists found differences in hourly movement within the day and differences in movement within the hour during different seasons.[19]

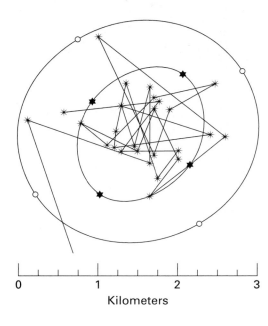

Figure 15–8 Data from field observations of mule deer with a telemetry unit. (Courtesy A. Reeve.)

The capability of telemetry to monitor physiological conditions, day and night habitat selection, and movement patterns without disturbing the animals makes telemetry a valuable tool for field biologists. Data on movement tell the manager where the animals are at various times. Decisions about the type of hunt and removal of different age groups rest on knowledge of movement patterns. Blocked migration routes resulting from roadways or construction can cause major changes in a big-game population.

Determining Age

Managers often need a reliable technique to determine the age of members of big-game species. One method commonly used is the cementum annuli technique.[20] The dark-staining annual layers present on the cementum of some animals' teeth are effective signs of aging (Figure 15–9). The first incisors are used for estimating the age of ungulates, whereas canine teeth are generally used to calculate the age of carnivores. Teeth are usually removed from the bodies of dead animals and transported to laboratories, where they are sectioned, placed on a microscope slide, and stained.

The major annuli appear as narrow, dark-stained bands separated by broad, lesser-stained bands. The age of the mule deer is one year more than the number of dark annuli in the first permanent incisor. Aging of other species must be done with a knowledge of the replacement chronology of the tooth being used.

When a sectioned tooth is viewed through a microscope, the annuli can be clearly seen and counted (Figure 15–9). However, all of the annuli in any one section should be examined carefully and counted, since there may be areas where reabsorption and

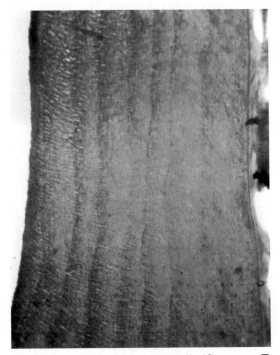

Figure 15–9 Cementum annuli. (Courtesy T. Moore, Wyoming Game and Fish Department.)

tooth repair have occurred, resulting in few annuli in that spot. The annuli are the result of slower growth rates of the cementum. This usually occurs during the winter; however, the rate of cementum deposit may be reduced during a rut, and it is not uncommon to find false annuli in males as a result.[21]

MANAGEMENT

Information concerning life-history patterns can be used in managing the populations and habitats of big game. Once the goals are established, the size of a population can be manipulated in consonance with habitat, habitat use, hunting, and other mortality factors.

Carrying Capacity

One of the important jobs in managing big game is estimating a habitat's carrying capacity, knowledge about which is essential in determining optimal yield. The carrying capacity, as we have noted, is the number of healthy animals that can be maintained by a habitat on a given unit of land.[22] For big-game herds that migrate, the smaller carrying capacity of either the summer or the winter habitat is what limits the population.

Carrying capacity can change from year to year because of changes in climatic conditions and the availability of food. Thus, there is no simple way to manage for carrying capacity, but there are a number of factors that can be watched as indicators of change. A large winter die-off usually means that the population is exceeding the habitat's carrying capacity. Most important is the availability of food: The type and quantity of food required for each animal and how much is available without destroying the area can be calculated. The formula to use is[22]

$$A = \frac{B \times C}{D}$$

where A = number of animal-days an area can support
B = food resources, in grams of available food in the area
C = amount of metabolizable energy contained in the food
D = amount of metabolizable food energy required by each animal per day

After determining the size of a population by one of the census or surveying techniques, biologists can use the preceding formula to figure out how many animals should be harvested on the unit of land. The formula also allows a better evaluation of management efforts to increase or maintain carrying capacity.

There are so many generalities, however, in any formula for carrying capacity that it is important to look carefully at the special characteristics of each area. Sometimes simplified models, such as ONE POP (see Chapter 5), can be used to indicate the number of animals that can be harvested.

Habitat Improvement

Managers are often asked to develop a habitat-maintenance or habitat-enhancement program for big-game species. Some animals, such as mountain lions, do not do well when there is an influx of people, so that habitat improvement means simply increasing isolated areas. In contrast, many ungulate species can be maintained even when large numbers of people come to the area.

Range improvement, which affects all species in an area, can be achieved by chemical, fire, mechanical, and biological means. (See Chapter 7.) The conversion of a dense stand of trees to a single type of vegetation can be beneficial to wildlife when doing so creates more edge. It is well documented that edge habitat is used by many types of wildlife more than its share of the landscape would lead one to expect.

Burning and mechanical control can remove woody vegetation and promote greater grass cover. Results of these methods on wildlife are similar to those obtained from spraying herbicides. Mule-deer forage increases when some trees are left; indeed, the diversity of wildlife generally remains high when strips of the original vegetation are retained. Burning the forest opens areas and increases edge, thereby increasing food for bear.[23]

Range sites are fertilized only when the increased production of forage pays for the cost of fertilization in the first year. A significant increase in elk use resulted on nitrogen-fertilized sites the first year, but there was no significant carryover in the second.[24]

Proper management of grazing continues to be the best and most cost-effective way to improve range conditions. Such management benefits wildlife by ensuring adequate food, cover, and water. Year-round grazing is the most detrimental to wildlife, since the understory vegetation then has no period of rest. All yearlong grazing is continuous, but not all continuous grazing is year long: The grazing period may be continuous only for a season of the year. When cattle are forced to turn to browse as the more desirable vegetation is grazed out, their dietary overlap with deer increases.

Competition between Big Game and Domestic Cattle

Competition for forage between wild ungulates and domestic livestock is a complex, controversial, and continuing problem. Some species, such as antelope and cattle, seek different plant life. But there is evidence that ungulate populations are limited by weather[25], hunting[26], the quality and quantity of forage[27], and livestock grazing.[28] It is likely that populations are limited by a combination of factors. Biologists working with wild horses point out that, for competition to exist, horses and cattle must overlap spatially, must be using a resource that is in short supply, and must reduce each other's population and ultimately, fitness below levels that would prevail in the absence of the other.[29]

Because of the social values of wildlife, the public-land administrator is obliged to allocate range forage for the support of big game. Prior use by a single-resource user (for example a rancher allowed to graze cattle on public land) should not obscure the fact that public needs, being of paramount importance, should be supplied first. The single-resource user must be made to understand that his or hers is a secondary right, to be exercised only as long as it does not conflict with primary needs.[30] It has become clear, with the Multiple Use and Sustained Yield Act of 1960 and similar legislation in 1964 and 1976, that wildlife should receive at least equal consideration with livestock in the management of public lands.

Allocation of forage has traditionally taken two forms. One, the *animal-unit-month* (AUM) method, consists of modifying procedures used for domestic livestock to allocate forage for wildlife. The other method, involving an analysis of key areas and species, consists of marking and measuring key plant species in a critical range and then deciding how those plants shall be used.

An AUM is the amount of forage required by an animal unit for one month of grazing. Animal units are an attempt to compare how different species of animals use the range. AUMs are determined by a combination of factors, such as the density, palatability, production, and proper use of forage plants. In converting livestock AUM figures to those for wildlife, range managers using the AUM method of making forage allocations take into consideration differences in body weight, surface area, and amount of dry forage consumed per day between livestock and big-game animals. Based on body weight alone, 9.62 antelope, 5.82 deer, or 1.88 elk consume forage equivalent to that consumed by one cow. Standard equivalents, based on a 435-kg (1,000-lb) cow, are as follows: cow, 1 AUM; bull, 1.25 AUM; yearling steer or cow, 0.60 AUM; two-year-old steer or cow, 0.80 AUM; mature horse, 1.25 AUM; sheep, 0.20 AUM; lamb, 0.15 AUM; white-tailed deer, 0.15 AUM; mule deer, 0.20 AUM; antelope, 0.10 AUM.

The AUM method has several shortcomings. The first and most important springs from the fact that a cow is not a deer: Straight weight conversions fail to allow for body weight–surface area relationships, different metabolic rates, different food habits, and different feeding-site selectivity. Other difficulties in using AUM conversions include a lack of information on forage requirements for wildlife species and uncertainty in estimating wildlife populations. Nor do AUM conversions always recognize the major biological and environmental factors that regulate wild populations, such as the quality of forage, weather patterns, predators, and age structure.[31] Wyoming and Nevada use the AUM system. In Wyoming, the system is modified slightly; the number of livestock AUMs is multiplied by a seasonal wildlife–livestock competition percentage and divided by the conversion figure derived from body weight. Other factors, such as the impact of fences and availability of water, are also taken into consideration.[32]

In newer methods of forage allocation, the amount of nutrition in the available forage supply is divided by the forage requirement of the species. The quotient is the number of animals that can be supported over a unit of time.[32] These methods, grouped as "optimal forage allocation methods," use the following information: delineation of winter-range boundaries and dates of use for each big-game species, quantity of big game forage present, nutritional quality of the forage and its digestibility and crude protein content, availability of forage during the winter, and winter nutritional requirements of the species.

Such a system has problems, too. One is the cost of collecting so much data over a number of years. Another is the variability of all data sets: A severe winter may cause animals to remain longer on winter range and cause an earlier starting date on that range; the quantity, quality, and availability of forage are also heavily dependent on the weather. Finally, the nutritional requirements of a big-game animal are related to the stress on the animal, and this varies from day to day and season to season. Incidentally, computer simulation is now being used to implement similar methods.[31]

In view of the increasing numbers of big game on the western range and the practice of multiple use of land, management must be well thought out although it need not be difficult. The nutritional requirement of the individual animal is not the factor that needs the most attention; the most important thing is what the range can carry without loss of quality. Trend data already present, along with site productivity, will tell the manager if too many animals are on the range, and adjustments can be made accordingly. A good rule to follow is to plan as if every year will be the worst on record; then an upward adjustment can easily by made if conditions warrant. Finally, utilization of 40 to 50 percent of the forage available ordinarily will leave enough forage and cover for wildlife use and vigorous plant growth.

Bison and Elk in Yellowstone National Park

Yellowstone National Park is the nation's first national park created for the benefit and enjoyment of the people and for the preservation and retention of its resources in their natural state. The park is approximately 2.2 million acres and part of the Greater Yellowstone ecosystem, which is approximately 20 million acres. In the northern part of the park, there are 16,000 to 20,000 elk. About 3,500 bison live in the park most of the year. Bison and elk carry brucellosis, which is transmitted among animals primarily through exposure to infective reproductive material. For this reason, cattlemen do not want infected bison or elk grazing in the same pasture with their cattle. As herds of bison and elk increase in the northern part of Yellowstone, they move off parklands and onto cattle-grazing areas. No one knows the reasons the bison are moving. Some say that overgrazing is a cause, while others think that snowmobile trails in the park provide easy avenues for bison, in particular, to move easily. Conservation and some native American groups oppose any form of bison control, while ranching groups want controls. If cattle contact brucellosis, they must be slaughtered, since they cannot be transported with the disease. When people contact the disease, it is called undulating fever.

Working groups have developed a management plan to keep the number of bison and elk within a prescribed population limit. The plan calls for removing any bison that wanders out of the park if it tests positively for brucellosis. This action has caused much debate and focus from the news media. The controversy will undoubtedly continue and even escalate as people reduce natural wildlife habitat to suit their own purposes.

Hunting and Management

We come now to the question of how hunting is used to manage big game, or how big game are managed for hunting. Common goals in big-game management include *maximum production of animals, sex-ratio manipulation, production of trophy animals,* and *population manipulation.* Reaching each of these objectives requires analyzing the size of a population, as well as its age composition, sex ratio, and reproductive biology. To help reach goals through hunting, managers can manipulate the hunting season and quotas and use such tools as winter feeding, disease control, and habitat manipulation.

Maximum Production.   Managers may want to maximize the production of big-game animals in areas where there are many people wanting to hunt. Highly desirable species or animals near metropolitan areas are examples. To achieve maximum production, managers need to determine what number of males and females in the population can produce the largest number of offspring—that is, what density of animals provides the highest net reproduction (Figure 15–10). This determination can be made with the assistance of the logistic growth equation.[33] (See Chapter 3.) Habitat conditions must enter into this calculation. For instance, year-to-year differences in the availability of food may alter the values annually. Decisions on maximum production of herds can best be made in the light of long-term sampling data. For maximum production, animals that do not contribute significantly to the net reproduction must be removed. This means harvesting males at a higher rate than females in *polygamous* animals.

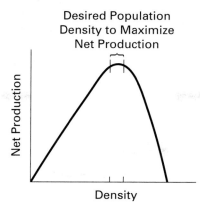

Desired Population
Density to Maximize
Net Production

Net Production

Density

Figure 15–10 Optimal population size to maximize net reproduction. (From Ref. 33.)

In elk populations, hunting regulations have brought about maximum production when access to the elk is not difficult and there are many hunters. Hunting seasons, which vary annually according to the size of the population, could include an "antler only" period or hunts for antlerless elk, if appropriate, to reach the right population mix for maximum production.[33]

Sex-Ratio Manipulation.   Managers often find that a certain ratio of males to females is desirable in maintaining a desired population size. This may be the size that biologists estimate can be supported by the habitat or that minimizes damage claims while maintaining the big-game species. A higher male–female ratio generally reduces population growth.

Harvesting both sexes generally results in a satisfactory sex ratio. In ungulates with antlers, early hunting pressure often favors males. Later hunts for elk or other species that migrate often result in the removal of more females. Removal of males or females can be further controlled by the number and types of permits issued.

Trophy Animals.   Some game and fish agencies provide hunting of elk, moose, deer, or bighorn sheep for trophies. Trophy animals, being older, require larger ranges than females or yearlings, so that the number of animals the habitat can support is reduced.

Two methods are used to produce trophy animals. The first is restricting the take to males with a certain characteristic, such as a four-point antler in bull elk or three-quarter curl in bighorn sheep. The second is to reduce hunter pressure by limiting the number of permits through a quota-drawing system. Even though most hunters say they prefer the lower probability of taking a trophy animal to the higher probability of taking a nontrophy animal, management for trophy animals is not common. However, rough topography and dense habitat often results in de facto trophy management for those willing to spend the time and effort necessary to take an animal from such areas.

Population Manipulation.   Managers may want to increase or decrease the size of a population while maintaining a hunting experience. This goal can be reached through controlling the length of the hunting season, limiting the number and/or increasing the cost of permits, and manipulating the sex ratio of a population.

When an increase in population is desired, a limited number of permits to take males may be issued during a restricted season and/or in an area. This limitation allows the female ratio to increase in polygamous populations. Sometimes, permits issued by the season are closed when a specified number of animals have been removed. A reduction in the size of a population can be achieved by extended or special seasons. Permits can be sold at reduced prices for females and young, and areas that are difficult for hunters to access can be improved.

Seasons and Regulation.

It is often difficult to determine how to set seasons for harvesting big-game species so as to give the hunter the best experience possible. For example, the largest portion of the elk harvest occurs during opening weekend of the elk season.[34] This has been so for controlled hunts in Arizona and general hunts in Colorado and other Rocky Mountain states. Regardless of the hunting season, the hunting pattern is consistent, with opening weekend accounting for 49 to 67 percent of the total harvest. Not surprisingly, highly accessible areas are used more by hunters than are less accessible areas. This information can be documented using the percentage composition of big game harvested by area.[33] It is important to distribute hunters in order to provide a satisfactory hunting experience. Some areas open the season on Monday, to allow local people to hunt before visitors arrive on the weekend. Selective hunting permits for each herd is another device for distributing hunters. Some states divide permits between in-state and out-of-state hunters, thereby controlling access to an area and reducing the number of hunters in and about urban areas.

Harvest Assessment.

An effective management program for big-game species entails some form of harvest assessment. By *harvest assessment,* we mean a follow-up to find how animals were taken, their age and sex, and how successful each hunter was. Information on takes is often entered into a computer so that agencies can determine the percentage taken from each herd. Ways of following up include field checks and sample surveys—that is, sending survey forms to people who have received permits. (Often, people will not answer these questions or will answer inaccurately.) Some states provide data-gathering cards, which must be returned by the hunter at the end of the hunting season. The information requested on the cards may include the number of days spent hunting, locations of hunts, amount of game harvested, and sex and age of the animals harvested. It is common to have check stations, with agency personnel in attendance, at main exits from hunting areas. This, of course, is one of the most accurate ways of getting information on takes.

Disease.

Disease takes a heavy toll on many game species. Most game and fish agencies use some criteria to evaluate and minimize the impact of disease on big-game species. Wyoming, for example, annually traps a number of bighorn sheep to collect information on disease. All trapped animals are given inoculations to prevent the spread of disease, as well as to remove those in excess of the estimated carrying capacity of the habitat.

Diseases that attack big-game species are carried by viruses, bacteria, parasitic protozoa (one-celled organisms), flatworms, roundworms, and ectoparasites (parasites on the external body). All either introduce toxic material that interferes with the animal's normal physiological processes or drain the animal's energy, making it susceptible to other disease-causing organisms or increasing the chance of death by predation.

Field managers can reduce the incidence of disease by normal management procedures. Preventing overcrowding and maintaining an adequate habitat lessens the probability that disease will spread. A winter feeding program in time of severe snow helps animals keep up their energy, but is a very controversial subject. A successful winter feeding program must include food of nutritional value, must be carried out during the entire winter, and is very expensive. Furthermore, disease can be spread when the animals share pasture or come in contact with domestic animals. Removal of diseased domestic livestock and of habitats that harbor disease vectors (transmitters) is an effective management tool.

Managers should be able to recognize the indicators of disease. A weak, inactive antelope, deer, or sheep that does not respond normally to people, particularly in late summer and early fall, can be a victim of *blue tongue*, an insect-carried virus transmitted by livestock. If livestock owners are experiencing outbreaks, managers should be particularly alert. Acutely ill animals must be removed to prevent infection of others. The assistance of veterinarians is necessary.

Brucellosis is a bacterial disease that affects many mammals, including people. Common in cattle in some parts of the country, it causes abortion in the second half of pregnancy.[35] It appears to be contracted by the animal's licking infected material. When there are excessive abortions, particularly in the later stages of pregnancy, brucellosis should be suspected (Figure 15–11). Some cattle ranchers' associations have set up a national program to control the disease through immunization, which is difficult, although possible, in game species.

In zoos and game ranches, attempts are made to protect animals from disease. Some biologists feel that protecting animals from disease may be weakening the population. The lack of genetic variability in cheetahs is cited as a possible reason for the spread of feline infectious peritonitis in the population. Some people believe that if an endemic infection exists in the host population, the population may become more resistant to the disease.[36]

Diagnosing a disease often requires laboratory facilities not immediately available to the field manager. Wildlife veterinarians must be called to assist when disease is suspected. While field managers cannot ordinarily be diagnosticians, they can learn

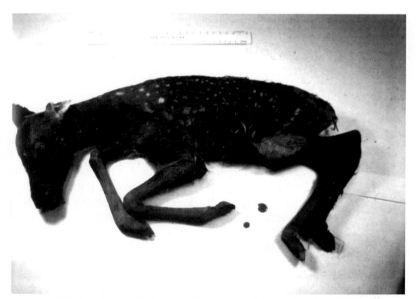

Figure 15–11 Elk calf aborted because of brucellosis. (Courtesy T. Thorne, Wyoming Game and Fish Department.)

to detect abnormal behavior. They should learn to recognize major symptoms and should understand the basics of disease transmission.

Wildlife sometimes act as carriers of diseases that can infect people. Fleas found on many species of mammals have long been known as carriers of plague. During the 1980s, ticks living on deer were found to carry a bacterium that causes Lyme disease. The tick, which has a two-year life cycle, uses intermediate hosts, including songbirds and mammals. Adult ticks obtain the bacteria from the blood of animals, particularly deer and small rodents that are infected. When the tick moves to a new host, including humans, and obtains a meal of blood, it injects bacteria into the host.[37] In people, early detection and treatment with antibiotics can be successful. Unfortunately, the disease can mimic many others and be difficult to detect.

DESCRIPTIONS OF SPECIES

Deer

In the United States, there are two native species of deer: mule and white tailed. (Some taxonomists feel that some subspecies are really species.) The black-tailed deer, which is common on the Pacific slopes, is now considered a subspecies of the mule deer. They can interbreed, but occupy, with some exceptions, different geographic areas. Their habitat preferences are similar: Both occupy mixed forests, forest edges, and foothills in the western United States.

The white-tailed deer overlaps with the mule deer in parts of the west and extends throughout the eastern United States. It is found in open field, brushy areas, and open wood. Land-clearing practices in the eastern United States have substantially increased the habitat for this deer.

White-tailed deer, which biologists divide into as many as 30 subspecies in North America, mate in the fall. The doe is in heat for only about 24 hours during several 28-day periods. The gestation period is about 200 days; thus, most fawns are born in late spring. Does can breed during their first year, but commonly do not until their second year. The breeding season is called the rut. Fawns remain bedded in an isolated area during their first days and move only short distances during their first few months.

Some records indicate that these deer can live 20 years; however, bucks taken by sport hunters are usually not older than 4 1/2 years, and many are only 2 1/2. Deer are the major big-game species sought by hunters in the United States. In much of the eastern United States, the white-tailed deer is the only big game hunted. These deer are the sole big game seen by most people.

Deer eat a variety of vegetation: woody plants, small bushes, trees, shrubs, and some ground plants, although usually not grasses. In the east, white-tailed deer migrate if they live in areas where food is less available in the winter. In other areas, the herds are nonmigratory. Mule deer have both migratory and nonmigratory population. Management generally involves maintaining habitats through clearing and keeping the population in balance with the availability of food.

Elk (Wapiti)

Elk (Figure 15–12) generally summer in high, open mountain pastures of the northwestern United States, New Mexico, and Arizona. During the winter they move into lower, wooded forests. Most states with elk populations have a hunting season that, although it may last only a few days or weeks, usually brings a good deal of money in to outfitters,

Figure 15–12 Elk. (Courtesy of the U. S. Soils Conservation Service.)

motels, and restaurants in the affected areas. Poaching, market hunting, and changing land-use practices caused a major decline in the elk population during the 19th and early 20th centuries, but the population today is rather stable, thanks to hunting quotas.

Elk generally begin their rutting season in mid-August and continue until November, although the period varies somewhat in different regions of the country. Conception ranges from September 15 to November 4, with about 75 percent coming between September 26 and October 10. Elk calves are born in early June and join the herd with the mother several weeks later. Cows are very protective of their young during the first year of life.[38] Males are sexually mature by 15 months, but they rarely breed at that early age. Yearling cows generally do not have young. Elk have a relatively complex social structure: Bulls set up harems, but older cows also play a leadership role.[38] The species has a low birthrate, with normally only one birth per female at a time. Elk generally do not live much beyond nine years, and most males die by their seventh year.[38]

Management of elk habitat means making a source of food available. Heavy winter mortality generally indicates too many elk for the range. Since most elk migrate between summer and winter ranges, the migrating route must be kept open. Loss of winter range because of human activity or mineral exploration can be a problem.

Caribou

With long, inward-curving antlers, these majestic animals are found in the northern part of the hemisphere. They prefer tundra and taiga and get into the coniferous forests of mountain regions, where lichens grow. They are considered to be of the same species as the reindeer of Europe and Asia. Caribou are an important source of food for many people in the far north.

Breeding occurs in the latter half of October. Caribou produce one offspring at a time, starting when they are three or four years old. Largely because of wolf predation in much of their range, a high proportion (sometimes 95 percent) of the young do not survive their first year.

Caribou migrate long distances in groups of as many as 10,000. They move into the tundra and sedge meadows in the spring and summer, during which time they eat any vegetable and some parts of plants. For the fall and winter, they move into open coniferous forests and windswept mountains, where lichens are available under the snow. Management involves keeping hunting and predation in balance. Maintaining areas where lichens are available under snow cover is also important.

Moose

This large ungulate is found in spruce forests, swamps, aspen groves, and willow habitats of North America. The Rocky Mountains, Great Lakes states, New England, and Alaska constitute its range in the United States.

The rut occurs in the fall of the year. Most cows are in heat only 24 hours, with intervals of 20 to 22 days, and have a single calf each year between their 4th and 10th years. Twins do occur.

The moose usually moves among several small home ranges during the year. Unlike some other members of the deer family, moose are solitary animals. They are the object of heavy wolf predation in some parts of their range.

Since moose prefer willow or early-succession habitats, often near water, habitat can be a factor limiting their increase. Activities such as logging make these habitats less suitable because of water pollution and fire. However, logging, which sets back succession in a forest, can be beneficial. Management therefore involves habitat maintenance, together with harvest and predation control.

Pronghorn

Pronghorn antelope are swift animals of the Rocky Mountain, Great Basin, and southwestern states. They are found in grassland and brush habitats, often where sage grows.

Antelope are an important game species in many western states. Their use of open habitat makes hunting success as high as 90 percent. Although some people do not feel that antelope hunting is a true hunting experience, many take advantage of this opportunity to obtain a big-game species. The antelope population grows rapidly if habitat is available. Most females breed during their second year. They mate in the fall, when bucks try to obtain a harem of does. This period lasts from two to three weeks. Does give birth to one or two fawns in the spring, at which times females are usually isolated. It is not uncommon for most females in a herd to give birth within a few days of each other. Females and young remain in the herd through the summer. Fawns often fall prey to coyotes, bobcats, and golden eagles.

During the late fall and early winter, antelope move to winter ranges, usually windblown ridges where sage is available. In the spring, they move to summer ranges to dine on the early grass shoots. Sage becomes an important food in the dry summer. Browse, forbs, and grasses are important in some parts of the antelope range.

Management of pronghorn involves the prevention of barriers, such as interstate highways, that block seasonal movement and hinder maintenance of winter habitat. Harvests must be geared to keep antelope numbers within the habitat's carrying capacity.

Bighorn Sheep

Because they inhabit remote mountain areas, bighorn sheep are seldom seen by people. They are found in the Rocky Mountain and southwest states, in alpine regions near or on rocky cliffs (Figure 15–13).

Bighorns are known for their fall rut, in which rams have butting contests, which can be heard for miles. Photographers like to catch these sessions. The mating season

Figure 15–13 Bighorn sheep. (Courtesy E. Arnett.)

comes between late summer and January, varying with latitude. Ewes give birth to a single lamb, beginning in their third year. The lamb is well protected during its first few weeks. Yearlings, ewes, and lambs live together in the summer. In the fall, the rams rejoin the herd, having formed a separate herd in the summer. Bighorn sheep migrate to high mountain meadows in the summer and return to traditional valleys in the winter. The winter range often becomes a limited habitat when people have utilized some of the area for development.

Sheep are very susceptible to the lungworm parasite, which spreads rapidly in the large, concentrated winter flocks. The parasite has an especially adverse impact on the sheep population when a harsh winter limits the availability of food. Sheep are, for the most part, a trophy animal: rams are highly prized by many hunters.

Management of bighorn sheep involves maintenance of the population within the carrying capacity of the habitat, primarily the winter range. Access to water can be an important survival factor, particularly in the arid southwest. Competition from cattle has become a limiting factor in some parts of the range.

Mountain Lion

The mountain lion, a large cat, sometimes called a cougar, is usually found in steep mountain areas with cliffs and scattered openings in trees. Its range is the western United States, the Rocky Mountains, and a few isolated areas of the Gulf coast and south.

Mountain lions are sought by trophy hunters. The most common method of hunting is by using dogs, which can pick up the scent of a lion, follow it, and tree it. Then shooting a lion is easy; the problem is getting in and out of lion country (Figure 15–14).

Mountain lions can have their first litter as early as 20 to 21 months of age, although most females do not reach sexual maturity until 30 months. A lion is in estrus from 4 to 12 days, with a 14-day interval between cycles, until conception occurs. Females can resume the estrus cycles immediately following the birth of a litter, but they sometimes do not do so for nearly a year. The gestation period is only 90 to

Figure 15–14 Mountain lion. (Courtesy of the U. S. Bureau of Land Management.)

96 days, and litters range from one to six kittens. Mountain lions can have litters in any month of the year, but spring is the most common season.[39]

When they are one or two years old, the young leave the mother to set up a territory. Males will defend an area of 65 to 90 km² (25 to 35 sq mi) against other males. Young males may be killed if they come into an area defended by an older male. Each male activity area can contain from two to four smaller female activity areas. Mountain lions rarely live beyond 12 years. Adults weigh 34 to 125 kg (75 to 275 lb).

Mountain lion management generally has been associated with predator programs, because the lions prey on domestic sheep and cattle. When ranchers use high-mountain areas to graze their animals, the attacks become more frequent. (See Chapter 22.) Biologists recognize that some behavior traits are involved in the attacks on domestic animals: Apparently, some lions are more likely than others to attack sheep and cattle. Generally, however, domestic cattle make up less than 10 percent of the mountain lion's diet. Habitat management, harvest quotas, and removal of predators are the primary techniques to use for this species.

Bear

Most taxonomists agree that there are three species of bear in North America: the black bear, the grizzly bear, and the polar bear. All are found in the United States. The black bear—which has several color variations, including brown—is the most common. It has been forced from much of its range by the influx of people. In the eastern United States it is found in deciduous and hardwood forests, as well as around swamplands. Western black bear, some of which are cinnamon colored, live in forests, wooded hills, and mountains.

Like polar bear and grizzly bear, black bear breed in April or May, depending on latitude. The cubs of the black bear are born from November to February, while the fe-

Figure 15–15 Grizzly bear. (Courtesy of the National Park Service.)

males are in a den. A female usually has two or three cubs, but larger litters are known. The cubs usually stay with the mother for a year and a half.

Except for females with cubs, black bear usually travel alone. Congregations do occur, however, around garbage dumps. Although they seek high-protein foods, bear eat a wide variety of items, so that the type of food is probably not a limiting factor.

Black bear move around their range during the season, seeking vegetation with a high moisture content. In the summer, if precipitation is scant and shortages of fruit occur[40], they will feed on domestic sheep, small mammals, and insects. In the winter, they will take elk or moose.

Management of the black bear revolves around the harvest, which is used both to keep the bear within the limit of its habitat and to reduce predator damage. Grizzly bear (Figure 15–15), a threatened species, require a large expanse of wilderness. The closing of Yellowstone Park garbage dumps in recent years has reduced the food supply for grizzlies.

SUMMARY

When people say "wildlife," they usually mean big-game animals: moose, elk, antelope, deer, sheep, goats, mountain lions, and bear. These animals are harvested by hunters and viewed for aesthetic reasons. Understanding the biology, life history, and habitat needs of the animals is important in developing management programs. To gain insight into big-game populations, managers must often gather data through censusing.

Competition between wild ungulates and livestock is commonly discussed, but seldom sufficiently documented. Most wildlife use different plants or parts of the range than livestock use. Managers need to develop range integration patterns to maintain the best range for all animals. Some use the animal unit month (AUM) method, which determines animal equivalencies based on the density, palatability, and production of forage plants and on use features of the range. Other forage allocation methods put greater emphasis on nutrients.

Hunting is a management tool. Management objectives that can be furthered by hunting include maximum production of game animals, sex-ratio manipulation, harvesting for trophies, and manipulation of the size of a population.

All managers need to be informed concerning the diseases of big game. By controlling disease, managers can make more big game available for hunting, which, in reality, is a desirable alternative in population control.

DISCUSSION QUESTIONS

1. How accurate should big-game censuses be?
2. What role does disease play in big-game management?
3. Discuss some of the uses of radiotelemetry in managing big-game species.
4. How should wildlife managers determine carrying capacity for big game? How should the concept be used?
5. Compare the impact of managers' work on species whose young disperse with that on species whose young are accepted into the social structure. With which group is it easier to increase the population? For which group is habitat manipulation more important?
6. Why do we need to understand the behavior and social structure of big game to manage these animals?
7. How are harvest regulations used in managing big game?
8. What is an AUM? How is it used?
9. Compare management functions in maximizing production and hunting for trophies.
10. Describe different movement patterns among big game, and give reasons for their occurrence.

LITERATURE CITED

1. Stelfox, J. G. 1971. Big Horn Sheep on the Canadian Rockies: A History, 1800–1970. *Canadian Field Naturalist* 85:101–21.
2. Gilbert, D. L. 1978. Evolution and Taxonomy. In J. L. Schmidt and D. L. Gilbert (Eds.), *Big Game of North America: Ecology and Management,* pp. 1–10. Harrisburg, PA: Stackpole Books.
3. Hornocker, M. G. 1970. *An Analysis of Mountain Lion Predation upon Mule Deer and Elk in the Idaho Primitive Area.* Wildlife Monograph 21. Bethesda, MD: Wildlife Society.
4. Cowan, I. M. 1974. Management Implications of Behavior in the Large Herbivorous Mammals. In V. Geist and F. Walter (Eds.), *The Behavior of Ungulates and Its Relation to Management.* Morges, Switzerland: International Union for Conservation of Nature and Natural Resources.
5. Adams, A. W. 1982. Migration. In J. W. Thomas and D. E. Toweill (Eds.), *Elk of North America: Ecology and Management.* Harrisburg, PA: Stackpole Books.
6. Craighead, J. J., G. Atwell, and B. W. O'Gara. 1972. *Elk Migrations in and near Yellowstone National Park.* Wildlife Monograph 29. Bethesda, MD: Wildlife Society.

7. Short, H. L., W. Evans, and E. L. Boeker. 1977. The Use of Natural and Modified Pinyon Pine–Juniper Woodlands by Deer and Elk. *Journal of Wildlife Management* 41:543–59.

8. Krefting, L. W. 1974. *The Ecology of the Isle Royal Moose.* Technical Bulletin 297. Minneapolis, MN: Agricultural Experiment Station, University of Minnesota.

9. Klein, D. R. 1968. The Introduction, Increase, and Crash of Reindeer on St. Matthew Island. *Journal of Wildlife Management* 32:350–67.

10. Caughley, G. 1970. Eruption of Ungulate Populations, with Emphasis on Himalayan Thar in New Zealand. *Ecology* 51:53–72.

11. Nixon, C. M., L. P. Hansen, and P. A. Brewer. 1988. Characteristics of Winter Habitats Used by Deer in Illinois. *Journal of Wildlife Management* 52:552–55.

12. Bookhout, T. A. 1990. *Wildlife Techniques Manual.* Washington, DC: The Wildlife Society.

13. Smith, R. B., and F. G. Lindzey. 1982. Use of Ultrasound for Detecting Pregnancy in Mule Deer. *Journal of Wildlife Management* 46:1089–92.

14. Philo, L. M., E. H. Follmann, and H. V. Reynolds. 1981. Field Surgical Techniques for Implanting Temperature-Sensitive Radio Transmitters in Grizzly Bears. *Journal of Wildlife Management* 45:772–75.

15. Jacobsen, N. K. 1978. Telemetered Heart Rates as Indices of Physiological and Behavioral Response of Deer. *Pecora IV Symposium Proceedings,* pp. 248–55. Washington, DC: National Wildlife Federation Scientific and Technical Series 3.

16. Cook, R. S., M. White, D. O. Trainer, and W. G. Glazener. 1971. Mortality of Young White-Tailed Deer Fawns in South Texas. *Journal of Wildlife Management* 35:47–56.

17. Goldberg, J. S., and W. Hass. 1978. Interaction between Mule Deer Dams and Their Radio Collared and Unmarked Fawns. *Journal of Wildlife Management* 42:422–25.

18. Biggins, D. E., and E. J. Pitcher. 1978. Comparative Efficiencies of Telemetry and Visual Techniques for Studying Ungulates, Grouse, and Raptors on Energy Development Land in Southeastern Montana. *Pecora IV Symposium Proceedings,* pp. 188–93. Washington, DC: National Wildlife Federation Scientific and Technical Series 3.

19. Kammermeyer, K. E., and R. L. Marchinten. 1977. Seasonal Changes in Circadian Activity of Radio Monitored Deer. *Journal of Wildlife Management* 41:315–17.

20. Gilbert, F. F., and S. L. Stolt. 1970. Variability in Aging Maine White-Tailed Deer by Tooth Wear Characteristics. *Journal of Wildlife Management* 34:532–35.

21. Anon. 1982. *Handbook of Biological Techniques.* Cheyenne, WY: Wyoming Game and Fish Department.

22. Mautz, W. M. 1978. Nutrition and Carrying Capacity. In J. L. Schmidt and D. L. Gilbert (Eds.), *Big Game of North America: Ecology and Management,* pp. 321–48. Harrisburg, PA: Stackpole Books.

23. Anderson, S. H. 1982. *Effects of the 1976 Seney National Wildlife Refuge Wildfire on Wildlife and Wildlife Habitat.* Resource Publication 146. Washington, DC: U.S. Fish and Wildlife Service.

24. Skoulin, J. M., D. J. Egerton, and B. R. McConnell. 1983. Elk Use of Winter Range Affected by Cattle Grazing, Fertilizing, and Burning in Southeastern Washington. *Journal of Range Management* 36:184–89.

25. Peek, J. M. 1971. Moose–Snow Relationships in Northeastern Minnesota. In A. O. Hauger (Ed.), *Proceedings, Snow and Ice in Relation to Wildlife and Recreation Symposium.* Ames, IA: Iowa State University.

26. Longhurst, W. M. 1957. The Effectiveness of Hunting in Controlling Big Game Populations in North America. *Transactions of the North American Wildlife Conference* 22:544–69.

27. Mautz, W. W. 1978. Sledding on a Bushy Hillside: The Fat Cycle in Deer. *Wildlife Society Bulletin* 6:88–90.

28. Mackie, R. J. 1978. Impact of Livestock Grazing on Wild Ungulates. *Transactions of the North American Wildlife Conference* 43:462–77.

29. Denniston, R. H. 1981. *University of Wyoming Feral Horse Study: Habitat Preference and Use.* Laramie, WY: University of Wyoming.

30. Smith, A. D. 1958. Considerations Affecting the Place of Big Game on Western Ranges. Bethesda, MD: Society of American Foresters, pp. 188–92.

31. Cooperrider, A. Y., and J. A. Bailey. 1980. Simulation Approach to Forage Allocation. *Forage Allocation Symposium,* Albuquerque, NM.

32. Grieb, J. R. 1979. Allocation of Range Resources for Wildlife. In *Council on Environmental Quality Symposium: Rangeland Policies for the Future.* Washington, DC: U.S. Government Printing Office.

33. Mohler, L. L., and D. E. Toweill. 1982. Regulated Elk Populations and Hunter Harvest. In J. W. Thomas and D. E. Toweill (Eds.), *Elk of North America: Ecology and Management,* pp. 561–97. Harrisburg, PA: Stackpole Books.

34. Boyd, R. J. 1970. *Elk of the White River Plateau, Colorado.* Technical Publication 25. Denver: Colorado Division of Game, Fish and Parks.

35. Thorne, E. T., N. Kingston, W. R. Jolley, and R. C. Bergstrom. 1982. *Disease of Wildlife in Wyoming.* Cheyenne, WY: Wyoming Game and Fish Department.

36. Scott, M. E. 1988. The Impact of Infection and Disease on Animal Populations: Implications for Conservation Biology. *Conservation Biology* 2:40–56.

37. Ricciuti, E. R. 1988. Something Scary Lurks Out There. *Audubon* 90:88–93.

38. Boyd, R. J. 1978. American Elk. In J. L. Schmidt and D. L. Gilbert (Eds.), *Big Game of North America: Ecology and Management,* pp. 11–36. Harrisburg, PA: Stackpole Books.

39. Russell, K. R. 1978. Mountain Lion. In J. L. Schmidt and D. L. Gilbert (Eds.), *Big Game of North America: Ecology and Management,* pp. 207–26. Harrisburg, PA: Stackpole Books.

40. Lindzey, F. G., and E. C. Meslow. 1977. Home Range and Habitat Use by Black Bears in Southwestern Washington. *Journal of Wildlife Management* 41:413–25.

16
SMALL MAMMALS

People regulate many small-mammal populations. Taxonomically, the group includes carnivores, lagomorphs, rodents, and marsupials (Figure 16–1). For management purposes, we divide them into animals desired for fur (badger, beaver, marten, mink, and muskrat); furbearers that are declining in many regions and have been afforded special protection (fisher, otter, and wolverine); small predators (coyote, fox, weasel, lynx, and bobcat); and small game (rabbit and squirrel). General distribution and management techniques are discussed in this chapter. Animal damage control problems, particularly as they relate to the predators, are addressed in Chapter 22.

DISTRIBUTION

Small mammals are found in almost all types of habitat in all parts of the world. The habitat classifications discussed in Chapter 8 can be used as focal points for looking at the distribution of the various species. Many of the current distributions are quite different from the original patterns because people have altered these animals' habitats.[1] The river otter, for example, originally ranged over much of the North American continent, but today its distribution is considerably reduced, and in much of its range it is very rare (Figure 16–2). Fox squirrels, in contrast, still range widely and are even expanding their range, since they do well in woods near farmland.

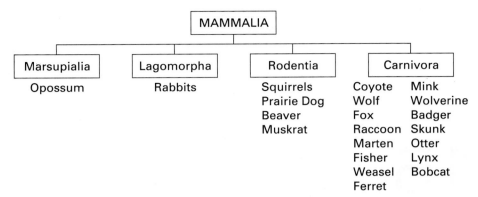

Figure 16–1 Taxonomy of small mammals.

Aquatic

A number of prized furbearers are dependent on aquatic habitats. Beaver, muskrat, and river otter all spend much of their time in streams and lakes. When these water habitats are altered by chemicals or organic wastes, the animals may be forced to move or die. Each species, of course, has a tolerance level that differs from that of other species. Beaver, for example, can withstand substantial change and are usually not disturbed unless the habitat is so altered that siltation occurs. In some streams and ponds, beaver, using adjacent open water or land habitat, live very well with people.

Beaver can actually be a major force in the reclamation of silted waterways or eroded streams. Where vegetation has been altered through such activities as logging and grazing, water runoff can become intense, carrying a great deal of the soil away in the form of silt and forming gullies with rivers or streams in the bottom. These areas lose their streamside vegetation because of a lowered water table and become suboptimal for fish, waterfowl, and wildlife. With a little help, beaver placed into such areas begin to cause major changes in the habitat. For example, in the high prairies of western Wyoming, streams had formed furrows up to 9 m (30 ft) deep in the flat plains. The streamside vegetation of lush shrub and willow, with aspen at some elevations, was all

Figure 16–2 Distribution of the river otter.

but gone. Biologists introduced beaver, along with several truckloads of aspen and, in some cases, tires, to help support the dams (Figure 16–3). Protection against heavy grazing by livestock was also needed in some areas. As the dams backed up water, the silt load dropped, in one case from 33,500 kg (33 tons) per day to 4,000 kg (4 tons), a 70-percent reduction. The water table was raised so that the roots of grasses and herbs could obtain water. Willows began to sprout and stabilize the stream bank and later cooled the water by shading. A riparian habitat returned, as did the fish and waterfowl.[2]

Muskrat are found in a variety of waterways in back country, suburbs, and creeks and drainages associated with highway systems. They also use farm ponds, swamps, and other stagnant waters and are able to tolerate some pollution. They are known to burrow into stream banks and dirt dikes, causing a collapse of the dirt structure.

River otter are found in wetlands, where disturbances are minimal. They will tolerate human beings, but do not adapt to them as well as beaver and muskrat do. The removal of logs, excessive boating, and the disposal of wastes in rivers cause the otter to go elsewhere. Some animals, such as mink, depend on the wetland habitat for both food and shelter and prosper in direct relation to the amount of wetland habitat available.

Terrestrial

Each terrestrial vegetation complex or geomorphological variation provides elements that support small-mammal populations. Large timber, extending through much of the boreal forest, provides habitat for fishers and marten, which do not like open areas. Squirrels occupy a variety of forest habitats. The fox squirrel, which prefers transitional areas between forests and fields, has extended its range from the eastern United States into the prairie, where greenbelts and fencerows with trees provide habitat. Gray squirrels, in contrast, thrive in more mature, dense woodlands in the east.

The prairie provides a habitat for the prairie dog and some colonies of ground squirrel, which subsist on the same plant material as livestock (Figure 16–4). Because

Figure 16–3 Old tires help to reinforce beaver dams. (Courtesy of U.S. Bureau of Land Management; photo by B. Smith.)

Figure 16–4 Prairie dogs are part of the prairie ecosystem. (Courtesy of U.S. Fish and Wildlife Service.)

of this competition and the fact that cattle occasionally break their legs in their holes, these species are often the object of poisoning or other control measures.

Badgers are fossorial (burrowing or digging) mammals of the western grasslands and deserts. Rabbit like areas with edge habitat and multiply rapidly when brush piles are left after clearing vegetation. Lynx and bobcats favor sparsely treed areas with cliffs and rock outcrops. Coyote, on the other hand, seem to adapt well to changes created by people.

CENSUS AND SURVEY TECHNIQUES

Population density surveys of small mammals are particularly difficult because the animals are found in such a variety of habitats. Complete counts, or even population indices of large areas such as a state, are usually not feasible. Counts of some colonial animals, such as prairie dogs, can be made by means of the mark-recapture technique. Rabbit, squirrel, beaver, muskrat, mink, fishers, and other small mammals can also be counted by that technique. Animals are usually trapped by means of live traps. Metal ear tags can be used to mark the animals; sometimes toes are clipped. Radiotelemetry can now be used on many small mammals. Such techniques have been particularly useful in following small mammals to determine their habitat use and the impact on them of changes to their habitat.

Population estimates of river otter in Louisiana were obtained by the use of a mark-recapture technique involving radioactive tracers. Otter were caught and injected in their hip muscles with a radioactive zinc solution. Otter scats were then collected along otter trails and rest and latrine sites. By comparing marked versus unmarked scat, investigators made estimates of the size of the otter population.[3]

A number of indirect techniques are also used. Beaver, which live in colonies, are frequently counted by the number of colonies per unit length of stream. Such counts can be made with aerial surveys. The size of the population is then projected on the basis of the average number of animals per colony, a number derived from ground counts. Muskrat are counted by the number of houses (or burrows) along streams. Aerial surveys are an effective means of making such counts. Information on burrows of fossorial animals has been obtained by radio probes. Several types of fiber-optic devices, flexible tubes through which researchers can probe into burrows and see, have also been used to study fossorial-animal habitats.

Pelt-count comparisons from yearly summaries are used as an index of the size of some populations. These data are useful in drainage areas, forests, or other distinct geographic areas. In some parts of the country, mink, muskrat, and river otter population trends are also estimated by pelt counts. Scent posts, on which animals deposit excretory material, are used to determine the presence of some furbearers.

MANAGEMENT

The group of animals that we are discussing varies widely, which means that management techniques must also vary. Techniques generally fall under four headings: *habitat manipulation, harvest* or *removal, reintroduction,* and *education.*

Habitat Manipulation

Habitat management for small mammals ranges from general to specialized. Several species of small mammals have very specialized needs. For example, fishers (Figure 16–5) use mature forests, river otter prefer downed vegetation near the river, and eastern flying squirrel need snags or hollow trees. Other species can adapt quickly to a variety of habitats. For instance, coyote do well when wolves are removed and open fields exist, and raccoon like farms, parks, and suburban areas, where people leave garbage or farm produce. Habitats must be evaluated for each species that is under management consideration. Preserving or creating the desired habitat for specialized species is necessary for their survival.

Figure 16–5 Fishers use mature forests. (Courtesy of the Wyoming Game and Fish Department.)

Removal

Removal techniques are used for many pest species. Sometimes, removal is a continuous process; at other times, removal occurs only when animals reach a nuisance level. Thus, ground squirrel and prairie dog colonies are selectively poisoned in the prairie habitats. Many regions have continuing programs for removing coyote and nutrias. Techniques include poisoning, trapping, and shooting. Since most animals that cause economic damage reproduce and recolonize rapidly, the clearance is only temporary. Habitat-manipulation programs, which require extensive research, could pay dividends.

Many furbearers, along with rabbit and squirrel, are harvested by people. Regulations controlling the hunting season and take can be effective. Areas can be closed until populations have reached desirable levels. When animals become pests, bounties are paid, but bounties are usually not effective unless some form of habitat manipulation is instituted to discourage the pest. Populations of animals are sometimes removed when they are known to carry a disease. This has happened to skunk, raccoon, and squirrel populations when they have become carriers of plague and rabies.

Reintroduction

Reintroduction (or introduction) is a very difficult management technique. As with biological control, the manager must understand all the needs of the species and determine whether the new habitat will meet those needs. It is important to recognize that this technique can disrupt the system. Managers should carefully monitor the animals to determine their success or failure. Future reintroduction programs are more likely to be successful when the reason for failure are known.

Reintroduction of animals can take one of two forms. First, animals can be moved from one location to another (translocated). In translocation, certain dangers exist and should be guarded against. Predators on people may threaten the newly released animal. Key habitat needs must be met. Photoperiods or climatic variables may cause difficulties. Species such as fishers, marten, and river otter have been translocated with varying degrees of success.[4]

The second form of reintroduction consists of removing animals from the wild to a captive-breeding facility, where, presumably, they will reproduce. The young can then be placed in selected field locations. The dangers noted for translocation are present here, too, and in addition, animals raised in a captive-breeding facility may have acquired degrees of domestication. Weasels, river otter, and marsh hare have all been raised in captive-breeding facilities and released. (See Chapter 21 for more on captive breeding.)

Education

An educated public can make a manager's job much easier. By educating the public in the values of beaver, for instance, biologists have been able to utilize this natural engineer in habitat-improvement projects. People who know something of the life history of furbearers and their habitat needs are not likely to engage in indiscriminate shooting and overharvesting of these animals. Knowledge of the food habitats of some small mammals can convince the public of their value. Landowners can benefit from knowing the habitat requirements of desirable and undesirable species. Urban residents can understand how their living habitats may attract animals such as raccoons.

DESCRIPTIONS OF SPECIES

In this section, we describe a group of small mammals that are of interest to wildlife managers. Some are of considerable economic value, while others cause economic damage or are nuisances.

Opossum

The opossum has thrived with the expansion of the human population. Originally found only in the southern states, the opossum has expanded its range north into Canada and west of the Rocky Mountains (Figure 16–6). The only marsupial found in North America, the opossum has pouched relatives that are much more common in Australia. Opossums eat almost any type of food and can live almost anywhere. They prefer the flesh

Figure 16–6 Opossum mother carrying litter on her back. (Courtesy of the U.S. Fish and Wildlife Service; photo by F. M. Blake.)

of dead animals and insects and will take mice, shrews, and birds when available. They are common in farming areas of the eastern United States and are most active at night. They use areas of 6 to 16 hectares (15 to 40 acres) and appear less able to inhabit areas away from human dwellings because of a lack of food. The male has a double-headed penis and the female a double womb. Most opossums have two litters a year, some three. There can be 17 or more young per litter, although litters of 6 to 10 are more common. Gestation is only 13 days. The young are incubated in the pouch and travel on the mother's back when they are old enough. Opossums live about four years.

Hunting opossums by dog is common in some parts of the country, but the opossum has expanded its range, indicating that hunting does not keep the animals in check. Although the animal is not really a nuisance, some people are bothered by it, so managers must often institute removal practices.

Rabbit

A number of species of rabbit are hunted in much of the country; both rabbit and hare are important food for some carnivores and raptors. For example, golden eagles prey extensively on white-tailed jackrabbit. Rabbit reproduce very rapidly. The eastern cottontail breeds from February through December, usually having three or four litters of from one to nine young each. The female mates again shortly after giving birth. The breeding time, age at first breeding, and litter size vary considerably among species and sometimes within species that have geographic variations.

A number of management techniques have been tried with rabbit, from clear-cutting to leaving habitat edge. Rabbit require cover, and once this is available, their population grows rapidly. Downed logs, woodlots, fencerows, and small refuges are all ideal for rabbit. Removal programs, including poisoning, have been undertaken in areas where rabbit are pests. In some potato-growing areas, jackrabbit have been clubbed to death in large numbers, an action that has brought strong protests from conservation groups.

Squirrel

There are a number of different squirrel populations throughout the United States. These small animals are found in public parks, on farms, and in rural areas and woodlands. In some parts of the country, they have become a popular sport animal. Squirrel are very adaptable and tend to control themselves through dispersal of the young.[5]

Most squirrel management involves some form of habitat manipulation. In farming, woodlots and snags provide ideal habitats. Woodlots adjacent to meadows or fields are especially attractive. Ground squirrel are considered pests in areas where they disturb agriculture or affect cattle. In such cases, poisoning programs are often instituted. In Michigan, wildlife managers worked with landowners to manipulate habitats and control fox squirrels. Fire or heavy grazing tends to destroy the matty vegetation in woodlots; thus, fewer squirrel are found there. In cases of very dense understory, cattle open the area for fox squirrel populations. Trees planted in clumps, creating an island effect, will often attract some species of squirrel.

Prairie Dog

Blacktail prairie dogs are found in the short-grass prairie and sagebrush of the plains and Rocky Mountain states from Montana and North Dakota to Arizona and New Mexico. Whitetail prairie dogs' range is the central Rocky Mountains region of Wyoming, Utah, Colorado, Arizona, and New Mexico. The Utah prairie dog is endangered. Blacktail prairie dogs live in colonies or towns of several thousand animals, with a complex, interlinking burrow system. Their social structure, based on family groups or coteries, is also quite complex.

Blacktail prairie dogs breed in February or March and have their young about 6 weeks later. The parturition period is about 10 weeks. Although prairie dogs can live as long as eight years, they fall prey to many carnivorous animals, predatory birds, and some snakes. While they are less active in the winter, they do not truly hibernate; rather, during the severest part of the winter, they go into a period of torpor in which the body temperature lowers slightly. Prairie dogs have a number of calls. For example, one member, when alarmed, will give a call and flick its tail, and the other dogs will dive for safety.

Management of prairie dogs consists chiefly of a removal program. Ranchers, who do not like prairie dog colonies' disrupting their range, solve the problem with poison and gunshot. Unfortunately, a reduced prairie dog population means less prey for other animals, and the poison also means a destroyed burrow system as animals die. Burrowing owls and black-footed ferrets use burrows as homes. Ferrets are also dependent on prairie dogs as prey.

Beaver

Beaver are large rodents that are important in maintaining ponds and streams, which provide homes for a variety of wildlife. The beaver dams block water, preventing excessive erosion. Beaver, which nest either in stick houses or bank holes, are active primarily at night, when they fell trees, construct their houses, and pull a supply of food under water in areas where a frozen surface prevents winter food expeditions. They prefer the relatively flat terrain of fertile valleys. Aspen groves near waterways make ideal beaver habitat. Beaver are monogamous animals. They breed once a year and have litters of up to nine, but usually of four or five. They live in colonies, most of which contain the two adults and the young from one or more previous years. The females are sexually mature in their third year.

In some areas, beaver have been declining as a result of indiscriminate hunting, heavy trapping, and water pollution. A study made in Alaska indicates that some form of territoriality keeps them far enough apart that their food source remains adequate.[6] In some areas, the removal of predators allows beaver to increase despite trapping. Management tactics are designed to keep people from disturbing or killing the animals, which set up housekeeping and continue in areas for many years if the waters are not polluted and there is an available food supply. Reintroduction is a very feasible management option.

Muskrat

Both numerically and in dollar value, the muskrat is the most important furbearer in the world.[7] It is found throughout North America, except in Mexico and parts of the southeastern and southwestern United States. The habitat for this species includes the standing and slow-moving waters of streams, marshes, and lakes. Muskrat like areas with dense vegetation, but are adaptable to human water developments, such as canals and ponds. The muskrat can build two types of permanent shelter: a burrow or den built into the sides of lakes, ponds, or rivers and a house or hut constructed of herbaceous plants and mud in shallow standing water.

One study revealed that the muskrat prefers long streams with shores and cliffs for burrows and covers for shelter from water currents.[8] Such areas usually provide a readily available supply of food. Apparently, it is not so important to be alongside forests or fields, but it is important to have cliffs where the animal can dig. In the study, no burrows were found where the banks were less than 2.2 m (7.2 ft) or the slope was less than 10 percent. Coves where water conditions, bottom substrate, and vegetation usually differed from those in the river itself were favored, as were branches of the main channel. Also popular were rivers with islands, which, of course, increased the shorelines for burrows.

The diet of this animal is variable, but it relies chiefly on plants, especially aquatic plants, for which it digs on the bottom. The diet also includes some animal food, such as clams, crayfish, and fish. People generally do not observe muskrats, but in areas where the grassy bottom of waterways can be seen, muskrat runs are often visible. Traps placed in these runs are frequently successful (Figure 16–7). The important aspects of management of the muskrat include maintaining stream banks, vegetation, and unpolluted water. A Canadian study showed that muskrat populations did not decline following seismic activities, even though some animals were hurt during the process.[8]

Coyote

Coyote are now considered a pest species in many areas of the United States. They are known to kill sheep, and they reportedly attacked a person in southern California. The coyote is classified as a carnivore, although it will eat many different types of food. Food analyses in Oregon indicate that the animal eats a variety of small rodents, rabbits, large ungulates, and fruits. The diet changes by season in response to the availability of prey.[9] Coyote move both alone and in groups of two, three, and four. They move a short distance each day, ordinarily not more than 2 km (1.25 mi).

Data show that coyote and wolf habitats do not overlap.[10] One study revealed that wolves kill coyotes; in any event, coyotes avoid areas where wolves are active. But

Figure 16–7 Muskrat runway. (Courtesy M. Boyce.)

wolves require large areas, with little human activity, so that when fences are built and ranches established, wolf populations decline. These features do not bother coyotes, and the loss of wolf habitat turns out to be the creation of coyote habitat. Most management effort vis-à-vis coyotes has been to reduce the population, but it has not been very effective. There is a need to develop more effective controls. (See Chapter 22.)

Wolves

Gray wolves used to be found throughout a large part of North America, with a few remnants coming down into the Great Lakes states and Montana. They are still found in Canada and Alaska, but in much smaller numbers. Red wolves are found in a small part of Texas and Louisiana. This species of the Canidae family occupies a large home range. Development and other human disturbances reduce their prey and subdivide their habitat.

A study in the Superior National Forest showed a population declining because of a reduction in the white-tailed deer population.[11] Pup starvation was followed by a lower production of pups. In areas with a high density of wolves, members of adult wolf packs were the most secure members of the wolf population because they protected each other from external threats. As food became scarce, however, the pack offered less security, because individuals became involved in intraspecific strife for the limited food. A detailed study of the wolf social system on Isle Royale has been published.[12]

The gray wolf has from 5 to 14 young between April and June. The young, which feed on meat regurgitated by the adults, remain with the adults from one to two years. The basic management strategy for gray wolves appears to be to ensure forest habitat in areas where the animals are able to remain isolated from human disturbance.

Because the gray wolf is listed as an endangered species, recovery efforts are under way in parts of its range.[13] These recovery efforts are bringing livestock owners and conservation groups into direct conflict, as the reintroduction of wolves into Yellowstone National Park occurs. The question being raised is how far we should go to promote and protect an endangered species.[14] Conservation organizations are setting up a fund to compensate landowners for loss of livestock. Politicians, however, are now in the picture; thus, members of Congress from the states near Yellowstone (and some from afar) are either for or against reintroducing wolves into the park.

Fox

There are four species of fox in North America: the red fox, kit fix, gray fox, and arctic fox. All these carnivores in the Canidae family have doglike characteristics. The swift fox (a subspecies of the kit fox) is rare in much of its range (Figure 16–8). Red and gray foxes are common species found throughout the United States. The red fox is a relatively small animal, usually under 7 kg (15 lb), with many color variations. The white tip on the tail distinguishes it from other species of fox. It prefers settled open country with occasional trees or edge habitat. The gray fox, common in brushy and woody habitats in the west and south, is often found in areas where people are living. The arctic fox, found only in Greenland and the far northern parts of Alaska and Canada, lives in the tundra above the tree line. The red, arctic, and gray fox are important economically for their fur.

Red fox mate in winter or early spring; the females produce from one to nine kits after a 53-day gestation in a maternity den. As with wolves, the kits' first solid food is meat regurgitated by adults. The kits begin to disperse at four months.

Figure 16–8 Swift fox.

Fox are active primarily in the late evening and at night. They do not mind human activity and can often be found near or under homes. They are hated by farmers because they steal chickens. They feed on a variety of animal and plant material, the bulk of their diet consisting of small mammals, birds, insects, and fruits. Fox adapt quickly to alterations in their habitat. A study of replanted strip mines in Illinois showed that red fox populations increased rapidly and did better in the stripped area, once the diversity of vegetation returned, than in the intensively farmed area. Trees planted in the reclamation program provided particularly good red fox habitat.

Management techniques include providing brushy habitat, scattered trees, and, for the red fox, an added edge effect. But managers most frequently use removal procedures. Little effort is made to manage the arctic fox. Presumably, its habitat will remain suitable unless extensive human disturbance destroys its prey base.[15] A study in the Prudhoe Bay, Alaska, area showed that oil and gas development increased the chances for survival of the arctic fox, apparently because garbage from the food that was brought in provided scavenging material.[8] An increase in rabies transmission became noticeable as a result of increased fox–fox, as well as fox–human, contact. Population fluctuations were more pronounced in disturbed than in undisturbed areas.

Raccoon

The raccoon is one of the few well-established omnivorous feeders in the country. It ranges throughout the United States, except in the Great Basin and parts of Montana. Its range extends south into Mexico and Central America. The raccoon is primarily a species that inhabits forested areas; however, it does get out into the open range when there are scattered trees. It prefers trees along wetlands, where it can find water and a variety of foods.

Raccoon coexist easily with people—so much so, that they can become a nuisance. In a study near Cincinnati, Ohio, raccoon were found to be very adaptable to human settlement as long as they could find streams with trees along the side to climb into and rest.[16] They have comparatively small home ranges of about 5.1 hectares (12.6 acres) and travel primarily between rest sites and feeding areas. One of the reasons for their adaptability to human settlement is that raccoon can eat a variety of garbage items. It is not uncommon for them to rummage through trash cans in picnic areas or suburbs. They also like crayfish and crabs when available.

Raccoon breed once a year, having from one to seven offspring. A male may breed with several females. Except during mating, raccoon are mostly solitary. Females, which can breed in their first year, use hollow trees, culverts, or caves as homes for their young.

Management efforts for the raccoon are generally in response to complaints from suburbanites who find the animals in their garbage and yards. Canine distemper, a viral disease, affects raccoon in some parts of their range, causing them to act as if they are dizzy or drunk. Care must be taken in handling these animals because they are susceptible to rabies, although less so than dogs. Trapping and removal procedures are necessary if they are to be taken out of an area. Raccoon are also a pest species around farms. In the midwest, corn is a common food of raccoon, but they also eat grapes, figs, melons, and other types of fruit.

Marten

The marten, or pine marten, is a carnivore found in Alaska, Canada, and northern parts of the continental United States, down into the Rocky Mountains. Marten are closely related to weasels and mink and are valuable furbearers. Thus, they come under furbearer regulations in some states.

The marten is a swift tree climber and can be found in climax forested areas in its range. Since marten eat small mammals, particularly voles and squirrels, they are attracted to forests or small forest openings with understory vegetation.[17] A study of their habitat in Maine indicated that when extensive clear-cutting occurred, they did not utilize the open area, but inhabited forests with scattered openings.[18] The study, using radiotelemetry and snow tracking, revealed that uncut soft wood mixed with strands of coniferous forest were heavily used by martens in both summer and winter. The home range, as determined by live trapping and radiotelemetry, was found to be between 5 and 10 km². The animals used cavities in stumps or logs for resting in the winter and the crowns of conifers for resting in the summer (Figure 16–9).

Marten habitat is affected by impacts on the forest itself. Clear-cutting, fire, and development all reduce food and cover. Crown fires, however, do not seem to destroy the marten habitat.[19] Marten populations have declined in parts of the range because of the ease with which the animal can be trapped. Marten use dens in tall, hollow trees, underground in rock piles, or in tree roots. They mate in the summer and have two to four young the following spring. The young disperse in late summer or fall.

Marten have been successfully reintroduced into some areas. One such program is in a northern Wisconsin hardwood forest.[18] As a result of the program, the biologists have recommended that marten be trapped live after the summer breeding season and moved as quickly as possible. The reintroduction program should continue for several years. In all such programs, adequate food should be available, to reduce movement away from the release area.

Fisher

The fisher is a solitary predator found in the northern part of the United States and into Canada (Figure 16–10). It eats snowshoe hare, mice, porcupine, and, occasionally, other carnivores and berries. Fishers have been known to live in zoos as long as 10 years.

Figure 16–9 Pine marten. (Courtesy of the Los Angeles Zoo.)

Figure 16–10 Distribution of the fisher.

During the last part of the 19th century and early part of the 20th century, the number of fishers decreased markedly, partly because of trapping and logging.[20] Fishers are one of the easiest animals to trap. Before the 1920s there were no trapping regulations, and the prices paid for pelts were high. Between 1910 and 1930, an excellent fisher pelt would bring about $150, and some as much as $345.[20] In 1979, after serious inflation had started, the Hudson Bay Company paid $410 for an excellent pale female pelt. Prices in recent years have continued to rise, and it appears that the demand for fisher pelts will hold.

Fishers occupy forested habitat, where they make dens for raising their young (Figure 16–5). Mating data, which come mostly from fur farmers, show that mating occurs in the spring and that gestation exceeds 325 days. Females can bear young at one year. Fishers are found in forested areas with some openings, such as transmission-line corridors, pipeline corridors, or abandoned logging roads. Management involves isolating the animals from human activity and controlling their harvest.

Weasel

Weasels are found in many parts of the United States. The long-tailed weasel inhabits almost all parts of the country, parts of Canada, and south into Mexico and Central America. The short-tailed weasel is more common in the northern part of the United States and into Canada, and the least weasel occupies a range that dips down around the Great Lakes states and continues up into Canada and Alaska. Weasels are primarily carnivorous, taking small mammals such as prairie dogs and ground squirrels alive.

The short-tailed weasel feeds heavily on meadow voles and other small mammals, while the long-tailed weasel will take some of these species, but will also pick up chipmunks, ground squirrel, rabbit, some birds, and even some reptiles. The short-tailed weasel, sometimes called the ermine because its coat turns white in the winter, prefers early successional communities and avoids forested habitats. Males tend to use extensive shrub communities, while females prefer the more open grassy areas.[21]

Long-tailed weasels prefer more advanced successional stages and can be found in forested habitats. There is a fair amount of overlap in the range of the species in the edge and ecotone areas near field and forest. Weasel populations can be maintained even near human development; difficulties arise from *extensive* human development or loss of food. Then weasels can become pests, and removal measures are in order.

Black-Footed Ferrets

The discovery in 1981 of a black-footed ferret population in Meeteetse, north-western Wyoming, showed that this mustelid, thought to be on the brink of extinction, survives in isolated areas of the country (Figure 16–11). The ferret is now listed as an endangered species. Before the Meeteetse discovery, the last confirmed sighting of a wild ferret was in 1979 in South Dakota. There are few early records of the black-footed ferret, partly because of the animal's nocturnal habits and skittish nature (Figure 16–12). Reports from early trappers indicate that the Plains Indians used ferret skins during their ceremonies. The first recorded reference to the black-footed ferret was in 1851, when the naturalist John James Audubon and the Reverend John Bachman described it as a species after examining a single skin given them by a Wyoming trapper.

Historically, the ferret's range corresponded with that of the prairie dog, extending north across short-grass prairies from Texas and Arizona to the Canadian provinces of Alberta and Saskatchewan. Ferrets were typically found within prairie dog towns, since they apparently relied on prairie dogs almost exclusively for prey and burrows in which to live and raise their young.

As the cattle industry grew in the 19th century and prairie dogs became competitors with livestock for limited forage, an extensive extermination program was instituted. The successful control of the prairie dog also caused the ferret to decline, and extensive plowing of the grassland decreased the habitat for both animals. It is difficult to assess the extent of the decline of the ferrets because nobody knows how abundant they had been. The only population formally studied was in South Dakota in 1964. It was observed for more than 10 years, but then it disappeared, possibly wiped out by disease.

The fact that there is little information about the ferret's biology and life history makes management efforts difficult to plan and assess, but maintaining prairie dog towns and minimizing human disturbance are emerging as important methods of man-

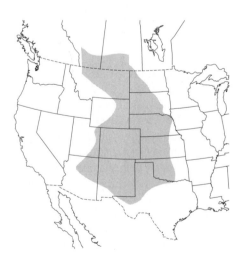

Figure 16–11 Historic range of the black-footed ferret.

Figure 16–12 The rare and endangered black-footed ferret, now part of a successful captive-breeding program. From six animals brought into captivity, biologists currently have more than 300. Reintroduction into the wild has been moderately successful; however, more than 90% of the animals released become prey. (Courtesy of the U.S. Fish and Wildlife Service; photo by L. C. Goldman.)

agement. It is quite likely that climatic conditions and vegetation play a role, but the optimum climate and vegetation are not clear.

Mink

With the exception of some of the drier areas in the southwestern United States, mink are found throughout a large part of this country and Canada. They are usually rich brown and may have light spots on their bellies. Normally, they are found in the wild in wetland habitats along streams and lakes, using logjams and downed vegetation to create dens during a considerable part of the year. They spend about 60 percent of their active time at night.[22] The home range for adult males can be up to 5,000 m (16,400 ft) along the streams. For females, the range is up to 2,800 m (9,180 ft). Males sometimes wander some distance from water, apparently to look for food.[23] Mink prefer muskrat, but will take such other prey as rabbit, mice, frogs, and birds. Mink mate in late February through April. The young are born in the spring in a fur-lined den. The litter (around four animals) remains together until the fall.

Harvest regulations are needed, since extensive trapping operations have eliminated the mink population from a number of streams in the country, and human encroachment on wetlands has had a destructive effect. Mink are so prized for their fur, that entrepreneurs have established mink farms. Habitat enhancement is the best technique for maintaining mink populations. Improvements made for fish and waterfowl also benefit mink. These include preventing stream pollution and managing the riparian habitat. Transect counts show mink activity increasing by as much as 53 percent in improved wetland habitats. They appeared not to respond to an increase in trout biomass, but sought crayfish that increased in abundance with improvement to the mink habitat.[24]

Wolverine

The wolverine, a relatively large furbearer, can weight up to 13.6 kg (30 lb). Except for the bushy tail, it looks like a small bear with big feet, so it often startles people. Living chiefly in high mountain forests in Canada and Alaska, it extends into a small area of the Pacific Northwest and Rocky Mountains. Its fur, being resistant to ice buildup in winter, is used to trim parkas.

Wolverine live primarily on carrion, often feeding on dead elk, deer, and domestic cows. They are adapted to eat frozen as well as fresh meat.[25] Possessed of a very keen sense of smell, they can sniff out food buried underground or in snow and can cover large areas when searching for food. Data indicate that the males defend territories as large as 2,000 k^2 (500,000 acres).[26]

In the north, wolverines are often considered pests, so that little has been done for their conservation. They have been known to raid trapper lines, food caches, and cabins. Wolverines are commonly trapped incidentally with other furbearers. Large, fairly isolated forests are a necessary habitat for wolverines. Studies show that human activities, such as logging and camping, do not have a major impact on the wolverine population. When large clear-cuts are made, wolverines stay away, but they use small openings in forests, which provide for the movement of other animals, to search for food.

Badger

Although morphologically specialized to prey on fossorial rodents, the badger is a very adaptable animal: It also eats birds, reptiles, insects, and some vegetation (Figure 16–13). It is more common in open shrub land, but also lives in forest edge and sometimes in forests with openings. Considered a furbearer in most parts of the country, it is covered by state harvest laws for furbearers. Trappers take badgers for their pelts from areas with sufficient numbers to allow trapping.

Figure 16–13 Badgers are adapted to dig (fossorial).

An Idaho study of the impact of human beings on badgers showed that roadkills and shooting were the main causes of mortality.[27] Badgers usually live from 3 to 5 years, but are known to have lived as long as 14 years. Generally, first-year males do not breed, but the females do. Conception usually occurs in July or August. Males maintain a home range of more than 600 hectares (1,450 acres), which may include more than one female. Females have a considerably smaller home range.[28] The young tend to disperse from the area of birth.

Management efforts consist primarily of education: People need to be made aware of the importance of badgers in removing rodents and arthropods. Since human activity causes the major impact on the badger population, there appears to be little need for management efforts other than education in areas of heavy human activity.

Skunks

Skunks are small, omnivorous animals found throughout most of the United States. Striped skunks are at home in practically every habitat in every state, living in dens and often beneath abandoned buildings. They can be seen wandering around on cloudy days and at sunset. They eat a variety of fruits, berries, insects, earthworms, other small invertebrates, and some rodents. They sport many color variations, from almost black to almost white.

Spotted skunks are also found throughout a good portion of the country, but they are not common in some of the more northerly states and the northern part of the east coast. They eat a variety of invertebrates, eggs, and, sometimes, small birds. The hog-nose skunk and the hooded skunk are found in the Pacific southwest and extend down into Mexico and parts of Central America.

In a country where millions of dollars are spent every year on human deodorants, it is not to be wondered that the skunk is not favored. In addition, the animal can carry rabies. Thus, removal procedures are the order of the day when skunks invade suburban areas or campgrounds in large numbers. They can be kept away from buildings by repellents—mothballs (paradichlorobenzene) are effective. Screens can prevent them from getting under buildings. Proper fencing will keep them from chicken coops or apiaries. (Skunks like honeybees.) Removal of insects from golf-course grasses is useful. Despite their bad reputation, skunks do help keep small-rodent and insect populations in check.

Otter

River otter used to inhabit most waterways of the United States and Canada, with the exception of areas in the southwest and the Great Basin. Since their distribution coincides closely with that of mink and muskrat, many studies compare the three populations. The otter's highly prized fur is rich brown above and silvery below. Its webbed feet and wide tail base make it an excellent swimmer. Otter have been subjected to shooting, habitat reduction, and extensive trapping. In some states the otter is a protected animal, with special laws concerning their removal.

Although they use mountain lakes and streams in the fall, otter prefer valley habitats and are more common in valley streams than in valley lakes, reservoirs, or ponds. During the summer, they use mudflats, open marshes, swamps, and backwater sloughs. Their use of lakes, reservoirs, ponds, and unobstructed streams is greater during the winter. Logjams create ponds and rest areas throughout the year.[29]

Otter feed primarily on fish, but apparently also eat other foods such as crayfish, amphibians, insects, birds, and mammals.[30] Otter mate in early spring, shortly after the birth of the last litter. Males usually leave not long afterward, but many return when the young are partially grown. The young usually disperse at about eight months.

A major requirement for otter seems to be undisturbed, unpolluted streams, although some adapt to low-level pollutants. Extensive downed vegetation near stream edges obstructs feeding, denning, swimming, and play areas. Thus, managers can create otter habitat by removing some downed material, leaving enough to create some impoundments, and by maintaining the riparian habitat along the sides of streams.

Lynx

Lynx are found in the Pacific northwest, around the Great Lakes, in the northern Rocky Mountains, throughout a large part of Canada, and up into Alaska. They have a cyclic population, peaking about every 10 years. (See Chapter 3.) Along the southern edge of their range in the United States, they may suddenly increase from extremely rare to abundant in a few months,[31] apparently as a result of dispersal from their population center. Lynx prey to a large extent on snowshow hare, which is partly the reason for the cyclic population, but they also take deer, squirrel, mice, and grouse. It is estimated that a lynx requires about one hare per day.[32] In winter, lynx can catch and kill fox that become bogged in snow (the padded feet of the lynx allow it to run swiftly across snow); at other times, they chase fox from the area.

Lynx young (two per litter, normally) are usually born in early summer. They remain near the den site—under tree roots, in a log, or under a rock—for a few weeks. They may stay with their mother for as long as a year, at which time they become sexually mature. They can live between 11 and 15 years, but in natural conditions the life span is probably much less.[33]

Lynx prefer climax boreal, mature coniferous, and mixed forests. The most common trees in their habitat are balsam, fir, black spruce, white birch, yellow birch, and red maple. These areas are also commonly used by the snowshoe hare (Figure 16–14).

Management of the lynx population involves keeping areas in an undisturbed state, allowing the snowshoe hare population to increase. Human activity tends to disrupt lynx activity, and controlled sport hunting assists in stabilizing the lynx population.

Bobcat

Bobcats are found throughout a large portion of the northern, western, and southern United States. The cyclic nature of prey populations influences the bobcat population, which has declined drastically because of loss of habitat and trapping. At one time, bounties were paid to remove bobcats; now they are a game species in many states.[34] Most hunting is done with dogs.

Bobcat kittens are usually born in the spring in rock fissures or cracks that are used for dens. The kits remain with the mother through the summer while she teaches them to hunt. Bobcats are carnivores (Figure 16–15). They prefer cottontail rabbit and eat woodrats, but can switch to other prey, such as small mammals and birds. Rarely have they been reported to feed on ungulates. They appear to be somewhat selective once they find prey that they prefer. They have been reported to take poultry, young pigs, and lambs, but they do not like to be near human habitation, so attacks on domestic animals are not common. Bobcat populations can best be protected by maintaining isolated areas. Encouragement of an adequate prey base is also important.[34]

Figure 16–14 Lynx. (Courtesy of the Wyoming Game and Fish Department.)

Figure 16–15 Bobcat. (Courtesy of the Wyoming Game and Fish Department.)

Chap. 16 / Small Mammals

SUMMARY

Small-mammal populations constitute a varied group of furbearers, predators, small game, and some rare animals afforded special protection. Species from groups of animals are found in most habitats of the United States. Aquatic habitats have beaver, muskrat, and river otter, some of which are important furbearers. Management of this group centers on keeping aquatic systems free from pollution or heavy human impact. Beaver have been introduced successfully in areas where heavy stream erosion has occurred. They have assisted in raising the water table and controlling river siltation.

Each terrestrial vegetation complex provides elements that support varied groups of terrestrial small mammals. Forested areas are homes for pine marten, fishers, wolverine, bobcats, and lynx. Forest edges have squirrel, weasels, and coyote. Grassland species include prairie dogs, ferrets, wolverine, and rabbit. Skunks and raccoon are found in field and forest edge, but are also attracted to suburban areas.

Because of people's varied feelings toward small mammals, management approaches are often varied. Habitats must be preserved to keep prized furbearers. Removal programs are instituted for prairie dogs, coyote, and sometimes skunks when they are perceived as nuisance species.

DISCUSSION QUESTIONS

1. Describe some techniques that can be used for censusing muskrat populations.
2. What data should be obtained before and during a reintroduction program?
3. Explain why erosion lowers the water table. How can beaver help raise the water table?
4. Why is removal an ineffective management technique for most animals?
5. Describe methods of managing mink.
6. Can you manage wolverines by managing their food? Explain.
7. How can the nuisance aspect of beaver be controlled?
8. Is hunting a good means of regulating rabbit populations? Why or why not?
9. Discuss the importance of the river otter to the ecosystem.
10. What species appear to increase as people settle in an area? What characteristics of these species give them the ability to coexist with people?

LITERATURE CITED

1. Hall, R. E. 1981. *The Mammals of North America.* 2 vols. New York: Wiley.
2. Smith, B. H. 1980. Not All Beaver Are Bad; or, an Ecosystem Approach to Stream Habitat Management with Possible Software Applications. *Proceedings, 15th Annual Meeting, Colorado–Wyoming Chapter, American Fisheries Society.* Laramie, WY.
3. Shirley, M. G., R. G. Linscombe, N. W. Kinler, R. M. Knause, and V. L. Wright, 1988. Population Estimates of River Otter in a Louisiana Coastal Marshland. *Journal of Wildlife Management* 52:512–15.
4. Berg, W. E. 1982. Reintroduction of Fisher, Pine Marten, and River Otter. In G. C. Sanderson (Ed.), *Midwest Furbearer Management: Proceedings, Symposium of the 43rd Midwest Fish and Wildlife Conference.* Wichita, KS.
5. Thompson, D. C. 1978. Regulation of a Northern Grey Squirrel Population. *Ecology* 59:708–15.

6.   Boyce, M. S. 1981. Habitat Ecology of an Unexploited Population of Beavers in Interior Alaska. In J. A. Chapman and D. Pursley (Eds.), *Worldwide Furbearers Conference Proceedings 1,* Frostburg, MD: pp. 155–87.

7.   Brooks, R. P., and W. E. Dodge. 1981. Identification of Muskrat Habitat in Riverine Environment. In J. A. Chapman and D. Pursley (Eds.), *Worldwide Furbearers Conference Proceedings 1,* Frostburg, MD: pp. 113–38.

8.   Westworth, D. A. 1981. The Effects of Oil Exploration on Muskrat Population in the Mackenzie Delta. In J. A. Chapman and D. Pursley (Eds.), *Worldwide Furbearers Conference Proceedings 1,* Frostburg, MD: pp. 1698–1727.

9.   Toweill, D. C. and R. G. Anthony. 1988. Coyote Food in a Conifer Forest in Oregon. *Journal of Wildlife Management* 52:507–12.

10.   Fuller, T. K., and L. B. Keyth. 1981. Non-overlapping Ranges of Coyotes and Wolves in Northeastern Alberta. *Journal of Mammalogy* 62:403–5.

11.   Mech, L. D. 1977. Productivity, Mortality, and Population Trends of Wolves in Northeastern Minnesota. *Journal of Mammalogy* 58:559–74.

12.   Peterson, R. O. 1977. *Wolf Ecology and Prey Relationships on Isle Royale.* Scientific Monograph Series 11. Washington, DC: National Park Service.

13.   Anon. 1982. *Northern Rocky Mountain Wolf Recovery Plan.* Denver, CO: U.S. Fish and Wildlife Service.

14.   Carey, J. 1987. Who's Afraid of the Big Bad Wolf? *National Wildlife* 25:4–11.

15.   Yearsley, E. F., and D. E. Samuel. 1980. Use of Reclaimed Surface Mines by Foxes in West Virginia. *Journal of Wildlife Management* 46:729–34.

16.   Hoffman C. O., and J. J. Gotschang. 1977. Numbers, Distribution, and Movement of a Raccoon Population in a Suburban Residential Community. *Journal of Mammalogy* 58:623–36.

17.   Hargis, C. D., and D. R. McCullough. 1984. Winter Diet and Habitat Selection of Marten in Yosemite National Park. *Journal of Wildlife Management* 48:140–46.

18.   Davis, M. H. 1982. Post-release Movements of Introduced Marten. *Journal of Wildlife Management* 47:59–66.

19.   Koehler, G. M., and M. G. Hornocker. 1971. Fire Effects on Marten Habitat in the Selway–Bitteroot Wilderness. *Journal of Wildlife Management* 41:500–505.

20.   Powell, R. A. 1982. *The Fisher: Life History, Ecology, and Behavior.* Minneapolis: University of Minnesota Press.

21.   Simms, P. D. 1979. North American Weasels: Resource Utilization and Distribution. *Canadian Journal of Zoology* 57:504–20.

22.   Linscombe, G., N. Kinler, and R. J. Aulerich. 1982. Mink. In J. A. Chapman and G. A. Feldhammer (Eds.), *Wild Mammals of North America,* pp. 629–43. Baltimore: Johns Hopkins University Press.

23.   Melquist, W. E., J. S. Whitman, and M. G. Hornocker. 1980. Resource Partitioning and Coexistence of Sympatric Mink and River Otter Populations. In J. A. Chapman and D. Pursley (Eds.), *Worldwide Furbearers Conference* 1:187–220.

24.   Burgess, S. A., and J. R. Bider. 1980. Effects of Stream Habitat Improvement on Invertebrates, Trout Populations, and Mink Activity. *Journal of Wildlife Management* 44:871–80.

25.   Hornocker, M. G., and H. S. Hash. 1981. Ecology of the Wolverine in Northwestern Montana. *Canadian Journal of Zoology* 59:1286–1301.

26.   Wilson, D. E. 1982. Wolverine. In J. A. Chapman and G. A. Feldhamer (Eds.), *Wild Mammals of North America.* Baltimore: Johns Hopkins University Press.

27.   Messick, J. P., and M. G. Hornocker. 1981. *Ecology of the Badger in Southwestern Idaho.* Wildlife Monograph 76. Bethesda. MD: Wildlife Society.

28.   Lindzey, F. G. 1978. Movement Patterns of Badgers in Northwest Utah. *Journal of Wildlife Management* 42:418–22.

29. Melquist, W. A. 1981. Ecological Aspects of a River Otter Population in Central Idaho. Unpublished Ph.D. thesis, University of Idaho.

30. Toweill, D. E., and J. E. Tabor. 1982. River Otter. In J. A. Chapman and G. A. Feldhamer (Eds.), *Wild Mammals of North America.* Baltimore: Johns Hopkins University Press.

31. Mech, L. D. 1980. Age, Sex, Reproduction, and Spatial Organization of Lynxes Colonizing in Northwestern Minnesota. *Journal of Mammalogy* 61:261–67.

32. Parker, G. R. 1981. Winter Habitat Use and Hunting Activities of Lynx on Cape Breton Island, Nova Scotia. In J. A. Chapman and D. Pursley (Eds.), *Worldwide Furbearer Conference Proceedings* 1:221–48.

33. Rue, L. L. 1981. *Game Animals.* New York: Van Nostrand Reinhold.

34. Jones, J. H., and N. S. Smith. 1979. Bobcat Density and Prey Selection in Central Arizona. *Journal of Wildlife Management* 43:666–72.

17

WATERFOWL

Waterfowl constitute one of the major groups of sport animals in North America. Along with big-game mammals, they have been the major object of wildlife management. Not only are waterfowl species sought by hunters, but their beauty makes them attractive on ponds in and around living areas and in parks. Many people enjoy just going out with binoculars and looking at waterfowl species, in both urban and country areas. Their annual spring and fall migrations concentrate waterfowl species in several major areas of the country. Bird-watchers visit these areas, and during the fall, hunters drive to them.

In this chapter, we discuss the historical development of waterfowl propagation, hunting, and management in the United States; the migratory behavior and biology of waterfowl; hunting as a substitute for natural mortality; lead shot and its implications for waterfowl mortality; the impact of human disturbance, drought, loss of wetlands, and disease on waterfowl production; and management programs on behalf of waterfowl, including regulations, prevention of disease, habitat manipulation, and predator control.

TAXONOMY

Waterfowl in North America are of the family Anatidae. This group of birds is divided into two subfamilies and eight tribes, a classification based in part on their use of habitat and feeding (Figure 17–1). Management efforts thus vary somewhat, depending on the type of group. Swans, the largest of the waterfowl, are easily recognized because their necks are longer than their bodies. In the United States, all three breeding species of swans are white and are relatively rare, found in only a few locations. Geese, intermediate in size

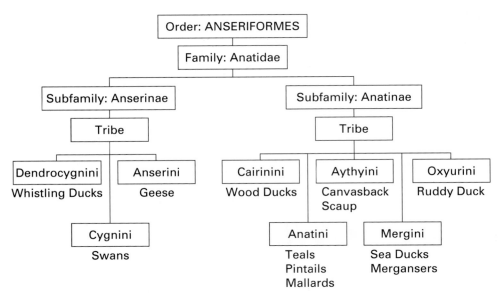

Figure 17–1 Taxonomy of waterfowl.

between swans and ducks, are common throughout North America. As in the case of swans, males and females look alike; their necks are heavier and longer than the necks of most ducks, and their legs are farther forward than the legs of both ducks and swans, an adaptation, apparently, for grazing. Whistling ducks, or tree ducks, have a limited distribution in the south. Some are arboreal. Fulvous and black-bellied tree ducks look somewhat like geese, and, as with swans and geese, the plumage of the sexes is very similar.[1] They do not dive.

The subfamily Anatinae is divided into five tribes of ducks, in part on the basis of their anatomy, which probably evolved as a result of feeding strategy. Surface-feeding ducks—*puddle* and *dabbling* ducks—are found in ponds, lakes, and rivers, where they feed on water plants and other vegetation. They take some invertebrates and fish. This group includes the common mallard, one of the species most sought after by hunters, and the pintail, gadwall, widgeon, shoveller, teal, and black duck. *Wood duck* males are colorful birds found in lakes and streams around woodlands. Females are of a duller color than the males. Wood ducks are of a different tribe than the dabbling ducks, although some taxonomists lump the two together. Normally, wood ducks nest in trees.

Bay ducks and sea ducks, often called "diving ducks," are a single subfamily of expert divers, with legs set far back. They are found in many bays and rivers and, particularly during the migratory season, inland in some streams. Their diet consists chiefly of invertebrates and aquatic vegetation. This group includes the redhead, canvasback, ring-necked duck, scaup, goldeneye, and buffle-head. Sea ducks are large and short necked. They dive deeper than the bay ducks and so are able to take more mollusks in their diet. They are found along coastal and inland waters. Such species as the harlequin ducks, eiders, and scooters are in this group. Mergansers are placed in the same tribe (mergini) as sea ducks. They are larger than other ducks and slower in taking off from the water; however, they have a rapid flight pattern. They eat fish commonly caught while the ducks are diving.

In the United States the ruddy duck, common on lakes and ponds, is a member of the *stiff-tailed* tribe. Smaller and chunky, with tails that stick straight up, ruddy ducks dive for their food, which consists mostly of plant material.

The history of waterfowl propagation and hunting, like that of big game, is tied to the development of the country. Early explorers moving westward were amazed at the large flocks of migratory waterfowl. But by the late 19th century, the decline of wetland habitats had begun to take its toll of the waterfowl. This was intensified by exploitation of the populations through hunting as great numbers of ducks were removed for eastern markets. Between 1955 and 1989, the breeding populations of waterfowl in the United States dropped by 16 percent.

In the east, the wood duck declined because of the impact of settlements along the Atlantic coast. Its former range coincided with the deciduous forests, which gave way to the fields and farms of early colonists. The ducks were associated, to a large extent, with beaver, since ponds provided food and snags created by dammed water served as nesting cavities. As nesting cavities were eliminated, a further decline in the wood duck population set in.[2]

Change in Habitat

The prairie pothole region (Figure 17–2) in the central part of North America provides nesting grounds for about half the continental mallard population, as well as many other waterfowl species. It also provides habitat for migrants in the spring and fall. During the past 200 years, the region has changed from a nearly pristine wilderness to a land of intensive agriculture. This transaction has altered the relationship between ducks and their environment. Although wolves used to take some waterfowl, red fox and coyote, which have become common, prey heavily on them. Red fox take an especially heavy toll of females, because they are easy to reach when nesting. Apparently, this is the reason for the higher number of male mallards in the region.[3]

Although other habitats are important to waterfowl, the wetlands of the prairie pothole region are the most used habitat for feeding and nesting. Loss of habitat, of course, has been a major contributor to the decline of waterfowl populations. Improvement in wetland habitat and protection from inundation by people and predators can help maintain the waterfowl population and a normal reproduction level.

Figure 17–2 Prairie pothole region.

In areas where water fluctuation is great, birds must be able to find the food necessary for survival. Water levels have been studied in North Dakota in relation to the feeding cycle of blue-winged teal.[4] The types of food consumed by teal before and after water-level changes were compared. During high water, the teal had a diet high in snails, which were taken on wetland habitat. As the wetlands dried, the teal diet shifted to midge larvae, consumed on semipermanent lakes. Even during years with average precipitation, some wetland areas dry up, so birds must be able to move to other areas and find food in temporary, seasonal, or semipermanent wetlands. At the same time, they must be able to avoid predators.

When the food supply decreases, waterfowl spend more time searching for food. The summer feeding behavior of lesser snow geese in Canada is illustrative (Figure 17–3). As the crude protein content of ungrazed vegetation begins to decline in August, snow geese and their goslings spend as much as 17 hours a day feeding.[5] Waterfowl that live in less than optimal areas often need larger home ranges. For example, mallards living in the prairie pothole region in North Dakota maintain a larger home range than those in central Minnesota.[6]

Management by the Federal Government

Pressure on the federal government from concerned conservation groups has been partly responsible for management efforts such as the Migratory Bird Treaty Act, which, among other provisions, prescribed regulations for waterfowl hunting. The Migratory Bird Stamp Act was passed when the public became aware of the heavy loss of wetland habitat during the late 19th and early 20th centuries in the midwest. By 1920, the loss of nest cover had resulted in a lower waterfowl population, and heavy droughts into the 1930s increased the concern.

The Migratory Bird Stamp Act was one of a number of efforts to maintain and restore waterfowl habitat. Another was the development of the National Wildlife Refuge Program. The Stamp Act provided funds for the purchase of refuges set aside primarily for waterfowl in the midwest. Congress later shifted the financing of the refuges from the general tax revenue to a tax on waterfowl hunters. Each time a hunter purchases a license, a stamp is purchased from the federal government, and stamp funds are earmarked for waterfowl management (Figure 17–4).

Figure 17–3 Snow geese. (Courtesy of the U.S. Fish and Wildlife Service.)

Figure 17–4 Barrow's goldeneyes have benefited from the Migratory Bird Stamp Act.

In the 1950s, the U.S. Fish and Wildlife Service and the states jointly determined that 5 million hectares (12.5 million acres) of waterfowl habitat needed to be placed under state and federal control to maintain waterfowl populations that then existed.[7] The federal share of the total was 3.2 million hectares (8 million acres), including 1.4 million hectares (3.5 million acres) already owned. By the end of 1976, about 769,000 hectares (1.9 million acres) had been purchased in fee title or easement by the government. From 1977 to 1986, the Service's waterfowl habitat acquisition program was guided by Waterfowl Habitat Acquisition Concept Plans, which identified and ranked, within 15 broad geographic areas, significant habitats of national importance to all waterfowl. A total of 728,000 additional hectares (1.8 million acres) was targeted for acquisition in the 15 areas, including 395,000 hectares (975,000 acres) in the three areas covering the prairie pothole region.[8]

In 1980, a new federal resource planning system evolved that focused on those in-dividual species and populations most in need of attention. Nine species of waterfowl (the Pacific white-fronted goose, cackling Canada goose, Pacific brant, mallard, black duck, pintail, canvasback, redhead, and wood duck) were determined to have serious habitat-related problems. To better address the habitat needs of these species, the government de-veloped a new acquisition strategy—the Ten-Year Waterfowl Habitat Acquisition Plan. This document identified the national habitat protection needs of the nine species within 11 habitat acquisition priority categories. Approximately 1.1 million hectares (2.7 million acres) were targeted nationwide for acquisition in these categories from 1986 to 1995.

The North American Waterfowl Management Plan was completed in 1986.[9] The Plan recognizes 34 waterfowl areas of major concern, including 21 in the United States. For the period 1986–2000, the North American Plan established an acquisition goal of some 769,000 hectares (1.9 million acres) in the six areas of highest priority in the United States, including 445,000 hectares (1.1 million acres) in the prairie pothole region. Thus, acquisition and management are a major approach by the federal gov-ernment to manage waterfowl species.

After 10 years, proponents of the management plan summarized their accom-plishments. Not only did the plan support a major increase in waterfowl, but signifi-cant steps to improve habitats were taken. Several joint ventures were in progress: The San Francisco Bay area program, the Great Lakes program, and a number of flyway joint venture programs were in place to improve waterfowl habitat. These projects in-volved joint work with state and local governments. Plan supporters pointed out that their efforts help improve not only waterfowl numbers, but in fact, all wildlife.

Flyways

By 1936, four waterfowl flyways had been identified by intensive analysis of waterfowl banding (Figure 17–5). The North American continent is divided into these four regions for migratory-bird management, and a waterfowl biologist is assigned to each of the flyways. Through the years these biologists have developed a variety of survey techniques to determine usage of the flyways by waterfowl and have systematically gathered data for each flyway. As a result, information is available on the number and species of birds utilizing each flyway for migratory stopovers, breeding, and wintering.

In January 1947, aerial surveys were undertaken over winter waterfowl grounds in Mexico, primarily in the coastal areas and later inland. As the information about winter, migratory stopover, and breeding areas increased, biologists were able to make more substantial management recommendations.[10]

Today each flyway has a council with representatives from the federal government, all states in the flyway, and, at times, private conservation organizations. Each council meets at least once a year to develop management plans for species under its jurisdiction. The plans establish goals for the number of birds and hunting days, define problems associated with species management, and make recommendations for im-

Figure 17–5 Administrative waterfowl flyways in North America.

provement and acquisition of habitats, as well as for reintroduction programs. They also spell out survey and research programs to be used in reaching these goals. Since these plans are developed through collaboration of the federal government and the states, all political subdivisions try to follow the procedures. Planners and developers find the plans helpful in analyzing habitat alterations in the flyways that might affect waterfowl.

Flyways represent geographic boundaries that can be used for administrative purposes, but waterfowl generally follow narrow corridors in these areas. By means of recovery of banded birds and radar surveillance, biologists can define the areas where the heaviest migrations of each species occur. Some of the corridors cross more than one flyway; also, different populations of a species sometimes use different corridors (Figure 17–6). Because of this irregular use of flyways, managers must evaluate more than just geographic boundaries. Currently, the U.S. Fish and Wildlife Service, in cooperation with the Canadian Wildlife Service and a number of private organizations, conducts annual breeding-ground surveys and takes winter inventories. Data are gathered by flyway and are combined to help managers draw up regulations.

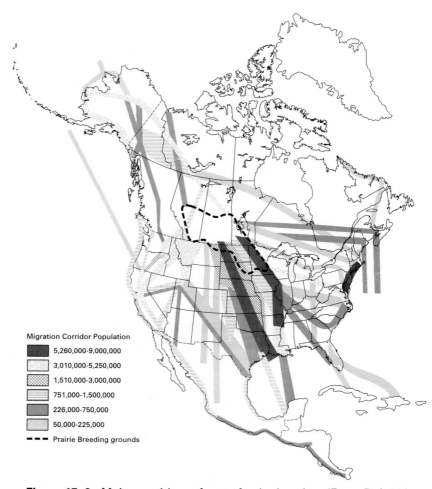

Migration Corridor Population

■	5,260,000-9,000,000
░	3,010,000-5,250,000
▨	1,510,000-3,000,000
▤	751,000-1,500,000
▦	226,000-750,000
▨	50,000-225,000
- - -	Prairie Breeding grounds

Figure 17–6 Major corridors of waterfowl migration. (From Ref. 11.)

In the spring and fall, many waterfowl that nest in North America migrate thousands of miles, others only a few hundred miles. Many so-called resident species go south and are replaced by a more northerly population that migrates into the vacated areas. Most people, not realizing that there are different populations of the same species in their area all year, think that they are seeing residents. Some populations of wood duck and hooded mergansers do remain throughout the year.[1] (Figure 17–7).

For the most part, waterfowl make very long migratory flights. Pintail ducks banded in Alaska have been found 8,050 km (5,000 mi) away, in Guatemala. Some populations of the same species winter in a number of different areas. For example, tundra swans that nest in the northern Arctic and Alaska migrate south into the United States. The western population, which nests along the northern Alaskan coast, winters in the Central Valley of California. The eastern nesting population migrates to Chesapeake Bay on the eastern seaboard (Figure 17–8).

Causes

Biologists speculate on a number of reasons that migration occurs. Some feel that birds return to their ancestral home each winter. Food appears to be a reason for waterfowl migration. As the harsh winters set in on the northern part of the continent, waterfowl migrate south until food becomes available. Canada geese, for example, are likely to remain in an area until winter conditions make it difficult to find food. Then they fly south until food is available, and when winter reaches the new location, they continue south. In years of particularly mild winters in lower altitudes, the geese are not likely to migrate at all.

What else triggers migration? Probably, a number of things act together. For some birds, the length of the day, or photoperiod, appears to be important. Apparently, the photoperiod acts somehow on the endocrine system of these birds and controls their

Figure 17–7 Hooded merganser.

Courtship

Many species of waterfowl are closely related genetically. In fact, some species can crossbreed and produce viable offspring. We see this with some hybrids in city parks. In the wild, distinct courtship patterns often keep the different species from interbreeding. Elaborate courtship displays involve specific movements that provide clues from prospective mates. Some species initiate courtship on winter grounds and then migrate to a nesting area. Ruddy ducks, on the other hand, begin courting after they arrive on the breeding grounds. The males usually arrive a week or so before the females. The male (drake) courts a prospective mate on the water by beating the underside of his bill against his neck. This produces a rapid tapping sound, and as air is forced out of the breast feathers, a ring of bubbles appears. Ornithologists aptly refer to this as bubbling. The drake generally displays before a number of females. Once he is accepted by a female, he swims around with her before mating. The pair then disperses to a small pond to construct a nest and lay eggs.

Boon, L. 1997. The Amazing Dollar Duck. *Ducks Unlimited* 61:12–13.

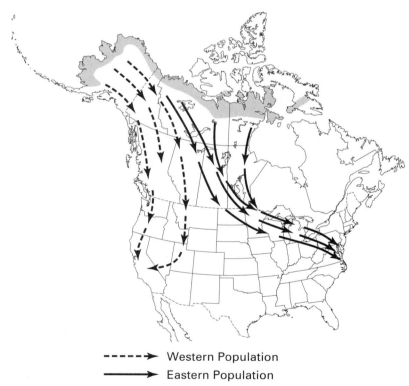

- - - - -▶ Western Population
———▶ Eastern Population

Figure 17–8 Tundra swan migration routes.

biological clock. This, in turn, controls the buildup of fat reserves in the body, providing the energy needed for long flights. Fat is important because it furnishes twice as much energy per unit of weight as does protein. When food is short in an area, birds need more time to prepare for migration.[11]

Studies also indicate that for some species, aspects of weather can trigger migration. In Canada geese, which go looking for food, migration is correlated with wind and temperature changes; heavy migration occurs under or near the west side of a ridge in a high-pressure area when winds are favorable.[12] Snow geese usually begin migrating northward from Texas within three to five days after the maximum daily air temperature first reaches 29°C (84°F) and remains at 18°C (64°F) or above for four days. Departures are not related to minimum temperature, surface wind, atmospheric pressure, relative humidity, sky cover, or precipitation.[13]

Navigation

Waterfowl tend to return to the same breeding area year after year. Some species return even to the same nest site if it has not been disturbed. (This behavior is called "site tenacity.") Apparently, the birds imprint on the area (i.e., recall the specific area) in which they were born.

Waterfowl have evolved the ability to use different clues as guides in migration.[14] Some make extensive use of the landscape in recalling areas where they have been before. Young birds migrate with older birds the first year and apparently always have some way of remembering rivers, ridges, ponds, and other landmarks so that they can return along the same route. Celestial bodies, the earth's magnetic field, and polarized light patterns, in combination with landmarks or alone, also appear to guide some species of waterfowl.

Timing

Although we have made the general statement that waterfowl migrate northward in the spring and southward in the fall, the times of year can differ. Some species begin their fall migration in July, others well into the winter. The time of day also varies. Many ducks and geese migrate during the day, then feed, and rest during the night. But some waterfowl migrate at night. Hunters sitting in their blinds frequently observe ducks and geese in flight throughout the night. In many parts of the country, it is not uncommon to hear Canada geese calling during the night in both spring and fall.

NESTING SUCCESS

Most waterfowl species are **indeterminate layers**—that is, they continue to lay eggs until the nest is filled, at which time some stimulus terminates egg laying. If eggs are removed from the nest, the bird may continue to lay. The clutch size—number of eggs per nest—varies within and between species. Mallards commonly lay 11 to 15 eggs, while geese lay 3 to 8. Most young can walk, swim, and find their own food as soon as they are hatched and dry. These young are called **precocial,** compared with **altricial** young, which are helpless and require food from the parents. Waterfowl young remain with the parents throughout the summer months and frequently migrate in the same flock as their parents, although the latter varies with species. In early to middle summer, on lakes and ponds of North America, it is quite common to see two adult birds with a string of young following. At this time, the young are vulnerable to predation and a variety of diseases.

Predation

The predatory loss of waterfowl eggs and young can be considerable. Most of the time, waterfowl nests on islands are isolated from predators, but changes in the water level sometimes create access routes for predators. The presence of humans can cause an increase in rats, which eat birds' eggs.

Drought

Throughout the prairie region, droughts occur at irregular intervals. These are difficult times for waterfowl production. In a severe drought in 1958, only 8 percent of southwest Manitoba's lesser scaup breeding population nested.[15] Droughts not only cause physiological changes in the birds' bodies, but alter the habitat and create different balances in the ecosystem (Figure 17–9).

There is evidence that when the prairie potholes in the north-central United States and southern Canada suffer from drought, waterfowl populations move farther north to breed. This puts added stress on some arctic habitats. Drought intensifies the problem of habitats reduced in area by lessening wetlands. It is very important, then, that managers decrease bag limits in drought years. Major reductions in hunting occurred in the late 1980s in response to the severe drought then.

Oil and Chemicals

Petroleum products that find their way into wetland habitats have an adverse impact on both the reproduction of waterfowl and the activity of the young. When mallards ingest crude-oil products, their eggs are lighter and smaller than normal.[16] Ingestion also delays the onset of laying, inhibits egg production, lowers hatchability, and reduces the thickness of the eggshells.[17] Hatchability is lowered when eggs come in direct contact with petroleum products. Young birds fed low levels of petroleum products have depressed growth rates during the first eight weeks of life and undergo physiological changes that could affect their migration and survival later.[18]

Since both breeding and winter wetlands are subject to contamination by petroleum products, energy development, and waste disposal, managers must educate the public in the effect of these types of contamination on waterfowl production. Further damage occurs when feathers become coated with oil products. Rehabilitation measures are usually taken following large oil spills. The oil is removed from the waters, and waterfowl are collected, cleaned with detergents, and then released. There are as-yet unanswered questions about the birds' survival.

Figure 17–9 Shovelers have been harmed by drought.

Waterfowl management includes (1) harvesting regulations, (2) infectious disease control, (3) habitat management, and (4) predator control.

Hunting

Waterfowl provides both sport and food for the American hunter. Studies show that regulated hunting of waterfowl reduces other causes of mortality, such as disease. When game agencies have shortened hunting seasons, the survival rate and size of the breeding population have not increased. In black ducks, changes in hunting regulations affected male and female survival differently. Differences were also found in regional survival rates for black ducks when hunting regulations were changed.[19] While hunters are thought to take approximately one-fifth to one-fourth of the mallard population in an average year, hunting accounts for only one-third to one-half of the total deaths.[20] This means that a variety of environmental factors limit the population when hunting does not occur.

Managers need to discover the population level below which hunting can have an adverse impact on the total continuing population. The threshold level can vary geographically and annually because of drought, habitat alteration, or pollution, and so must be determined seasonally for each nesting area. Properly regulated, hunting can be used not only as a compensating mortality factor but also as a stabilizing factor.

Regulations. Since waterfowl are migratory species, their hunting regulations are the responsibility of the federal government. This responsibility is carried out primarily under the U.S. Fish and Wildlife Service, but is shared with the various state governments through the four flyway councils. There are a number of regulations for waterfowl. The Migratory Bird Treaty Act requires a closed season for wild ducks, from March 10 to September 1. Flyway councils generally set framework dates for duck hunting of October 1 through January 20.[21] Within the framework, states can choose a season that coincides with an abundance of ducks. It is possible to split the allowed time and have two hunting seasons if states want to take advantage of different peaks in duck abundance. States also may petition flyway councils for special seasons or special takes if they feel that a population should be reduced to within the carrying capacity of the habitat. It is also possible for states to have two or more areas with different hunting seasons. This will occur in states where the distribution of waterfowl varies because of geographic differences. Shooting hours are generally established by treaties; with few exceptions, they extend in the United States from a half hour before sunrise to sunset.

Waterfowl harvests can be regulated. Traditionally, a *fixed bag limit* defines the number of ducks allowed per day. The number may vary by species and/or sex. The take of ducks is based on management objectives. When populations need protection, the take may be reduced to numbers less than the combined bag limit (*restriction*). If a population needs to be reduced, hunter take can be increased to a number greater than the normal bag limit (*bonus*).

The *point bag regulation,* another technique, assigns lower point values to ducks needing less protection and higher point values to ducks needing greater protection. The bag limit under the point system is calculated by adding the point values of all the ducks shot. In recent years, the bag limit has been 100 points.[21]

Surveys of breeding and wintering waterfowl are used by biologists to determine the season and bag limits. The data are compiled by the U.S. Fish and Wildlife Service, which utilizes a number of statistical and modeling techniques to ascertain the status of the different populations around the country. These techniques are based, to some extent, on mark-recapture information or on population trends. If a population appears to be doing well, biologists generally recommend the same, or only slightly increased, hunting pressure. If the data indicate that a population is declining, the bag limits will probably be decreased and areas closed to hunting. Information required to compile hunting regulations includes estimates, made after the previous hunting season, of breeding-population levels, production, harvest, and mortality rates. Data used for the estimates include banding returns, harvest takes, and wing surveys (Figure 17–10). Some areas require hunters to send in a wing of each bird taken. Biologists have conducted extensive banding programs to assist in discovering movement patterns, survival rates, and the effect of hunting on different age groups and sexes. The banding data are sent to the U.S. Fish and Wildlife Service in Laurel, Maryland, which keeps data on all banded birds.

Lead Shot.   One of the problems associated with waterfowl hunting has been poisoning by lead shot. It is estimated that in the United States, about 2 million ducks, or 2 to 3 percent of the fall and winter population, die each year from lead poisoning as a result of swallowing spent shotgun pellets when feeding on bottoms.[1] These pellets, when swallowed, pass to the gizzard, where they are converted to a soluble form and absorbed into the bloodstream. The lead causes a reduction in the oxygen supply to all tissues, thus interfering with the body's ability to break down sugar and other carbohydrates. It also disrupts the production of hemoglobin, resulting in anemia. This imbalance of the blood chemistry causes the duck's liver and heart to function improperly. The symptoms are apparent loss of weight, wing droop, refusal to eat (because of paralysis of the digestive system), a tendency to seek isolation and cover, and loss of ability to walk.

Because biologists feel that lead poisoning affects the total number of birds available for hunting, and because they are concerned about human ingestion of lead from ducks, evaluations of the impact of swallowing lead pellets from shotguns are under way. A proposal to require steel shot has caused a protest from hunters who object to steel shot because of its higher cost and reduced penetration. In fact, the controversy has led to heated debates before Congress and in public meetings. This disagreement is intensified because of the danger of lead shot to some endangered bird species, such as the whooping crane. Checks are being made to see whether steel shot will reduce

Figure 17–10   Waterfowl wings. (Courtesy C. Patterson.)

mortality in waterfowl species. Thus, in 1989, waterfowl hunters were required to use steel shot in zones in 46 states. Copper and various alloys, as well as plastic buffering, are also being tested. (The fine plastic particles help prevent deformation of the spherical shot pellets, giving a better pattern than that of the more expensive metal shells).

Disease

Diseases can have a major impact on waterfowl populations, even when they are spread by only a few members of the population. Deaths by disease may go unnoticed unless a massive outbreak occurs. Thus, it is important to recognize disease early. Doing so requires keen observation, experience, and a knowledge of factors associated with the habitat that induce the growth of disease organisms. Once managers notice symptoms in a few individuals, they should ask for the help of disease laboratories and veterinarians and immediately isolate the individuals affected.

One bacterial disease that affects waterfowl is avian cholera (Figure 17–11), an acute infectious disease often picked up by wild waterfowl from contaminated water or from contact with infected domestic chickens and turkeys.[22,23] The cholera bacteria are transmitted through the water that birds drink and can also be contracted by inhalation. The disease can progress rapidly through waterfowl populations when they concentrate during migration, especially if the birds are suffering from poor nutrition. Control consists of removing dead or diseased birds from the area as soon as possible. The disease is not readily transmitted to people.

A toxin produced by an anaerobic bacterium causes botulism, which can infect human beings when it grows in food that has not been thoroughly preserved such as canned items. When botulism strikes ducks, they become somewhat listless and have difficulty holding up their heads, a condition described by the popular term *limberneck*. Outbreaks of botulism have been known to destroy massive numbers of ducks. The dead ducks need to be disposed of properly, since flies lay their eggs in carcasses and spread the disease through the bodies of the maggots that hatch and infest other waterfowl. The bacteria seem to live in water or marshlike areas late in the summer, when water levels are low and there is little circulation. Draining these areas can help control the disease. It is possible for birds affected with botulism to recover if they are placed in isolation pens and given fresh water.[1]

Several viral diseases, including duck plague, crane herpes, and avian pox, are reported in some species of waterfowl.[23] Parasitic diseases can also harm waterfowl

Figure 17–11 Pintail dying from avian cholera. (Courtesy T. Thorne, Wyoming Game and Fish Department.)

populations. Leucocytozoon is a parasitic disease that affects the blood of birds. It is commonly transmitted during the breeding season by black flies from adult birds, which apparently act as reservoirs, to young birds. This parasite causes an anemia that results in the death of the young. The Canada goose population on the Seney National Wildlife Refuge in the upper peninsula of Michigan appears to be susceptible to leucocytozoon. Approximately every four years, a large portion of the gosling population is virtually destroyed at the age of two to three weeks. Currently, there appears to be little that managers can do to prevent this cyclic loss of goslings.

Habitat Management

Waterfowl require a number of habitats for survival. Figure 17–5 shows the major migratory routes for most waterfowl. These areas are important habitats that birds need for resting and feeding during migration. They also require a winter habitat in which water remains open and food is available. Some birds migrate to the southern part of the United States, others to Central and South America.

Breeding habitat is important, too. The birds must have various habitats for courtship, nesting, and raising their young. Each species of waterfowl requires a slightly different combination of these types of habitat. In managing waterfowl, we must address all of the birds' needs throughout the year. Losing one habitat can mean loss of a population. To a large extent, habitat management involves structural manipulation to provide nesting areas, food, and protection from predators. Biologists have found that manipulation of both aquatic habitat and vegetation helps maintain waterfowl. In Manitoba, Canada, the responses of breeding dabbling ducks to their food source, aquatic microinvertebrates, were compared with responses to modifications in the cover–water ratio. Emergent vegetation and open water were regulated in three sites in the proportions of 30-percent cover vegetation and 70-percent open water, 50:50, or 70:30. The greatest density and species diversity occurred in the 50:50 plots, apparently in response to changes in the invertebrate populations.[24] A similar study was made in Iowa, where increased water levels produced an approximate 50:50 interspersion of plant cover and open water. The density and species diversity of dabbling duck breeding pairs increased on these plots.

Management strategies in wetlands should consider the effect of pond densities on the abundance of both breeding ducks and their young. The temperature, type of soil, drainage, precipitation, and evaporation determine the plant associations that directly or indirectly affect the breeding distribution and abundance of some species of ducks.[25] Waterfowl also need densely vegetated areas to hide in when molting. Some species of waterfowl prefer upland areas for nesting. Studies of blue-winged teal indicate that areas where grasslands are maintained by burning or grazing produce more teal than areas where the grass is simply allowed to grow.

Pond size was important in duck utilization in western North Dakota. As pond size increased, total pair and brood use per pond also increased. While pairs were found on ponds as small as 0.04 hectare (0.1 acre), broods were generally not observed until ponds were larger than 0.4 hectare (1.0 acre) in surface area. Ducks appeared to prefer older ponds with grassy shorelines. This information is relevant to plans that encourage waterfowl production.[26]

When succession is controlled, desired waterfowl can be maintained. A shallow lake of 35 hectares (86 acres) was created in northern Sweden by flooding a sedge meadow. Within four to five years after flooding, the food supply apparently increased to such a level in the succession sequence, that duck populations peaked. Although there was a decrease in the populations after that period, it was apparently because of other changes in the maintenance of the water areas.[27]

Artificial islands have become important in maintaining waterfowl populations. Islands are sanctuaries for both nesting waterfowl and other species of birds. In some areas, special structures, such as driftwood nest enclosures, have been placed on the islands to provide cover and promote nesting activity (Figure 17–12). Artificial islands are often constructed on water impoundments to provide alternative nesting areas when habitat around the edge has been lost. Researchers have found that the density of nesting ducks is greater on small islands that are relatively far from shore and have a fairly high vegetation cover.[28] Rectangular islands appear more attractive to waterfowl because they have a greater perimeter than circular, elliptical, or square ones. The greater the ratio of water–land edge to landmass, the more attractive the insular habitat. Also, rectangular islands require only limited surveying and are easier to build. In some prairie pothole regions, relatively small islands have proved to be good waterfowl nesting areas. Islands can be placed in small impoundments, with little or no aquatic vegetation around the edge. They afford protection from predators.

Nesting and food supply are closely tied together. The annual food supply can affect the number of eggs and the time the eggs are laid. It is known that what food breeding females will eat is determined in part by their reproductive status and the time of year. It appears that they need a greater supply of protein before nesting and migration.[29] Nesting structures have been used successfully to compensate for a lack of natural nest sites and habitats capable of supporting waterfowl. Goose nesting platforms can be placed on or in water that has food. Often a tire placed on a flat platform attached to a pole will attract Canada geese, and if the platform is out in the water, they

Figure 17–12 Artificial islands can provide nesting cover and freedom from predators, allowing a waterfowl population to reproduce.

are generally protected from predators as well as from human intruders. (People can stand along the shoreline, however, and watch the nesting progression). Next boxes have been used to attract a number of species. Goldeneyes will use boxes with black interiors and relatively large entrance holes. Wood ducks use both wooden and metal nesting boxes placed in the middle of impoundments, away from predators. The recovery of wood ducks from the declines in population discussed earlier was due, to a large extent, to nest boxes.

The number of cattle grazing around ponds and marsh also influences waterfowl production. When rest-rotation grazing (no grazing for a season) occurs in fields with ponds, waterfowl production generally increases the following spring. When grazing occurs in late fall, waterfowl production the next spring is generally down.[30] Fire and mowing do not appear to have a major impact on puddle duck production when they occur in the fall of the year, prior to nesting.

Fences prevent cattle from intruding on heavy waterfowl nesting. Excluding cattle from riparian habitats in grassland areas having relatively few nesting sites has also been successful. Special attention needs to be given to the timing of activities: Tilling and mowing must be timed so as not to disturb nesting or essential cover.

Migratory Habitat.　Management of habitats for migrating birds is very important even though they are used for a short period of time. Impoundments must be maintained to provide an adequate food supply during migratory stopovers. These areas can be utilized for both fishing and hunting. Maintenance of natural vegetation along the shore helps protect an area from predators that might remove birds stopping for a short time.

Winter Habitat.　Wintering habitats are different for each species, and some species, such as the Canada goose, have a number of populations, each of which winters in a different region. Often, the birds are crowded into relatively small land areas during the winter, a situation that adds to stress. Birds that winter in colder latitudes generally lose body weight, making them more susceptible to disease and cold. In severe winters, there can be large die-offs. Some wildlife refuges have been established in areas of heavy winter concentrations of waterfowl. A number of wintering waterfowl areas are popular with bird-watchers, whose presence does not appear to affect the birds adversely.

Stocking.　Artificial stocking has also been used to maintain waterfowl populations. Canada geese, mallards, wood ducks, gadwalls, and redheads have been used to restock locations where waterfowl populations have declined. As with the introduction of any other species, it is important that artificial restocking programs be undertaken only after a thorough evaluation of the location is made in relation to the species. Restocking from game farms has not been particularly successful in establishing breeding populations. Most game-bird farms release birds just before the hunting season, to increase hunter success. Generally, the release of wild strains has brought the greatest success. When birds are released at an early age and forced to remain in the area for a time, they apparently migrate, but return the next spring because of imprinting. Moving nests with eggs has met with only moderate success in wood ducks.[31] In moving boxes with newly hatched birds or eggs about to hatch, researchers in the northeast found that more than 50 percent of the eggs or broods were later abandoned.

Canada geese have probably been one of the best success stories in restocking programs. Some of these geese were established in the Rockefeller Refuge in Louisiana

Bentonite Mining and Waterfowl Habitat

Volcanic ash from erupting volcanoes now in the Yellowstone National Park region spewed eastward millions of years ago. In eastern Wyoming and western South Dakota, the ash produced deep deposits, upon which additional soils have collected in layers. Bentonite from this ash has since become a valuable product for use in such items as ice cream and kitty litter and as a cleanup material for toxic waste.

Many mining companies now remove the topsoil and scrape layers of bentonite from below. When they are finished removing the bentonite, they recontour the depressions and stabilize the banks by planting grass and shrub seeds. New data collected by biologists show that these depressions collect water during the winter and often retain it throughout the dry summer. The more than 300 ponds in bentonite-mined areas are quite attractive to waterfowl and other wildlife. Working with the mining companies, biologists guide the reclamation work of recontouring and seeding to attract waterfowl. The ponds now attract breeding waterfowl. Game and fish departments stock some ponds with fish to provide a fisheries resource. All species of wildlife in these areas are increasing because of the year-round water supply.

McKinistry, M. 1993. Evaluation of Wetlands Created or Reclaimed as a Result of Mining in Northeast Wyoming. Master's thesis, University of Wyoming.

during the early 1960s. The objective was to establish a population of birds for sport hunting. The birds quickly became acclimated to the warmer temperatures, and the flocks grew and nested extensively.[32]

Refuges as Waterfowl Management Areas. Duck hunters often view refuges as essential to the perpetuation of their sport, but as interest in nongame bird-watching grows, others feel that the refuge system should not be utilized by hunters. There are those who view refuges simply as duck factories that should produce a quota of birds per acre, birds per day, or birds per duck-stamp dollar. These attitudes must be considered and attempts made to reconcile them at each wildlife refuge.

Refuges maintained for waterfowl usually need frequent extensive work if they are to maintain the proper habitat. Levees and other water-control structures and roads to facilitate patrolling, management, and hunting are needed. Refuges tend to dictate the distribution of waterfowl, so that it is important for a well-managed network of refuges to be developed to achieve a good distribution of waterfowl over a wide area.

Biologists are finding that a more varied and low-key feeding program may be the best method of managing refuges for waterfowl. A number of agricultural crops are attractive to, and nutritious for, the birds. In an appropriate marshland, foods can often be grown with forage yields approaching those of agricultural crops. A good stand of smartweed or pondweed can yield a high poundage of seed per acre. Alkaline bulrush is one of the more productive waterfowl foods in western marshes. It is impossible to write a generalized formula for food management, but variety is much more likely to meet nutritional requirements than is monoculture. Many refuges, with the major objective of managing waterfowl, plant grain to attract the birds. Some of the more difficult management problems with Canada geese have arisen because the geese become used to this easy forage and ignore their natural diet.[33]

Predator Control

In the prairie pothole region, uplands with tall, dense nesting cover can produce nesting success of 70 to 90 percent with predator management. Without predator management, such cover produces success rates of 50 to 56 percent. Because of the limited cover in the pothole region, upland nesting ducks concentrate their nests, possibly making them more susceptible to predators. Recruitment of young can be very low where red fox are abundant. Skunks, weasels, raccoon, and coyote will also remove eggs and young from nests. Snapping turtles may take large numbers of young birds. These and other predators can frequently be controlled with structures such as islands and nesting platforms. Trapping and removal programs are also effective. Some places have been closed to the public so that bird-watchers, in approaching the nests, do not open paths that can later be used by predators. At any rate, in areas where predators are particularly high and access is easy, some kind of predator control program should be instituted.

DESCRIPTIONS OF SPECIES

Swan

The tundra swan, formerly called the whistling swan, is the most common swan in the United States (Figure 17–13). Trumpeter swans are found in some areas of the west and midwest, and mute swans inhabit the northeast and east coast. Mute swans are found in many parks. Tundra swans breed along the arctic coast in northern Canada and migrate to the eastern Chesapeake Bay and west coast. Nesting usually occurs near lakes, large rivers, and estuaries of the far north in Canada and Alaska. Swans build a nest of moss and grasses on an elevated hummock in calm water. The three to five eggs commonly laid are vigorously guarded, as are the young.

Swans eat a variety of aquatic vegetation. They prefer to feed in shallow water, where they can scrape plant material from the bottom. Predators do not appear to be a major problem in most of the range, because swans attack other animals with a vengeance. Habitat modification and pollution are the major threats to swan populations. The polluting of estuary water in the winter grounds, destruction of migrating stopover areas, and invasion of breeding areas by mineral developers all threaten swan welfare and survival.

Figure 17–13 Tundra swan. (Courtesy of the U.S. Fish and Wildlife Service; photo by R. Erickson.)

Canada Goose

The Canada goose, the most common goose in the United States, is divided into 11 races, based on size and habitat use. It is common to hear flocks migrating over cities and countryside during the fall and spring of the year.

Canada geese breed in a diversity of habitats in the United States and Canada. They prefer some water nearby, nesting in marshes and on cliffs, islands, and elevated platforms in lakes. The average clutch size of the Canada goose is five to six eggs, but the variation is great. The female usually incubates while the male is nearby. When the young hatch, they are led from the nest by the male and female. They usually like to go to water, but some semidomestic birds head for lawns, parks, and golf courses. Geese seem to take other pairs' young. It is not uncommon to see one pair with a following of 15 to 25 young, other pairs having lost some or all of their young to that group (Figure 17–14).

Geese eat a variety of grain crops, aquatic plants, and some insects. Corn and wheat fields attract them. Migration seems quite variable: The birds in the north migrate annually, but some of the birds that have adapted to parks seem not to migrate at all. Managers are often called on to deal with nuisance birds that invade parks and suburban lawns. Removal appears to be the only means of dealing with these individuals.

Mallard

The mallard breeds along any sort of body of water, including irrigation ditches and stock ponds. The females usually nest on the ground under trees or shrubs, but have been known to use old crow or magpie nests. Mallards generally lay eight or more eggs. The female leads her brood to water shortly after hatching. The birds eat a variety of vegetation, both wild and cultivated, and sometimes insects.

Some of the more domestic variants and southern birds appear not to migrate, but most mallards do. They are late migrants in the fall, but are among the earliest birds to travel to their breeding grounds in the spring. Since mallards adapt easily, management involves maintenance of isolated wetland habitats during the breeding season. The

Figure 17–14 Canada geese are now found in many urban areas.

availability of nest cover is very important in maintaining mallard breeding populations. People's disturbing nests often opens areas for predators to enter, and high predator populations affect mallard survival.

Canvasback

Canvasbacks are diving ducks that breed in the reeds of freshwater lakes, salt bays, and estuaries of the northern United States, Alaska, and Canada. (The prairie pothole area is popular.) Females use down-lined basket nests, and some have been reported to breed in their first year. They generally have more than eight eggs. The young feed in and among the reeds on plant material and will also take insects. The birds winter along much of the coastal regions of the lower 48 states. Some winter in inland waters.

Predators take a large number of canvasbacks. In some years, nesting success is very low. In addition, canvasback nests are a favorite drop nest for redhead duck eggs. (Redheads sometimes parasitize a number of waterfowl species in this manner.) Once the eggs hatch, the canvasback generously raises the redhead young. Habitat management involves the maintenance of isolated wetland areas for nesting, an effort that will probably also help reduce predation.

Bufflehead

Buffleheads are small ducks. The males have a large white patch running from their eyes to the back of their head. They breed in Alaska, Canada, and parts of the western United States and winter in lakes and ponds, as well as parts of the coastal areas of the lower 48 states. Buffleheads generally do not breed until they are two years of age. They like to nest in small tree cavities fairly close to water; flicker cavities are a favorite. Females lead the young to water, sometimes a far distance over land. They feed on aquatic insects, aquatic plants, and some fish.

Since buffleheads are secondary cavity nesters, they depend on other birds to do the excavating. Several other species of cavity nesters compete with the bufflehead for the cavities. Some birds will enter bufflehead nests and puncture their eggs. Water free from excessive pollution is required for brood rearing, migration, and wintering.

Merganser

Three species of mergansers are found in North America: hooded (Figure 17–7), common and red breasted. All three breed around water, mostly in the northern United States and Canada. The hooded species uses a free cavity or stump, the common species uses a hollow tree (sometimes nesting on the ground under brushes and rushes), and the red-breasted species nests under bushes, trees, and roots. The three species winter in different waterways along the coast or inland.

The common merganser, which is the most numerous throughout the country in the winter, is discussed here. It prefers hollow trees near water. When cavities are used, predation is minimal. The common merganser normally lays 6 to 12 eggs. A day or two after the chicks hatch, they jump out of the nest at the encouragement of the female. Several broods may join together. The birds do not usually breed until their second year. Courtship activities are followed by the dispersal of pairs to widely scattered sites. The male usually leaves the female shortly after incubation begins. Next boxes are sometimes used.

The common merganser's food is chiefly fish, so that pollution-free waters are important. During migration, which can occur day or night, mergansers seem to flock to the same stopover areas. In the winter, they use open ponds or lakes to look for food.

Management of the common merganser consists of maintenance of waterways with snags nearby.

Ruddy Duck

Ruddy ducks are small, chubby birds with short necks and stiff tails. They nest in Canada and the northern United States. A large number use the prairie pothole area.

The birds use emergent vegetation near the water, usually lakes or ponds. They build nests in the water with dead vegetation; some nests actually float on decaying vegetation. The female lays 5 to 15 eggs at a time and sometimes dumps them in another marsh bird's nest. Both male and female may accompany broods when feeding. Winter habitat includes inland lakes and coastal bays, estuaries, marshes, and some rivers. Management centers on maintaining marsh and wetland habitats for nesting (see p. 359).

SUMMARY

Waterfowl are popular sport animals and the delight of many animal watchers. In the United States, waterfowl management has been tied closely to the development of wildlife management. The Migratory Bird Stamp Act was instrumental in getting funds to purchase land to manage and maintain waterfowl.

Treaties have established the federal government as the principal management group. The government shares this responsibility with states through flyway councils; flyways form the basis of waterfowl management in the United States.

Management centers on harvest regulation, disease control, predator control, and habitat management. Most waterfowl migrate between breeding and winter ranges. Thus, any management effort must consider breeding, migratory, and winter habitats. Adverse impacts on waterfowl reproduction have come from predation, drought, pollution and loss of wetland.

Management of waterfowl areas therefore involves a combination of techniques. Essentially, the area must have an adequate food supply to attract the waterfowl. Isolated areas with adequate cover for nesting are very important. These areas must be evaluated in light of the desired species' needs. In some cases, short grasses or long grasses should be maintained to attract the species. Finally, it is very important to keep the areas free from disturbance by humans and predators during the breeding season.

DISCUSSION QUESTIONS

1. How can we determine that hunting substitutes for other mortality factors in waterfowl?
2. Why is it important to know something about the breeding biology and habitat requirement of a species before reintroduction programs begin?
3. Does a program that results in the removal or extermination of predators to maintain waterfowl populations run counter to the principle of maintaining a natural system? Explain.
4. If hunting of a waterfowl population were not allowed, do you feel that the number of eggs and young fledged would decrease? Why or why not?
5. Can you suggest or visualize an example in which a waterfowl population has become or could become a nuisance species? What type of control measures would you recommend if this occurred near urban areas?

6. Is it appropriate that wildlife refuges be maintained solely for the purpose of keeping our waterfowl population? Discuss.

7. What might be the reason that some segments of our waterfowl populations which were formerly migratory are now year-round residents?

8. Do you feel that lead shot should be outlawed? Explain why, taking into consideration the fact that steel shot is more expensive and appears not to be as effective in killing waterfowl.

9. Discuss the important components of managing waterfowl populations. How can these components be combined with the objectives of a private landowner who has a cattle operation?

10. What are some of the techniques a manager can use to (a) prevent disease, (b) control disease, and (c) reduce disease in waterfowl populations?

LITERATURE CITED

1. Bellrose, F. C. 1980. *Ducks, Geese, and Swans of North America.* Harrisburg, PA: Stackpole Books.

2. Trefethen, J. B. 1975. *An American Crusade for Wildlife.* New York: Winchester Press and the Boone and Crockett Club.

3. Johnson, P. H., and A. B. Sargent. 1977. *Impact of Red Fox Predation on the Sex Ratio of Prairie Mallards.* Resource Report 6. Washington, DC: U.S. Fish and Wildlife Service.

4. Swanson, G. A., and M. I. Myer. 1977. Impact of Fluctuating Water Levels on Feeding Ecology of Breeding Blue-Winged Teal. *Journal of Wildlife Management* 41:426–33.

5. Harwood, J. 1977. Summer Feeding Ecology of Lesser Snow Geese. *Journal of Wildlife Management* 41:48–55.

6. Dwyer, T. J., G. L. Krapu, and D. N. Janke. 1979. Use of Prairie Pothole Habitat by Breeding Mallards. *Journal of Wildlife Management* 43:526–31.

7. Ladd, W. N. 1978. The Future: U.S. Fish and Wildlife Service Migratory Waterfowl Habitat Acquisition Program. *Transactions of the North American Wildlife and Natural Resource Conference* 43:226–34.

8. U.S. Department of the Interior. 1996. *Progress Report for Waterfowl Habitat Protection: Prairie Potholes and Parklands.* Denver, CO: U.S. Fish and Wildlife Service.

9. Canadian Wildlife Service and U.S. Fish and Wildlife Service. 1986. *North American Waterfowl Management Plan.* Washington, DC: U.S. Fish and Wildlife Service.

10. Saunders, G. B., and D. C. Saunders. 1981. *Waterfowl and Their Wintering Grounds in Mexico. 1937–1964.* Resource Publication 138. Washington, DC: U.S. Fish and Wildlife Service.

11. Bellrose, F. C. 1982. Waterfowl Migration Corridors East of the Rocky Mountains in the United States. In J. T. Ratti, L. D. Flake, and W. A. Wentz (Compilers), *Waterfowl Ecology and Management: Selected Readings,* pp. 1136–69. Washington, DC: The Wildlife Society.

12. Blokpoel, H., and M. C. Gauthier. 1980. Weather and the Migration of Canada Geese across Southern Ontario in Spring 1975. *Canadian Field Naturalist* 94:293–99.

13. Flickinger, E. L. 1981. Weather Conditions Associated with the Beginning of Northward Migration Departures of Snow Geese. *Journal of Wildlife Management* 45:516–20.

14. Lincoln, F. C. 1979. *Migration of Birds.* Circular 16. Washington, DC: U.S. Fish and Wildlife Service.

15. Rogers, J. P. 1982. Effect of Drought on Reproduction of Lesser Scaup. In J. T. Ratti, L. D. Flake, and W. A. Wentz (Compilers), *Waterfowl Ecology and Management: Selected Readings,* pp. 404–13. Washington, DC: The Wildlife Society.

16. Vangilder, L. D., and T. J. Peterle. 1980. South Louisiana Crude Oil and DDE in the Diet of Mallard Hens: Effect on Reproduction and Duckling Survival. *Bulletin of Environmental Contamination and Toxicology* 25:23–28.

17. Vangilder, L. D., and T. J. Peterle. 1981. South Louisiana Crude Oil and DDE in the Diet of Mallard Hens: Effect on Egg Quality. *Bulletin of Environmental Contamination and Toxicology* 26:328–36.

18. Szaro, R. C., M. P. Dieter, G. H. Heinz, and J. F. Ferrell. 1978. Effects of Chronic Ingestion of South Louisiana Crude Oil on Mallard Ducklings. *Environmental Research* 17:426–36.

19. Krementz, D. G., M. J. Conroy, J. E. Hines, and H. F. Percival. 1988. The Effects of Hunting on Survival Rates of American Black Ducks. *Journal of Wildlife Management* 52:214–26.

20. Anderson, D. R., and K. P. Burnham. 1978. Effect of Restrictive and Liberal Hunting Regulations on Annual Survival Rates of the Mallard in North America. *Transactions of the North American Wildlife and Natural Resource Conference* 43:181–86.

21. Anon. 1988. *Final Supplemental Environmental Impact Statement: Issuance of Annual Regulations Permitting Sport Hunting of Migratory Birds.* SEIS 88. Washington, DC: U.S. Fish and Wildlife Service.

22. Thorne, E. T., N. Kingston, W. R. Jolley, and R. C. Bergstrom. 1982. *Diseases of Wildlife in Wyoming.* Cheyenne, WY: Wyoming Game and Fish Department.

23. Friend, M. 1987. *Field Guide to Wildlife Diseases.* Resource Publication 167. Washington, DC: U.S. Fish and Wildlife Service.

24. Kaminski, R. N., and H. H. Prince. 1981. Dabbling Duck and Aquatic Microinvertebrate Response to Manipulated and Wetland Habitat. *Journal of Wildlife Management* 45:1–15.

25. Bellrose, F. C. 1979. Species Distribution. Habitats and Characteristics of Breeding Dabbling Ducks in North America. In T. A. Bookhout (Ed.), *Waterfowl and Wetlands: An Integrated Review,* pp. 1–16. La Crosse, WI: La Crosse Printing Co.

26. Lokemoen, J. T. 1973. Waterfowl Production and Stock-Watering Ponds in the Northern Prairie. *Journal of Range Management* 28:37–42.

27. Danell, K., and K. Sjoberg. 1982. Successional Patterns of Plants, Invertebrates, and Ducks in a Man-Made Lake. *Journal of Applied Ecology* 19:395–409.

28. Giroux, J. 1981. Use of Artificial Islands by Nesting Waterfowl in Southeastern Alberta. *Journal of Wildlife Management* 45:669–79.

29. Drobney, R. D., and L. H. Fredrickson, 1979. Food Selection by Wood Ducks in Relation to Breeding Status. *Journal of Wildlife Management* 43:109–20.

30. Gjersing, F. M. 1975. Waterfowl Production in Relation to Rest-Rotation Grazing. *Journal of Range Management* 28:37–42.

31. Capen, D. E., W. J. Crenshaw, and M. W. Coulter. 1982. Establishment of Breeding Populations of Wood Ducks by Relocating Wild Broods. In J. T. Ratti, L. D. Flake, and W. A. Wentz (Compilers), *Waterfowl Ecology and Management: Selected Readings.* Washington, DC: The Wildlife Society.

32. Chabreck, R. H., H. H. Dupuie, and D. J. Blesom. 1974. Establishment of a Resident Breeding Flock of Canada Geese in Louisiana. *Proceedings of the Annual Conference of the Southeast Association of Game and Fish Commissioners* 78:442–55. New Orleans.

33. Baldassarre, G. A., and E. G. Bolden. 1984. Field-Feeding Ecology of Waterfowl Wintering on the Southern High Plains of Texas. *Journal of Wildlife Management* 48:63–71.

18

SHORE AND UPLAND BIRDS

Shore and upland birds include species in five different orders (Figure 18–1). Some of these species are seldom seen by people, some are hunted, and others are considered nuisances. Some of the species nest in colonies, others in isolation. Some nest on rocky islands, some near the shore, and others in island forests. Why is such an assemblage lumped together as shore and upland birds? The reason is that most of these birds were at one time hunted or sought after because of their eggs or feathers.

Many of the species discussed in this chapter are migratory; some come under the management control of flyway councils. (See Chapter 17.) Migratory behavior has an important effect on management approaches. Some of the species, mostly of the order Galliformes (grouse, pheasant, quail, and turkeys), are nonmigratory and therefore under the jurisdiction of state management agencies. General management approaches, game-bird farms, census techniques, and characteristics of each group of birds are discussed.

MANAGEMENT APPROACHES

Management of shore and upland birds again brings the manager into contact with two kinds of people: hunters and bird-watchers. Some birds are harvested and are therefore the object of removal regulations. All species in these groups are popular with bird-watchers and photographers. Some people spend a great deal of money just to see the birds.

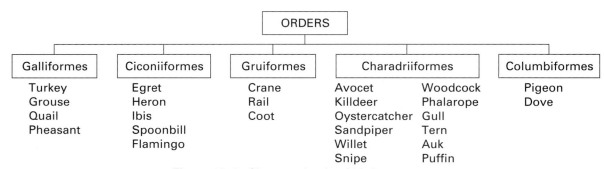

```
                              ┌──────────┐
                              │  ORDERS  │
                              └──────────┘
     ┌──────────┬──────────────┬──────────┬──────────────────┬──────────────┐
┌──────────┐ ┌──────────────┐ ┌──────────┐ ┌──────────────┐ ┌──────────────┐
│Galliformes│ │Ciconiiformes│ │Gruiformes│ │Charadriiformes│ │Columbiformes│
└──────────┘ └──────────────┘ └──────────┘ └──────────────┘ └──────────────┘
  Turkey       Egret          Crane     Avocet       Woodcock   Pigeon
  Grouse       Heron          Rail      Killdeer     Phalarope  Dove
  Quail        Ibis           Coot      Oystercatcher Gull
  Pheasant     Spoonbill                Sandpiper    Tern
               Flamingo                 Willet       Auk
                                        Snipe        Puffin
```

Figure 18–1 Shore and upland bird taxonomy.

Habitat Management

Shore and upland birds have many habitat needs. Shorelines, marshes, and wetland habitats are important for many of these species, whose continuance is best assured by protection of these areas from grazing, extensive recreational use, and development. The birds have developed a variety of breeding habits and specialized uses of habitat; presumably, these contribute to the survival of the species. When habitats are altered or the population structure is changed, the avian populations are affected. Thus, managers must consider the breeding habitats of each population in making any management decisions.

Colonial Nester Breeding Habits. A number of birds nest in colonies—some on the ground, some on rocky cliffs, and others in trees. Coastal rocky sites and oceanic islands provide nesting places for many seabirds (Figure 18–2). Murres and auks commonly form large colonies. Guillemots, puffins, and cormorants breed in and around rocks, although not in such large numbers.

Figure 18–2 Cliffs provide nesting sites for many seabirds. (Courtesy of U.S. Fish and Wildlife Service; photo by V. B. Scheffer.)

Gulls and terns are also colonial nesters. The kittiwake, an oceanic gull, breeds on the rocky coastal oceanic islands of the arctic and the east and west coasts of northern Canada. Some colonies have 10,000 or more birds nesting. Many gulls and terns breed in colonies along the grassy flats or sandy shores of the ocean or inland lakes. The birds require some degree of isolation during mating.

Coastal areas worldwide are home to seabirds, but these areas, when subject to development or pollution, can no longer support the animals. Eggs are often deposited in shallow depressions or on rock ledges. The mere presence of people can be disastrous.

A number of gull species migrate inland. Bonaparte's and Franklin's gulls travel north from the Gulf states and east coast, and the California gull travels from the Pacific coast eastward (Figure 18–3). Year after year, the gulls return to the same nesting site, generally along some inland lake or marsh. A number of terns breed on inland lake islands and in marsh areas. Some species construct elaborate nests; others just make shallow depressions in the sand or grass or in floating masses of vegetation.

Herons and ibises nest as colonies in trees (Figure 18–4), where the nests can be so dense that branches break off. Ibises also nest on the ground. Some colonies are mixed, while others contain only one species. They are found both inland and along the coast. Mixed colonies are common in some areas of the eastern seaboard. Some colonies can survive very close to developments, but are destroyed when the trees are removed or indiscriminate shooting occurs.

Isolated Nester Breeding Habits. The marshes and wetlands of the coastal and inland waterways provide a wide variety of some of the best nesting habitats for shorebirds. Sandhill cranes nest on high spots near water or in open fields. Many species of rails nest in marsh grass. Some require special features, such as slow-flowing water

···· Migratory Route

▨ Wintering Ground

Figure 18–3 Migratory route of the California gull, one of the few bird species that migrates from west to east and back.

Figure 18–4 White ibis.

or drowned debris. Curlews, willets, snipes, and plovers, among others, use the more open grass near waterways for nesting. Their nests are often hard to find.

In inland areas, quail, pheasant, and turkeys like the edge areas between field and forest. Grouse species nest in a variety of inland habitats. Woodcock nest in the earliest stages of deciduous forest in the east, preferring edge areas. They are common along the edge of transmission-line corridors in the eastern deciduous forests. Like grouse, doves nest in a variety of habitats, from grasses and edges to forest trees. The semidomesticated rock dove, or pigeon, likes ledges of buildings, but its droppings detract from the aesthetics of many buildings, making this bird the object of damage-control management.

Marbled Murrelet

Marbled murrelets are small seabirds in the order Charadriiformes that breed in the northern Pacific Ocean. The species range from the Bering Sea south to the coast of Northern California. For the breeding season, the birds fly to the disjunct pockets of coastal old-growth forests, where they nest on nearly perpendicular branches. In Alaska, about 5 percent of the murrelet population nests on the ground in areas where there are predators. The nesting birds are secretive and often very difficult to detect, except when they fly to the ocean for fish. In order to survive, this seabird must have two very distinct habitats, the open ocean and old-growth forests. Human impact on both areas is a source of concern for the welfare of the marbled murrelet population. Cutting down old-growth forests decreases the birds' nest sites, and oceanic oil pollution reduces their food supply and damages their feathers. Gill nets used by fishing companies pose yet another hazard to the birds. After studying murrelets for a number of years, biologists now know about the dual-habitat use of this species. Current knowledge can be used to help manage the population better.

Migratory Habitat. Like waterfowl, many shorebirds congregate in large numbers and migrate in the spring and fall. The waterways used for congregating are generally known and should be protected. Stopover areas are very important to migratory shorebirds. Research has shown that flocks will stop at the same coastal area, inland, or marsh each year. Some of the inland lakes, being on wildlife refuges, are now protected.

There are nine major stopover areas on the east coast of the United States that are used extensively during fall migration. Transoceanic migrating flocks remove large numbers—as much as 90 percent—of rapidly reproducing invertebrates from the sand on the shore to build up the fat supply needed for the long flight. Since so much of the coastal area is subject to development and extensive human use, the only apparent management option for such stopover areas is to set them aside in public or private refuges.

Winter Habitats. Wintering habitats for shorebirds are generally in coastal marshes of the southeastern, western, and Gulf states. Many birds migrate to the coastal areas of Central and South America. The impact of pesticides and habitat fragmentation on species using those areas is currently causing concern.

Population Management

Most efforts to manage the populations of shore and upland birds center around hunting regulations and permits for commercial takes. A number of species of upland birds are popular with hunters. Grouse, turkeys, quail, mourning doves, and woodcock are sought after. Since many species hide in grass, shrubs, or forest edges, bird dogs are popularly used to hunt shore and upland species that have open seasons. States control resident species, while regulations pertaining to sandhill cranes, woodcock, and mourning doves are set as a result of consultation between federal personnel and flyway councils because these bird species migrate from state to state (Figure 18–5).

Birds that impinge on human activities are the object of management to prevent damage. Gulls and pigeons are two species that have become nuisances. Garbage dumps have attracted thousands of gulls, the increased food supply enabling more of them to survive. The most effective deterrent seems to be simply covering garbage as soon as it is dumped. The location of a garbage dump next to the Patuxent Wildlife Research Center in Maryland in the late 1970s led to concern that the gulls might transmit disease to the endangered species housed at the center. Although this has not proved to be so, the lakes and ponds of the center now have a large number of gulls. Any long-term changes due to alteration in the energy-flow patterns will take some years to be noticed.

Pigeons have been controlled with avicides. Public opinion, however, may prevent their use in some places. One of the better deterrents is to construct new buildings without convenient nesting ledges. It is difficult to avoid having pigeons in many large cities that have massive old buildings.

Game-Bird Farms

A number of states and some individuals maintain game-bird farms to rear pheasant, quail, or grouse from egg to adult. The operation of such a farm is usually quite like a chicken-ranch operation. Managers of game-bird farms have sometimes been plagued by disease problems: Some bacterial and fungal infections can destroy confined birds. The birds are usually released several days before the hunting season, but pen-reared birds often do not give the hunter a satisfying experience, because they do not fly well and are not wary. Most states find that game-bird farms are a very expensive method

Figure 18–5 Many people enjoy hunting game birds. Here, hunters have captured ring-necked pheasant that live on the edge of grassy fields. (Courtesy M. McKinstry.)

of providing a hunting experience. The license fees usually do not even come close to covering the cost of operating the farms.

CENSUS AND SURVEYING METHODS

Because of the large variety of species and habitats, many census and survey techniques are used. Counts made at different times of the year will render different information. Colonial nesting species can be counted when nesting by either aerial or ground counts. Gull colonies can be counted by photographs. Species that exhibit courtship behavior at specific locations, such as the sage grouse, or prairie chicken, can be counted at leks. The congregation of some birds on lakes or other stopover areas before or during their fall migration can be photographed, and winter counts can be made where the species congregate in a small range.

Birds that congregate to mate in the early spring are easier to count than species that nest in isolation. The breeding-bird roadside count is effective for some species, but is biased in the case of those not normally associated with roadsides. Call counts and tape-recorder playbacks are used to census such species as rails, which are very difficult to observe.

The U.S. Fish and Wildlife Service conducts annual dove-call counts and woodcock-singing ground counts. At designated listening points on specified roads, calling doves and singing or actively courting woodcock are counted. The locations of routes in relation to the forest succession stage can influence woodcock call-count results.[1] Data from the dove and woodcock counts are made available by the service to assist in establishing regulations.

DESCRIPTIONS OF SPECIES

Seabirds

While seabirds are not members of the five major orders discussed in this chapter, some comments about them are appropriate at this point. More attention has been paid to some of these birds since recent offshore developments. Many species can be seen soaring in large flocks over oceanic waters. They are favorites of those on seabird boat trips. Many of the species are colonial nesters. The black petrel builds burrows on coastal cliffs and hills.

The habitats and biology of seabirds are not well known, nor are their movement patterns fully understood. Management efforts involve spotting nesting areas and keeping them intact and free from disturbance. Oil spills and other major water pollution must be prevented if food sources are to be maintained and feathers kept free from oil (Figure 18–6).

Pheasant, Grouse, and Turkeys

Members of the order Galliformes, such as pheasant, grouse, quail, turkeys, prairie chickens, and some of the partridge species, are managed primarily by state agencies. Different types of habitat are used by different species of these gallinaceous birds.

The pheasant, which has been introduced into many parts of the country, is a popular sport animal. The ring-necked pheasant is particularly common where fencerows and vegetation along roadways, fields, and transmission lines are left.[2] These types of edge habitat, fencerows, and fallow fields are fewer now than in previous decades, because small fields have been combined to make large fields that are easier to work with today's large machines. Thus, game birds have decreased in numbers, but pheasant can be attracted to areas by permanent cover in the form of either naturally occurring vegetation or seeded legumes and grasses. Shallow lakes and marshy areas

Figure 18–6 Tufted puffins are colorful seabirds. They nest on steep cliffs along the ocean and fly out to the ocean to feed.

with cattail and bulrush provide good areas for food and cover.[3] Management agencies should work with the county road commission or the state highway department to maintain these areas along roadways for pheasant populations.

Grouse are particularly susceptible to changes in the successional sere. Thus, maintenance of different species of grouse requires specific knowledge of the type of habitat each requires. Sharp-tailed grouse, for example, favor the early successional sere. As succession proceeds and dense vegetation and trees encroach on open fields, these grouse decline. Burning or clear-cutting areas improves the habitat and generally results in an increase in population. Ruffed grouse prefer edge where trees are common—especially, aspen saplings and alder in central Wisconsin or areas with a variety of dense herbaceous vegetation.[4] Management programs should be designed to obtain good interspersion of forest types and different-aged aspen and alder.

Sage grouse prefer large expanses of sage with edge, but studies show that removal of sage does not stop the birds' activity unless the area is destroyed (Figure 18–7). Dense sage communities can be made more desirable to sage grouse by opening up some edge. Sage grouse leks are commonly used year after year. Winter habitat is also important. Concentrations of grouse are usually found where the snow accumulation is less than 15 cm (6 in), which means that removing ridges, cliffs, and other barriers to drifting snow can be detrimental.[5]

Quail, which are popular game in some parts of the country, are divided into a number of species, associated with different types of cover. In general, though, they prefer cover within a canopy of overhead shrubs. The California quail thrives in chapparral that has interspersed open areas.

In some northern states, the gray partridge thrived, then declined, and now appears to be increasing. In some areas of former pheasant range now cleared for farming, gray partridges have become remarkably abundant. This probably is the result not of active wildlife management, but of a change in land use and, possibly, in weather conditions.

Figure 18–7 Sage grouse males performing on a lek in order to attract females to mate with them. One dominant male can mate with many females in this mating system. Leks are very traditional and, when destroyed, often lead to the demise of the birds that used them.

Farming has a major impact on most gallinaceous birds. During the early 1920s, for example, the greater prairie chicken was found on many of the prairie lands of the central United States. As farming efforts intensified in the decade 1930–1940, a decline in the prairie chicken population occurred.[6] But when the prairie chickens disappeared, the introduced ring-necked pheasant increased. Then, as field edge areas became less numerous, the pheasant declined.

Several programs have been initiated by wildlife-management agencies to improve gallinaceous bird habitat in farmlands. A program in Nebraska was designed to benefit ring-necked pheasant, bobwhite quail, and white-tailed deer by the creation of new habitats on marginal cropland, using grass and leguminous cover. Landowners were compensated to establish cover and were paid an annual fee to maintain the land. Managers have also tried to preserve existing cover on private lands, including wetlands, grasslands, and woods. They have recommended ways to increase cover, have promoted crop-rotating systems, and have demonstrated the importance of patchy habitat.

Artificial feeding has been successful in keeping the survival rate of gallinaceous birds high during severe winters. In a Minnesota study, a population of wild turkeys that had corn available suffered no weight loss and a mortality of only 10 percent. Another population, which lived on natural food, had substantial weight loss and a mortality of 60 percent. No differences were found between the survival rate of the two populations during mild winters.[7]

Management efforts for gallinaceous birds are local. The species desired, the stages in successional sere, and the amount of disturbance the birds will tolerate must all be taken into consideration.

Bobwhite Quail

In the east, one cannot help but be entertained by the call of the bobwhite quail. Bobwhites are found from the Rocky Mountains and Great Plains eastward, but in few western areas. They favor edge habitats, parks, and even suburbs. During the winter, bobwhites live in groups or coveys, which break up in the spring. Males find a singing post—a fence post will do. Nesting takes place amid dense grass along fencerows or in forest edge. The clutch is usually large: As many as 40 eggs have been reported. There may be several broods a year.

Bobwhites feed on seed stubble, grain left in fields, and fruit from shrubs. They roost in a close circle, with heads facing outward. Roosts are usually in sheltered spots under shrubs or in dense vegetation, but the bird is also known to roost in shrubs and trees. Bobwhites are relatively easy to manage, since they adapt well to people and developments. Keeping down predators, discouraging dogs and cats from interfering with nesting and bothering the young, and educating the public on the bird's beauty are all effective management approaches.

Long-Legged Wading Birds

Birds of the order Ciconiiformes include some of the most magnificent wildlife. For centuries, plume hunters have sought their feathers for women's hats and clothing. As we have noted, the first wildlife refuge in south Florida was established to protect wading birds from these plume hunters. The order includes herons, bitterns, wood ibises, ibises, roseate spoonbills, and flamingos. In the United States, the roseate spoonbill and flamingo are at the northern edge of their range (Figure 18–8). Since these birds find their food in water or on the bottom of waterways, pollution poses a great danger. The exception in feeding patterns is the cattle egret, recently arrived in the United States, which follows cattle to glean invertebrates turned up by grazing.

Figure 18–8 Range of flamingos and spoonbills in the United States.

Most of the species nest near the water in colonies. Some colonies have an immense number of bird species. A. C. Bent describes one such Florida colony in the early part of the century.

We had toiled all day, dragging our skiffs over miles of mud flats, poling them through several lakes and laboriously pushing and hauling them through the tortuous channels of sluggish streams, choked with roots and fallen tree trunks, in the almost impenetrable mangrove swamps of extreme southern Florida. The afternoon was well spent when we emerged on the open waters of Cuthbert Lake and saw ahead of us the object of our search, a mangrove island, about a mile distant, literally covered with birds. It was a beautiful sight as the afternoon sun shone full upon it; hundreds of white and blue herons, and a score or two of beautiful "pink curlews" could be plainly seen against the dark green of the mangroves, like feathered gems on a cushion of green velvet. As we drew nearer the picture became more animated, we could see the birds more clearly and we began to realize what a variety of birds and what a host of them the far famed Cuthbert rookery contained. The taller trees in the center of the island were dotted with the great white American egrets, perhaps 300 or 400 of them watching us from points of vantage; on the mangroves below them, among the hundreds of white ibises, we could see about 75 or 100 of the rare roseate spoonbills; the outer edges of the mangroves, growing in the water, were black with Florida cormorants and anhingas; and everywhere were flocks and clouds of Louisiana and little blue herons. The egrets and the spoonbills were the first to leave; the former rose deliberately, long before we were within gunshot range, and flapped lazily away on their broad white wings; the latter were equally shy, flying around the island, circling to a considerable height and then flying straight away, with their necks outstretched and their feet extended, in long lines or in wedge-shaped flocks; we watched them longingly as they faded away in the distant sky with the blush of sunset flowing through their roseate wings. Then hundreds of white ibises were rising from the mangroves with a mighty roar of wings and scores of cormorants were flopping off the outer branches into the water. When fairly in their midst, the air seemed full of the smaller herons, flopping up ahead of use, drifting around the island and floating over us; and mingled with them were circling water turkeys, soaring turkey vultures, and hovering fish crows, ready to pounce on unprotected eggs.

We landed on the island and found it much like other islands of its class in southern Florida; it was not over two acres in total extent, with not over an acre of dry land in the center. The dry land was covered mainly with black mangroves, mixed with some white button-woods; it was surrounded by a wide belt of red mangroves growing in the mud and water, which was 3 feet deep at the outer edge..[8]

Management efforts in behalf of waders consist of maintaining nesting sites free from disturbance and keeping water and marsh areas free from pollution. The normal level of water must be maintained: Each species has evolved characteristics—the bill in particular—to probe the water or mud below the water for food, so that a fluctuating water level can be upsetting.

Great Blue Heron

The great blue heron, common in much of the United States, nests in colonies, or *rookeries,* as colonies of great blue herons are sometimes called in the states, of from two to thousands of nests, which are used year after year. The rookeries are a scene of noisy activity. Bird-watchers do best not to enter them without disposable headgear. The nests are in trees, generally near a body of water, but if the trees are destroyed, the birds will attempt to nest on the ground and will use platforms for resting (Figure 18–9).

The heron is migratory in parts of its range, wintering with nonmigrating birds along the southern coastal states. When they return to their nest sites, the females normally lay four eggs, with both birds assisting in the incubation. The young are fed a diet of regurgitated fish, which leaves its signature in the rookery air. Adult herons feed primarily on fish, but also take some insects. It is not uncommon to see the long-legged bird standing in the water quietly and then scooping up fish and accompanying insect life.

Management efforts for herons center around teaching the public not to disturb these large birds. Waterways must be kept relatively free from pollution so that the birds can obtain food. Natural vegetation along the breeding and winter wetlands helps maintain habitat. Areas where rookeries are found must be isolated. The removal of nesting trees can be detrimental, since the birds tend to return to the same site year after year.

Cranes and Rails

Cranes and rails are members of the order Gruiformes, which also includes limpkins, gallinules, and coots, the latter two sometimes lumped under waterfowl. A number of species in this order are hunted; others are extremely rare and, in some cases, legally

Figure 18–9 Great blue heron nest in a rookery. Rookery sites are concentrations of a few to a few hundred nests.

classified as endangered. Birds in the order Gruiformes are closely associated with aquatic and wetland habitats for nesting and feeding and during migration.

Sandhill Crane

Sandhill cranes are divided into six subspecies. Three (the lesser, Canadian, and greater cranes) are migratory; the other three (the Mississippi, Florida, and Cuban cranes) are not. The three nonmigratory subspecies are listed as endangered.[9] The lesser and Canadian subspecies, which nest in Alaska and northern and central Canada, are relatively unstudied, so that data on their migratory use of habitat and their population are lacking.

The birds can defend relatively large territories—up to 405 hectares (1,015 acres)—but some defend areas as small as 2.5 hectares (6.25 acres). Their known habitat requirements are (1) water, (2) a feeding area—meadows, cultivated fields, open woodland—and (3) privacy. They will abandon their nests if unduly disturbed. Normally, the clutches contain two eggs and are incubated in from 28 to 31 days. Hatching takes from 24 to 36 hours; however, the young bird (called a *colt*) is up walking about the next day. The parents take the young into upland fields or meadows for feeding and return to the marsh every night to roost. Cranes become quite antagonistic toward their young of the previous year at the beginning of the breeding season. They eat a great number of insects, but also feed on grain crops.

The migratory members of the greater subspecies normally arrive in their breeding areas shortly after the snow disappears in the northern United States. They often move in family groups, but can also be seen flying in a crowd. A group of sandhill cranes moving down a valley catching the updraft creates a bizarre sight and an unusual sound. They fly in a huge forward-moving circle and sound like an old truck cranking along the road, dragging rusty chains. Cranes migrate primarily during the daytime and rest in a variety of stopover areas at night. Once they arrive at the breeding ground, they set up their territories and start courting. They leave the summer area in August or September and fly to their winter area. It is during fall migration that they are hunted.

A number of states allow hunting. Seasons are set in collaboration with flyway representatives (Chapter 17). Generally, the birds are shot at dusk when they enter a field to feed. While the states issue permits to hunt sandhill cranes, free federal permits are also issued in the central flyway. These permits are accompanied by questionnaires that show that 60 percent of the hunters are successful in bagging one or more birds.

Hunters at times mistake the endangered whooping crane for the sandhill crane. Since most states and provinces curtail hunting when whooping cranes are in the area, some ranchers and others interested in hunting do not report the appearance of whooping cranes.

Management needs of cranes are met mostly by habitat manipulation. A study of the eastern population of greater sandhill cranes indicated that they had increased in recent years, but drainage of wetland habitats and the crane's intolerance of disturbance from waterfowl hunting had forced the population into reduced areas. Refuge management criteria, particularly in the Jasper Pulaski Fish and Wildlife Area in northwestern Indiana, were established. It was found that water less than 20 cm (8 in) deep and an absence of human disturbances were needed. The birds did not avoid roosting where woody vegetation encroached upon the shoreline; in fact, human activity was better tolerated when roosts were surrounded by trees than when they were open and visible from roadways or other access areas. Hunting from a half hour before sunrise until noon or sunset every other day caused the cranes to avoid roosts.[10,11]

Census techniques can be used in breeding grounds, wintering areas, and areas of concentration of the birds during migration. Photographing concentrations is practical. The lesser and Canada sandhill crane populations have been studied by means of both fixed-wing aircraft following transects and hunters' surveys, in which questionnaires were mailed to holders of cranes to determine harvests. The returns provided some information as to the age distribution and number of birds taken.

Rails

Rails are divided into a number of different species and subspecies throughout the United States: yellow rails, black rails, clapper rails, king rails, Virginia rails, and sora. Rail subspecies are often localized and difficult to distinguish. In some cases, a distinction between subspecies of clapper rails can be made only with birds of the different subspecies in hand. Several subspecies, including the light-footed and Yuma clapper rails, are classified as endangered or threatened. The important distinguishing characteristic here is that they are found in different habitats. Rails also live in marsh or watery areas covered with dense aquatic vegetation, which makes them difficult to reach and study. There are both migratory and nonmigratory rails. Because of their secretive nature, the life history data on some species and subspecies are incomplete.

The life-history pattern generally follows the example of the Virginia rail.[12] The birds appear to be monogamous, but whether or not they hold a territory is not clear. The species nest in sedge and cattails of freshwater marshes, although some of the clapper rails commonly nest in salt water and in brackish areas. The nests are often placed on hummocks slightly above the water level, so a rise in the level during the breeding season can destroy them. Rails have large clutches, laid in the early spring. The young are precocial, leaving the nest shortly after hatching. They appear to be able to swim immediately and are capable of following their parents around and probing for food. Predator take is high during this early period.

Many states set hunting seasons for rails. Most of the time, however, the take of rails is low, since most people do not go out looking for them, but take them incidentally in their quest for waterfowl or other birds. Although small rails do not provide much food, their elusive nature makes them a challenge to some hunters.

Habitat management, again, appears to be the key to maintaining the population. Good management of marshes, so that water levels do not fluctuate greatly during the nesting season, is important. Early successional stages of marsh and cattail areas are more favored than dense, impassable later stages. Some water movement and downed cattails will allow the birds to move in an out of forage. A study in Missouri found that impoundments could be made attractive to rails by planting suitable vegetation on the sides. However, when waterfowl were attracted to impoundments for managed rails, the rails did not utilize these areas as frequently as when waterfowl were not present.[13] Pollution of waterways by heat, human wastes, or toxic material has a major impact on rails. Much information must still be gathered concerning their migratory behavior and movement patterns.

Shorebirds

The term *shorebirds* signifies mostly the order Charadriiformes: oystercatchers, sandpipers, avocets, stilts, phalaropes, gulls, and alcids. There are some 52 species of shorebirds in the United States. Many of these birds were once classified as game birds, with hunting seasons assigned them. Currently, the common snipe and the American woodcock are the only shorebirds that have a hunting season.

Shorebirds, most of which are migratory, generally nest in marshes or near the water. The nests are usually distinctive, although often difficult to find. Shorebirds typically utilize the edge of the water, but woodcock are found inland. The varying bill structures of shorebirds are adapted to utilize different sections of the water: Some probe the mud bottom, some utilize the bottom fauna near the surface, and others catch small fish or other organic material floating in the water. The biology of the birds varies from species to species. Some of the species have been studied extensively, but little is known about others, particularly those that migrate to the far north. The fall and spring migration delight bird-watchers, who come out in great numbers to witness them.

Shorebird management is primarily that of maintaining adequate habitat. Where waterways are polluted, marshes are drained, or human activity increases, the shorebirds are likely to be driven from the area. Introduction of predators to isolated islands and other land areas can result in major changes in the populations, since the native populations have no protective mechanism against predators. Some isolated islands are very important to migratory species, providing food and rest areas before long migratory flights. Although migratory stopover areas are used for only a short time by shorebirds, they are critical to their survival. Refuge areas and impoundments along the coast can help greatly in managing these species.

Many inland nesting areas of shorebirds have been destroyed by urbanization, ranching, and pesticides. Formerly, many of the shorebirds, such as willets, curlews, and sandpipers, were found in great numbers in the prairie states and up into the grasslands of the high plateau. Shorebirds still occupy grasslands; however, they are not so easy to locate. Curlews and willets, for example, are found in high prairie grasslands where an adequate supply of water exists. These birds, which are known by various local names, probe in standing waters and small streams that flow from mountain regions onto the plains area. Since they nest in open fields, they are affected by cattle, as well as by processes such as dragging, in which ranchers pull chains over their fields early in the season to turn up the soil. If these activities are properly timed, shorebird nesting, as well as crane nesting, can continue in the area without negative impact.

Common Snipe

The common snipe, found in many portions of the United States, usually nests in marsh areas near waterways. It is well known for its aerial courtship flight.[14] The birds are found both in brackish marshes as low as sea level and at altitudes of over 3,000 m (10,000 ft). Direct counts of the birds during migration and at some wintering areas are possible. Banding studies involving mark-recapture and territorial response involving the playback of recorded calls from the breeding areas have also been used. As with other shorebirds, management of marsh areas is important, since disturbance or destruction of marshes causes a decline. Because snipe are so small, most hunting of them does not occur directly, but is incidental, although the actual hunter take is not well documented (Figure 18–10).

Long-Billed Curlews

Before 1870, long-billed curlews nested commonly on prairielike habitats across North America. On the western grasslands, early explorers saw curlews nesting by the hundreds, from Montana to Texas. But in the last third of the 19th century, extensive hunting virtually exterminated the species in parts of the United States. Their numbers continued to decline across the continent during the first 30 years of the 20th century. By 1947, however, the trend had been reversed. Populations increased because hunting was stopped and scattered grasslands were allowed to recover. Too, the curlews were adapting to newly created "artificial" habitats: annual grasslands and irrigated lands (Figure 18–11).

Curlews arrive at their breeding grounds in April. There the males and females go through a courtship flight together. The birds nest in small depressions on high spots in the grassy prairie. The nests are very hard to find even with the female sitting, since she blends into the grass. Four eggs are the normal curlew clutch. Male and female

Figure 18–10 Common snipe. This small shorebird is hunted in many parts of its range. (Courtesy K. Downs.)

share incubation duties: When one bird is incubating, the other is not far away and will come quickly if an alarm is raised. Curlews feed on small invertebrates. They like to probe the bottoms of shallow waters for crawfish, small crabs, snails, and worms. They also take a variety of insects from grasses. Curlews migrate to coastal states from middle to late summer on into the fall.

Management efforts center around educating people on the value of the birds as consumers of undesirable invertebrates. Habitats for breeding need to be protected from ranching activities such as fertilization, grazing, and plowing.[15] The preservation of small patches of natural prairie is helpful. Nests and young are very susceptible to predation by small mammals and reptiles, and protection of areas from human intrusion reduces predation. Maintenance of pollution-free wetlands and estuaries with shore vegetation is the key to managing these birds' migration and winter habitat.

Have You Heard a Snipe Winnow?

For about six weeks each spring in marshy wetlands and bogs, common snipe begin an unusual courtship and territorial defense display. The males generally arrive a few weeks earlier than females and begin winnowing. As the females arrive, this activity increases. At sundown, the males (and sometimes the females) fly high into the air and then plunge downward, moving their tail feathers to create the winnowing sound. Snipe can winnow anytime of the night or day, but are clearly more active during the short period just around sunset. They display this behavior most often on moonlit nights when it is cold ($-5°C$ to $5°C$) and there is little or no wind. Nesting soon follows.

Figure 18–11 Long-billed curlew. These birds probe the mud along coastal and inland waters for small invertebrates. They nest in grassy fields near water, where they scrape a small depression and lay eggs in a grass nest.

Woodcock

The woodcock is a squat shorebird that has adapted to inland habitats. It is found primarily in the eastern United States. Because it is a popular sport bird, a great deal of information has been accumulated about it. The woodcock has a brown protective coloration that blends into the dry-leaf pattern of the forest edge floor; most people do not find woodcock without looking for them carefully. The birds feed on earthworms and other invertebrates that inhabit the soil.

Migration north from the wintering ground generally occurs in January or February. Most birds arrive at their breeding ground in late March and April. Woodcock perform unusual courtship displays at dawn and dusk on their singing grounds. The displays consist of flights lasting from 40 to 60 seconds over the singing ground, during which time the male performs acrobatics, accompanied by twittering of wings and vocal chirps.[16] The flights are followed by a ground display, during which the male utters a series of "peent" calls. This repertoire is repeated from 10 to 20 times over a period of 30 to 60 minutes each morning and evening during the courtship season. The male mates with one or more females on the singing grounds during this period.

Most nests are located within a few meters of a brushy field edge. However, nests are found anywhere from open fields to middle-aged hardwood stands. The nest consists of well-formed cup on the ground and usually contains four eggs; incubation lasts about 20 days. When habitat is adequate and there are few predators, nesting success can be up to 75 percent. The chicks are precocial and can fly shortly after hatching.

The woodcock is restricted primarily to the forested regions of the eastern half of the United States, eastern Oklahoma, Texas, and parts of Arkansas. The breeding and winter ranges overlap considerably. With improved telemetry, biologists are beginning to gather information on the birds' movements. In the southeast, woodcock are more commonly found in the higher elevation piedmont, where the habitat is more suitable, than on the coastal plains. Forests just preceding their climax stage are poor woodcock habitat, because they have few openings. An effective management tool is a fire that burns quickly and does not destroy all the trees, but creates a patchy environment in the forest. Transmission-line rights-of-way, pipeline rights-of-way, and other openings in the forest also encourage woodcock populations.

Aspen are popular sites for woodcock in Wisconsin. Well-drained upland nest sites near the brushy edge of loosely stacked pole timber are preferred. Clear-cuts at the edge of the habitat are highly attractive to woodcock for feeding and night roosting. In a study in Maine, biologists found that male woodcock used new clear-cuts as singing grounds soon after such areas were cut in the spring.[17] These areas are also good for trapping and banding. The use of clear-cuts by woodcock can be extended for several years by bulldozing vegetation from trails once a year. A continuing high demand for aspen pulpwood could assist in maintaining woodcock numbers. Because the woodcock is such a popular game bird and hunting pressure is relatively high, state and federal agencies have devoted a large proportion of upland game-bird research funds to this species. That does not mean a large dollar amount, of course: Most of the federal money goes for waterfowl.

Doves

Mourning doves are popular with hunters because they are a more open country bird and are easier to find than woodcock and snipe. They are in the order Columbiformes, together with the rock dove, or pigeon. There are several species of doves, including the mourning dove and white-winged dove. The precise habitats required have not been described; however, the birds commonly appear in edge habitats and appear to prefer some kind of signing posts. Modifications in existing forestry, range, and agricultural practices to increase edge would probably help maintain the dove population. The estimated size of the population of mourning doves ranges from 350 million to 600 million birds. The species has a very long nesting season, from March to September of each year.

The pigeon has become semidomesticated and is a problem in many areas of the country. Management of the bird generally falls to urban wildlife biologists and consists chiefly of reducing or exterminating the population.

SUMMARY

Shore and upland birds include an array of species that, for the most part, require wetland habitats. Some of the birds are harvested; all are enjoyed aesthetically. Each species has its own habitat requirements. Those that nest in colonies require isolated areas with adequate structure.

Nonmigratory species, chiefly the Galliformes, are managed by state agencies; migratory species fall under federal regulations. Some states have established game-bird farms to increase the number of birds harvested by hunters, but most of these operations are not cost effective.

Biologists use a variety of indirect methods to determine population trends of shore and upland birds. Counts can be made in colonies, on migration stopover areas, or in leks. The population trends of doves and woodcock are determined by annual call-count surveys.

DISCUSSION QUESTIONS

1. What can managers do to maintain populations of shorebirds on prairie land when ranchers and farmers use these areas extensively?

2. What are the most important habitat needs of shorebirds?

3. Which shorebirds prefer edge? Are these populations increasing or decreasing?

4. Would an open hunting season on gulls be a good way to reduce their populations? Explain.

5. Most shore and upland birds are surveyed by trend methods. What are population trends? Why and how are they used?

6. Why do you think that colonial nesting exists in some shorebird populations?

7. How is hunting game birds similar to, and different from, hunting big game?

8. Why is public education important in managing the great blue heron population?

9. How do fluctuating water levels affect shorebirds?

10. Why do you suppose relatives of shorebirds, the Ciconiiformes, have developed such colorful plumage in the process of evolution?

LITERATURE CITED

1. Keppie, D. M., W. R. Watt, and G. W. Redmond. 1984. Male Woodcock in Coniferous Forests: Implications for Route Allocations in Surveys. *Wildlife Society Bulletin* 12:174–78.

2. Snyder, W. D. 1974. *Pheasant Use of Roadsides for Nesting in Northern Colorado.* Special Report 36. Denver: Colorado Division of Wildlife.

3. Guthery, F. S., and R. W. Whiteside. 1984. Playas Important to Pheasants on the Texas High Plains. *Wildlife Society Bulletin* 12:40–43.

4. Kubisiak, J. F. 1978. *Brood Characteristics and Summer Habitats of Ruffed Grouse in Central Wisconsin.* Technical Bulletin 108. Madison, WI: Wisconsin Department of Natural Resources.

5. Dalke, P. D., D. B. Pyrah, D. C. Stanton, J. E. Crawford, and G. F. Schlatterer. 1963. Ecology, Productivity, and Management of Sage Grouse in Idaho. *Journal of Wildlife Management* 27:811–41.

6. Madsen, C. R. 1981. Wildlife Habitat Development and Restoration Program. In R. T. Dumke, G. V. Burger, and J. R. March (Eds.), *Wildlife Management in Private Lands,* pp. 209–17. Madison, WI: Wisconsin Chapter of the Wildlife Society.

7. Porter, W. F., R. D. Tangen, G. C. Nelson, and D. A. Hamilton. 1980. Effects of Corn Food Blots on Wild Turkeys in the Upper Mississippi Valley. *Journal of Wildlife Management* 44:456–62.

8. Bent, A. C. 1926. *Life Histories of North American Marsh Birds.* Bulletin 135. Washington, DC: Smithsonian Institution.

9. Lewis, J. C. 1977. Sandhill Crane. In G. S. Sanderson (Ed.), *Management of Migratory Shore and Upland Game Birds in North America.* Lincoln, NE: University of Nebraska Press.

10. Lovvorn, J. R., and C. M. Kirkpatrick. 1981. Roosting Behavior and Habitat of Migrant Greater Sandhill Crane. *Journal of Wildlife Management* 45:842–57.

11. Anon. 1988. *Final Supplemental Environmental Impact Statement: Issuance of Annual Regulations Permitting Sport Hunting of Migratory Birds.* SEIS 88. Washington, DC: U.S. Fish and Wildlife Service.

12. Zimmerman, J. L. 1977. Virginia Rail. In G. S. Sanderson (Ed.), *Management of Migratory Shore and Upland Game Birds in North America,* pp. 46–56. Lincoln, NE: University of Nebraska Press.

13. Rundle, W. D., and L. H. Fredrickson. 1981. Managing Seasonally Flooded Impoundments for Migrant Rails and Shorebirds. *Wildlife Society Bulletin* 9:80–87.

14. Fogarty, M. J., and K. A. Arnold. 1977. Common-Snipe. In G. S. Sanderson (Ed.), *Management of Migratory Shore and Upland Game Birds in North America.* pp. 189–210. Lincoln, NE: University of Nebraska Press.

15. Cochran, J. F., and S. H. Anderson. 1987. Comparison of Habitat Attributes at Sites of Stable and Declining Long-Billed Curlew Populations. *Great Basin Naturalist* 47:459–66.

16. Owen, R. B. 1977. American Woodcock. In G. S. Sanderson (Ed.), *Management of Migratory Shore and Upland Game Birds in North America,* pp. 149–88. Lincoln, NE: University of Nebraska Press.

17. Dwyer, T. J., G. F. Sepik, E. L. Derleth, and D. G. McAuley, 1988. *Demographic Characteristics of a Maine Woodcock: Population and Effects of Habitat Management.* Fish and Wildlife Research 4. Washington, DC: U.S. Fish and Wildlife Service.

19
FISHERIES

FISHERIES RESOURCES

The aquatic biota that makes up fishery resources is quite diverse. Most of the time, we think of harvested fishery products as trout, walleye, pike, salmon, tuna, cod, and herring. Most of the fish that we eat are in the vertebrate classes Chondrichthyes and Osteichthyes. Managers of fisheries also include mammals, such as whales, dolphins, seals, and sea otters, in their domain. Crustacea, such as shrimp, lobster, crayfish, and crabs, which are heavily exploited for human consumption, are also considered fishery resources, as are mollusks—that is, clams, oysters, squid, and octopuses. Some oriental fisheries specialize in managing aquatic plants that are harvested for human consumption. Actually, fisheries management goes beyond the harvested products, because the aquatic organisms influence the way other species of fish, mammals, and invertebrates live, thus affecting the health of the ecosystem.

In this chapter, fish species and some invertebrates are called fisheries resources. Fisheries-management techniques in the open ocean, coastal waters, inland lakes, and rivers are compared. It is important to keep in mind that while we are reading and discussing fisheries management in the perspective of our culture, many other cultures seek and utilize a different variety of fisheries resources than we do. Therefore, management must not be only for desired populations, but also for a healthy aquatic ecosystem.

ECONOMIES OF, AND DEMAND FOR, AQUATIC PRODUCTS

Most nations of the world utilize fish for human consumption. Worldwide, fish products supply some 10 percent of the world's protein.[1] The demand differs considerably in different parts of the world. In Japan, for example, annual consumption of fish products approaches 40 kg per person, and the weekly menu for a typical family includes several types of fish, invertebrates, and sea plants.

The ocean waters, which cover 75 percent of the earth's surface, have been a major source of food for centuries. Since about 1950, sophisticated technological methods, such as telecommunications, echo sounding, aerial communications, temperature monitors for locating fish, and electrical impulses and light for attracting them, have been used to ensure large catches. There is extensive information today about fish migration and schooling behavior. Modern fishing fleets include processing ships, which prepare catches for consumers right at sea. Peru, Japan, the Soviet Union, and China lead the world in the use of ocean resources (Figure 19–1).

Between 1955 and 1970, the biomass extracted from the oceans increased from 310 million to 700 million kilograms (31 million to 70 million tons). This catch remained fairly stable between 1970 and 1975. Experts estimate that the oceans could provide more than 1,100 million kilograms (110 million tons) of biomass per year without destroying the fish populations if currently unutilized species were harvested. Nearly half the biomass currently removed is used for purposes other than directly as food, such as manufacturing fish and oil meal, which indirectly reaches the tables of the western nations. Furthermore, much of the catch is not considered edible. There are more than 21,000 known species of fish, but fewer than 50 are sought by commercial fisheries. The species taken commercially are usually large or appear in large schools, making them easy to locate and remove. Oceanic fishing efforts have led to overfishing, depleting stocks and diminishing catches.

The economics of fisheries can be divided into *commercial* and *recreational* use. Commercial fisheries involve an array of people who fish for their livelihood. There is the person who owns one boat and hires a crew member to go out in the ocean and bring in a daily catch, to be sold to a local processor. There is the large canning company, which owns or leases many ships and hires captains and crews to obtain a huge supply

Figure 19–1 Commercial vessels take large catches from the oceans. (Courtesy C. A. Morgan.)

of fish to can. Then, in some nations of the world, there are national fishing fleets, which operate away from home ports for many months at a time. The economics of all of these is based on supply and demand. If demand for a fishery product goes down, the commercial fishery operation, including the canning industry, can lay people off and let their shipping fleets lie idle until the economy picks up. The individual person who owns one ship may be more severely affected by an economic recession or change in demand for fishery products. There have been a number of shifts, however, in the use of individual boats in the last few years. For example, along the east and west coasts of the United States, a number of former fishing boat operators use their boats for wildlife-watching excursions. On the west coast, watching whale migrations has become a popular spectator sport.

A study of some of the so-called delicacy foods, such as lobster and crab, indicates that the supply–demand curve does not necessarily apply to those foods: They are bought by people who apparently are not concerned about price increases.[2] This means that, as more people demand such foods, exploitation of the resource increases, and finally, the only way to supply the luxury demand may be through aquaculture.

Recreational fishing continues to be popular with people as they spend more time and money on the sport. (See Chapter 14.) The economics of recreational fishing includes more than expenditures on the sport itself. States with recreational fishing benefit from increased tourism, with people spending at hotels, motels, and campgrounds. The locals want to ensure that the state game and fish departments provide an adequate source of fish stock to continue to attract sport-fishing enthusiasts, as well as maintain selected waters for trophy fish.[3] The tourist industry is not concerned about the total amount spent by a state or federal agency, as long as it keeps local businesses flourishing. Oceanic recreational fishing, which has also expanded in recent years, yields good income from boat rentals. More people are fishing along the coast, and many are buying small boats for daytime fishing trips.

BIOLOGY OF FISH

Stocks and Strains

Fishery scientists use the term *stock* and differentiate between it and the biological term *population*. As we noted earlier, a population is a group of interbreeding individuals that live in the same area. Because populations are breeding units, the gene makeup within one population tends to be different from that in other populations. Biologists have been successful in identifying fish populations by analyzing the percentage of different blood proteins in each.[4]

Stock means a group of fish or other aquatic animals that can be treated as a single unit for management purposes.[5] Management does not require that genetic units be treated separately, only that the stock being managed form an identifiable unit. Thus, in an inland pond with four species, each species could be managed as a separate stock, so that population data would be necessary for each stock. Populations of anadromous fish might be managed as a separate stock in each spawning stream. Environmental conditions may differ in large bodies of water, so that growth and reproduction conditions for the same population may differ. Consequently, a given fish stock could be only a segment of a lake's population, or it could be the entire population.

Fishery managers also look at different strains of fish species that have separate genetic adaptations. Often, hatcheries raise different strains. For example, strains of rainbow trout that have various behavioral and biological characteristics are placed in different lakes and streams.[6] These strains can interbreed with other rainbow trout; however, local strains are often better able to survive in the local environment.

Reproduction

Fish have a relatively high reproductive potential, with females of species capable of egg production in the millions. Checks generally operate to maintain a balance: Not all females reproduce, not all eggs survive to hatch, and not all young mature. Usually, fish that simply distribute eggs but provide no parental care produce a large number of eggs. Mortality due to environmental conditions and predation can be more than 99 percent during the first few weeks of life. Fish that lay eggs and guard their young generally have fewer eggs, with increased survival rates. Thus, the average survival to maturity from fish spawning 2 million eggs ight be no higher than the survival from those spawning 2,000 eggs. The processes of egg laying and survival are adaptive mechanisms that have evolved in the different species.

Fish lay their eggs or spawn in areas where physical conditions are optimal, allowing the best survival of the fish. Some anadromous fish migrate many miles from the ocean into streams or lakes to spawn. Spawning sites are often those with rocky bottoms, free of silt or other pollution.

Habitat

Like other species of wildlife, fish require the proper biological and physical conditions to survive. Often, fish have different needs for distinct periods in their life history: Spawning may require one habitat, young fish (fry) another habitat, and adults still another. (See the discussion of southern kingfish in Chapter 9.) Accordingly, management can involve maintaining a variety of habitats to perpetuate a desirable species of fish.

Temperature. Some fish thrive in warm water, others in cold. Water temperatures that exceed 24 to 26° C (75 to 78°F) for an extended period are considered warm. Largemouth bass, perch, green sunfish, catfish, and a variety of smaller fish do well in such temperatures. Cold-water streams—generally speaking, those that do not exceed 24 to 26°C—are inhabited by salmonids (salmon and trout), grayling, and whitefish (Figure 19–2). Locating industrial facilities with warm effluents along streams or removing waterside trees and other vegetation can have a major impact on fish because of consequent changes in the water temperature.

Lakes can also be divided into warm and cold. Cold-water lakes have salmonids; warm-water lakes have black bass, sunfish, perch, walleye, pike, and some striped bass. Some lakes are two layered, with warm-water fishes in the upper layer (*epilimnion*) and cold-water fishes in the deep water (*hypolimnion*), which is generally characterized by a low concentration of oxygen during summer.[6] The fisheries manager can draw temperature regions. For example, in Lake Michigan, spottail shiners live at 6 to 14°C (43 to 57°F), trout at 13 to 22°C (55 to 72°F), yellow perch at 10 to 16°C (50 to 72°F), and alewifes at 8 to 22°C (46 to 55°F).[6] Obviously, fish vary in the temperature range they can tolerate. Managers must know these tolerances and must understand the biological interactions. Temperature must be considered in many situations—for example, in removing fish by poisoning so that desirable fish can be introduced, in creating new ponds in outdoor areas, or in setting up fish cultures.

Latitude and altitude are the major factors separating the temperature categories. Low latitudes and altitudes in the United States are the major environments for warm-

Figure 19–2 Trout are a cold-water species.

water fish. Cold-water fish are commonly found in mountain streams. Most midrange species are inhabitants of lakes in the northern latitudes.[7]

Salinity. Salinity is a direct or indirect limiting factor for many species. Some species living in estuarian regions can move across several salinity gradients, but most estuaries have species specifically associated with each gradient. Along the coast, where an insufficient change in water occurs between bays on lagoons and the ocean, heavy mortalities can occur. When fish move back and forth between fresh and salt water, their body chemistry changes. In a saltwater habitat, salt must not be excreted, and in a freshwater habitat, water must not intrude. Fish in inland streams and lakes have been affected by the return flow from field irrigation, which has high salinity.

Oxygen. Aquatic organisms in all bodies of water require oxygen. The chief cause of oxygen depletion today is the discharge of organic wastes into our waterways. Bacteria act rapidly on organic wastes, utilizing the oxygen needed for fish to survive. Die-offs can result. Winter kills below the ice in lakes and streams are also caused by lack of oxygen: As light is shut off by the ice and snow, plants—including the planktonic algae—consume more oxygen than they produce, depleting the supply. Too, the decay of organic matter on the bottom of lakes forms methane, hydrogen, sulfide, carbon dioxide, ammonia, nitrogen, and other gases that may be toxic or that may contribute to oxygen depletion.

Light. Light has as subtle effect on the distribution of fish within lakes, streams, and oceanic bodies. As lakes become more turbid, sight feeders can no longer find their food, and prey may be unable to elude predators (Figure 19–3). In Clear Lake, Iowa, the black bass, a sight feeder, decreased in number as turbidity increased, yet walleyes, which do not rely on sight for capturing food, did not decline. Below the light zone in the ocean, which extends to a depth of about 61 m (200 ft), fish have some form of luminescence, either to be recognized or to attract prey. Some species move between

Figure 19–3 Dace have been harmed by silt in some streams.

light zones during day and night periods. Light is one of the factors that influence the daily movement of trout. (Food, temperature, and oxygen are others.) Trout approach diffused light and avoid bright light.

In lakes, vision-oriented predators are most active during the day, with peaks of feeding in the early mornings and evenings, when invertebrates are most available. Zooplankton often tends to move up in the morning to the light zone and move down from it in the evening. These colonies of invertebrates are then eaten by small fish.

Space.   Areas that a fish can use for food or spawning are obviously critical to the survival of the fish. Some lakes cannot maintain certain species of fish because they have inadequate spawning areas or because the total area is too small for foraging. Very deep lakes often do not have productive bottoms and so lack the food supply needed to support many fish species. In the process of eutrophication, the composition of fish species changes. Lakes with shallower bottoms and more vegetation are usually much more productive. Because they are approaching the end of their successional sequence, however, they must be managed carefully so that the bottom will not be destroyed or the lakes filled with too much organic matter. The density of substrates in such lakes is also very important to spawning and growth.

Chemistry.   Toxic material that is dumped into a lake can have an impact on fish. Chemicals can change the pH of the water to such levels, that the fish are either killed outright or can no longer reproduce.

Changes in turbidity of the water also affect fish. An increase in erosion, causing siltation of the water, can bring about physical changes in the resident fish by changing their visual clues, decreasing their oxygen uptake, and making food less available.

A combination of different habitat characteristics is generally responsible for the presence of fish. In Wisconsin, biologists found changes in fish assemblages as a function of gradient from bog ponds to small oligotrophic lakes. This variation was the result of (1) the size and heterogenity of the habitat, (2) productivity in relation to pH, and (3) the concentration of oxygen in the water.[8]

Competition.   Competitive interactions are very important in shaping a community of fish. Competition can help communities become more diverse. On the other hand, competition, particularly from introduced fish, can alter a community's balance. Some species destroy or spoil the habitat. For example, a carp species destroys the aquatic vegetation that serves as cover for other species. Competition also can occur for spawning areas: Some salmon disrupt the bottom, destroying other species' eggs in the process. Overpopulation can cause a major problem among competing species.

Some species of tilapia have been widely released into U.S. waters and are being promoted as a part of the expanding aquaculture effort. Detrital feeders native to the southeastern tip of Africa, they are capable of living and reproducing in everything from fresh water to double-strength seawater. This fish, which sometimes reaches a weight of 2.7 kg (6 lb), is prolific, is fast growing, and has been used worldwide for food production.

Tilapia are established in fresh water in southern Texas, in both closed and open water systems, and in southern Florida, Arizona, and southern California. Breeding populations have also been established in brackish water along the east coast of Florida, in the Cape Canaveral area, and along the southern coastal areas of California. The mean lowest lethal temperatures for this species is 9.5°C (49°F) in fresh water and slightly less in salt water. Tilapia populations are expanding rapidly in the brackish waters along the southern coast of Florida and in the Salton Sea of California. The future impact of these fish on the commercial and sport fisheries in coastal areas, although unknown, could be significant. In the areas inhabited by tilapia, within a short period of time after their introduction, they have become a dominant species even where a normal native predator fish population has existed. In some Texas reservoirs, tilapia have eliminated most other fish. In recent years, a large sport fishery has developed around the tilapia in the Salton Sea.

Research is under way to produce a sterile population of this fish for aquaculture, which would negate the possible ecological problems should the fish escape or be released into the nation's open waters. Continued establishment of the species in nontarget areas could jeopardize its use for aquaculture.

Predation.   Predation can limit the abundance of fish or be a means of providing catchable fish in lakes. The removal of large predator fish by intensive fishing does not generally result in better fishing. On the contrary, the removal of predators often permits the survival of too many young fish for the habitat to feed. The result is overcrowding of the waters by slow-growing fish unable to reach a reasonable size. The balance produced between carnivorous fish and their prey can result in sustained yield.

The balance to be achieved in a community can be illustrated by the pond-fish culture. Largemouth bass, the principal carnivorous species, feeds on insects part of its life, but to reach a size in excess of 0.125 kg (0.3 lb), it must also have small fish. Bluegills have been found to provide bass with the food necessary for them to reach a desirable size. To achieve the proper balance of bass and bluegill, bass must feed not only on excess bluegills, but also on small bass. Thus, managers must construct ponds with a habitat that can support the spawning of both species, and the ponds must be stocked in such a way as to maintain the predator–prey relationship.[9]

Habitat can influence predator–prey interactions. Scientists have found that densities of vegetation influence the foraging behavior of largemouth bass: Bass in ponds with sparse vegetation had a more specialized prey diet than did fish in dense vegetation.[10]

The introduction of a new predator into a lake can drastically reduce the population of fish in the lake. The classic example is the introduction of the sea lamprey into the Great Lakes. Lake trout were nearly exterminated in Lake Michigan as the lamprey moved into the area: Only eight lake trout were caught in 1955. It was estimated that, as a result of the introduction of the lamprey, the annual production of lake trout fell from 2.7 million kg (0.27 ton) to less than 45 kg (100 lb) in just over a decade. An intense and costly program was then initiated to eliminated the sea lamprey. The life-history studies indicated that the most vulnerable period in its life occurred in streams. The lamprey's spawning migration took place in the spring and early summer during a relatively short time, from four to five months. Under ideal conditions, the adult lamprey could be stopped by either mechanical or electrical barriers installed below suitable spawning areas in the stream. The disadvantage of this procedure was that major reduction in parasitic stocks may require several years because of different age classes of larvae already present in the stream and that had to pass through the parasitic phase before the life cycle was interrupted. Also, adult lampreys that were spawning could move into streams that did not have barriers.

In the interest of restoring lake trout, managers first installed mechanical shockers on Canadian and U.S. streams and rivers that flowed into three of the upper lakes and added more later. Alternating-current electrical fields were set up to prevent the movement of lamprey and other fish upstream, while permitting the free downstream passage of floodwater, ice, and debris. Lampricides were used during the late 1950s and 1960s. Today, this unique control measure has reduced the number of sea lampreys effectively and permitted reestablishment of the lake trout sport fisheries and industry. Work continues on some large rivers and lakes in the vicinity that also have lamprey problems. Currently, the sea lamprey is controlled almost exclusively with chemicals; however, because of budget cutbacks in the 1980s, the lamprey population is increasing. It has been difficult to evaluate the total cost, but estimates by the Great Lakes Fisheries Commission, which was responsible for the program, indicate that it was very high.[11]

SAMPLING

Determining which fish or other aquatic organisms are present in a body of water, as well as what their age, structure, health, and other characteristics are, is important in management. These data may be helpful in setting regulations, evaluating the impact of a management program, and assessing the general health and structure of the ecosystem.

How do you set up an aquatic sampling program? First, decide on your objectives—what you want to learn from the sampling and how accurate you need to be. (See Chapter 4.) Second, design a sample system. Here, you need to consider the sample size required to achieve the desired accuracy and precision.

Many devices and techniques are used to sample aquatic populations. Nets are used to sample fish that are swimming. Traps or nets can be placed in selected locations to determine the movement patterns of some fish, such as anadromous species.[12] Spawning behavior and habitats are difficult to sample. After the spawning areas have been identified, selective sampling techniques, including stratified sampling, must be used.[13]

Gill nets of various mesh sizes can be used to catch pelagic fish. Some gill nets, selective for fish that can be frightened into them, are modified to have interior traps, in which the fish are caught (trammel nets).[12] Seining nets strung between two poles can be used to encircle fish in shallow water. Here, technicians hold each pole and move slowly through the water, pulling the nets and encircling the fish.

Some fish and other aquatic organisms are caught by entrapment devices. Nets are placed on the bottom of the aquatic habitat, and the animals swim or crawl into these units. Fish that are attracted by bait, such as catfish, buffalofish, and carp, can be caught in such a device.[14] Traps placed in the water are commonly used to capture bottom animals, such as crayfish or lobster. Sometimes, small fish (e.g., minnows) are trapped. There are many types of traps, from nets to plastic or metal enclosures.

Chemicals can be used as a sampling technique in a small body of water or arm of a lake. This is a method of obtaining an unbiased sample if the body of water is small or the arm of the lake is representative (in fish species or density) of the larger water body of interest. Generally, some means of blocking the spread of the chemical to other parts of the water is employed. Fish can be picked, counted, and weighed in the fashion of a census. There are chemicals on the market that are selective for fish species. Some chemicals cause fish to either rise to the surface or sink. Other chemicals cause the fish to die and remain on the bottom. All impacts of chemicals must be evaluated before a sampling program is initiated.

Electroshock techniques can be used to obtain a complete sample of fish. A number of different methods are included in this category; all use the principle of an electric current moving through the water to shock the fish (Figure 19–4). Most fish recover in from 30 seconds to two minutes. After they are counted, they can be released or collected.[15] A survey of anglers' catches can be an indicator of how many fish are present in a body of water.

Acoustical techniques are now being developed for fish sampling. They are based on the principle that sound waves in the water are bounced back by targeted fish.[11] Obviously, these techniques have the limitation that other objects also cause the sound waves to return.

Some biologists want to discover what bottom fauna there are in streams, lakes, and marine waters. Most of the sampling devices grab a portion of the soil on the bottom, and the biologist then projects a count from the organisms present. These techniques are somewhat crude, and it is important to design bottom sampling carefully in order to get the best data.

Figure 19–4 Electroshocking is used to collect a sample of fish. (Courtesy W. Hubert, U.S. Geological Survey.)

The traditional objective of most commercial fisheries has been to *maximize sustainable yield*—basically, an economic objective. As in other business areas, this objective soon changes to one of maximizing profit—a quite different thing, although still an economic objective.

When the profit motive is short range, the effect on the resource base can be disastrous. Even when the profit motive is long range, the commercial operation is almost inevitably less than optimum insofar as the public interest is concerned. This is where the fisheries manager comes in. Commercial fisheries can be a valuable resource themselves, but only if the public interest is protected in a highly skilled manner.

The fisheries manager performs a number of functions: regulation of fisheries, habitat protection, rehabilitation, and improvement of the fisheries habitat. Each of these categories require somewhat different activities in different aquatic systems.[1] We will look at the different categories in the oceanic, coastal, lake, and stream ecosystems.

Oceanic Fishing

Regulations aimed at the use and protection of oceanic fish generally come under international treaties. As we have seen in Chapter 10, a number of treaties have provisions related to oceanic fish resources. For many years, the oceans were exploited, resulting in the disappearance of many species. Now, major problems result from extensive pollution: Reports of tar globules and other waste materials floating in the open ocean are numerous. Some improvements in habitat have been attempted between the coastal zone and the oceanic areas. For example, since fish tend to gather around underwater structures, debris anchored to the bottom can be a simple attracting device. Artificial reefs on the bottom of the ocean also attract fish. After the Alabama Department of Conservation placed a number of automobile bodies on the Gulf floor, snapper fishing became quite good, even though in three to five years the automobiles disintegrated. Similar attempts have been made using quarry rock, old tires, and wooden posts, some of which also tend to disintegrate.

Coastal Ecosystems

Coastal ecosystems are regulated by both national and state laws. The Fisheries Conservation and Management Act of 1976 was a major step toward regulating and protecting these ecosystems. The act established a 200-mile fishery zone off the coast. Fees charged are based on the type of permit and amount of catch (in pounds) desired. Currently, the eight regional fishery management councils are identifying spawning sites critical to the development of fish and devising means of protecting them. It is presumed that a number of research activities will be carried on to facilitate the maintenance of the most productive fishery resources in the areas (Figure 19–5).

Coastal ecosystems are particularly susceptible to pollution. Dumping of human waste and toxic materials into these areas has been going on for many years. The Global 2000 report indicated that these coastal waters, which make up only 10 percent of the oceanic area, are responsible for 99 percent of the total oceanic fish production. Sewage, industrial waste, litter, and petroleum products have short-term effects, but synthetic organic chemicals, heavy minerals, and radioactive materials have a long-term impact. Some of the coastal waters along the eastern seaboard have been closed to both human recreation and fishing activity because of contamination.[16]

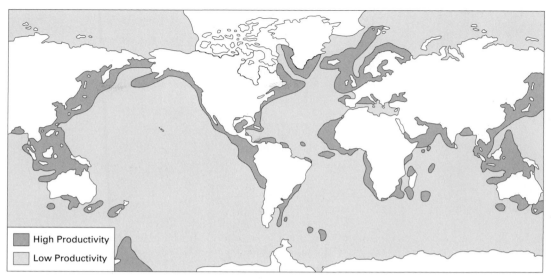

Figure 19–5 Areas of resource productivity in coastal waters.

Inland Fisheries

Lakes. Inland lakes, reservoirs, and ponds—inland standing waters—are often called *lentic habitat*. Regulation over, and protection of, these areas is under state laws, unless the lakes cross state boundaries or, in the case of the Great Lakes, are operated under the Great Lakes Fisheries Commission. Lakes have been subjected to a great deal of human and chemical waste pollution in this country. Reservoirs, which are really half-lakes, have contributed considerably to fresh water and require the same management as that of other types of fisheries resources. Generally, management aims at optimizing recreational fish values for the sportsperson. In other words, stocking is done so as to maintain communities of desirable fish. This requires understanding competition, predator–prey interactions, and the physical potential of the lake for its various fish species. Some agencies employ a form of zoning, managing some waters for trophy fish through carefully regulated harvest and allowing intensive fishing of a single species in other waters.

Management techniques to rehabilitate or improve an area include manipulating the population by selectively removing undesirable species or stocking desirable species. Excessive numbers of carp, suckers, gars, and gizzard shad can be reduced in small bodies of water by poisoning, netting, or trapping. In large lakes, these fish are harder to control. When undesirable fishes become overabundant and difficult to thin, the only recourse is to destroy the entire fish population and restock the body of water with desirable fish.

Reservoirs have been utilized to initiate fishery programs. With the construction of a new reservoir, the fishery manager can stock an entire community. Actually, many reservoirs quickly establish populations of fish without the necessity of stocking them.

A number of habitat-management programs have been initiated to maintain fisheries' populations in reservoirs and lakes. When reservoirs are constructed, bottoms are often cleared of trees and other vegetation. If some trees are left around the side, fish and other wildlife can find protection from predators when the water inundates the trees.[17] Brush and tire shelters have been placed on the bottom of reservoirs to provide areas where fish can find food and shelter. Concrete or other permanent structures are also useful.

Managers can meet with agencies that control the water flow in reservoirs to prevent excessive fluctuations in waterway movement. If dams are used to regulate the water flow each month of the year, excessive flooding or drought can be avoided, thereby preventing damage to the fishery resource. Small ponds in reservoirs upstream from dams can help avert sudden changes in the depth of the reservoir when hydroelectric facilities begin operation.

A fisheries problem that has become severe in recent years in many lakes, particularly in the eastern United States, is that of acid rain (Chapter 9). A change in the pH or chemical contents of a lake can destroy some or all of the fish. The means by which this occurs is not completely understood, but alteration of the food supply or physical habitat plays a role. The disappearance of fish from an acidic lake may be of more importance to the ecosystem than the direct effect of pH and the toxic materials on the individual.[18] Outside the United States, the addition of chemicals is the most widely used method to neutralize acid. The least expensive and most available neutralizing agent is limestone ($CaCo_3$). In Sweden, some 900 lakes and rivers have been limed to change the pH of the waterways and improve fisheries. Currently, research is under way to determine whether lakes in the United States that have been affected by acid rain could be neutralized by chemicals.

Streams. Streams come in a variety of habitats, from open-flowing water to narrow riffles (Figure 19–6). Open water that flows slowly can be very similar to a lake, while small streams can become isolated pools during dry periods. The physical conditions in each of these waterways are quite different, providing habitats for different fish.

Streams or rivers are generally characterized as belonging to different zones.[19] The terminology varies in different parts of the world, but in the United States, high mountain streams, which begin the water flow down to the sea, are said to be in the *erosional zone*. This zone is characterized by rocky bottoms, swift, usually cold water, long riffles, and small pools. The fish in the erosional zone—including trout and small, bottom-dwelling forms of sculpins—tend to be streamlined and active. In the *intermediate zone* are the long middle reaches of tributary streams. This zone typically has a moderate gradient, warm water, and about equal amounts of shallow ripples and deep rocky or muddy bottom pools. The streams here run and undercut the bank. The fish of this zone include minnows, suckers, smallmouth bass, and some catfish.

The *depositional zone* occurs in the warm turbid and sluggish low reaches of stream systems, where bottoms are muddy and beds of aquatic plants are common. The fish here are such bottom feeders as carp, suckers, sunfish, shad, and some predators, including bass. This zone may grade into the estuarian zone, which contains a mixture of fresh and salt water.

A number of studies show how habitat management can result in improved fisheries in streams. In the Ten Mile Creek region west of Denver, Colorado, highway construction in 1971 was destroying what was left of the creek. To prevent total destruction of the fisheries resources, the Colorado Division of Wildlife designed a new channel, which was dug around the area.. The channel was as narrow as possible, to concentrate the low water flow of the fall spawning season and to prevent the formation of anchor ice in the winter. The channel was designed with a vertical slope, to reduce the width of the clearance and increase the chances that the stream would undercut banks and so provide cover for the fish. Since the new channel lay almost entirely in the coarse gravel deposits, these steep slopes could be expected to weather down to a natural appearance without serious erosion of the bank.[13] Fish samples taken in the pre- and post-construction periods indicated that brook trout increased from 17 to 34 percent of the population of the stream.[20]

Artificial overhead bank devices have been successful in improving stream fisheries. A number of techniques have been used, including building rock walls with overhanging banks supported by oak pilings.[17] Placing trees and shrubs along banks has proved to be an effective way to maintain water temperatures. In a study conducted in northwest Montreal, the increases in brook trout in an improved section and an unimproved section of stream were compared. The improvement techniques included building small dams of rocks and logs at various locations, thus producing deep pools interspersed with ripples. Cover in the form of logs, stumps, and rafts of alder and ash lashed together was placed at locations close to abundant food. Large rocks placed randomly in the pool gave trout visual isolation, reducing intraspecies aggression as the population increased. The control section was unaltered, except for the regulation of water flow. The habitat-improvement techniques resulted in an increase in the trout population and biomass of more than 200 percent within two years. Crayfish biomass also increased in the improved sections, with resulting greater mink activity. Initially, it was thought that the mink came for the fish, but it turned out that they were eating the crayfish.[21]

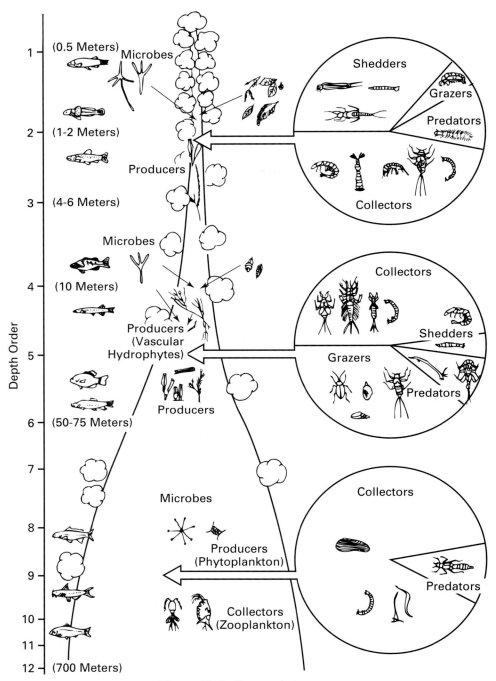

Figure 19–6 Types of streams.

In another study, in Oregon, gabions—baskets full of rocklike material—were used to improve the aquatic habitat. This was in streams flowing from the coastal range, in which natural spawning rocks had disappeared. The gabions were placed across the stream perpendicular to the current, and gravel was placed upstream behind the structures. Small pools that formed around the gabions improved passage for salmon and steelheads. Spawning areas around the gabions also improved fish production in the area.[22]

ANADROMOUS FISH

Fish that migrate from the sea or lakes to spawn in rivers are called *anadromous*. Many sought-after sport and food fish are anadromous, including a number of salmon and steelhead trout. In the Great Lakes, the alewife, Atlantic salmon, and sea lamprey are anadromous. The alewife is a herringlike fish that increased in the Great Lakes partly because its predators were destroyed by the sea lamprey. Extensive management efforts to control the alewife and lamprey have brought back some of the popular sport and commercial fishes, including the anadromous steelhead trout, which migrates to streams from the Great Lakes.[23] Current stocking programs have been partially successful in establishing the Atlantic salmon in the upper Great Lakes.

Dams and impoundments have been particularly troublesome in the maintenance of anadromous fish populations. Massive dams along such major rivers as the Columbia, between Oregon and Washington, have caused a decline in salmon runs. Biologists working with engineers have designed fish ladders that allow salmon to return to the streams where they were born and spawn (Figure 19–7).

By means of their olfactory sense, salmon can, three to five years later, find their way back to the area where they were born. Thus, maintenance of stream habitat is very important. Logging, extensive grazing, and dumping of waste materials into these small tributaries can seriously alter the habitat. Gravel bottoms that are disrupted by human activity, siltation, or excessive runoff can also be detrimental to anadromous fish. Managers must encourage people who use habitats around mountain tributaries to create a band of natural vegetation so that water temperature, oxygen content, flow, and stream bottom remain unaltered.

While the riparian areas must be preserved for the return of spawning salmon, the salmon in turn provide a vital role in maintaining the riparian community. Minerals like nitrogen are considerably higher in spawning areas, contributing to the health of the ecosystem. Fish carcasses provide food for insects and plankton, which in turn provide food for other fish and terrestrial organisms. Many nutrients have been brought from the ocean. This whole complex of aquatic–terrestrial interaction is very important for the health of forests that surround streams.[24]

Most Pacific species of salmon die after they deposit their eggs. Atlantic salmon can spawn more than once. Biologists are using the two-chambered Whitlock–Vibert egg incubation box, which can be placed in headwater streams, to protect salmonid eggs (Figure 19–8). An upper chamber acts as an incubator. Fry pass through slots in the floor to a nursery chamber, which protects them until they emerge. Siltation still causes problems, but the devices have better hatching and emerging success than that of gravel-planted embryos.[25] The Whitlock–Vibert box has been placed in streams by conservation groups assessing fisheries management. Regulations are also very important in the management of anadromous fish. Regulated seasons, restrictions on fishing gear, limits, and the protection of juveniles prior to migration downstream are all effective means of maintaining sport and commercial anadromous fisheries.

Figure 19–7 Fish ladders provide an opportunity for salmon to return to their birth site and spawn.

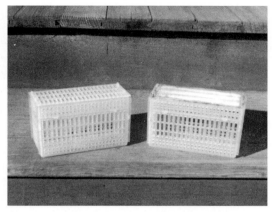

Figure 19–8 Whitlock–Vibert egg incubation box to protect eggs from predators. (Courtesy R. Barnhart, U.S. Geological Survey.)

DISEASE

Because freshwater aquatic organisms live in a relatively stable environment, disease caused by other organisms can spread among members of the desired population. Managers must learn to recognize fish diseases to prevent or control them. Diseases in fish are caused by six groups of organisms: ectoparasites, fungi, endoparasites, bacteria, viruses, and organisms that produce toxins in the water.[26]

Diseases are generally contracted either by direct contact or through an intermediate carrier. Fish, like other organisms, have an immune system. When the environment is disturbed, this immune system is more liable to be disrupted. Since most aquatic organisms are highly regulated by their environment, they can easily be disrupted by stress or alterations in habitat, which can be particularly troublesome in hatcheries or fish cultures.

Whirling Disease

Whirling disease, discovered 20 years ago, is now found in more than 20 states. It is caused by a parasite that invades the cartilage of its fish host. The parasite often destroys the cartilage around the organ responsible for the fish's equilibrium in the head of salmon and trout. This causes the fish to whirl around as though it were chasing its tail. Since the disease eventually kills its host and contaminates waterways, fisheries managers try to control whirling disease by preventing contaminated fish from being released. The life cycle of the parasite is as follows: Spores from infected fish are released into the water when the fish dies or when feces of fish-eating predators get into the water. These spores are ingested by mud-dwelling worms, in which they transform into a grapple-shaped form over a period of three to four months. Other fish become infected when they eat the grapple form or an infected worm. The parasite then enters the fish's bloodstream, travels to the cartilage, and begins a new cycle.

Alteration can cause diseases in natural habitats. Lowering the level of a lake to start a mosquito control program can result in crowding and a more rapid spreading of ectoparasites. The food supply changes with the water level, so that weaker fish may be unable to survive. Organic wastes introduce disease organisms into the aquatic system, as do vectors. Some parasites are associated with particular strains of fish (Figure 19–9). For example, populations of ocean perch can be distinguished from each other by their copepod parasite. Age classes can also be separated by disease organisms.

What can managers do to avoid or control disease? In natural habitats, managers should make regular assessments. Periodically, fish populations can be evaluated in the laboratory. Waste materials should be properly treated before they are put into the water. Managers need to keep informed about parasites and other disease-causing organisms. Many such organisms go through life cycles that involve intermediate hosts. When we know the process, we can sometimes pinpoint a stage vulnerable to attack. In extreme cases, managers may need to remove an entire diseased fish population. This is often done by poisoning. A healthy population can then be introduced into the aquatic system.

In an artificial environment, the fish are normally taken from other stock or from the natural habitat. They can be isolated before they are introduced into the artificial habitat, to remove diseased fish. Proper water conditions are very important in artificial ecosystems.[27] The fish in these systems may be crowded or under stress, making an adequate supply of oxygen and a proper diet even more important than usual. Fish in artificial habitats must be monitored closely for disease; periodic laboratory checks of selected individuals are appropriate.[26]

Vaccines can control some bacterial diseases, such as "red mouth" in trout. When steelhead survival was threatened in 1976, the California Fish and Game Department vaccinated a number of juveniles against two types of bacteria. The return rate of vaccinated adults to the hatchery was 19 percent above that of the nonvaccinated group, and no difference between the groups were found in the size and sex of the returning fish.[28]

Disease prevention obviously involves careful planning, controlled environments, and monitoring. The biological knowledge and experience of a manager are most important.

Figure 19–9 Trout parasitized by an anchor worm. (Courtesy D. Mitchum, Wyoming Game and Fish Department.)

In 1985, the U.S. Fish and Wildlife Service produced 138 million fish at a cost varying from $0.58 per kilogram ($1.29 per pound) of striped bass to $0.06 per kilogram ($0.13 per pound) of channel catfish in the west.[29] Over half the product consisted of various species of trout. The cost of producing rainbow trout in the west was $0.98 per kilogram ($2.16 per pound). About one-third were anadromous fish; the remainder were warm-water fish, including bass, bluegills, sunfish, channel catfish, and walleyes. These fish were released for sport and commercial fisheries. By 1999, the amount of fish produced in hatcheries remained the same; however, the price to produce a rainbow trout increased to $1.05 per kilogram ($2.32 per pound), still a great expense to produce a fish. Then there are extensive state hatchery programs. Truly, hatching fish is big business.

Effective hatchery management presupposes a knowledge of the many physical and biological factors that affect fish growth and development. Special strains of fish are often developed for stocking nearby streams and lakes. Some hatcheries run a purely put-and-take fishery program; others supplement natural fish populations.

Hatcheries in the United States were developed as an outgrowth of European fish culture practices. Early in this century, the U.S. Fish and Wildlife Service distributed hatchery-reared fish throughout the United States in railroad cars (Figure 19–10). Personnel had to maintain the tanks in transit. Several methods of transporting fish are available, so that the fish will not be put under undue stress. There are appropriate methods for each species (Figure 19–11).

Early attempts at cultivating hatcheries were not successful because the biologists did not examine sufficiently the physical and biological conditions of the water into which the fish were released. Managers today carefully consider types of habitat, food, predators, and all physical conditions before fish are introduced. Another problem faced by hatchery managers is acclimatization of the fish. Sometimes hatchery-reared fish become so accustomed to being fed at a certain time of day or so used to the

Figure 19–10 In the early 1900s, fish were transported in railroad cars. (Courtesy of the National Archives.)

Figure 19–11 Transporting fish from hatcheries to stocking areas. (Courtesy N. Ward, U.S. Fish and Wildlife Service.)

presence of human beings, that they are unable to survive in another routine and environment. Hatchery management should include follow-up studies of the survival of released hatchery-reared fish. If natural fish populations are providing an adequate recreational base and balance in the system, hatchery fish are better put elsewhere.

Fish are usually stocked as fingerlings, or fry, but there have been successful programs with fertilized eggs and adults. Stocking of fingerling fish is often supplemented by catchable fish when the demand for fishing is great. Put-and-take fisheries specialize in placing fish for immediate exploitation. By maintaining stock over many generations, they have developed strains with rapid growth rates. But not all of these can compete successfully in the wild. Some of the more successful stocking operations take the eggs from wild stock each year, fertilize them, hatch them near the receiving waters, and put them back into the water from which the brood fish were collected.

Stock-fish survival can be affected by the stress of being hauled and the abrupt difference in quality between the hauling and the receiving water. Hatcheries have attempted to develop hybrid fishes that are more likely to survive under natural conditions and are less susceptible to disease and attacks by predators. Stocking programs are successful in lakes where conditions are good for the survival of adults, but not for the young or for eggs. For example, fish may have to be stocked annually or every few years in lakes with inadequate spawning beds.

AQUACULTURE

Aquaculture is rearing plants or animals in water under controlled conditions—fish farming and underwater agriculture combined. Aquaculture involves establishing and controlling an environment, whereas natural-systems management involves manipulating the habitat. In aquaculture, the amount of biomass produced per hectare is very high compared with that of natural aquatic life. Aquaculture includes fish raised for sport, commerce, bait, ornament (goldfish and tropical fish) and plant species cultured for food and drugs. Pearl cultures are an industry in some parts of the world.

Aquaculture is more than 4,000 years old, apparently beginning in China following the development of the silkworm industry.[30] The carp was one of the first fish cultured. Aquaculture spread throughout southern Asia, incorporating species to suit local tastes. It started in the Roman Empire with culturing of pearls. Later, carp became important, probably as trade with Asia increased. Aquaculture techniques and some species, including carp, were brought to the New World when it was settled by Europeans. Oyster production began in the United States in the 1850s.[29] By 1900, there were trout farms in many states. Later, bait fish and ornamental fish became important objects of aquaculture in some regions. In the 1950s, a commercial channel-catfish industry developed in the southeast; today there are more than 2,000 fish farms in 13 states.

Crayfish have been successfully cultured in rice ponds after the rice is harvested. These ponds, although relatively shallow, provide cover and vegetation for shelter and food. Catfish are usually grown in ponds 120 to 150 cm (47 to 59 in) deep that slope from one end to the other (Figure 19–12). Often, artificial ecosystems are created in special facilities, such as indoor tanks or outdoor ponds. In some instances, the natural system is fenced in, much as livestock are corralled on a range. Marine species, such as oysters, can be kept in wire cages submerged in estuarine waters.

In any form of aquaculture, there are certain objectives and critical needs: a continuous and preferably cheap production of young, prevention of disease, maximization of rate and efficiency of growth, and proper control of temperature and oxygen. Ornamental species that can withstand only a few degrees change in temperature must be reared indoors.

Because most aquaculture occurs in artificial systems, diseases can spread rapidly as a result of inadequate oxygen, improper pH, or an improper diet. As we have noted, treatments are available for some diseases. Sometimes, entire cultures must be destroyed and facilities sterilized to remove the disease organism. In either case, the expense is considerable. Therefore, proper preparation and maintenance of the facility is one of the best and most economical approaches to reducing the incidence of disease. Growth enhancement can be successful only if the manager has extensive knowledge of an organism's diet. In aquaculture, there has been much research in artificial diets, which are effective if properly constructed and if costs are held to a minimum.

Figure 19–12 Catfish ponds are very common in parts of the country.

Fish Farming

Fish farming is becoming more and more popular in the United States. Many forms of fish, including trout and catfish, are raised on farms. Also, many invertebrates, such as crayfish, crabs, lobster, clams, and oysters, are raised in specially constructed ponds. Fish farmers attempt to create a special ecosystem in which the desired crop can be raised. In doing so, they are using some of the principles of ecosystem management and ignoring others. Since fish farming usually involves raising just one species, no predators are introduced. Controlled temperature, water quality, and pond construction are important. Food is often provided, since there are few primary producers. The fish species is usually re-stocked annually to replace those that are harvested. The most effective fish farm managers take advantage of natural conditions such as springs to raise their crop. People who apply the principles of ecosystem management often have the greatest success with the lowest costs.

Is aquaculture the answer for food in the future? Certainly, intensive agriculture can produce more food, but the energy input may be expensive and a drain on other systems. Experiments by those engaged in aquaculture indicate that there are possibilities of increasing production in some facilities. For example, most aquacultures in the United States produce a single species (monoculture). In other parts of the world, polyculture, the rearing of two or more species, is practiced. With better understanding of the food chain, some additional forms of polyculture may be possible. Still, energy limitations apply to all systems. Effluents from power plants and wastewater treatment facilities are being tested as possible aquaculture media. If feasible, this will give these facilities an interesting and useful by-product. Food produced through aquaculture is not always acceptable, of course, to all cultures. All in all, while aquaculture cannot be viewed as a panacea, it can make a significant contribution to a solution of one of the world's great problems: providing enough food for everyone.

DESCRIPTIONS OF SPECIES

Paddlefish

The largest inland fish in the United States, reaching 73 kg (160 lb), is the paddlefish (Figure 19–13). Currently found in large bodies of water in the Mississippi Valley, it has also been taken from bay waters. The paddlefish's nearest living relative is in the Yangtze Valley of China. As its name implies, the paddlefish has a long snout, which can be more than one-third of its body length. From the side, the beast resembles a shark. It swims slowly or remains quietly in slow-flowing waters that have abundant zooplankton, on which it feeds.

The habitat of the fish is large, slow-moving waters. The Mississippi River, with its oxbows and backwaters, is an ideal habitat. The fish requires extensive gravel bars, with fluctuating water levels, on which to spawn. Since 1900, paddlefish have declined as a result of overfishing and loss of habitat. Stream channelization and lake drainage have been particularly detrimental to the paddlefish. Maintenance of habitat is therefore the major management task in behalf of this large fish.

Figure 19–13 Paddlefish. (Courtesy T. Gengerke.)

Alewife

The alewife, a member of the herring family, enters rivers on the east coast and the Great Lakes early in the spring in order to spawn in slower moving water. This anadromous species became abundant in the Great Lakes—excessively so in Lakes Ontario, Erie, and Michigan—when some of the larger fish were killed when the sea lamprey was introduced into the lakes. As a result, it was washed up on shore in great numbers, causing an unpleasant odor in many beach areas. There are also landlocked populations of the species.

Management efforts to reduce the population failed. When the lamprey was brought under control, some of the Great Lakes' salmon increased and used the alewife as a forage fish. Thus, through predator–prey interactions, the alewife was kept in check. Some biologists feel that the fact that the alewife population is small now prevents further growth of the salmon population, for which the alewife is a source of food.

Cutthroat Trout

A popular sport fish in many western streams and lakes is the cutthroat trout, of which there are 13 recognized subspecies. Some of them are on federal or state endangered- or threatened-species lists. The cutthroat will hybridize with rainbow and golden trout. Some anadromous forms occur on the west coast, intermixed with nonmigratory forms.

River habitats for cutthroats are characterized by clear, cold, silt-free water; rocky substrate in riffle run; and areas of slow, deep water, with dense vegetation on the banks. Lake habitats are characterized by clear, cold, deep water, usually oligotrophic. Management efforts center on maintaining the habitats in their natural state. The dispersal of human wastes on other pollution activity disrupts the trout habitat, and alteration of the vegetation on the side of streams is detrimental. To keep prize fish, it is necessary to establish regulations against excessive removal of large fish, thus effecting a distribution of different-sized classes.

Channel Catfish

The channel catfish is found throughout a large part of the United States in large streams with low or moderate gradients. The adults like large pools of water with such cover as downed logs. Through introductions, the channel catfish has expanded its range in recent years from the more central United States to the east and west.

The fish feed on bottom material, such as small fish, insects, crayfish, and some plants. There seems to be little selectivity in their diet. Catfish spawn in hollow logs or cavities in the banks of streams. The eggs are deposited as a gelatinous mass in a small nest. The male guards the fry in the nest for about a week after they hatch.

The channel catfish is a very popular sport fish. Some farms raise the fish commercially. Management is relatively easy if streams are kept free of toxic materials. Channelization or bank-clearing projects disrupt the catfish habitat.

Northern Pike

The northern pike is distributed throughout Alaska, a large part of Canada, and into the north-central and eastern states. It is found in small lakes, in the shallow, more-vegetated parts of large lakes, in marshes, and, to a lesser extent, in rivers.

Pike generally are solitary fish. They lay their eggs in the spring, spreading them over submerged vegetation and then abandoning the eggs. Sometimes, spawning occurs under the ice. The young can grow very rapidly, reaching a length of 45 cm (18 in) by nine months. Sexual maturity occurs in three to four years. The fish can live more than 24 years. Pike are restricted in part by the shortage of shallow-water habitat. Furthermore, many pollution-producing activities affect pike feeding and spawning areas in the United States.

Bluegill

In the United States, the bluegill is a popular sport fish taken by young and old alike. This species of the sunfish family has become more widely distributed in recent years. Originally restricted to east of Texas and the Mississippi, it is now found in most of the warm waters of the United States and in other parts of the world (Figure 19–14).

Bluegills do not get very big—only about 24 cm (9 in) long. They are gregarious and move around in aggregations through clear, slow-moving water. They eat small insects and crustaceans. The female lays eggs in a nest constructed as a depression by the male. The preferred substrate is gravel, but other substrates will do. The male guards the nest until the eggs hatch.

Pacific Herring

The Pacific herring is a very popular commercial species. It spawns along the California coast, with large populations spawning in Tomales and San Francisco Bays. Spawning season is from November to June, with larger numbers spawning in December,

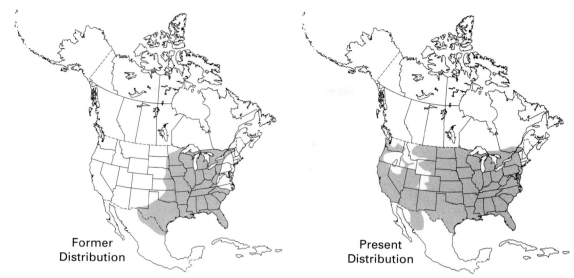

Former
Distribution

Present
Distribution

Figure 19–14 Distribution of the bluegill, a popular sport fish.

January, and February. Generally, the males and females enter the bays several weeks before spawning and remain in schools in the deep channels. Apparently, external stimuli such as water temperature initiate spawning, although the trigger mechanisms are not well understood. The fish spawn on vegetation and rock surfaces in the bays. The offshore distribution of the Pacific herring is not well documented; therefore, its habitat needs are not known.

Pollution, as well as heavy human use of beaches and estuaries, has an adverse impact on herring. These areas need to be kept clear for survival of the eggs and young. Excessive turbidity resulting in the settling out of sediments may hinder spawning and incubation. Because herring roe are very popular in Japan and the number of roe is highest just before spawning, fishing is heavily concentrated then. California has set herring stock quotas, to prevent overfishing and allow preservation of the resource.[31]

SUMMARY

Fisheries resources include fish, some marine mammals, and invertebrates harvested for human consumption. Fish are managed for both commercial and recreational use. They constitute an important component of the food web. Fish generally spawn many more eggs than hatch or survive. Different species require specific physical characteristics of their habitats. The biology of fish is also very important. Competition and predator–prey interactions play an important role in the growth of fish communities.

When biologists want to sample fish, they need to understand the biology of the organisms. Nets are appropriate for pelagic species, while traps and bottom samples can be used for bottom-dwelling species. Electric shock and poison are useful for total counts in an enclosed body of water.

Populations can be managed in different habitats through regulating harvests and protecting, enhancing, and rehabilitating habitats. Control of disease is important in maintaining good fishing. Hatchery fish are released into lakes and streams, but this put-and-take operation for sport fishing is rather expensive. Currently, many aquatic species are being raised by fish farming (aquaculture), but the future scope of aquaculture will probably be limited by various energy impacts.

DISCUSSION QUESTIONS

1. How do regulations serve as a means of management in fisheries?
2. How can we determine the true value of hatchery-reared fish? Of a hatchery program? Should we try to put a price tag on this activity?
3. Can human wastes be used effectively in aquaculture? Discuss the good and bad points.
4. How can we determine the harvestable limits of a fish population?
5. What forms of habitat improvement can benefit stream fisheries?
6. What techniques can be used to prevent disease in fish?
7. Are habitat-improvement techniques feasible for anadromous fish?
8. What is maximum yield? What causes change in maximum yield? How is the concept used in fisheries management?
9. What conditions are important for fish survival?
10. How can marine-fish populations be managed?

LITERATURE CITED

1. Lackey, R. T., and L. A. Neilsen (Eds.). 1980. *Fishery Management*. New York: Wiley.
2. Russell-Hunter, W. D. 1970. *Aquatic Productivity*. New York: Macmillan.
3. Aldrich, E. C. 1997. Angling for Answers. *Wildlife Journal*. March/April, pp. 121–23.
4. Tyler, A. V., and V. F. Gallucci. 1980. Dynamics of Fish Populations. In R. T. Lackey and L. A. Nielsen (Eds.), *Fishery Management*, pp. 111–49. New York: Wiley.
5. Lackey, R. T., and W. A. Hubert. 1977. *Analysis of Exploited Fish Populations*. UPI-SG-76-04. Blacksburg, VA: Virginia Polytechnic Institute and State University.
6. Everhart, W. H., and W. D. Youngs. 1981. *Principles of Fishery Sciences*, 2d ed. Ithaca, NY: Comstock.
7. Stickney, R. S. 1979. *Principles of Warmwater Aquaculture*. New York: Wiley.
8. Rahel, F. J. 1984. Factors Structuring Fish Assemblages along a Bog Lake Succession Gradient. *Ecology* 65:1276–89.
9. Swingle, H. S. 1950. *Relationships and Dynamics of Balanced and Unbalanced Fish Populations*. Bulletin 274. Montgomery, AL: Alabama Agricultural Experiment Station.
10. Anderson, O. 1981. Optimal Foraging by Largemouth Bass in Structured Environments. *Ecology* 65:851–61.
11. Smith, R. R., and J. J. Tibbles. 1980. Sea Lamprey (*Petromyzon merinus*) in Lake Huron, Michigan, and Superior: History of Invasion and Control, 1936–1978. *Canadian Journal of Fisheries and Aquatic Sciences* 37:1780–1801.
12. Bennett, G. W. 1971. *Management of Lakes and Ponds*. New York: Van Nostrand Reinhold.
13. Backiel, T., and R. L. Welcomme (Eds.). 1980. *Guidelines for Sampling Fish in Inland Waters*. Eifact/T133. Rome, NY: Food and Agricultural Organization of the United States.
14. Hubert, W. A. 1983. Passive Capture Techniques. In L. A. Nielson and D. L. Johnson (Eds.), *Fisheries Techniques*, pp. 95–111. Bethesda, MD: Fisheries Society.
15. Platts, W. S., W. F. Megahan, and G. W. Minshall. 1983. *Methods for Evaluating Stream, Riparian, and Biotic Conditions*. General Technical Report INT-138. Washington, DC: USDA Forest Service.

16. Barney, G. O. 1980. *The Global 2000 Report to the President of the United States.* New York: Pergamon Press.

17. Nelson, R. W., G. C. Horak, and J. E. Olson. 1978. *Western Reservoir and Stream Habitat Improvements Handbook.* FWS/OBS 78156. Fort Collins, CO: U.S. Fish and Wildlife Service.

18. Haines, T. A. 1981. Acid Precipitation and Its Consequences for Aquatic Ecosystems: A Review. *Transactions of the American Fisheries Society* 110:669–707.

19. Moyle, P. B., and J. J. Ceck. 1982. *Fishes: An Introduction to Ichthyology.* Englewood Cliffs, NJ: Prentice-Hall.

20. Babcock, W. H. 1981. *Stream Degradation Studies: Ten Mile Creek Channelization, Colorado Fisheries Research Review 1978–80.* Review 10. Denver, CO: Fisheries Research Section, Colorado Division of Wildlife, pp. 66–70.

21. Burgess, S. A., and J. R. Biden. 1980. Effects of Stream Habitat Improvements on Invertebrates, Trout Populations and Mink Activity. *Journal of Wildlife Management* 44:871–80.

22. Anderson, J. W., and J. J. Cameron. 1980. *The Use of Gabions to Improve Aquatic Habitat.* BLM Technical Note 342. Washington, DC: U.S. Department of the Interior.

23. Bley, P. W., and J. R. Moring. 1988. *Freshwater and Ocean Survival of Atlantic Salmon and Steelhead: A Synopsis.* Biological Report 88(9). Washington, DC: U.S. Fish and Wildlife Service.

24. Levy, S. 1997. Pacific salmon bring it all back home. *BioScience* 47:657–60.

25. Barnhart, R. A. 1980. Current Status of the Whitlock–Vibert Egg Incubation Box. In *Trout Unlimited, Proceedings of Wild Trout H,* pp. 10–14.

26. Pritcher, T. J., and P. J. B. Hart. 1982. *Fisheries Ecology.* Westport, CT: AVI.

27. Brown, E. E., and J. B. Gratzek. 1980. *Fish Farming Handbook.* Westport, CT: AVI.

28. Kerstelter, T. H. 1980. Increase in Ocean Survival of Freely Migrating Steelhead Vaccinated against *Vibro anguillarum. Transactions of the American Fisheries Society* 109:287–89.

29. Division of Program Operations—Fisheries. 1986. *Artificially Propagated Fish for National Hatchery Programs.* Washington, DC: U.S. Fish and Wildlife Service.

30. Avault, J. W. 1980. Aquaculture. In R. T. Lackey and L. A. Nielsen (Eds.), *Fishery Management.* New York: Wiley.

31. Barnhart, R. A. 1988. *Pacific Herring.* Biological Report 82(11.79). Washington, DC: U.S. Fish and Wildlife Service.

20
NONGAME

Traditionally, wildlife managers have been trained to maintain wildlife for harvesting. Until the public became ecologically aware in the late 1960s and early 1970s, the funding dictated that, except for endangered species, managers concentrate on the game species. As J. S. Gottschalk said in a 1975 symposium on the management of forest and range habitats for nongame birds: "In a sense, one could characterize wildlife management policies over the world, as a matter of fact, as being superficial. They deal largely with end products and ignore the vast and complex matrix of plant and animal life which in the long run supports not only fish and wildlife but man himself."[1] Now, it is difficult to see how managers can deal with wildlife without heeding our present knowledge of population, community, and ecosystem management. Although we may set bag quotas or recommend criteria for increasing or decreasing a game population, we must also recognize how the recommendations affect other populations.

Nongame is an administrative term for a subset of wildlife species that are not hunted, harvested, or intentionally removed by human beings. *Nongame* can also be construed as including game species that must be managed as nongame, as in national parks. Some states are developing extensive programs for nonhunted wildlife and calling them *nonconsumptive*-wildlife programs. As a result of bringing the public into wildlife issues, state wildlife-management agencies are increasing their constituencies.

Nationwide, there is a strong public interest in managing nongame. In 1980, Stephen Kellert surveyed American attitudes toward wildlife and wildlife-related activities. (See Chapter 14.) The results showed, among other things, that during the preceding two years, 14.5 percent of the respondents hunted, 44.4 percent fished, and 25.2

Plant Community	Possible Management Feature
Deciduous Forest	Size
Coniferous Forest	Configuration
Riparian	Distribution of Vegetation
Grassland	Adjacent Land Use
Desert	Human Impact

Figure 20–1 Management features of selected environments for nongame.

percent spent time bird-watching. In addition, 46 percent visited zoos and 42 percent spent time photographing animals.[2] In 1996, the U.S. Fish and Wildlife Service found that 63 million adult Americans actively participated in such nongame activities as feeding, watching, or photographing wildlife.[3] Clearly, the public is becoming more aware of our wildlife resources and spending more time on activities that are wildlife related. Managers can therefore expect more questions about nongame and more interest on the part of the public.

Each community has its own combination of wildlife species that are where they are because of the abiotic and biotic characteristics of the region. These animals form an intertwined network whose links are altered when people encroach on the system. Thus, managing nongame often involves managing the community; it is difficult to restrict attention to the attributes of individual species (Figure 20–1). This may indeed be the most positive note about nongame management: *It includes all wildlife.*

In this chapter, we examine several approaches in nongame management and discuss nongame wildlife mechanisms in relation to types of community. Because of the interest in wildlife as both nongame and consumptive species (for removal by falconers), we include a section on raptors. Needs peculiar to managing wildlife in urban settings are also discussed.

NONGAME AND PEOPLE

Most nongame activities can be classified as primary or secondary. Primary activities include those in which people start out to have a wildlife experience. Thus, driving to a fish hatchery, visiting a wildlife refuge, and going on a nature trail are all primary wildlife activities. So are such activities as planting garden plants to attract wildlife and putting in bird feeders or birdbaths. Those who seek out wildlife to photograph also are primary wildlife users (Figure 20–2).

Secondary wildlife activities are those which enhance another activity. For example, hikers who see a moose or bear increase their total enjoyment of the trip. When traveling by car, seeing wildlife species such as deer, antelope, prairie dogs, and kestrels enhances one's enjoyment and aesthetic pleasure. People often stop to photograph wildlife or just to look. Wildlife displays at rest areas, as well as signs along waterways and trails, all add to a pleasurable experience for people.

Wildlife managers must consider resident and nonresident wildlife users when considering primary and secondary wildlife activities. Nonresidents may come many miles to photograph a prairie dog, alligator, or mountain sheep. Residents are often interested in having parks and home plantings that attract wildlife. All in all, managing nongame wildlife means managing wildlife for people. At the same time, people must be managed so that they do not destroy the wildlife or are not hurt by the animals.

Figure 20–2 Many people enjoy watching nongame birds such as this yellow-bellied sapsucker.

CURRENT STRATEGIES IN COMMUNITY MANAGEMENT

The objectives in any nongame-management program are maintaining species or communities and minimizing human impacts on wildlife. Instrumental to these objectives is the accumulation and evaluation of data on the biology, ecology, and distribution of all species and communities so that management strategies can be developed. Some popular nongame-management approaches rest on data collection and synthesis systems that emphasize wildlife habitat associations. To model "change," these approaches describe the relative abundance of species associated with different habitats or the successional stages of vegetation. Thus, change is equated with replacing one habitat and its associated species with a new habitat and *its* associated species.

Managers can and do manage for individual nongame species when appropriate, and the needs of some individual species are discussed in this chapter. However, it is often more desirable to manage for communities of wildlife. Three field approaches to managing wildlife communities are managing for the richness or diversity of species, usually through fostering a diverse habitat; managing for indicator or featured species; and life-form management. Species richness and featured species were discussed in Chapter 2.

Diversity

Natural ecosystems consist of mosaics of structural types in which a diversity of wildlife has evolved. This diversity of species is believed to have numerous values, including a function in maintenance of the ecosystem. Logically, then, we should preserve diverse habitats to maintain a diversity of wildlife and the integrity of ecosystems.

Habitat diversity can occur at different levels: within the habitat, between habitats, and over a large geographic area. (See Chapter 6.) Biologists have found that some nongame species actually select discrete patches of different types of habitat within a plant community. Ecotones appear to increase the richness of nongame species within a large geographic area (Figure 20–3). Although some species found in urban areas select edges and disturbed sites, these species are very likely to persist without special management attention. Thus, if we manage for maximally diverse types of vegetation, species with specific requirements will be lost. While diversity sounds like a worthy goal, the requirements of individual species also need attention. To provide adequate patches within a habitat, how much of it is needed for the purpose, where and in what arrangement, and of what quality?

Because diversity can be viewed from a site-specific to a geographic perspective, many methods of managing for diversity of wildlife exist. One scheme for geographic diversity consists of leaving parts of each successional sere of the area under management. In addition to maintaining each sere with active management procedures, managers can evaluate proposed impacts on the habitat in the light of alterations in each sere.

Indicators

The major concern in the use of indicator species is selection of the proper indicators or featured species. In the National Forest Management Act, the U.S. Forest Service defines an indicator species as "those which are believed to indicate effects of management activities on a number of other species." In other words, a nongame indicator species represents the nongame community of species. The habitat requirements developed for any indicator species-management program are likely to be much broader than that animal's actual niche, so that, by design, the niches of many species may be covered (Table 20–1).

Even with a wealth of background biological information, problems can exist. While indicator species generally represent communities, most are selected because they have narrow tolerances to environmental factors (steno species). As a result, a change in the environment causes a change in the steno species. Although the selected indicator species are not likely to become extinct, others may, or ecosystems may change undesirably if the entire concept or the initial species selected was faulty. The question is, then, whether a few species truly represent the needs of entire communities. This species–ecosystem approach is conservative as to individual species, but not automatically so as to communities, despite the underlying theory.[4]

Figure 20–3 Ecotones increase the number and diversity of species through the junction of two habitats.

TABLE 20–1

Possible Nongame Indicator Species

Species	Characteristics
Northern pike	Shallow northern lakes with emergent vegetation
Cave salamander	Limestone caves or spring
Desert iguana	Dry desert with rocks and creosote
Red-breasted nuthatch	Mature conifer forests
Red tree vole	Douglas fir of western United States

It is unlikely that we could pick so-called indicator species that would represent virtually all types of habitat. For instance, one style of indicator-species management could be based on rarity.[5] The criteria for species to be monitored would then be (1) a small geographic range, (2) high habitat specificity, and (3) either low numbers or very localized populations. Although this approach makes no claim to being able to safeguard all representative ecosystems, it is capable of revealing declines in types of general habitat.

Life-Forms

Every community has an array of animal species, and it is sometimes difficult to develop comprehensive community-management criteria. Forest Service biologists have suggested that a life-form approach can help managers work with groups of species associated with components of the plant community.[6] They suggest classifying the animals of a community into groups based on specific combinations of habitat requirements for reproduction and feeding. The result is groups of animals not necessarily taxonomically or morphologically similar (Table 20–2). It is then possible to relate each life-form of the animal community to life-forms of the plant community (Table 20–3). The life-form idea is similar to the guild concept, in which all animals that remove similar food from a similar site in the environment are in the same guild. By focusing on changes in the plant community, managers can predict the impact on species of a particular animal life-form.

TABLE 20–2

Classification by Habitat Requirement

Life-Form	Reproduces	Feeds	Example
A	In water	In water	Bullfrog
B	On the ground	In bushes, trees, or air	Common nighthawk, Lincoln's sparrow, porcupine
C	Primarily in conifers	In trees, bushes, or air	Golden-crowned kinglet, yellow-rumped warbler, red squirrel
D	In hole excavated in tree	In trees, bushes, ground, or air	Common flicker, pileated woodpecker, red-breasted nuthatch
E	In burrow underground	On ground or under it	Rubber boa, burrowing owl, Columbian ground squirrel

TABLE 20–3

Animal–Plant Relationship by Habitat

Plant Community		Successional Stage*			
		Grass	0–10	11–39	40–79 Years
Quaking aspen	(Animals reproducing)	0	0	0	3
	(Animals feeding)	1	1	1	3

*Numbers in each successional stage refer to the number of animals in lie-form D (Table 20–2) that use the

As in the diversity-management concept, care must be taken that no specific requirement of a species be omitted. Too, the requirements of migratory stopover species or different needs for each life stage of an animal's cycle must be included.

NONGAME HABITATS

Nongame animals, like all other wildlife, have specific habitat needs, which can be examined from both species and community points of view. It is important to understand the role that nongame wildlife plays in communities.[7] In the process of energy movement in a natural system, nongame species feed on plants and other animals. If a pesticide kills many insects in an area, the impact may be pronounced on both birds and small mammals. Birds directly affect some plant communities through the consumption of seeds. Dispersal is achieved when seeds pass unharmed through a bird's digestive system and start growing in areas where they are excreted. This process is particularly helpful in restoring damaged habitat.

Habitat Size

Major changes known to affect nongame communities include changes in community size, changes in structure (which can involve changes in successional sere), alteration of moisture, and disturbance of unique habitat features such as riparian zones. A number of studies indicate that the size of a habitat is particularly important in nongame bird communities. There is other evidence to show that small mammals also rely on habitats of a certain size. In an eastern deciduous forest, biologists found that the fragmentation of a contiguous forest totaling more than 5,260 hectares (13,000 acres) into plots of 40 hectares (100 acres) over 25 years resulted in a change in bird species despite little change in the internal character of each 40-hectare plot.[8] In this study, species of long-distance migrants, such as broad-winged hawks, yellow-throated vireos, worm-eating warblers, ovenbirds, and hooded warblers, disappeared from the breeding population of the study's sites. Other species, including the acadian flycatcher and scarlet tanager, had fewer breeding pairs than before. There was a positive correlation, in other words, between continuous forest area and the number of migratory birds breeding in the area.

Other data have confirmed the fact that the management of nongame communities requires an understanding of the size of those communities. The size requirements differ in different parts of the country. In the Blue Mountains of Oregon, 34 hectares (84 acres) were sufficient to maintain nongame communities, but in western Maryland, areas greater than 1,300 hectares (3,200 acres) were required to support a full complement

TABLE 20–4

Forest Subdivision Has a Major Impact on Nongame Birds

	Hectares of Eastern Deciduous Forest	
	5,300	40
Yellow-billed cuckoo	Present	Present
Wood thrush	Present	Present
Yellow-throated warbler	Present	Absent
Red-eyed vireo	Present	Absent
Worm-eating warbler	Present	Absent
Ovenbird	Present	Absent

of breeding-bird communities[6] (Table 20–4). The differences in these studies are due to differences in the forest communities in the regions. The highly migratory bird communities in the eastern deciduous forest require much larger tracts of land than do the less migratory communities in the west.

Habitat Structure

Habitat structure—that is, the layering pattern of vegetation and the size of the vegetation community—appears particularly important for nongame communities. Clearing of brush, burning of understory vegetation, and an increase in grazing pressure all affect the structure of the community and alter the wildlife species present. Edge—the area between two habitats—is structured differently than the surrounding community and supports different species. There is a correlation between breeding birds and structural components.[9] Different nongame species use each layer and occupy different microclimates within the habitat. For example, wood thrushes are found on the ground and the lower shrub layer of a forest community, blue-gray gnatchers are found near the top, and cardinals prefer edge.[9]

Habitat Features

Habitat features include ledges, slab rock, and shallow caves that are used by bats, raptors, small mammals, and reptiles. Each habitat has its own unique featured attraction for wildlife (Figure 20–4); for example, desert cliffs have a number of nongame species. Beyond the large features, each component of a habitat is uniquely attractive to certain species of wildlife. Sometimes these features are difficult to measure. So, although we might not identify why a seemingly insignificant feature is important, past experience tells us that it possibly is. As we learn more, its value may be discovered. This idea goes back to the concept of habitat size: Although changes can occur in a habitat, it is important to leave *components* of the natural habitat in order to maintain the community of wildlife that is there. If it is necessary to make large-scale changes, the habitat must be re-created (Figure 20–5).

Habitat Types

Some habitats, such as the riparian, appear to be particularly important to nongame; witness the fact that more than 136 bird species use riparian habitats in the west.[10] A few breeding birds nest only in riparian vegetation, and many species that live in nearby grass-

Figure 20–4 Dead logs are important to these raccoon. (Courtesy of the Los Angeles Zoo.)

lands use the riparian habitat. Some small mammals have been found to be susceptible to alteration of the riparian habitat. The apparent reason for the importance of this habitat is the diversity in food and structure near arid and semiarid environments.

Nongame wildlife species utilize the habitat differently during different times of the year. During the breeding season, birds set up territories in habitats that meet their

Figure 20–5 Marmots live in snags in forest-land.

specific requirements for nesting and raising young. In the fall and winter, resident birds form flocks and move around much larger areas, so that specific habitats are not so important. The limitation in winter is the availability of food. Disturbances of the habitat during the time when nongame species, particularly birds, are moving into an area and setting up their territories can disrupt the entire nesting cycle.

NONGAME PROGRAMS

As a professional group, nongame managers evolved from among biologists—particularly those in the U.S. Forest Service and U.S. Fish and Wildlife Service—who were charged with implementing nongame laws. Outside government, the environmental movement spawned countless private organizations. Some of them set excellent precedents in public education, political action, and, in the case of the Nature Conservancy, holistic resource inventorying and preservation, starting as far back as the 1960s.

While these developments were under way, state resource agencies were reluctant to extend game-species management to include nonconsumptive wildlife, largely because of funding restrictions. The 50-year precedent of "user" funding made wildlife management read "game." Through the Pittman–Robertson and Dingell–Johnson programs, hunters and fishing enthusiasts had paid to have their resources managed. Nongame management required a much broader perspective of wildlife resources.

In 1980, Congress passed the Fish and Wildlife Conservation Act to provide funding and technical assistance to states for nongame conservation. The act was in part a response to a report prepared for the Fish and Wildlife Service which showed that Americans spent close to $13 billion a year on binoculars, birdseed, and travel to areas where nongame wildlife could be found. Unfortunately, in the political climate in the years immediately following passage of the act, no funds were appropriated to carry out the activities prescribed in it. In 1988, the U.S. Fish and Wildlife Service prepared a document called *Nongame Bird Strategies*.[11] This document outlined objectives and tasks that the Fish and Wildlife Service intended to accomplish in the next five years. Hopefully, this is a step in the recognition of the need for management of nonhunted species by the federal government. Again in the late 1900s, *Teaming with Wildlife* was an effort to gain funding for nongame species (See Chapter 14).

State Programs

Rhode Island claims to have had a nongame program since the 1950s; at any rate, 38 states have initiated official, staffed nongame programs in the last 15 years.[12] If endangered-species work is included, the list of states doing some sort of nongame work grows to 50 as of late 1990. Half of these states do endangered-species work. The states with the largest funding and staffs are Colorado, Oregon, Minnesota, New York, New Jersey, Washington, Michigan, and California.

While many states' nongame programs focus on single species, whether they are "endangered," "sensitive," "of special interest" (for example, raptors), or "featured," some state programs, such as Alaska's and Kentucky's, are more holistic. These typically cover plants, invertebrates, and vertebrates. Missouri has the most comprehensive wildlife conservation program in the country. In that state's "Design for Conservation," nongame is not considered or managed separately from game, plants, or other resources.[13] State programs are developing comprehensive inventories to be used in monitoring populations and communities. Public outreach, urban wildlife management, and planning to meet long-range objectives are frequently major goals.

> **Reptiles in the Pet Trade: Are They Nongame?**
>
> People have always wanted pets. In the United States, pet stores have recently seen an increased demand for reptiles such as snakes and lizards. In 1995, the United States imported 840,000 green iguanas, up from 92,000 in 1985. While some people have continued to enjoy these high-cost, high-maintenance pets, animal shelters and zoos have been inundated with requests to take the creatures off owners' hands. At the same time, natural populations in Africa, Asia, and South America have been depleted almost to the point of extinction.
>
> Many of these nongame pet species are coming from the United States. Rattlesnakes and horned lizards are now popular. People are searching the areas of the country where these reptiles are found and selling them to pet stores for high prices. Some enterprising individuals are selling mated prairie dog pairs around the world for substantial fees.
>
> So although we now have nongame species being removed from the wild in the same manner as hunted species, we are not managing or limiting the numbers that can be taken. If some of the nongame pet species are to survive, controls must be instituted to regulate the numbers that can be taken in relation to the number produced.

To illustrate the rapid growth of the nongame movement, in 1975, although 36 states reported activities directed specifically toward nongame fish and wildlife, only 17 had a full-time biologist for this work. Half of these projects concerned endangered species, and half involved research.[14] Three years later, 46 percent of the projects in the southeastern United States focused on endangered species, with 26 percent on a single species, the red-cockaded woodpecker. In 1988, all 50 states funded nongame programs at some level. In that year, a survey by the National Wildlife Federation found that more than $43 million dollars were spent by the state on nongame. These dollars included monies spent on endangered species as well as on nongame species.[11]

Funding

While monies from hunting licenses remain an important source of support, Endangered Species Act monies have funded a large share of nongame research to date. In 1981, some 47 states and three territories used Pittman–Robertson funds to benefit nongame directly or indirectly. State nongame programs have tapped a variety of funding sources. The most successful programs are not totally dependent on federal monies. Some sources of funding are as follows:

1. *The Endangered Species Act of 1973.* Section 6 of this act provides for state–federal endangered-species cooperative programs on a two-thirds or one-third matching basis.
2. *State sales tax.* Missouri passed a constitutional amendment in 1976 increasing its sale tax by one-eighth of 1 percent, with funds going to the Department of Conservation.
3. *Sale of wildlife stamps, patches, T-shirts, and other items.* These sales programs have been tried in several states, but with little success.

4. *Sale of personalized auto license tags.* The state of Washington has funded its nongame program from these sales for nine years.

5. *General appropriations.* A 1980 survey by the National Wildlife Federation indicated that 19 of 50 states polled used state dollars for nongame. Only 13 states reported using hunting or fishing license revenues directly for nongame.

6. *Pittman–Robertson funds.* Although many states use these monies for nongame, constraints include competition with game-management budgets and the restriction to work on birds and mammals only.

7. *Voluntary nongame income-tax checkoffs.* After Colorado originated this type of funding in 1977, more than 30 states followed suit and now receive nongame funding from state-income-tax checkoffs. New Jersey generated $403,000 from an investment of $43,500 in 1982.[15] A number of other states have been unsuccessful in passing checkoff legislation. Reasons given for these failures include competition from other state agencies for funding, protests by state revenue departments concerned about their own workload and complicated tax forms, vetoes by two governors as setting a bad precedent for special interests, and inadequate public understanding and support.

8. *Tax on selected items.* In 1990, Colorado considered user tax on some items, such as bird feed, binoculars, and some golfing equipment, for nongame funding. Since that time, those funds have assisted with nongame funding.

Case Study: The Colorado Nongame Program. The Colorado nongame program was initiated in 1972, when the state legislature allocated hunting license fee monies to set up one nongame biologist position, which provided for 50-percent game work within the Small Game Section of the Division of Wildlife. In 1973, the Colorado legislature passed a Nongame and Endangered Species Act declaring as state policy the "management of all nongame wildlife for human enjoyment and welfare . . . to ensure their perpetuation as members of the ecosystem."[16] The legislature also passed a bill providing for the sale of conservation stamps, a program that netted $5,600 in 1974, but was dropped in 1978. In addition, the Colorado Nongame Advisory Committee was formed in 1973, with 10 appointed committee members providing biological expertise and administrative and political support for the nongame program.

In 1974–1975, a general fund appropriation of $86,000 financed the addition of a staff mammalogist and an ornithologist to work on the recovery of endangered species. Throughout these early years, virtually all programs focused on endangered species. Appropriations for 1975–1976 climbed to $135,582, permitting expansion of the programs to include endangered fish.

A step forward in the nongame program came in 1977, when an advisory council member suggested an income-tax checkoff. The Colorado Nongame Tax Check-off Bill passed unanimously that year (Figure 20–6). This bill defined nongame as "any wildlife species which is endangered, threatened with extinction, or not commonly pursued, killed or consumed either for sport or profit." By that definition, nongame species account for 80 percent of the 783 species of Colorado's birds, mammals, reptiles, fish, mollusks, and crustaceans.[16]

Although checkoff funds were not available until 1978, funds from the Federal Endangered Species Act, Pittman–Robertson, and Dingell–Johnson totaling $584,000 were received for the first time in 1977. With this boost, the program added part-time staff and new projects. In 1978, more than 90,000 taxpayers contributed $350,000 in the first tax checkoff year, followed by $500,000 in 1979 and $650,000 in 1980. An

TAX AND CREDITS	15	**COLORADO TAX** from the tax table. Part-year residents and nonresidents enter tax from line Q. Form 104PN. ■ **15**		.00
	16	Alternate minimum tax from Form 104AMT .... • **16**		.00
	17	Total of lines 15 and 16 .... **17**		.00
	18	Personal Credits from line 4, Part I, Form 104CR .... • **18**		.00
	19	Enterprise Zone Credits from line 12, Part II, Form 104 CR • **19**		.00
	20	Total of lines 18 and 19, but not more than line 17 .... **20**		.00
	21	Net tax, line 17 minus line 20 .... **21**		.00
PRE-PAYMENTS $	22	**COLORADO INCOME TAX WITHHELD from wages and winnings** .... • **22**		.00
	23	**ESTIMATED TAX payments and credits**; extension payments; and amounts withheld on nonresident real estate sales ..... • **23**		.00
	24	Total of lines 22 and 23 .... **24**		.00
	25	If line 24 exceeds line 21, enter your overpayment **25**		.00
	26	Amount on line 25 to be credited to 1997 estimated tax • **26**		.00
VOLUNTARY CONTRIBU-TIONS		**ENTER THE AMOUNT, IF ANY, YOU WISH TO CONTRIBUTE TO:**		
	27	The Colorado Nongame and Endangered Wildlife Fund .... • **27**		.00
	28	The Colorado Domestic Abuse Fund .... • **28**		.00
	29	The Colorado Homeless Prevention Activities Fund • **29**		.00
	30	The United States Olympic Committee Fund .... • **30**		.00
	31	Action Older American Volunteer Program .... • **31**		.00
	32	The Colorado Child Care Improvement Program ...... • **32**		.00
	33	The Drug Abuse Resistance Education Fund .... • **33**		.00
REFUND OR AMOUNT OWED	34	**Total of lines 26, 27, 28, 29, 30, 31, 32 and 33** .... • **34**		.00
	35	Line 25 minus line 34. This is your **REFUND** .... • **35**		.00
	36	If amount on line 21 exceeds amount on line 24, enter amount you owe. Include amounts entered as voluntary contributions on lines 27 through 33, if any. Include penalty (96) $ • _____ .00 and interest (97) $ • _____ .00 if applicable .... • **36**		.00

MAKE CHECK PAYABLE TO COLORADO DEPARTMENT OF REVENUE. TO ENSURE YOU RECEIVE CREDIT FOR YOUR PAYMENT, WRITE YOUR SOCIAL SECURITY NUMBER ON YOUR CHECK. PLEASE DO NOT SEND CASH.

SIGN YOUR RETURN

Under penalty of perjury, I declare that to the best of my knowledge and belief, this return is true, correct and complete. Declaration of preparer (other than the taxpayer) is based on all information of which the preparer has any knowledge.

Your signature	Date	Paid preparer's name, address, telephone # and signature
Spouse's signature (if joint return, BOTH must sign)	Date	

MAIL YOUR RETURN TO: COLORADO DEPARTMENT OF REVENUE 1375 SHERMAN STREET DENVER, CO 80261

Figure 20–6 Checkoff form for use in funding nongame wildlife programs in Colorado.

aquatic specialist, four regional biologists, and a five-person research staff were hired with these funds.

By 1980, program goals had evolved beyond simply reversing the decline of 25 threatened and endangered species to include keeping other nongame species from sliding into the threatened and endangered categories. Other goals included encouraging such nonconsumptive uses of wildlife as bird-watching and nature study.

The Colorado program still has the preservation of endangered species as its top priority. But reintroductions, the development of urban wildlife habitats, and educational projects are ongoing. An inventory of state resources has been under way for a number of years. Now attention is turning to the massive job of monitoring all species and habitats for years to come.

Although inventories and research are filling gaps in biological and ecological knowledge, great unknowns will probably always exist. For Colorado, the pressing issue is how to proceed with ecosystem management. The state is facing this gap between knowledge and need by centering its management around key species that are perceived to be ecological indicators.

RAPTORS

Raptors are birds of prey, such as hawks, falcons, owls, and vultures (Figure 20–7). Because they are large, highly visible, and at the top of the food chain, they often need special management. Raptors are harassed because they occasionally take livestock and eat wildlife species sought by hunters. Actually, raptors are a very important part of the ecosystem and are of great economic value. When human activity causes a decrease in the raptor population, a number of mammal populations, including rabbit, small rodents, and prairie dogs, usually increase, and, of course, an increase in these populations reduces the food available for livestock and wild ungulates. Some raptors, particularly vultures and bald eagles, remove dead animals, a function that helps maintain balance in the ecosystem.

Pesticides

Because of raptors' high position in the food chain, they have been affected by pesticides during the past 25 years. Many species have declined to levels where they are endangered or of concern and are listed as birds of high interest. In some parts of the

Figure 20–7 Great horned owls found throughout much of North America are year-round residents. (Courtesy T. MacLaren.)

Flies and Red-tailed Hawks

Each spring, red-tailed hawks establish nests in many trees around the Jackson Hole area of Wyoming. The large, visible nests are usually 20 to 40 feet up in a tree. Some of the trees are isolated in a field of sage or grass and so can be seen from a distance. In this area of cold winters and wet springs, hatchings of the black fly are to some degree controlled by weather. When the spring is cool, the black flies hatch late, and when the spring is warm, the hatch can occur earlier. Early-hatching flies often bite nestling hawks, sucking blood and introducing the parasite *Leucocytozoon*, which causes anemia and organ damage. Young birds sometimes die from the parasite when they fledge too early by jumping from the nest because of extreme irritation from the biting flies. As the chicks get older, their feathers protect them from the bites. So here we have an interesting combination of events that affect red-tailed hawks' nesting success each year. When the flies hatch later in the spring, the birds have enough feathers covering them to provide protection. Cool springs with late hatches of flies mean that more red-tailed hawk young survive.

Smith, R. N. 1994. Factors Affecting Red-tailed Hawk Reproductive Success in Grand Teton National Park, Wyoming. Master's thesis. University of Wyoming.

---

nation, entire populations have disappeared. An example is the peregrine falcon in the east. Bans on some of the pesticides that are more persistent in the environment have improved these birds' chances of survival and reproduction.

Impacts of People

Raptors are affected by an increase in human population. When people move into areas frequented by raptors or used for nesting, the birds are often disturbed and will not raise their young. If the disturbances are extensive, as happens with some mining activities or urban development, raptors may leave the area and not return.

Human disturbance of raptors is particularly harmful during the breeding season. Ospreys were disturbed whenever people approached their nests in Arizona.[17] They failed to use the nests during the year following a timber harvest, but returned the next year when human activity was reduced. Four of 13 red-tailed hawk pairs abandoned their nests following the detonation of explosives near Rio Blanco, Colorado. The other birds seemed extremely nervous immediately after the blast. No second-year study was made.

Raptors generally require specific types of sites for nesting. Vultures, for example, use rocky cliffs, areas also used by prairie falcons, and golden eagles (Figure 20–8). Goshawks prefer dense evergreen forest, but will use clumps of trees along canyons and moist draws. Sharp-shinned hawks, also found in evergreen forests, nest in other forested areas, particularly near streams. The Cooper's hawk uses dense aspen stands.

Riparian sites are also popular. Swainson's hawks nest in trees along moist canyons. They commonly return to their nests year after year. Marshy habitat is ideal for ground-nesting marsh hawks. Ospreys nest in riparian habitats and wintering bald eagles roost in trees. The trees also provide homes for many owl species not commonly seen during the daytime. The great horned owl, among others, uses nests constructed by other birds or builds its own nest on niches in ledges or cliffs.

Figure 20–8 Golden eagle nest site on a cliff. This "other" eagle is found in grassland and desert habitats. (Courtesy T. MacLaren.)

Human activity often results in decreasing rabbit, small-bird, and rodent populations, which provide food for a number of raptors. When this occurs, the raptor population declines and may not return to its predisturbance level. The most important consideration in managing raptors is maintaining their habitat, especially in the face of human impact. It is important that, before any form of human disturbance occurs, planners be aware of the species of raptors present. Once this information is available, data on the birds' life history can be gleaned from the literature. This includes information on home range, specific types of nesting substrate, and prey base.

Management Techniques

Habitat. There are a number of tested techniques for managing populations. If their native habitat can be left undisturbed, raptor populations often will remain in an area even when human activity occurs around the fringes. When habitat has to be disturbed, plans can be made to avoid disruption during the breeding season of most species. It is important to reseed as soon as possible after vegetation is removed. In one instance, reseeding after the removal of junipers from a native sagebrush–grassland habitat, increased the prey base of rodents for at least two years. People involved in developing important wildlife areas should consider restricting public access and should educate employees in the importance of raptors.

When constructing or altering any habitat, those responsible should consider a number of things. It is known that raptors are often killed by contact with high towers and transmission lines. Towers should have lights at night and some form of color marking during the day. Placing transmission lines below the tree line will also help. Wires should be placed at levels below the tops of the poles so that birds do not hit their wings when taking off from a perch on a pole. Fences and utility lines should be placed as far away from marshy areas as possible. Roadways do not pose as great a hazard to raptors as to ground-dwelling animals, so the concern here is the extent to which they prey base may be reduced.[1] But there is also some concern that raptors may be attracted to roadside structures and be killed when traffic becomes heavy.

Structures. Artificial structures can be used to manage raptors. Outside Saguache, Colorado, perch sites have been placed near food sources for ferruginous hawks, red-tailed hawks, marsh hawks, and golden eagles. Artificial nesting platforms for ospreys and golden eagles have worked. In the Powder River Basin of Wyoming, golden eagles were moved successfully from natural nest sites in cottonwood trees to artificial nest sites some distance from planned mining activity (Figure 20–9). Preliminary work on the activity patterns and territories of the birds showed that they could not be moved out of their territories or too close to other nesting raptors.

Figure 20–9 Moving a golden eagle nest to mine for coal.

Agricultural Diversity. Hedgerows were once commonly placed between crops in the United States. As farming became more sophisticated, the hedgerows disappeared. More recently, organic farmers have found them helpful because they harbor beneficial insects. Many hedgerow plants have berries and seeds that attract a variety of birds and small mammals, providing an excellent prey base for raptors. Farmers have begun to re-create these strips of vegetation, selecting plants that attract helpful insects, birds, and mammals. The hedgerows have provided new opportunities to improve the diversity of wildlife.[18]

URBAN WILDLIFE

With an ever-higher proportion of people living in urban areas, wildlife management has taken on a new dimension that lacks the aura of outdoors and isolation. Managers of urban wildlife must deal with different perceptions on the part of the public.

Types of Wildlife

Most of the wildlife forms that inhabit the central city area or residential areas are edge species—cardinals, robins, raccoons, squirrels, and so on. These animals adapt quickly to changing habitats. As a habitat becomes more dense in suburban areas or areas where vegetation grows rapidly, the diversity of wildlife increases (Figure 20–10). Management of wildlife in urban settings often involves animal damage-control problems: People complain about English sparrows nesting in the eaves of houses or squirrel getting into attics. Chimney swifts can clog chimneys; raccoon get into garbage cans and rabbit into gardens.

Public Perception

Surveys show that many city dwellers are quite interested in wildlife. In a Missouri study, 26 percent of the people said that they enjoyed outdoor activities such as bird-watching.[19] Similar studies in New York and Colorado have found that the public is aware of wildlife issues and interested in wildlife activities.

One measure of the importance of an activity is the amount of money spent on it by participants. In a study of the economics of enjoying nongame birds, biologists found that urbanites spend money for birdseed, birdhouses and feeders, field guides, gift books, binoculars, cameras, and dues to professional wildlife societies. The survey found that 20 percent of households purchase an average of 60 pounds of birdseed annually.[20] People in cities are interested enough to spend money to see, photograph, or attract wildlife (Figure 20–11).

Methods of Management

Management generally involves attracting wildlife by creating necessary habitats and features or reducing wildlife by eliminating habitats or removing the animals. In urban settings, biologists can help design features that attract animals, such as wildlife refuges, parks, and greenbelts. Refuges and parks can include ponds, lakes, and marshes interspersed among forest and field. The areas must be isolated enough to provide refuge, and people must be routed so that they can view the wildlife, yet not disturb it.

If city planners are aware of ways in which wildlife can be attracted to use existing urban facilities, they can assist with wildlife management. For example, impound-

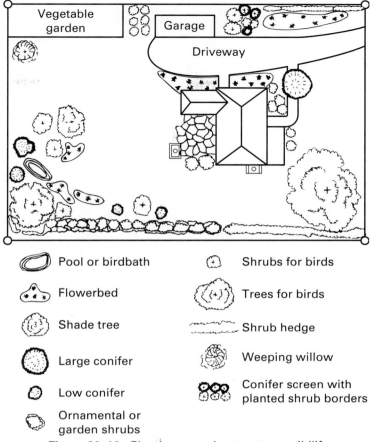

	Pool or birdbath		Shrubs for birds
	Flowerbed		Trees for birds
	Shade tree		Shrub hedge
	Large conifer		Weeping willow
	Low conifer		Conifer screen with planted shrub borders
	Ornamental or garden shrubs		

Figure 20–10 Planting a garden to attract wildlife.

ments created for runoff from storms can be designed for breeding and migratory birds.[21] Areas in parks can be isolated for breeding waterfowl, yet the public can view the young that later use the lakes. Around the home, the right kinds of trees and shrubs, ponds, birdbaths, and feeders and protection from dogs and cats can attract wildlife for people to enjoy.

Biologists consider animals' needs when making recommendations for attracting wildlife. For example, birds need a place to feed, sing, court, nest, and hide. The plants that homeowners select will influence wildlife. Yards that have only deciduous trees and shrubs can be improved by the addition of pines, junipers, cedars, and other evergreens that provide winter shelter. Fruit-bearing shrubs and trees will attract wildlife around the home. Knowing the food preference of some familiar birds is helpful (Table 20–5). Many wildlife groups, such as the Audubon Society and local garden clubs, have pamphlets about attracting wildlife. Through their extension services, state and federal agencies will supply pamphlets on birdhouses and plantings that attract wildlife.

People also want to know how to keep some wildlife away. Building designs that attract or discourage nesting birds and ratproof garbage cans are concerns of the urban wildlife manager. When populations of starlings, pigeons, squirrel, or even rats become troublesome, biologists are called on to set up removal programs. Poisoning programs are not desirable, because they leave dead animals. The best approach that biologists

Figure 20–11 Birds at feeder. Feeders are the most popular non-consumptive activity. People in the United States spend millions of dollars each year on food for bird feeders.

have found is to use preventive measures, such as removing garbage or sealing off house sparrow nesting sites, to keep animals away from structures.

One of the major roles of the urban wildlife manager is that of educator. Through news media, pamphlets, and public talks, managers bring to the public a greater awareness of both the benefits and beauty of wildlife. They can show people how to attract desirable species and avoid undesirable ones through such things as appropriate construction, proper disposal of garbage, and the use of shrubs.

DESCRIPTIONS OF SPECIES

Hellbender

In clear, fast-flowing streams and rivers of the east-central United States, anyone who goes fishing may see a large, flattened salamander that can reach a length of 75 cm (30 in). This gray salamander, the hellbender, with molting on its back, is strictly an aquatic organism.

Hellbenders are much used in biology experiments. They are sought by some people for food. Because of their size and habitat use, they are also known in folklore. Their diet consists of aquatic invertebrates, such as worms, snails, and crayfish. They lay their eggs in depressed nest cavities below rocks or logs. The males usually prepare the nests and guard the eggs until they hatch two to three months later.

TABLE 20-5

Food Preferred by Birds Attracted to Urban Settings

Plants	Grouse, pheasant, and quail	Band-tailed pigeons	Woodpeckers	Steller's and scrub jays	Chickadees, bushtits, and nuthatches	Robins	Thrushes and bluebirds	Cedar waxwings	Orioles and tanagers	Black-headed grosbeaks	Evening grosbeaks	Purple, Cassin's and house finches	Goldfinches and siskins	Rufous-sided and brown towhees	Juncos and sparrows
Blackberries	○	○	○	○	○	○	○		○	○	○	u		○	
Cascaras		○	○	u	u	◐	◐	u		u	○	u	u	○	u
Cherries	○	○	○	○	u	○	◐	○	○	○	●	◐		○	u
Crab apples	○	○	○	○	○	○	○	○	○	○	◐	○		u	
Dogwoods	○	○	○	○	u	○	○	○	u	u	○	○			u
Elderberries	○	○	○	○	○	○	○	○	○	○	○	○	○	○	○
Firethorns	u		u	u		○	○	○				○			○
Grapes	○	○	○		u	○	○	○	○					○	○
Hawthorns	○	u	○	u		○	u	○	u	○	○	○			○
Hollies	○	u	○	u		○	○	○				u		u	
Junipers	○		u			○	u	u			○	u			
Mountain ashes	○		u			○	○	○	○		○	u			
Russian olives	○			○		○	u	○		u	●				
Serviceberries	○		○	○	u	●	●	○	◐	○	○	u	○	◐	
Snowberries	○				u	○	○			u	◐	u		○	
Sunflowers	○	○	u	◐	◐					○	◐	◐	●	○	●

Note: Insects around plantings attract vireos, warblers, wrens, flycatchers, kinglets, swallows, and martins. Grasses and weeds attract seed-eating birds. Sugar-water feeders, in conjunction with flowers of salvia, evergreen huckleberry, red-flowering currant, and the columbines and fuchsias, attract hummingbirds. Other plants especially attractive to birds include madrone, cotoneaster, salal, highbush cranberry, huckleberry, blueberries, currants, gooseberries, maples, oaks, and other "mast" trees. Groups of birds usually have similar plant food habits, although individual bird preferences and extent of use may vary. Preference ratings for the plants were based on data from food-habit studies, but actual plant use by birds varies by season and situation.

○, Choice; ◐, Good; ●, Fair; u, used, amount not determined.

441

Once fairly common in many streams and rivers of the east, hellbenders are now on the decline, the victims of water pollution from toxic materials and debris. Dam construction has limited the number of flowing streams with rock bottoms, these animals' natural habitat. Management efforts are directed toward keeping streams and rivers free of pollution.

Indigo Snake

The indigo snake is a large, dark snake that can exceed 260 cm (102 in) in length. The body is blue black and the underside of the head orange, red, or sometimes green. This snake is found in the southeastern United States, in a disjoint range from Texas to Florida and Georgia. Pinewoods and palmetto stands, some hammocks, and grassland near water are the snake's habitat. In the east, from Alabama to Georgia, the indigo snake likes to use tortoise burrows. They eat birds, some small mammals, frogs, and even other snakes. The indigo snake startles people, who find them hissing, shaking their tails, and sometimes flattening their necks. They probably can live at least 30 years.

The massive developments in the south, along with the influx of people, have reduced the number of indigo snakes. The snake is collected by some as a pet. But people also gas the tortoise burrow in which the snakes find refuge to collect tortoises for eating. Management means keeping the snake's habitat relatively undisturbed. Education of the public on the role of this reptile in the community can prevent unnecessary killing. Indigo snakes assist in controlling amphibians, reptiles, and small mammals. The snake is endangered in part of its range.

Desert Horned Lizard

The desert horned lizard is a flat-bodied creature that reaches the diameter of a large jar lid about 13 cm (5 in). With scales on its side, back, tail, and head, the lizard has the appearance of a prehistoric reptile. This lizard lives in sandy, gravelly soil in rather arid areas; often, there are rocks or shrubs under which it can hide. It is found from Oregon to Arizona. There are closely related species in other arid regions of the country. The desert horned lizard can be found during the day as it seeks insects. It lays 10 or fewer eggs under rocks in the late spring or early summer.

The lizard is caught and kept in terrariums. The influx of people and development into the desert ecosystem has disrupted its habitat and food base. Management involves educating the public in the desert ecosystem and encouraging people to leave the system undisturbed and to let the wildlife alone.

Little Brown Bat

The little brown bat (*Myotes*) is widespread throughout the United States (except for some of the southern states) and up into Alaska (Figure 20–12). It likes to use the interiors of buildings and barns in the summer. During the winter it hibernates, usually in caves or mines. Some bats migrate hundreds of miles to their hibernating sites. Bats generally roost in large numbers in caves or other dark enclosures. Large deposits of their droppings (guana) accumulate. These deposits are high in nitrate content and are thus collected for fertilizers. Collection was, however, more common before the advent of synthetic fertilizers.

The females and young cluster in nursing colonies. The young are carried tucked under the mother's body. The bat feeds on insects, mostly flies and moths that it takes in flight. Its ability to locate insects by sonarlike sound waves is well known. The little brown bat, like many other bats in the United States, has been affected by pesticides. A reduction in its food supply, an increase in pesticide levels in insects, and loss of habitat due to the destruction of caves have been detrimental. Managers are sometimes called on to remove bats from homes or buildings because people fear the danger of rabies or have listened to superstitious folktales.

Tree Vole

On the Pacific Coast from California through Oregon is an unusually specialized tree vole—in fact, although some taxonomists would disagree, there are probably two species: red and dusky. These voles seldom come down to the ground. They build nests

Figure 20–12 Distribution of little brown bats.

with twigs and resin from the Douglas fir, Sitka spruce, or western hemlock. Their food consists of needles from the tree. There are separate nests for males and females. The voles are a favorite prey of spotted, saw-whet and long-eared owls. This species, which is found in old-growth forests, is declining because of logging operations. Managers need to maintain areas of undisturbed forest. It is important that clear-cutting operations not isolate the remaining stands of forest.

Prairie Falcon

The prairie falcon is a streamlined raptor with pointed wings and rapid flight. It is found from the Rocky Mountains westward. The birds winter in the southwest and Mexico, with breeding and winter ranges overlapping in the southwest. Management efforts involve the maintenance of isolated sites. In some states falconers receive permits to remove young, and harvest controls must be maintained so that not all the young are removed from one nest or one area.

The species is very adaptable, nesting from 60 m (200 ft) above sea level to tundra habitat in Colorado. The birds can breed at one year of age and fledge three to five young. Banding returns indicate a life span of four years or longer (Figure 20–13). Prairie falcon typically return to the same nest site, preferring either steep cliffs or rock outcrops, where they can nest in one of the crevices or holes sometimes 1.5 m into the cliff. These rock outcrops are also used as perch sites, from which the falcon can swoop down on ground squirrel and horned larks. Often, human intrusion into the area causes a decline in prey, resulting in poor nesting success.

Figure 20–13 Prairie falcon. This swift-flying falcon can reach speeds of over 100 km per hour. It feeds on small birds and mammals. (Courtesy D. Runde.)

Lewis Woodpecker

The Lewis woodpecker is a brightly colored woodpecker that breeds from the Rocky Mountain states westward and winters from Oregon southward. Individuals can be seen in much of the range all year, but birds from the north usually move south. The Lewis woodpecker is found in areas of scattered trees and open country and, at times, in forests. These birds like to perch on fence posts along fields. They excavate nesting places in dead trees, often those killed by fire. Management involves the maintenance of edgelike habitat with snags for nesting. The creation of edge habitat today often involves the removal of all snags and downed vegetation. These areas are not as good as areas with natural edge.

Ovenbird

The ovenbird is a warbler that commonly nests in the eastern deciduous forests. It builds a dome nest on the ground and uses perch sites in the shrub understory of the forest. It prefers moist (not wet) sites. The ovenbird winters in some parts of the southeast, but more commonly flies to the West Indies, Mexico, and South America.

Ovenbirds are one of many species being affected by the fragmentation of eastern deciduous forests. They need dense forest with open areas between the shrubs. Nests are normally in an open area on the forest floor or under shrubs with a fair amount of space above them. If a disturbance to the forest occurs, ovenbirds may abandon nests that have been started. They tend, however, to remain on the nest when an intruder comes nearby, flushing as almost a last resort.

Cardinal

The cardinal is a brightly colored bird found from the central plains states eastward and from southeastern California eastward through Arizona and Texas (Figure 20–14). A popular bird, it will come to feeders and nest in urban settings. It is an edge species and

Figure 20–14 Cardinal. This colorful seedeater is very common in the eastern United States. (Courtesy of the U.S. Fish and Wildlife Service; photo by L. C. Goldman.)

uses dense bushes in fencerows and along fields and isolated roadways for nesting. It builds its nest in bushes, vines, and small trees. Some cardinals migrate short distances. Most become somewhat nomadic in the winter. Urban wildlife managers might recommend dense shrubbery, such as blueberry, elderberry, poleberry, or spicebrush under beech, flowering dogwood, juneberry, and red cedar. Stationary feeders with seeds will attract cardinals.

SUMMARY

Nongame wildlife includes all species that are not hunted, harvested, or removed by people as well as game species in national parks and other preserves. Most nongame-management strategies involve some form of community management, such as managing for diversity, indicator species, or life-forms. Community management is management of all wildlife. Thus, when the minimum-sized habitat is maintained with the requisite factors, including structural components, all wildlife—both game and nongame—can exist.

Because most funding for wildlife work has come from fees assessed for hunting and fishing, most management efforts have traditionally been expended in behalf of game species. Many states are now developing nongame-management services. The management groups receive funds from many sources, including tax checkoffs, the sale of special wildlife items, personalized license tags, and state and federal appropriations. Some of the nongame management effort is in behalf of raptors, whose position on or near the top of the food chain makes them an important component of our natural system.

Urban wildlife is very important to many people. Managers are now helping city dwellers attract wildlife to neighborhoods by suggesting building designs and appropriate plantings. Urban wildlife management also involves the control of less desirable wildlife.

DISCUSSION QUESTIONS

1. How does nongame management differ from game management?
2. Why are raptors easy to see? Why are they important?
3. Should special funding be available for nongame management? Why or why not?
4. Distinguish among diversity, indicator species, and life-forms as nongame-management strategies.
5. Should nongame management efforts be directed primarily at endangered species? Explain.
6. What special background does an urban wildlife manager need?
7. What problems does human impact make for nongame-wildlife management?
8. Describe the techniques used to manage raptors.
9. What are the legal mandates related to the management of nongame?
10. What role should the federal government play in managing nongame? States? Private corporations?

LITERATURE CITED

1. Gottschalk, J. S. 1975. The Challenge of Practical Ecology. In D. R. Smith (Coordinator), *Proceedings, Symposium on Management of Forest and Range Habitat for Nongame Birds.* General Technical Report WO-1. Washington, DC: USDA Forest Service.

2. Kellert, S. R. 1980. *Activities of the American Public Relating to Animals.* Washington, DC: U.S. Fish and Wildlife Service.

3. U.S. Fish and Wildlife Service. 1988. *1985's National Survey of Fishing, Hunting, and Wildlife Associated Recreation.* Washington, DC: U.S. Fish and Wildlife Service.

4. Graul, W. D., and G. C. Miller. 1984. Strengthening Ecosystem Management Approaches. *Wildlife Society Bulletin* 12:282–88.

5. Niemi, G. J. 1982. Determining Priorities in Nongame Management. *Loon* 54:28–36.

6. Thomas, J. W. (Ed.). 1979. *Wildlife Habitats in Managed Forests: The Blue Mountains of Oregon and Washington.* Agricultural Handbook 553. Washington, DC: USDA Forest Service.

7. Peterson, S. R. 1980. The Role of Birds in Western Communities. In R. M. DeGraaf and N. G. Tilghman (Eds.), *Workshop Proceedings, Management of Western Forest and Grasslands for Nongame Birds.* Technical Report INT-86, pp. 6–13. Ogden, UT: USDA Forest Service.

8. Robbins, C. S. 1970. Effects of Forest Fragmentation on Bird Populations. In R. M. DeGraaf and K. E. Evans (Eds.), *Workshop Proceedings, Management of Northcentral and Northeastern Forests for Nongame Birds,* pp. 198–212. Technical Report NC-51. Washington, DC: USDA Forest Service.

9. Anderson, S. H., and H. H. Shugart. 1974. Habitat Selection of Breeding Birds in an East Tennessee Deciduous Forest. *Ecology* 55:828–37.

10. Tubbs, A. A. 1980. Riparian Bird Communities of the Great Plains. In R. M. DeGraaf and N. G. Tilghman (Eds.), *Workshop Proceedings, Management of Western Forest and Grasslands for Nongame Birds,* pp. 419–33. Technical Report INT-86. Ogden, UT: USDA Forest Service.

11. Office of Migratory Bird Management. 1988. *Nongame Bird Strategies.* Washington, DC: U.S. Fish and Wildlife Service.

12. Itekkers, J. 1982. List of Check-Off States at 16, Returns Aren't All In. *Nongame Newsletter* 1:1.

13. Wilson, J. 1982. In Missouri, Nongame Part of Design for Conservation. *Nongame Newsletter* 1:2–3.

14. Anon. 1975. *Current Investigations, Projected Needs and Potential New Sources of Income for Nongame Fish and Wildlife Programs in the United States.* Washington, DC: Wildlife Management Institute.

15. Applegate, J. E. 1984. Nongame Tax Check-Off Programs: A Survey of New Jersey Residents following the First Year of Contributions. *Wildlife Society Bulletin* 12:122–28.

16. Torres, J. 1982. History of Nongame Funding, Staffing in Colorado. *Nongame Newsletter* 1:6.

17. Adams, J. K., and V. E. Scott. 1979. Timber Harvest Modification around Active Osprey Nests. *Western Birds* 10:157–58.

18. Lewis, T. A. 1997. Using an Agricultural Relic to Create a Wildlife Haven. *National Wildlife* 35:18–19.

19. Wilier, D. J., D. L. Tylka, and J. E. Werner. 1981. Values of Urban Wildlife in Missouri. *Proceedings, 46th North American Wildlife Conference,* pp. 424–31. Washington, DC: Wildlife Management Institute.

20. DeGraaf, R. M., and B. P. Payne. 1975. Economic Values of Nongame Birds and Some Urban Wildlife Research Needs. *Proceedings, 40th North American Wildlife Conference,* pp. 281–87.

21. Adams, L. W., T. M. Franklin, L. E. Dove, and J. M. Duffield. 1987. Design Considerations for Wildlife in Urban Storm Water Management. *Transactions, 51st North American Wildlife and Natural Resource Conference,* pp. 249–59. Washington, DC: Wildlife Management Institute.

21
ENDANGERED SPECIES

THE SUBJECT DEFINED

Endangered species are species that have declined to such a level that their survival is questionable. The Endangered Species Act defines an endangered species as a species that is in imminent danger of extinction throughout all or a significant portion of its range. The act goes on to define as threatened species that are liable to become endangered in the foreseeable future. In the act, the word *species* is construed to include subspecies and races.

A species usually becomes endangered because its environment (habitat) has changed in such a way that it can no longer supply the species' needs. Most of the changes today are due to human use. For the most part, the endangered species is symptomatic of major changes in an ecosystem. Extinction as a natural phenomenon occurs gradually, usually over millennia. People speed up this process. Still, it is sometimes difficult to tell whether a species is endangered because of people or because of natural causes.

Hawaii is classic example of an ecosystem unbalanced by human beings. In the late 18th century, cattle, sheep, horses, goats, and pigs were introduced to the islands, were allowed to multiply, and eventually ran wild. Since the only native Hawaiian land mammal was a species of bat, native vegetation was vulnerable to destruction by these introduced herbivorous mammals. During the 19th century, herds of the animals moved into the forest, slowly destroying the habitat of native birds that had evolved as a part of the islands' delicate ecosystem. Unable to adapt to the different ecological conditions, many of the birds perished. Today, Hawaii is home for a high proportion of the endangered vertebrates of the 50 states, including many birds and the Hawaiian hoary bat.[1]

There are those who argue that declining species are a part of the natural process of evolution, and indeed, this may be true in the case of some endangered species, which have reached the pinnacle of their evolution. For the most part, however, the intrusion of human beings into an area has caused the decline. In a reasonably stable biosphere, the evolutionary rate and extinction rate are approximately equal: Extinction is normally linked with, if not caused by, a new species.[2]

By simply working on one endangered species or group of species, we are treating symptoms and not causes. When endangered species occur in an area, it is because changes are occurring in the habitat and ecosystem. The entire system must be examined and, if appropriate, treated. But conservationists have been successful only in getting legislation that deals with species. This is partly because most people do not think beyond the individual species to look holistically at the community and the ecosystem. It certainly makes more sense to develop management criteria for an entire community, including management of endangered species, and this may come with time. For now, wildlife managers must concentrate on the species in the community setting. In this chapter, we examine the need for managers to work with politicians, landowners, and other resource managers.

RATIONALE FOR SAVING SPECIES

The manager cannot spend much time with people without being asked about the reasons for saving endangered species and the use of public monies for such a purpose. Although managers may have ideas that should be very convincing, they must remember that not everyone has studied the workings of the ecosystem and that emotions can easily cloud issues and displace humanistic attitudes when one's personal livelihood or property seems threatened.

Some scientists point out, perhaps with an emotional overtone, that the chemical makeup of only a few of the world's species has been unraveled. Thus, we have identified only relatively few of the species, presumably, whose chemical makeup could be of great benefit to human beings in medicinal, industrial, and agricultural ways. If, inadvertently or intentionally, we allow a species to become extinct, we may be depriving ourselves of valuable products. For example, if the fungus known as penicillin had been wiped out, we would never have had the drug penicillin or the family of antibiotics that developed following the discovery of this class of compounds. Similarly, tropical plants are the source of alkaloids used in a variety of drugs to treat people for heart disease, cancer, and other ailments, yet today the extinction of tropical and subtropical floral and faunal species continues at a high rate.[3]

Researchers have discovered that snails and mollusks do not contract cancer. The discovery has set off a search for the chemicals that produce this immunity, in the hope that, when found, they can be used to prevent or alleviate cancer in people. No matter how small or obscure a species, there is no way of knowing that it could not be a direct aid to humankind. Each living species contains a unique reservoir of genetic material that has evolved over eons of time and cannot be retrieved or duplicated if lost. This genetic material is characteristic of the population and not of just a few individuals.

The role that each species plays in the ecosystem is also an important consideration in maintaining a diversity of species. As each species is eating or being eaten, it

Figure 21-1 Reintroduction of the wolf into the Yellowstone ecosystem has caused a great deal of controversy. Ranchers believe that they will lose more cattle, while some people feel that wolves belong in Yellowstone. Court decisions add to the controversy. If wolves increase in numbers, coyotes will decrease. In the field, people can distinguish between wolves and coyotes in that wolves hold their tails straight back, parallel to the ground, and coyotes hold their tails down. (Courtesy of Wyoming Game and Fish Department; photo by L. R. Parker.)

serves as part of a route for energy flow in the natural system. Each ecosystem has components that make it unique: The bison has a role in grassland; the black-crowned night heron is valuable to the salt marsh. When one species is removed, shifts occur in the energy-flow pattern. For example, removal of the American bison from much of its range undoubtedly led to an increase in herbivorous insects and rodents. And when the wolf population was removed from much of North America, there was no further predation of coyote, so that population expanded (Figure 21–1). People should remember that the species *Homo sapiens* is also subject to evolutionary processes. Scientists argue that the more we learn about biology and evolution in both nature and the laboratory, the better able we will be to understand and manage our own biology and evolution.[2]

CAUSES OF EXTINCTION

The extinction of species is a natural phenomenon, but human encroachment often greatly accelerates the process. Extinction occurs when a species fails to replace its numbers in the population growth equation: The value of d, mortality, is constantly higher than that of $b(b - d)$. (See Chapter 3.) The failure is generally caused by a stressful change or a new element in the environment. Extinction can be grouped into the following scheme, modified from Terborgh and Winter:[4] fragmentation of the habitat, loss of features such as a nest site or cover, and loss of genetic viability.

When the habitat of a species is destroyed, the species moves, adapts, or becomes extinct. The habitat on the Hawaiian Islands has been so altered, that some species can no longer survive. When the rocky substrate of western streams is filled with silt, the yellow-legged frog cannot continue to breed or find food.

Each species has a minimum critical area in which it can survive. Habitats can be so fragmented, that the size no longer accommodates the needs of the species. Generally, populations are in equilibrium with their habitat. The loss of a habitat can result in negative growth rates. When the habitat is destroyed, some species cannot adapt, and equilibrium may not be reestablished, as happened with the passenger pigeon and some migratory birds of the eastern deciduous forests.

When mining or logging occurs, nest sites in cliffs or trees can be destroyed. These areas also afford animals protection from prey and from the elements.

As the habitat becomes altered or subdivided, some animals may become isolated in small populations, with minimal genetic viability. These groups cannot always withstand environmental changes and therefore are unable to continue to produce viable offspring.

The alteration of ecological stability, including changes in the food web and loss of food sources, is probably due to loss of habitat.

The introduction of new species into an area often results in an ecological imbalance that may destroy or alter existing populations. As land bridges are formed between habitats, predator–prey cycles and competition between species change. For example, the coyote now interbreeds with the red wolf, and the result is a change in the genetic makeup of the wolf. Plans to reintroduce the red wolf into the "land between the lakes" in Kentucky and Tennessee may help save this species in parts of its range. Human or toxic wastes, such as sewage, pesticides, and acid rain, have changed the composition of species or rivers, lakes, and land communities. For instance, as we learned in Chapter 19, sea lampreys had a destructive effect on lake trout in the Great Lakes when the Welland Canal allowed their passage. Similarly, hunting pressure has caused the loss of the bison, and poisons are suspected in the decline of the black-footed ferret.

Gall Bladder Trade: A Threat to Bear Populations?

Bear gallbladders have long been a staple in traditional Oriental medicine. The demand for them has caused bear populations in China to drop to only 60,000 and in Japan to 10,000. In South Korea and Taiwan, bear have been totally decimated. In the United States, bear poaching has been on the increase as the gallbladder prices have risen to between $50 and $200 each. Buyers have been interested only in gallbladders they have sold in the Orient. Most of the time, the remaining parts of the bear have been left to decay. As the human population of the Orient has increased, the demand for gallbladders has also increased.

Will the demand for gallbladders cause the extinction of bear all over the world? Right now, the black bear population in the United States is in no danger. But the increased demand for gallbladders, together with the associated increase in price, could pose a threat to these bear in the future.

GENETICS OF SMALL POPULATIONS

When two organisms breed, their offspring represent a mixing of the genetic material of the two parents. This mixing is not often apparent in the offspring (the visible properties are called the phenotype), because it can be partially or fully masked by dominance, recessiveness, or other interactions between the genotypes of the parents. When some individuals are isolated from a population and interbreed over a number of generations, they tend to become unrepresentative of the total population's genetic material. The process by which the isolated, breakaway group develops genetic characteristics different from those of the source population is known as **genetic drift.** In general, smaller populations have less genetic variability. Such reductions in variability may be deleterious to these small populations. This means that they generally have less chance to survive when change occurs in the environment. Pesticides, for example, destroy large numbers of flies. Since the fly population is very large, a few flies had a genetic makeup that made them resistant to pesticides. These resistant flies are able to reproduce, so their resistant offspring can increase the fly populations to prepesticide levels. This recovery may not be possible in populations with limited genetic variability.

Therefore, the result of inbreeding in small populations often is that its members become more and more alike as they mate with individuals having a similar genetic makeup. Geneticists who worked with domestic animals suggest that populations can withstand some inbreeding, but are better able to survive with the introduction of new stock. Calculations indicate that there is a minimum size, depending on the species, at which a population can cope with inbreeding effects. The recommended minimum size for large mammals in the absence of introduced stock is 50 adults of breeding age.[5] The number (and sex ratio) may be higher in populations in which not all members are effective breeders (for example, those that maintain harems, that utilize lek behavior, or in which not all adults breed each year). The prescription of 50 adults for maintaining a viable population has been disputed when it is applied to populations in general. For critical species such as the Atlantic salmon, a minimum of 1,000 adults is suggested.

THE ENDANGERED SPECIES ACT

In 1966, the federal government, under the Endangered Species Act, began to list endangered species and designate habitats to be preserved. In 1969, the Endangered Species Conservation Act was passed. This act supplemented the 1966 act, providing a clearer definition of wildlife under protection. It has also been reviewed, amended, and renewed in subsequent years, including 1978, 1982, and the late 1980s. In the late 1980s, congressional stalling techniques prevented action on the renewal for over a year. In 1973, modifications of the act allowed the Department of the Interior to designate critical habitats. Endangered-species legislation gives the Secretaries of the Interior and Commerce the power to list species and initiate action that promotes recovery of the species.

Section 7

The 1973 act was indicative of a major change in the way people were viewing endangered species. Section 7, in which the Secretary of the Interior was given the authority to designate critical habitats of species, was most controversial. Here it is useful to compare Section 7 with the National Environmental Policy Act (NEPA). NEPA makes environmental quality a matter of national policy. In particular, it requires that federal programs give environmental objectives appropriate consideration, together with economic and technical considerations, in decision making. This consideration is to be demonstrated in the preparation of an environmental impact statement for most proposed federal actions that could affect the quality of the environment (Figure 21–2).

Section 7 goes beyond NEPA's requirements. It asks for a report on the possible effects of proposed actions specifically on endangered species. If, in consultation between agencies and the U.S. Fish and Wildlife Service, it is determined that there is a threat to an endangered species or its critical habitat, the project must be altered to remove the threat or be cancelled. In other words, rather than balance the endangered species consideration with others, the act requires flatly that there be no adverse impacts on endangered species.[6]

Cooperative Action

The Endangered Species Act was, of course, a major step forward in the role of the federal government in protecting wildlife. The Act was excellent as a means of listing species, but it was somewhat difficult to enforce. The 1973 Act gave the states and the federal government the enforcement powers needed. The most important provision of the Act is the requirement that states afford special protection to any species of wildlife determined by the Secretary of the Interior to be endangered or threatened. When the secretary has determined that an animal is endangered or threatened, the state or states affected become eligible to enter into an agreement with the federal government and receive federal financial assistance for up to two-thirds the cost of an approved program. That assistance, allocated among eligible states at the Secretary's discretion, is based on certain criteria specified in the Act. Special regulations are set

Figure 21–2 The Mt. Graham red squirrel was the subject of a Section 7 consultation between the Forest Service and Fish and Wildlife Service in Arizona. The squirrel population occupied a small habitat on a mountain where people wanted to construct telescopes.

up for federal endangered-species permits in states working under the agreement. States do not receive the right to permit others to take federally protected species without a federal permit.

The International Component

The Endangered Species Act of 1973 also has an international component. Besides encouraging the conservation of endangered species worldwide, the Act directs the president to implement the Convention on Natural Protection and Wildlife Preservation in the Western Hemisphere. (See Chapter 10.) In addition, it directs the Secretary of the Interior to encourage foreign nations to establish and carry out endangered-species programs of their own and authorizes both financial assistance and the loan of federal wildlife personnel. Finally, the act authorizes the Secretary of the Interior to conduct law enforcement investigations and prohibits the importation of endangered and threatened species.

The Endangered Species Act and its 1973 version give relatively broad powers to the federal government to manage wildlife habitats where endangered species are involved. The Act directs the government to become involved with other nations in preventing the extinction of endangered species.

The Listing Process

Species. The U.S. Fish and Wildlife Service may nominate a species for listing as endangered or threatened, and an individual or organization may petition to initiate the listing process. A petition may be filed with the Department of the Interior by anyone who has adequate data to support a proposed listing. The process begins with a letter to the Secretary of the Interior. When protection is considered necessary for a species during its evaluation for possible listing, it is placed under federal protection. For a species to be listed as endangered or threatened, evidence must be provided that its existence is in peril from one or more of the following: (1) the destruction or threatened destruction, modification, or curtailment of its habitat or range; (2) its overutilization for commercial, sport, scientific, or educational purposes; (3) disease or predation; (4) the absence of regulatory mechanisms adequate to prevent either its decline or the degradation of its habitat; and (5) other natural or human-made factors affecting its continued existence. To make its recommendation, the Fish and Wildlife Service follows what is known as a *rule-making* (or regulatory) procedure. This process is followed by all federal agencies in proposing regulations that will have the effect of law.

When the biological evidence concerning a species' status is not enough to justify a listing, the process may begin with the publication of a notice of review and solicitation of more information on the species from any source. This information, together with already synthesized data, is published in the *Federal Register,* a publication of the U.S. Congress. When the information is sufficient to warrant a consideration of listing the species, the Department of the Interior or Department of Commerce publishes a proposal in the *Federal Register* to list the animal or plant as endangered or threatened and to designate an appropriate critical habitat for the species.

At this and every other stage in the listing process, all interested persons are asked to comment on the proposal. Generally, a period of 60 days is allowed for public hearings to discuss the proposal. To make sure that all interested members are aware of the proposal, news releases and special mailings so inform the public, the scientific community, and other federal agencies.

Delisting or reclassifying occurs when a species is felt to have recovered sufficiently. The procedure is the reverse of the listing process. Each of the criteria for listing must be addressed, with evidence that it is no longer a threat to the species (Figure 21–3).

While the procedure for proposing to list an endangered species is clearly spelled out in the act, the exact biology of the animal is not. Often, the questions raised during a proposed listing concern how many animals there are and how that number is changing over time. These questions are frequently difficult to answer. For example, we do not have an agreed-upon population estimate of grizzly bear, a very large species that is listed as threatened. In reality, listing a species as endangered means that the population is so small that it is in immediate danger of extinction.

Several years ago, the ferruginous hawk was proposed for listing. There were no accurate counts of this raptor, and getting an estimate was difficult because the birds nested in remote areas. Furthermore, one pair could have had anywhere from two to six nests to choose from each year. An effort was undertaken to count more accurately the number of birds in an area.

Sometimes an isolated segment of a major population is proposed for listing because that segment is declining. Such efforts are now underway for sage grouse. Is it appropriate to list a population that may be declining in one area when the population as a whole may be doing well? While these biological questions are difficult to answer, we find some organizations being established for the sole purpose of finding and proposing species to list. The goal of these organizations is to force the federal government to list as many species as possible. Clearly, the Endangered Species Act poses a real dilemma for wildlife managers.

Habitat. Designations of critical habitats affect only federally authorized activities and are made primarily to help federal agencies locate endangered species and fulfill their responsibilities under the act. Critical habitat includes those areas of land, water, and airspace occupied by the species at the time of its listing that are required for its normal needs and survival:

1. Space for individual growth and growth of the population with normal behavior
2. Food, water, air, light, minerals, and other nutritional or physiological needs

Figure 21–3 The American alligator is an example of an animal delisted through federal–state cooperation.

3. Adequate cover or shelter
4. Sites for breeding, reproduction, rearing offspring, germination, or seed dispersal
5. Protection from disturbances in a location representative of the historic, geographic, and ecological distribution of the listed species.

Certain areas may be excluded from being designated as critical habitat if the Secretary of the Interior or Secretary of Commerce decides that the economic benefits outweigh the benefits of conserving the areas. Such areas are not to be excluded, however, if doing so would result in extinction of the species inhabiting them. Following the required period for public comment and public meetings on a proposal to list a species and its critical habitat, the information received is analyzed, and, based on the best available biological data, a final decision is published. The ruling generally becomes effective 30 days after its publication in the *Federal Register.* After a species is listed, its status is reviewed at least every five years to ensure that federal protection is still warranted.

Consultation. The U.S. Fish and Wildlife Service must be consulted by federal agencies when there is a possibility that an action or activity will affect an endangered species. There have been more than 20,000 consultations under Section 7 alone. Most of the consultations involving endangered species and developments have resulted in compromise.

Recovery Plan. To restore a protected species to its nonendangered status, the Fish and Wildlife Service develops a recovery plan for the species. Recovery plans are prepared by a knowledgeable person on a voluntary or contract basis through a public or private agency or by a recovery team appointed for that purpose.[7] Naturally, the elaborateness of recovery plans depends on the range and characteristics of the species. For migratory species such as the whooping crane or secretive mammals like the black-footed ferret, the plans can be quite complex.

Each recovery plan starts with background information on the species, its habitat, and its biological needs. The plan will cover possible manipulation of habitat, cleanup of habitat, transplantation, captive-breeding programs, acquisition of habitat, and recommendations to state, federal, and private agencies for changes in land-use practices. An implementation guide is developed, the overall plan is approved by the Department of the Interior of the Department of Commerce and initiated.

Candidate Species. The 1988 amendment to the Endangered Species Act allowed the U.S. Fish and Wildlife Service to spend money towards the recovery of plant and animal species it had identified as candidates for listing as endangered or threatened. The amendment placed species into one of three categories, based on the amount of data available on each species.

On February 26, 1996, the U.S. Fish and Wildlife Service revised its candidate list, eliminating the three categories and using just the category "candidate." The candidate species were only those for which the agency had enough information to warrant proposing an endangered or threatened listing. The revised list meant that the candidate species were in the early stages of the proposed listing process. The revised notice issued in 1996 identified 182 species as candidates for listing, whereas the old system listed nearly 2,700.

Habitat Conservation Plans. In the late 1900s, the Department of the Interior developed the idea of Habitat Conservation Plans (HCP) to work with private land-owners who have endangered species on their property. The HCP is an agreement between the landowners and a federal agency. It allows the landowner to proceed with habitat changes such as construction, logging, or other development in exchange for some conservation measures for the endangered species. The whole purpose is to balance the Endangered Species Act with private property rights.

Each HCP must be negotiated separately and special conservation measures for the species in question must be adopted. In some cases, habitats are improved; in others, habitats are set aside. Naturally, conflict still remains. Some feel that complete preservation is the only way to maintain endangered species. Others feel that preservation of endangered species isn't necessary. In 1999 the proposed reauthorization of the Endangered Species Act contained provisions for HCP's.[8,9]

MANAGEMENT

Management principles for a listed species are usually spelled out in the recovery plan. This section discusses some forms of endangered-species management.

Habitat

Central to the management of any species, including endangered species, is the maintenance or development of a proper habitat. Sometimes, the best way to bring this about is through habitat manipulation. For example, the water level of central Florida is crucial to the survival of the apple snail, the principal diet of the endangered Florida Everglade kite. Accordingly, dams and levees have been constructed to hold water at the level required to maintain the population of the snail. Here, habitat management contemplates a sort of double play, restoring one species by guarding against the threat to another.

The Kendall Warm Springs dace, a small endangered fish, inhabits a spring area and some short streams in the Bridger Teton National Forest in western Wyoming. The water, which flows along the north face of a small limestone ridge, is well mineralized and relatively warm (about 25°C, or 77°F). This habitat is protected by fences to prevent grazing and to keep people out. The use of soaps, detergents, or bleach in nearby waters is prohibited.[10]

Another endangered species, the red-cockaded woodpecker of the southeast, requires pine trees for nesting cavities. The trees are in open groves and are usually more than 70 years old. With the pulp industry harvesting smaller pine trees and disturbing the rotational cycle, the habitat for the red-cockaded woodpecker is declining. Efforts are being made to maintain pockets of undisturbed trees that can serve as homes for woodpecker colonies.[11]

Captive Breeding

Captive-breeding programs to maintain endangered species include those in zoos and others in which the species are bred in captivity and then released into the wild, using foster parents. Ideally, a population should have a minimum number to maintain a viable population—we mentioned 50 as the minimum for mammals. At times, however, captive-breeding programs are intended merely to perpetuate the species, when hope of a viable population seems remote. The panda is an example of such an attempt.

Captive breeding was used to reestablish the peregrine falcon. This species has bred in nearly every part of what is now the contiguous United States, extending back to the Pleistocene Ice Age. In the early 1950s, the breeding population throughout most of the northern half of the western hemisphere began a precipitous decline, which was related to the use of DDT. A study in 1975 led to the conclusion that peregrine falcons were no longer breeding in the east.

Once the species was declared endangered, a recovery plan was made calling for an inventory and priority ranking of suitable reintroduction sites throughout the east. A captive-propagation program was initiated from wild stock. The primary brood stock was raised at the Cornell peregrine facility at Cornell University, in New York. Peregrine stock was obtained from several sites in the world, including the arctic tundra, the Queen Charlotte Islands off the west coast of Canada, Scotland, and the Mediterranean. Techniques were refined to the point where as many as 200 peregrines a year could be produced.

Once produced in captivity, the birds must be released into the wild. "Hacking" is the process of gradually releasing young birds, giving them semi-liberty while still feeding them until they become accustomed to life in the wild. Hack sites were set up in different parts of the country (Figure 21–4). In many cases, volunteers staffed these sites to ensure that food was available in proper amounts. The birds would fly around, accustom themselves to the area, and return to the hack site, having imprinted on it.

Birds eventually were placed not only in forested areas where there are cliffs, but on tall buildings in the center of towns. One of the first successes was with birds placed on the Department of the Interior building in Washington, DC. The idea of placing birds on buildings in the centers of cities originated in the spring of 1979, when a female peregrine took up residence at the 33rd-floor level of the U.S. Fidelity and Guarantee building in downtown Baltimore, Maryland. The female laid some infertile eggs and began incubating them. After Cornell researchers removed her eggs and substituted young falcons from the incubator, she fledged two males and two females.

One captive-breeding program that has been successful in recent years is that of the black-footed ferret. Wild ferrets were found in 1981. Several years later most had succumbed to disease, and six were taken into a captive-breed facility. Those six formed the basis for a current population of over 300 animals. Quite a few have been

Figure 21–4 Hack site of peregrine falcon. (Courtesy C. Patterson.)

reintroduced into the wild. Survival seems to be low, as over 90% of the animals reintroduced into the wild have become prey for other animals. There is a note of success, however, as a number of females reintroduced into the wild have given birth to young.

For the most part, captive-breeding programs have been expensive, and not all reintroductions, translocations, and introductions have been successful. Less than half of the more than 1,000 cases of captively bred birds and only 5 of the 20 mammal reintroductions have been looked upon as successful.[12]

Double Clutching.
Another technique used to maintain endangered species is double clutching. Here, the first eggs laid are removed to induce the female to produce another clutch. The first clutch is raised in captivity and then released elsewhere. This method has been useful for some species of waterfowl and gallinaceous birds.

Cross-Fostering.
Double clutching can be combined with cross-fostering, a practice used in peregrine work as well as with other birds. Cross-fostering consists of putting the eggs of one species into nests of related species to be incubated and hatched, and the young reared. An intensive cross-fostering program is under way for the whooping crane population. By removing first clutches, biologists are on the way to developing a new population of whooping cranes. Eggs have been removed from birds at the Wood Buffalo National Park and placed in sandhill crane nests. The sandhill crane site selected was the Grays Lake Refuge in Idaho. Sandhill cranes nest at Grays Lake and winter at Bosque del Apache Wildlife Refuge in New Mexico (Figure 21–5). Eggs have also been placed at the Patuxent National Wildlife Refuge in Maryland.

Despite a number of setbacks, including bad weather and predation, as of 1984 there were approximately 28 birds in the Idaho population, with 19 fledged in the fall of 1983. Although it is too early to rate the cross-fostering idea in general, it appears that this recovery plan will help restore the whooping crane population.[13]

Zoos.
Captive breeding in zoos is a means of *maintaining* endangered species. The International Union for Conservation of Nature and Natural Resources (IUCN) keeps a census of rare animals in captivity, not only as a record of which zoos have which rare species, but also as part of a coordinated international breeding program in which intrazoo matings and exchanges are made. The census includes such information as the nearest places where particular species are kept and the number and sex of animals bred in captivity and currently on exhibit. Another record kept is the *stud book,* the main purpose of which is to facilitate the planned breeding of species recorded. Breeders use the stud book to prevent prolonged interbreeding, which may be a serious problem. Stud books contain such information as an animal's number, sex, date and place of birth, and date and place of death. Listed in the publication *International Zoo Yearbook* are annual updated censuses and study books for rare species of wild animals in captivity.

Currently, the recovery team for the California condor is attempting to extend the life of the condor by maintaining the birds in California zoos, where they hope to develop and maintain the genetic stock. Whether the birds will ultimately be released in the wild is not known. Zoos have an important role in the propagation of endangered species—in fact, there are some zoos whose stock consists almost entirely of endangered species. Species such as the Hawaiian goose, laysan teal, and swinhoe pheasant have benefited from captive breeding in zoos. This work of zoos will probably expand.

A number of breedings of endangered species have been attempted in the U.S. Fish and Wildlife Service Patuxent Research Center in Maryland. Not all of them have

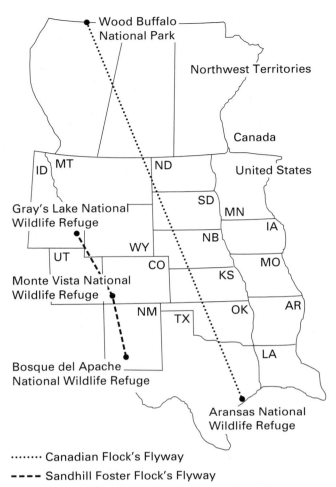

Wood Buffalo
National Park

Northwest Territories

Canada

United States

ID | MT | ND

SD | MN

Gray's Lake National
Wildlife Refuge

IA

NB

WY

UT | CO

MO

Monte Vista National
Wildlife Refuge

KS

NM | TX

OK | AR

Bosque del Apache
National Wildlife Refuge

LA

Aransas National
Wildlife Refuge

········ Canadian Flock's Flyway

- - - - Sandhill Foster Flock's Flyway

Figure 21–5 Migratory route of the whooping crane.

been successful; for example, the breeding of black-footed ferrets failed because of disease and possible loss of genetic variability. There is a possibility that natural areas or refuges could be established for endangered species. Management of such reserves would probably be intensive, since disease and predators would have to be controlled. In addition, most of the areas where reserves would be desirable are affected by tremendous human population growth.

Surrogate Species. Sometimes it is difficult to obtain information on the breeding biology of an endangered species. One useful strategy is to study surrogate species and use them as substitutes for particularly rare or endangered species. The surrogate species is generally a taxonomically close relative of the endangered species and in most cases is not itself considered rare or endangered. Surrogates are often used to determine which captive-breeding techniques can be developed or used successfully. Surrogates have been used in testing methods of obtaining and transporting stock, developing suitable pen facilities and caring procedures, and determining the medical and sanitary procedures necessary to prepare captive animals for independent existence. Surrogate breeding species include the Siberian ferret for the black-footed ferret, the prairie falcon for the peregrine falcon, the sandhill crane for the whooping crane, and the Andean condor for the California condor. (Since the late 1970s, the Andean condor itself has been declared endangered.)

Acquisition of Habitat

A major problem for most threatened and endangered species is the loss of their habitat. The federal government may acquire essential habitat lands as an emergency or last-resort measure if habitat is the key problem in the species' restoration. However, much endangered species habitat is managed by private conservation organizations, state agencies, and concerned individuals. There are many volunteer cooperative efforts to protect habitats without the need for acquisition. (See Chapter 10.) If federal acquisition is appropriate, special funds are used to purchase the critical habitat. Sometimes the essential land is donated. All acquired habitat becomes part of the National Wildlife Refuge System.

In the mid-1970s the U.S. Fish and Wildlife Service began acquiring nearly 4,050 hectares (10,000 acres), which make up the Mississippi Sandhill Crane National Wildlife Refuge, in Jackson County, Mississippi. This acquisition was essential to the survival of the Mississippi subspecies of cranes, which now number about 40. The land, bought from a developer who had planned to turn the area into a major residential development, is now the site of a refuge to maintain cranes' nesting, feeding, and roosting habitats.

INTERNATIONAL PROGRAMS

A number of international organizations are devoted to conserving the endangered species of the world. One is the World Wildlife Fund (WWF), which solicits support from private individuals. WWF has member organizations in many countries. Through a series of educational and research projects, it helps maintain endangered species in areas that are subjected to intense human pressure and habitat change. WWF has other programs, including some dealing with ecosystem development, international law, and worldwide conservation (Figure 21–6). Its overall aim is to create awareness of threats to the environment, to generate worldwide moral and financial support for safeguarding the living world, and to convert such support into action based on scientific research.[14]

Another international organization with concern for the environment is the International Council for Birds Preservation (ICBP). With its headquarters in Cambridge, England, ICBP has as its objective the preservation of rare and threatened species and the habitats on which they depend.

The International Union for the Conservation of Nature and Natural Resources (IUCN) is an older international organization that has encouraged international cooperation to conserve natural resources. The principal method used to meet its objective is the facilitation of cooperation between governments and national and international organizations in the conservation of wildlife and other natural resources. Particular concerns of the organization are the spread of public knowledge, education, scientific research, international draft agreements, and worldwide conventions for the protection of nature.

The Nature Conservancy is working in both the United States and foreign countries to preserve endangered species and biodiversity. This organization seeks to achieve partnerships between landowners and agencies. Its goal is to build strong partnerships and local support to promote the maintenance of plant and animal species in their natural habitats.[15]

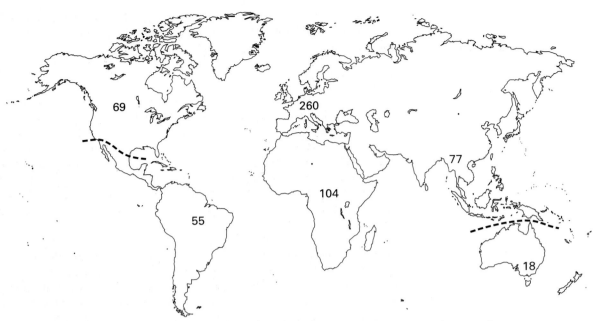

Figure 21–6 World Wildlife Fund programs to conserve endangered species are found on all continents of the world. As with other private conservation groups, World Wildlife funding comes from private donations.

International conservation bodies are under a number of constraints. The magnitude of the problem of enforcement is staggering. For instance, lack of personnel and equipment at the local level, especially in developing countries, has persuaded several of the nongovernmental organizations active in East Africa to use a substantial part of their funding to support antipoaching teams. At the governmental level, the United States has been a major leader in endangered-species research and management. In fact, most management techniques used for endangered species have been developed in the United States, with its large number of professional biologists.

In 1973, the United States hosted a conference of nations specifically interested in formulating an international approach to conserving endangered species. The conference produced an international treaty, the Convention on International Trade in Endangered Species (CITES). The delegates attending the meeting found marine species particularly difficult to deal with. CITES cited for action five species of whales that were already given complete protection by the International Whaling Commission (IWC). There was disagreement among delegates from different nations on the extent to which animal or plant parts and derivatives should or could realistically be a part of a protection plan. Disputes arose among representatives of wildlife producer and consumer nations and among those of countries in both groups either willing or reluctant to b e bound by the treaty's provisions.

One of the least costly, but most important and effective, modes of worldwide cooperation on endangered-species conservation is the exchange of technical information through access to published research and other publications. International cooperation of this type will become increasingly important as countries exhaust their limited resources.

The U.S. Fish and Wildlife Service plays an important role in the international endangered-species effort. Section 8 of the Endangered Species Act of 1973 authorizes the Department of the Interior to use foreign currencies for conservation programs in countries where such currencies are available. At present, only a few countries, including Burma, Guinea, India, and Pakistan, are eligible for the program. After the United States, India has more species (51) on the U.S. list than any other country. The United States–India joint program is designed to help India implement its wildlife objectives and to identify areas of cooperation that will benefit U.S. conservation programs.

The endangered species program worldwide is an effort of private and public groups. Like so much in wildlife work, a key to keeping the world's animal populations in a healthy state is a comprehensive program to educate people to preserve habitats.

BIOLOGICAL VERSUS POLITICAL CONSIDERATIONS

The endangered-species issue truly brings biologists into the political arena. No longer can biologists who have spent years in the field simply say they know that a population is increasing, decreasing, or suffering from some form of impact. Now their assertions must be documented in accordance with prescribed procedures and definitions established by lawyers and politicians. Some people regard this as an insurmountable obstacle, but it can be viewed as an opportunity to use biological knowledge and persuasion to obtain funding, legal assistance, and management authority.

Nevertheless, many wildlife managers become frustrated because public agency administrations and politicians do not automatically accept their skilled opinions as the final say. No longer is it enough for managers simply to point out the need for maintaining a waterway or wetlands because they support endangered species. Public demand for mosquito control may force the use of pesticides that result in the loss of fish and amphibians. The realities are that people do not like mosquitoes and are unaware of, or uncaring regarding, fish or frogs. Many who want to control mosquitoes with a pesticide pose the question, "What are we concerned about, people or animals?" Managers must then look for other ways to maintain that habitat to support the frog or fish population. One such way could be finding and encouraging the use of pesticides that are not toxic to the frogs and fish. Whatever the decision, the manager must be aware of the political overtones associated with conservation attempts.

The snail darter, a small fish that evolved in an east Tennessee river, is probably the classic example of a nationwide conflict among biologists, developers, politicians, and the public. (See Chapter 12.) In this case, conservationists sought an injunction against completion of the Tennessee Valley Authority's (TVA) Tellico Dam on the Little Tennessee River, claiming that the snail darter, a listed endangered species, would become extinct if its habitat were flooded. The plaintiffs pointed out that the Little Tennessee River was the only known habitat of the snail darter. (Since that time, the fish has been found in several other locations.) The case went all the way to the U.S. Supreme Court, which ruled that the protection afforded endangered species by Section 7 of the Endangered Species Act of 1973 was absolute. (The TVA had argued that congressional appropriation of funds for completion of the dam after listing the snail darter was evidence of congressional intent.) The controversy after the injunction was granted was in part responsible for the 1978 amendment to the act, which included a provision that the Tellico project be reviewed by a cabinet-level panel for possible exemption. The panel met, but did not act on the exception issue because its members concluded that benefits from the completed dam would not justify the cost of finishing

The California Gnatcatcher and the Endangered Species Act

The California gnatcatcher, a subspecies of the blue-gray gnatcatcher, is found from the coastal plains of southern California throughout most of Baja California in Mexico. In 1992, it was proposed that the California gnatcatcher be listed as an endangered species. The listing was based on data from a monograph of the subspecies published in 1988. When the listing was proposed, a number of landowners and developers undertook a major effort to prevent the listing because they felt that it would devalue their expensive property. Attorneys for the developers argued that the gnatcatcher should not be listed because it was not a legitimate subspecies. They argued that various components of Endangered Species Act were not followed in the listing process. Many scientists were asked to examine the data, the habitat, and the scientific methods involved. All of this controversy and the ensuing investigations resulted in the formation of the cooperative program called Natural Community Conservation Planning (NCCP). The NCCP has enabled all sides to plan collaboratively for the development and conservation of the coastal-sage scrub ecosystem in southern California. The key habitat for the gnatcatcher and other wildlife species has been protected, and the NCCP has allowed planned development with all parties knowing the rules. The result has been planned development within the natural community. This type of compromise has been a wonderful example of cooperative planning, especially in an urban setting. The NCCP is an example of a Habitat Conservation Plan.

Holing D. 1997. The Coastal Sage Scrub Solution. *Nature Conservancy* 47:16–24.

it. Subsequently, pressure from pro-development groups caused the members of the Tennessee congressional delegation to amend another bill before Congress to exempt the Tellico project specifically from the Endangered Species Act. The snail darter was reclassified as threatened in 1984 when additional individuals were found living in other rivers.

In other cases, compromises have been worked out. For example, in 1978 an injunction was granted against the completion of Grayrocks Dam in eastern Wyoming. Conservationists pointed out that the dam's use of water to generate power would reduce the water supply and so adversely affect a whooping crane habitat in the Platte River Channel 480 km (300 mi) downstream. The power companies set up a trust fund to purchase downstream water rights to replace water used by the dam. This action proved to be a satisfactory method of maintaining the habitat, so completion of the dam was allowed.

Because endangered-species legislation sets up specific criteria to guide managers in working with endangered species, the entire concept creates a certain apprehension in those who use or own land. For example, those exploring for oil and gas must spend a good deal of money surveying whether an endangered species exists in the area. Landowners with endangered species on their property are concerned that the land may be put in the public domain.

The example of the black-footed ferret in the north-central portion of Wyoming points up this problem. In the latter part of 1981, the ferret was discovered to be alive and doing rather well on a private ranch with a large cattle operation. The owner was also active in exploring for oil and gas, so that seismic activity was common on the ranch. With the discovery of the ferret and the influx of conservation-minded people, the owner became worried that the land might be taken away. Biologists set up strict conditions for the survey of the ferrets and the use of the land. It is likely that ferrets exist in other areas, but the landowners are reluctant to report them. We must find methods of compensating landowners and encouraging them to preserve endangered species—to some degree, for their own benefit.

The grizzly bear is another example. In this case, the conflict is more among biologists working for different agencies. Because the grizzly is spread across a number of states on federal, private, and state land, a number of agencies are studying it. This situation is confused further by the fact that different political groups have different management plans. They do not even agree on how many bear exist in different areas. Thus, it is very difficult to develop a coherent management program. The recovery plan solves some of these questions; however, it must be revised periodically. In addition, many people feel that the grizzly bear should be removed from the ecosystem because it sometimes attacks human beings.

DESCRIPTIONS OF SPECIES

The American Alligator: Recovery of an Endangered Species

The American alligator is a recovered endangered species. Although disagreement existed as to whether the alligator really was in danger of extinction, it was known that, because of excessive legal and illegal exploitation and habitat alterations, the number of alligators had been greatly reduced over the preceding 100 years. The need for better management was evident, and steps to improve management were taken by states in which the alligator was offered protection. Since the initiation of these protective measures by the Endangered Species Act of 1973, the number of alligators has increased significantly. In addition, wildlife managers and biologists have conducted research in many areas of alligator biology, helping to fill previous gaps and dispel misinformation. In some portions of its range, the alligator is again considered a huntable resource, with management now based to a greater extent on better scientific information.[16]

The alligator is a natural inhabitant of the 10 southeastern states. During the late 19th and early 20th centuries, alligators were taken from many areas, so that their range was reduced. There are records of alligators as far north as the Dismal Swamp in southeastern Virginia. Western boundaries have been set in the area of San Antonio in south Texas and around Dallas and Waco in north Texas. Climate appears to be the limiting factor on the northern edge of the range. An average minimum temperature of $-10°C$ (15°F) marks the reptile's northerly limits; to the south and east, saline waters appear to limit the distribution of the alligator.[17]

The alligator is somewhat secretive; it lives in a variety of habitats, including remote wetlands. Accurate estimates of its population density are difficult to make. Early explorers and residents of the southeast commented on the abundance of the reptile. There have been estimates of as many as 3 million alligators in Florida in the mid-19th century. But as alligator hides became popular, the number of animals decreased dramatically. Major dealers in the southeast processed 190,000 hides in 1929, but only 6,800 in 1943.

As the alligator population declined, there were localized attempts at protection. In Florida, alligators received some protection in 1944, but the action was heavily questioned by the public. Nevertheless, by 1969, most states in the southeast had enacted some legislation affording protection to the animal. Federal interest was evident in 1967, when the alligator was included on the list of rare and endangered species.

The alligator is found in swamps, bays, marshes, reservoirs, lagoons, ponds, rivers, canals, springs, and creeks. Encroachment by humans has both destroyed the creature's historical habitat by draining wetlands and created additional habitats in the form of reservoirs and canals. Males prefer open lakes throughout the year, while females go into the adjacent swamps during the summer nesting periods. Fall, winter, and summer ranges are usually included in the larger spring range.

Courtship takes place during April and May. The female builds nests and lays eggs from mid-June to early July. The nests are often located in shady areas near permanent water such as swamps or marshes or along streams. They are built of mud or sand and vegetation and are located, on average, 3 m (10 ft) from permanent water. Up to 80 white, hard-shelled eggs may be laid, although the average number is between 30 and 40. Hatching takes place in August, after an incubation of from 59 to 65 days. Hatching success has been reported to range from 45 to 58 percent.

In Louisiana, the alligator recovered sufficiently to sustain an experimental harvest in 1971. In 1975, alligators in three parishes of Louisiana were reclassified as threatened, and harvesting was again allowed, under strict regulation.

Throughout much of the southeast, the alligator has made a strong comeback. In 1977, alligators in all of Florida and parts of Georgia, South Carolina, and Texas were reclassified as threatened, and in that year nuisance control was initiated. More recently, experimental harvests have been conducted in some states, and efforts continue on the proposal to reclassify the alligator as threatened in all of its range.

Following the enactment of protective laws, for a number of years alligator management consisted of attempting to reduce poaching. As a consequence of the reclassification of alligators from endangered to threatened and delisting in part of its range, management of the reptiles required change. Under the reclassification, the broad goal of management is to keep the alligator from becoming endangered again, while allowing some commercial hunting and avoiding adverse impact on the habitat.

Whooping Crane

The endangered whooping crane, which numbers approximately 140 birds, migrates through the central United States (Figure 21–7). It reaches a height of 1.5 m (5 ft) and weighs up to 7 kg (16 lb). It becomes sexually mature at between four and six years of age. Whooping cranes come north to Wood Buffalo Park, Canada, in late April and early May. On arrival, they build their nests, chiefly of roundstream bulrushes. Normally, there are two eggs, but usually, because of sibling rivalry, only one nestling fledges. The eggs are incubated by both male and female, the female usually sitting during the night. Except for brief periods, incubation is continuous.

Whooping cranes are proficient at defending themselves, their nests, and their young against predators. The effectiveness of their defense is illustrated in two stories reported by Bent[18]:

> At another time I crippled one of the large white species by breaking a wing. As it was marching off rather rapidly, I sent a little rat terrier to bring it to bay. No sooner did the dog come up with it than it turned about, and quick as lightening drove its long sharp bill clean through him, killing him on the spot.

> An extraordinary tragedy was much talked of in my earliest days in the North. About 1879, there was a young Indian living near Portage la Prairie, Manitoba. In the spring, he went out shooting among the famous wild-fowl marshes of that section. A white crane

Figure 21–7 Whooping cranes in flight. (Courtesy of the U.S. Fish and Wildlife Service, photo by L. C. Goldman.)

flew low within range and fell to a shot from his gun. As it lay on the ground, wounded in both wing and leg, crippled and helpless, he reached forward to seize it. But it drove its bill with all its force into his eye. The brain was pierced and the young hunter fell on the body of his victim. Here the next day, at the end of a long anxious search, the young wife found them dead together and read the story of tragedy.

At hatching, the chicks weigh about 5 ounces and measure about 8 inches. They are precocial, but parents and young return to the nest at night for three or four days after the hatching. Then the chicks leave the nest permanently, although they are still brooded by the adults at night or in bad weather. During the first 20 days after the hatching, families usually remain within 1.8 km (1.1 mi) of the nest.

In early September, soon after the chicks have developed full flying ability, migration gets under way. The cranes begin arriving in Aransas National Wildlife Refuge in mid-November, with stragglers arriving by late December. They spend approximately six months at the refuge, where their principal diet is blue crabs and clams. Courtship displays (dancing) occur throughout the wintering period, increasing in frequency as spring approaches. Migration to Wood Buffalo National Park begins in early April, but stragglers remain at Aransas into early May.

The following appear to be the significant causes of the whooping cranes' being endangered:

1. Habitat destruction has restricted their range.
2. They are rigid in their migration routes and nesting areas, which prevents them from recolonizing a suitable habitat in their former range.[19]
3. Their late sexual maturity and small clutch size result in slow population growth.
4. The northern breeding range has an ice-free period of only four months, just enough time for completion of the breeding cycle. This means that if the first clutch is destroyed, there is no time to renest and produce a second clutch.

Yuma Clapper Rail

The Yuma clapper rail, one of seven subspecies of clapper rails found in the west, was declared endangered by the Secretary of the Interior in 1967 (Figure 21–8). Surveys based on call counts between 1969 and 1982 showed that the birds were distributed along the lower Colorado River in the United States and the Colorado River Delta of Mexico. Population estimates were 1,700 to 2,000 birds.

Figure 21–8 Yuma clapper rail. (Courtesy of C. Conway, University of Montana.)

The Yuma clapper rail breeds in freshwater marshes in the United State and brackish marshes in Mexico (Figure 21–9). It is one of the few clapper rail subspecies that breed in freshwater marshes. Its winter habitat is unknown, but biologists conjecture it to be the salt or brackish marshes of Mexico. In the United States, the Yuma clapper rail prefers mature cattail bulrush stands in shallow water near high ground. Shallow

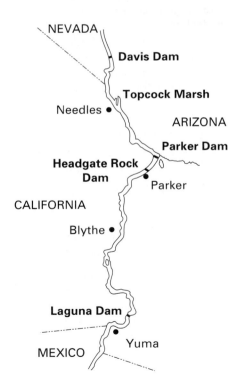

Figure 21–9 Breeding distribution of the Yuma clapper rail along the Colorado River.

Part 5 / Management Applications

water with hummocks and downed vegetation are ideal areas. Often, the areas where the rails breed have near row channels of flowing water. The rails take a variety of invertebrate species as their flood while breeding.

Between 1909 and 1942, human intervention through the construction of dams on the lower Colorado River created marsh habitat. Before that time, the free-flowing waters of the Colorado River, with its annual fluctuations in flow, apparently had not been home to Yuma clapper rails. Some 10 to 15 years after construction of the dams, rails were sighted in marshes created by the dams.

The key to maintaining or expanding the population of breeding Yuma clapper rails is maintenance of early successional stages of cattail marsh by creating shallow water with dredge spoils, altering channels, and using explosives in the lower Colorado River region of the United States. Doing this allows a mat of dead cattails to form in 0.3 to 0.7 m (1 to 2 ft) of water. Rails will then use these areas, since they can walk on the dead vegetation.[20]

People use the lower Colorado River area for camping, and their presence can adversely affect the rails. Too, the river waters must be kept at a flow level, preventing the marsh habitat from being destroyed by flooding or drying. High water levels in 1983 and 1984 appear to have reduced the nesting sites. Most important, biologists must determine where the Yuma clapper rail winters, so that they can prevent adverse impacts on that habitat.

Bowhead Whale

The bowhead whale, a baleen whale and one of the largest of the northern whales, occurs in small populations scattered throughout the upper limits of the northern hemisphere. In 1848, Captain Thomas Roys sailed into seas then unknown to whalers and discovered the great whaling grounds north of the Bering Strait. In the following years the news of this rich resource spread, and in 1852 more than 200 whaling vessels operated in the Bering Strait region. By 1866, the hunting pressure had put the bowhead population into a steep decline.

In 1880, the rising price of baleen stimulated the development of steam-auxiliary whaling vessels. These allowed bowheads to be exploited in all areas of their range. By 1889, steamers reached the summer feeding grounds off the Mackenzie River delta in Canada, and from then until 1915, the efforts of the whaling industry were concentrated in this area. It is estimated that between 19,100 and 21,500 bowheads were killed by all vessels from 1848 to 1915.[21] After 1915, with the collapse of the baleen market, few whales were taken commercially.

Subsistence harvesting of bowhead whales has gone on for more than 2,000 years. Eskimos were capable of taking between 45 and 60 whales annually, using traditional methods. During this time of traditional whaling, the entire animal was utilized—for food, utensils, weapons, and toys. Meat and blubber were the most important parts. With the advent of commercial harvesting, Eskimos were hired by the whaling stations and trained in the use of Yankee methods of whaling. This essentially eliminated traditional subsistence hunting until the end of commercial whaling in 1915.

With the end of commercial whaling, the Alaska Eskimos returned to subsistence whaling, but this time they were harvesting a severely depleted population and were armed with modern weapons. Between 1915 and 1969, the annual Eskimo bowhead harvest varied considerably, but did not exceed 23 and averaged only 10 whales yearly. The number of whales taken or struck and lost increased to 29 struck and landed and 82 struck and lost in 1977. This greater number matches the increase in the number of boats engaged in whaling, an increase resulting from the new prosperity of Eskimos employed on the construction of the trans-Alaskan oil pipeline.

The larger Eskimo harvest, coupled with an estimate that bowhead populations had declined between 7 and 11 percent of their original number, prompted the Scientific Committee of the International Whaling commission (IWC) to state that "any taking of bowhead whales could adversely affect the stock and contribute to preventing its eventual recovery, if in fact such a recovery is still possible." The committee recommended, and the IWC agreed, that "on biological grounds, exploitation of this species must cease."[22]

Very little is known about reproduction of the bowhead species. The Bering Sea herd appears to mate and calve mainly in April and May, during the spring migration. Gestation lasts about 12 months. It is unlikely that females breed more than once very two years.

The food habits of bowhead whales have not been well researched, although it is assumed that they feed primarily on zooplankton. They also take polycheate worms, gastropod mollusks, echinoids, and crustaceans—at least, these were found in one specimen's stomach. It is estimated that there are now between 1,000 and 2,000 bowheads, but the census is obviously very crude.

Each spring, the western Arctic bowhead whales migrate from the southern edge of the seasonal ice pack in the Bering Sea through the Chukchi Sea into the Beauford Sea. Some part of the population concentrates south of St. Lawrence and St. Matthew Islands. This migration begins with the breakup of the ice in the Bering Sea, generally sometime in April or early May. The breakup causes shifting currents and changes in wind and temperature in and around islands and land masses such as those between the Chukchi Peninsula and St. Lawrence Island.

Taking advantage of currents, bowheads begin their northward migration through the northwestern Bering Sea and western Bering Strait in April (depending on the ice conditions). They usually pass Points Hope and Barrow in middle to late April and continue through May. From Point Barrow, they travel northeast along an offshore route to the eastern Beaufort Sea, where the first whales arrive by early to middle May (Figure 21–10).

After spending the summer (June–September) feeding in the Beaufort Sea, including the vicinity of Banks and Hershel Islands, the whales begin their fall migration back to the Bering Sea. Records and recent observations indicate that most of the whales will have left Amundsen Gulf by the middle of September.

Management of the bowhead whale in the western Arctic requires habitat-protection measures together with control of the harvest. Continued exploitation by subsistence whalers and existing petroleum production in the summer feeding areas undoubtedly affect the bowheads. Offshore petroleum exploration and production east of Point Barrow, at Prudhoe Bay, Alaska, and in Canada's Mackenzie Delta and Amundsen Gulf, will have a further impact on bowhead populations. How great this effect will be is a matter of speculation. Indications are, however, that bowheads are disturbed by such activities and presumably could change their feeding areas.

Manatee

The West Indian manatee is one of four living species in the order Sirenia. Two subspecies of manatee exist in the world, the Caribbean manatee and the Florida manatee. The Florida manatee is a large, seal-shaped, gray or brown aquatic mammal with a flat, spatulate tail. It is hairless, except for bristly whiskers on its muzzle and scattered hairs on its back (Figure 21–11). The forelimbs are paddle shaped, with vestigial nails, and there are no hindlimbs. Being herbivorous, the manatee has only molar teeth. These

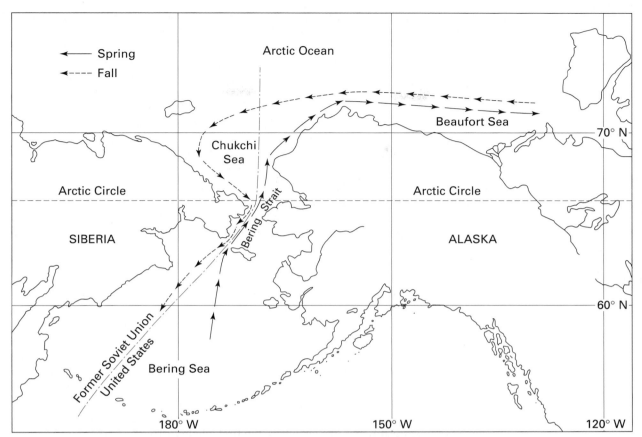

Figure 21–10 Migration route of the bowhead whale.

Figure 21–11 Female manatee with young. (Courtesy of the U.S. Fish and Wildlife Service.)

molars form at the back of the jaw and wear down as they move forward and replace other molars at the front of the jaw. This is thought to be an adaptation for eating forage mixed with sand. Manatees have poor eyesight, but are endowed with excellent hearing, their large earbones being fully developed at birth.

Manatees may grow to 2.3 to 4.1 m (7½ to 13½ ft) in length and weigh up to 203 to 608 kg (450 to 1,350 lb). They are able to swim at bursts of up to 24 km/h (15 mph) for short distances, but their cruising speed rarely exceeds 10 km/h (6 mph).[23] They are without natural enemies and thus have no means of defense.

Manatees do not appear to have strong cohesive social groups. The only long-term association appears to be that between females and their calves (for one to two years), although estrous females and their male consorts may remain together from a week to a month. Manatees feed from six to eight hours a day and eat a wide variety of plants. They grasp and tear with their lips, which are strengthened with lateral, horny pads, and pass the food back to the grinding molars. How much they eat in the wild is not known, but captives consume between 27 and 50 kg (60 and 110 lb) of vegetation daily.

Common habitats of manatees are slow-moving rivers, estuaries, and salt-water bays. The animals require channels 1.2 to 1.8 m (4 to 6 ft) deep for migration. Food must be available and warm water nearby for winter use.[21] A predictable movement to warm-water sites of the Florida Peninsula occurs during spells of 10 to 16°C (50 to 60°F) air temperature. Estimates of the size of the population range from 750 to 850.

Competition between people and the manatee for aquatic habitat has been a major detriment of the population. Changes in water temperature, disturbance of feeding areas, and alterations of springs and channels have reduced the area in which the manatee can survive. Explosives, pesticides, and dredging are all detrimental to the manatee. Accidental deaths from scraping by boat propellers have been frequent. The dams built in Florida waterways have caused periodic shifts in the water currents. Manatees have been sucked up against dams and immobilized or crushed in the locks. Less violent, but still disturbing, are the many people scuba diving or fishing.

Management of the endangered manatee involves maintenance of its habitat and reduction of accidents and harassment. Knowledge of the manatee's biology and a public education program are probably the keys to the survival of the species.

SUMMARY

Endangered species are species that are threatened with extinction through a significant part of their range. Threatened species are those that are liable to become endangered in the foreseeable future. Most endangered species are endangered because their habitat has been altered in some way. In other cases, the removal of individuals has been so great that the gene pool or genetic makeup of the population has been reduced to a level not permitting adaptation to a changing environment. Most approaches to managing endangered species treat symptoms, not causes. Endangered-species management can result in conflict between biologists and those interested in changing the landscape.

Specific steps establish a species as being endangered and designated under the Endangered Species Act. Once a species has been declared endangered, a recovery team is appointed to formulate and monitor a recovery plan. Recovery plans prescribe specific actions that should be taken to assist the species in its recovery. These actions include improving habitats, captive breeding, and acquiring habitats.

DISCUSSION QUESTIONS

1. Define *endangered species* and *threatened species.*
2. Is endangered-species management a good way to manage wildlife? Explain.
3. Discuss the history of endangered-species legislation.
4. Who has authority over the management of endangered species on forest service land, state land, and private land? Explain.
5. How does genetic variability influence the recovery of endangered species?
6. What is a recovery plan? How does it come into being? What is its purpose?
7. How does a species get listed as endangered? Delisted?
8. Should an endangered species be able to stop a major federal construction project such as the Tellico Dam? Why or why not?
9. Is money spent on captive-breeding programs well spent? Defend your answer.
10. What role does the United States play in international endangered-species management?

LITERATURE CITED

1. Anon. 1982. *Endangered and Threatened Wildlife and Plants.* 50 CFR 17.11 and 17.12. Washington, DC: U.S. Fish and Wildlife Service.
2. Ripley, S. D., and T. E. Lovejoy. 1978. Threatened and Endangered Species. In H. P. Brokaw (Ed.), *Wildlife in America,* pp. 365–78. Washington, DC: Council on Environmental Quality.
3. Anon. 1981. *Endangered Means There Is Still Time.* Washington, DC: U.S. Fish and Wildlife Service.
4. Terborgh, J., and B. Winter. 1980. Some Causes of Extinction. In M. E. Soule and B. A. Wilcox (Eds.), *Conservation Biology,* pp. 119–33. Sunderland, MA: Sinauer Associates.
5. Franklin, I. R. 1980. Evolutionary Change in Small Populations. In M. E. Soule and B. A. Wilcox (Eds.), *Conservation Biology,* pp. 135–49. Sunderland, MA: Sinauer Associates.
6. Bean, M. J. 1983. *The Evolution of National Wildlife Law.* New York: Praeger.
7. Foin, T. C., S. P. D. Riley, A. L. Pawley, D. R. Ayers, T. M. Carlsen, P. J. Hodum, P. V. Switzer. 1998. Improving recovery planning for threatened and endangered species. *BioScience* 48:177–184.
8. Lipske, N. 1988. Giving rare creatures a fighting chance. *National Wildlife* 36:(2) 14–23.
9. Luoma, J. R. 1988. Habitat conservation plans: Compromise on capitulation. *Audubon* 100(1):36–43.
10. Anon. 1982. *Kendall Warm Springs Dace Recovery Plan.* Denver, CO: U.S. Fish and Wildlife Service.
11. Jackson, J. A. 1978. Competition for Cavities and Red-Cockaded Woodpecker Management. In S. A. Temple (Ed.), *Endangered Birds: Management Techniques for Preserving Threatened Species,* pp. 103–13. Madison, WI: University of Wisconsin Press.
12. Kleiman, D. G. 1989. Reintroduction of Captive Mammals for Conservation. *BioScience* 39:152–61.
13. Erickson, R. C., and S. R. Derrickson. 1981. The Whooping Crane. *Proceedings, International Crane Symposium,* Sapporo, Japan. 1980.
14. Anon. 1996. *The World Wildlife Fund in 1996.* Washington, DC: World Wildlife Fund.
15. Anon. 1997. The Year in Conservation. *The Nature Conservancy Annual Report.* Arlington, VA: The Nature Conservancy.
16. Chabreck, R. H. 1980. *Status of the American Alligator in Louisiana and in Baldwin and Mobile Counties, Alabama.* Atlanta, GA: U.S. Fish and Wildlife Service, unpublished report.

17. Neill, W. T. 1971. *The Last of the Ruling Reptiles: Alligators, Crocodiles, and Their Kin.* New York: Columbia University Press.

18. Bent, A. C. 1926. *Life Histories of North American Marsh Birds.* Bulletin 135, Washington, DC: Smithsonian Institute.

19. Ward, J. P., and S. H. Anderson. 1990. Crane Collisions with Powerlines. In D. Woods (Ed.), *Proceedings of the 1988 Crane Workshop.* Tallahassee, FL: Florida Game and Freshwater Fish Department.

20. Anon. 1983. *Yuma Clapper Rail Recovery Plan.* Washington, DC: U.S. Fish and Wildlife Service.

21. Bockstoce, J. 1980. A Preliminary Estimate of the Reduction of the Western Arctic Bowhead Whale Population by Pelagic Whaling Industry: 1848–1915. *Marine Fishery Review* 42:20–27.

22. Tillman, M. F. 1980. Introduction: A Scientific Perspective of the Bowhead Whales Problem. *Marine Fishery Review* 42:2–5.

23. Hartman, D. 1976. Manatee. In J. M. Layne (Ed.), *Inventory of Rare and Endangered Biota of Florida.* Tallahassee, FL: Florida Committee on Rare and Endangered Plants and Animals.

22
ANIMAL DAMAGE

Wildlife species have always had some effect on human activities. In rare instances, wild animals, such as the grizzly bear in the United States or the tiger in India, have attacked and killed human beings. More commonly, wildlife species cause damage to commodities belonging to people—agricultural products, livestock, gardens, homes, and buildings; wildlife managers are often called on to prevent or control such damage. Predator control has come into the forefront in wildlife thinking partly because of emotions related to predation on sheep by coyote, mountain lions, bear, bobcats, and other animals. In the U.S. Department of Agriculture, there is an Animal Damage Control group that recommends methods of controlling predators.

ATTITUDES TOWARD ANIMAL DAMAGE

There are at least two ways of looking at damage caused by animals, or, simply, animal damage. From a philosophical point of view, animal damage is simply a normal event in the natural system.[1] One then considers, perhaps in admiration, the skills of the predator in seeking, finding, stalking, dispatching, and consuming its prey. For example, one sees the skill in falcons searching for their prey and in wolves looking for their meals. Some people feel that such interactions in the natural system should be allowed to occur and that we should not interfere with them. These people find certain ethical and aesthetic values in the entire natural process.

To the owners of ranches, farms, or homes, however, there is little beauty in a coyote's carrying off a sheep or starlings raiding a vineyard. Rather, they want to remove the wildlife species to minimize the losses. Owners of fish hatcheries besieged by herons want to do away with the herons to protect the fish. And when a grizzly bear attacks a person, there is often a public outcry to destroy the entire grizzly population.

The Brown Tree Snake of Guam

Although animal damage has taken many forms, people have usually thought of it when their livelihoods or property was damaged. An unusual form of damage has been occurring in the western Pacific Ocean on the small island of Guam. The brown tree snake, a native of nearby New Guinea, was accidentally introduced on Guam following World War II. The island's natural vertebrates were limited to animals such as birds and bats, whose ancestors flew to the island, and lizards, which originally arrived as eggs on floating debris in the ocean.

The brown tree snake is an efficient predator of ground and tree-dwelling animals. It appears to have a mildly toxic saliva, which is injected into the muscle tissue of a victim as the snake chews. Although the reptile is slow in reproducing, it is becoming very numerous because it does not appear to have any predators. As of 1996, only 3 of Guam's 12 native forest bird species, one of three native bat species, and 1 of 12 native lizard species remain. Many agricultural activities are suffering because the snake is destroying the natural balance of organisms on the island.

As the snakes eliminate their prey, they wander in search of new food. Now they are invading dwellings at night and biting children aged 2 to 11 while they are sleeping. The snakes appear to bite their victims a number of times as though to ingest them. The saliva does not kill the child, but makes him or her sick. Whether the snakes are invading people's houses because of the lack of food or because they are just wandering is not known. The brown tree snake is an example of how damage to an ecosystem affects many species and can directly affect human beings. This form of damage is very difficult to measure in dollar terms.

Rodda, G. H., T. H. Fritts, and D. Chiszar. 1997. The Disappearance of Guam's Wildlife. *BioScience* 47:565–74.

CAUSES OF ANIMAL DAMAGE

When people move into an area, they often disrupt the natural system in such a way that new predatory behavior sets in. Thus, the removal of wolves from a large part of North America, along with the heavy grazing of rangelands, resulted in an increase in the coyote population, and there are no apparent natural controls on that population, which occasionally prey on domestic sheep. Similarly, the removal of large expanses of natural prairie vegetation and its replacement with monocultures disrupts the habitat in such a way that insects, rodents, and some large ungulates begin to use farmers' crops for food. When acreages are converted into human-made ecosystems, the wildlife species still present in the area utilize the human-made systems as well as the natural system (Figure 22–1).

Figure 22–1 Elk crossing a road can be hit by a vehicle, causing damage to the vehicle and injury to the owner and the elk. (Courtesy B. Debolt.)

It is necessary to control some animals because they carry disease; for example, skunks, raccoon, bats, and fox may carry rabies. Exotic species often cause animal damage problems. (See Chapter 9.) For example, the introduction of starlings and the English sparrow into the United States, has resulted in major damage to some crops and buildings (Figure 22–2). And the gray squirrel, introduced into the Cape Peninsula in the Republic of South Africa during the last century for ornamental purposes, has become a major predator of nesting songbirds.[2] The case of the gray squirrel is of more than incidental interest: It shows the kind of considerations that are overlooked when wildlife are introduced into a new area. As we have noted, the introduction of any exotic species into a new environment without exhaustive ecological studies can lead to ecological disasters. The rat, introduced into many areas of the world, is a pest because it generally does not have natural predators in the new areas.

Figure 22–2 House sparrows frequently build their nests in unwanted places. (Courtesy C. A. Morgan.)

ANIMAL-DAMAGE PROGRAMS

During the settlement of the west, the pioneers saw quite early that large predators were able to disrupt their farming and ranching operations. So stockmen, ranchers, and farmers set out to destroy predators whenever they could. Wolves, for instance, were attacked so viciously that they were exterminated in parts of their former range. But other species become more abundant, so the U.S. government, through the Bureau of Biological Survey, searched for poisons to kill coyote, bobcats, jackrabbit, prairie dogs, and other mammals that were destroying or damaging property in the late 19th century. But this was a relatively minor effort until about 1920, when predator- and rodent-control programs began (Figure 22–3).

These programs have historically relied on poisons to prevent animal damage to any agriculture crop. The most controversial use of poisons has been against coyote. In fact, coyote control has been a focal point of animal-damage programs in the United States. From the 1880s to the middle 1930s, predator research was primarily a part-time attempt to improve strynchine-impregnated baits. In the late 1930s, thallium-treated bait stations on lambing grounds were introduced into the west, and by the 1940s, poison stations were placed throughout the Rocky Mountains in lambing areas. This was considered a step forward in reducing coyotes, particularly in remote areas. In

Figure 22–3 Coyote seem to adapt well to human-created habitats and are predators on some human livelihoods. Thus coyotes have been a focal point of animal damage programs.

1937, experiments with a device called the "coyote getter" were begun. It used a small charge in a .38 special cartridge to expel sodium cyanide powder. The device was later replaced by a spring-activated unit called an M-44, which, when bitten and pulled by an animal, delivers toxicants into the animal's mouth (Figure 22–4). Rotten meats are placed near the M-44 to attract coyote. There have been problems with the cartridge in the "coyote getter" and the M-44, and both have had only limited success. Various restrictions have been placed on the use of the M-44 with sodium cyanide, most of them relating to where the device can be placed.

In the 1940s and 1950s, studies were also made of the efficiency of thallium, compound 1080, anticoagulant bait stations, and strychine drop baits. In the 1970s, there was a controversy over the impact of compound 1080 on nontarget species. Steel traps were tested as a means of catching coyotes and evaluating their effect on the wildlife population.

Toxic sheep collars are constructed to take advantage of the fact that most adult coyotes kill sheep by biting them on the underside of the throat. When the first collars were used, in the 1920s, toxicant-filled syringes were attached to either side of the sheep's neck. Later collars contained 1080, sodium cyanide, and diphacinone.

Selected sheep have toxic collars placed on them and are tied to a stake in the area of high coyote predation. The main flock is then corralled into another area (Figure 22–5). Compound 1080 appears to be effective. Unfortunately, it requires the sacrifice of lambs and may have unknown hazards for other animals.[3]

By way of information, we should note that unlike chlorinated hydrocarbons, 1080 is a water-soluble compound (sodium monofluoroacetate) that is not subject to bioaccumulation in the food chain. The degree of toxicity of 1080 varies widely with species, and the effects appear not to be cumulative in victims of sublethal doses.

During the 1960s, several committees were formed to examine methods of animal-damage control. The first major committee was formed by A. S. Leopold (son of Aldo), under the direction of Secretary of the Interior Stewart Udall, to evaluate predator control. The committee's report recommended that government policy promote husbandry of all forms of wildlife, but that population control be local wherever a species was causing significant damage. This advisory committee concluded that in the west the most efficient measure for control of coyotes was the 1080 bait station.

Figure 22–4 M-44 used to attract and kill coyote. (Courtesy of G. Connolly, U.S. Department of Agriculture.)

Figure 22–5 Sheep collar containing coyote poison. (Courtesy G. Connolly, U.S. Department of Agriculture.)

In response to increased controversy over the use of toxic chemicals for predator control, the U.S. Department of the Interior and the Council on Environmental Quality sponsored a 1971 study of the entire predator damage situation in the United States. The committee conducting the study was formed by Stanley Cain, a noted wildlife biologist. This committee's report expressed concern about the nontarget effects of 1080. The report indicated that, because of the high value placed on all wildlife, predator control should be selective for individual predators taking livestock. As a result of this report and public pressure, President Nixon issued an executive order in 1972 banning the use of toxicants for predator control on federal lands or by federal agents.[4]

It was following this executive order that livestock owners, especially sheep ranchers, began to put political pressure on Washington to approve some form of toxicants for predators. The result of the ban, according to a number of biologists, was to transfer much of the responsibility for coyote control from trained personnel to landowners.[5] Some people felt that the pendulum was swinging in favor of predators, causing polarization of environmentalists and livestock owners.[6]

DOCUMENTATION OF LOSS

Animal Damage

In 1981, there were approximately 12.9 million domestic sheep in the United States. Less than 40 years earlier, in 1942, the stock sheep and lamb population had peaked at 49.3 million, an inventory exceeded only by the more than 50 million head during the 1800s. The many factors that contributed to this decline of the industry include poor markets for lamb and wool, a labor shortage, and predation. The extent of the impact of predators on the industry has been, and will continue to be, a subject of controversy, much of which stems from disagreement over the economic loss attributable to predators and over the prevention of predation.

Because it is not always easy to determine which animals have been killed by predators, predation losses are difficult to estimate. Carcasses can be destroyed, never found, or partially decayed, making the exact cause of death difficult to determine.

Nevertheless, there have been a number of surveys that have attempted to ascertain the amount of sheep loss from predation by animals, particularly coyote. The type of survey can influence the results. In a study conducted in southern Iowa, people were contacted by mail.[7] Respondents who had lost sheep by predation thought that 3 percent of their flocks had been killed by coyote and 1 percent by dogs. Some 60 percent of the respondents attributed their sheep loss to predation. Yet biologists who visited ranches where losses of domestic animals had been reported found that dogs had killed more sheep than coyotes.

In an Idaho study, biologists found that premature birth, starvation, and disease were the major cause of lamb deaths. The mean minimum predation loss during the year was approximately 2.9 percent, a figure based on observed and unaccounted-for losses. Coyotes accounted for the largest proportion, 93 percent of all predator-killed lambs and ewes.[8]

In the area around Yellowstone Park, the loss where sheep grazed in the vicinity of the park was 3.7 percent. Of that number, black bear accounted for 34 percent, grizzly bear 15 percent, coyote 6 percent, disease and poisonous plants 6 percent, and improper herding 39 percent.

Beginning in 1978, the U.S. Fish and Wildlife Service has attempted to control coyote in about 11 percent of the species' range. Surveys in ranges where federal biologists worked showed that coyote destroyed an estimated 4 percent of the sheep population annually.[9] Obviously, coyote do kill sheep, but it is obvious, too, that bear, mountain lions, and dogs also kill sheep and that the villains are different in different areas. Nevertheless, for most concerned people, the word association with "sheep loss" is "coyote."

Studies of the estimated expenditures to prevent coyote predation, based on personal interviews with ranchers, indicate that in 1983 small ranching operations (under 200 sheep) spent approximately $0.44 per head, medium (200 to 999 sheep) spent $0.37, large (1,000 to 5,499) spent $0.93, and operations of more than 5,500 sheep spent $1.65 per head, for an average of $1.06 per head.[10] A number of different techniques, including trapping coyotes, checking and corralling sheep, and using noises and M-44s (Figure 22–6), were utilized to prevent losses (Table 22–1). Costs varied by size of ranch because the practices utilized were different (Table 22–2).

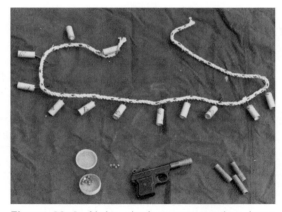

Figure 22–6 Noise devices are used to keep coyote from sheep. (Courtesy L. Jahnke, Wyoming Game and Fish Department.)

TABLE 22–1

Some Techniques to Reduce Predator Success

Predator tax	Doctor sheep
Ammunition	Guard dogs
Check sheep	Fence
Corral sheep	Phone calls
Nonuse of land	Electronic shepherd
Ground hunting	Pool fees
Added labor	Insurance
Aerial hunting	Cameras
Trapping	Radios and flashlights
Bounty	Poison
Denning	Get warden
Move sheep	Get trapper
Paid trapper	Bait lambs
Repellent	Spotlight
Flashing lights	Park pickup
Find sheep	Llama
Rifles	Buy back lambs
Hay	Intercom
Bells	M-44s
Propane guns	

It is obviously difficult, for various reasons, to compile the complete costs of animal predation. Some notion of the magnitude of these costs can be gained, however, from the following: In 1981, an estimated 14,800 sheep, valued at $370,000, and 71,900 lambs, valued at $2,609,970, were lost to predation in Wyoming alone.[11] These were direct costs only. Such figures can perhaps give us a better understanding of why ranchers and farmers feel as they do about certain aspects of wildlife management and control.

Agriculture Loss

Agricultural crops are another major victim of animal damage. Bird damage to crops is locally serious in the United States. Estimates are of more than $25 million annually in damage to ripening corn, $50 million to seeded corn, $12 million to sunflowers, $12 million to cherries, $6 million to sorghum, $5 million to rice, $4 million to grapes, and $1 million to lettuce.

Canada geese were found to graze in fields of fall and winter rye planted to reduce erosion and improve the quality of the soil in Connecticut. There the leaf biomass of rye was 535 percent in midwinter inside exclosures (exclusion fences) preventing

TABLE 22–2

Predator Control Practices by Size of Ranch

Small	Medium	Large	Very Large
Corral sheep	Check sheep	Added labor	Aerial hunting
Check sheep	Added labor	Check sheep	Added labor
Ammunition	Fence	Hay	Paid trapper
Ground hunting	Electronic shepherds	Paid trapper	Ground hunting
	Hay	Ammunition	Check sheep

geese from grazing. In the spring, rye biomass was 177 percent higher in the exclosure, showing that recovery was slower where grazing had occurred.[12] At the same time, of course, birds are very valuable to farmers. Gulls eat locusts, and blackbirds eat weevils, corn borer larvae, earworms, and root-worm beetles.

Other animals, too, are responsible for damage to agriculture. In the east, white-tailed deer often move into hardwood forest areas following logging and prevent regeneration by eating the new trees. Deer remove much of the vegetation from forests when the deer population is high. Many ungulates trample crops and vegetation.

Other Losses

Wildlife species cause other forms of damage, some of which can be evaluated. Bird-aircraft collisions cost more than $20 million annually in the United States and also take human lives.[13] Homes are damaged by birds' nesting and pecking.

MANAGEMENT APPROACHES TO DAMAGE CONTROL

Predator control or management is a part of wildlife management. It applies to carnivores, omnivores, herbivores, and scavengers alike. One of the great problems in animal-damage programs has been a lack of planning. Problem species generally exist because they bother people. As we indicated in Chapter 1, wildlife management can involve increasing, maintaining, or decreasing a population. All too often, a damage management program involves killing something. Success is thought of as the number of animals killed, not a reduction in damage. To decrease population numbers, it is necessary to look at population dynamics. Do we want to reduce all age classes or just one age group of animals? For example, in a population that is near the saturation level in its habitat, young animals may move out, causing damage problems. The social structure of the population should be known. Shooting the dominant male prairie dog may increase a population, not reduce it.

Managers must plan a program to reduce populations, carry it out, and evaluate it. New approaches must be initiated if the program is not meeting its objectives. Sometimes it is necessary to determine levels of damage with which it is acceptable to live. This could occur when the costs of management actions exceed the costs that are returned from the product or when public sentiment prevents complete removal of the animal producing the damage.

Many forms of population control are described in earlier chapters. One is predation, which occurs when one animal kills another, usually to eat. In meeting their own food demands, predators help reduce the number of prey when there is an overabundance of these animals. Predators also remove animals with poor survival characteristics, such as the weak or sick. Usually, the size of the predator population is determined by the size of the prey population. There are many predators, including grizzly bear, black bear, mountain lions, golden eagles, coyote, swift fox, and marten. Some predators kill sheep, cattle, or other domestic animals. The owners of these animals demand restitution and more stringent predator control measures. Most predator control is directed against coyote. Leopold mentions predator control as one form of wildlife management. Generally, predator control is employed when predators are too numerous to manage. Some ranchers want all predators removed, but they do not realize that other predators would likely increase in number to fill the gap.[14] Predator control programs work best when predators are numerous and prey, such as ungulates, are rare. Predator control is expensive and sometimes dangerous. Toxicants used to kill predators also kill other animals or harm the ecosystem. In some cases, control is even counterproductive: Coyotes, for example, recolonize areas left vacant by destroyed populations. Newcomers to the habitat have large litters, so coyotes return quickly. In some cases, improving the habitat is a good alternative to controlling predators. Livestock husbandry with the use of dogs or llamas can also be effective.

Choice of Control

In general, it is not wise to make predator control an independent objective. Rather, controlling predators should be conducted in relation to such broad management objectives as suppressing disease, and protecting wildlife and domestic animals. Management practices involving control require that (1) decisions pertaining to control not be arrived at independently of other decisions, (2) interdisciplinary efforts be planned, and (3) accurate data be available.[2] There is no single criterion for deciding when control should be practiced and no single standard for determining need. Decisions should be based on aesthetic, social, economic, ecological, political, and administrative perspectives.[2] Managers should regard animal damage as one aspect of the total wildlife-management program. For example, the problem of coyote predation cannot be solved simply by shooting the animals, for this type of removal cannot be complete, and the habitat in which the coyotes reproduce remains. Thus, the coyotes will continue to be a problem, both next week and next year. Furthermore, before managers begin a program of animal removal, they must consider the impact of planned actions on the system as a whole.

When instituting control measures, a manager should answer several questions: (1) Who will carry out the controls? (2) What are the primary areas in which the controls will be exercised? (3) What basic approach is appropriate? (4) What control methods should be used?[15]

The manager must also know the economics of the situation—the relation between costs and benefits. Actually, economics may *dictate* the type of control measures used. Renting an aircraft and buying ammunition to remove a few coyotes for a relatively short period of time would not be economically sound, but simple habitat manipulation might keep the coyote population at a level such that serious economic loss would be prevented.

Forms of Management

The management of animals causing damage can involve altering habitats, manipulating populations, or implementing biological controls. Chemicals can be used to destroy or repel animals. Physical methods, such as erecting barriers or altering the conditions under which the damage is occurring, are sometimes feasible. At other times, husbandry techniques or alternative forms of agriculture practices can be instituted.

Habitat Alteration. All animals causing damage problems find the habitat they are in suitable for their use. If it is possible to alter this habitat so that the nuisance species no longer finds it satisfactory, the damage should cease. Rats, for example, use debris on the ground for cover in some orchards. They also use garbage dumped in alleyways and between buildings in cities. Removing garbage and debris eliminates the food source and renders the habitat less suitable for the rats.

Population Management. Many forms of control involve some method of population management. One method is the harvesting technique. Bounty hunting is a possibility, but it has not been highly successful.[2] Actually, to have a successful harvesting program, it is necessary to know something about the biology of the animal causing the damage: what time of day it is active, when it breeds, how many young it usually has, how the young disperse, and how the animal uses different habitats at different times of the day or during different seasons of the year.

Trapping and Removal. The Kirtland warbler is of interest because of its endangered status. One of the reasons for its decline and subsequent listing as an endangered species is the high proportion of its nests parasitized by cowbirds (Figure 22–7). The brown-headed cowbird lays one or two eggs in the nest of the warbler. The cowbird

Figure 22–7 Cowbird eggs in warbler nest. Some warblers that nest near forest edge have cowbird eggs in nearly 100% of their nests. (Courtesy of M. Hamis.)

chick generally hatches sooner than the warbler chick, pushes the eggs or young warblers out of the nest, and is then raised by the foster parents. In the early 1970s, when the warbler population had declined to fewer than 300, the Fish and Wildlife Service and the Forest Service instituted a program to remove the cowbird. Cowbird decoy traps were erected, and from 1975 to 1981, a total of 24,158 cowbirds were removed from Kirtland warbler nesting areas. Statistics highlight the dramatic effect of the program. As the solid line in Figure 22–8 indicates, from 1931 to 1971, some 59 percent of warbler nests were parasitized; by 1972, the figure had dropped to 6 percent, and the average was only 3.4 percent over the 10 years of the cowbird control program. The dashed line shows the corresponding rise in Kirtland warbler fledglings per nest for the same period.[16]

Trapping and removal have also been used to move large mammals from areas where they are causing damage. One of the more extensive programs involves the relocation of black bear. National parks and national forests often have their personnel trap bear around campgrounds and take them to the back country. While disappointing people who like to see bear along the roadways, this action has reduced personal injuries.

In a western study, investigators found that the removal of litters of coyote pups from an area reduced predation by 88 percent, and when both pups and adults were removed, the reduction was 98 percent.[17] Note that, for such striking results, ease of access and considerable time in the field were necessary.

Use of Chemicals. A number of chemicals have been used to control predatory animals. Chemicals can kill, repel, or frighten the nuisance animal. The densities of predators of waterfowl in Canada, including skunks, raccoon, and ground squirrels,

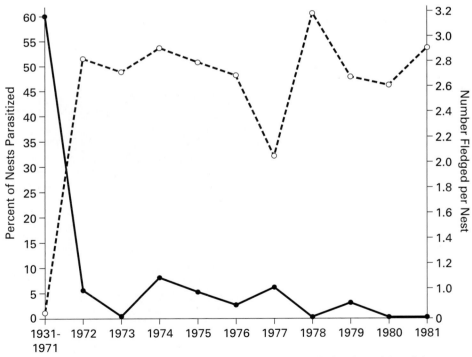

Figure 22–8 Changes in cowbird parasitism on Kirtland warblers following the institution of cowbird trapping.

were reduced by injecting some of the birds' eggs with strychnine. Predators destroyed only 16 percent of the waterfowl eggs in an experimental area where strychnine was used, compared with 34 percent of the eggs in a control area.[18]

Blackbirds have long plagued farmers in the midwest by descending in great flocks on the field and removing grain very rapidly. A number of chemical controls have been used against them.[19] The U.S. Fish and Wildlife Service, through studies of caged birds, showed that red-winged blackbirds, house sparrows, grackles, tricolored blackbirds, and brown-headed cowbirds can be controlled by using grains covered with toxic chemicals.[20] The birds eat the poisoned grain and die a short time later.

The selectivity and efficiency of M-44s, hunting, strychnine, and steel traps were evaluated with respect to their effect upon predators of sheep in Texas.[20] M-44 and hunting were found to be the most selective methods, each taking only target species. Strychnine and steel traps took a variety of animals, including game animals, raptors, rodents, songbirds, and reptiles. M-44 was probably the most efficient tool for taking coyotes as target species, while steel traps took bobcats more efficiently. Of course, a nontarget species in one area may be a target species in another area, so that care should be exercised in generalizing about the selectivity of predator controls.[21]

In the west, prairie dog and ground squirrel towns have been destroyed by poisoning. Ranchers want to remove these animals, which compete with cattle for forage and create holes that may cause livestock to fall and break their legs. Unfortunately, poisoning also kills some other species of wildlife or reduces their prey base. Thus, the decline of burrowing owls and black-footed ferrets has been related to the poisoning of prairie dogs and ground squirrels.

Methocarb (mesurol), a repellent, was evaluated as a spray treatment for reducing bird damage to ripening cherries in Michigan. More than 45 percent of the untreated cherries had been pecked by robins and grackles by harvest time. When the trees were sprayed with methocarb, the damage was reduced by 66 percent.[22] In California and Oregon, damage to grape crops by robins, house finches, quail, goldfinches, and other birds was reduced by applying chemicals. Chemicals such as lime, lye, powdered sulfur, and cayenne pepper have reduced rodent activity when placed in established runways.

Chemicals have been used as a frightening device to protect sunflowers and other crops from damage by birds. Corn particles can be treated with 4-aminopyridine and scattered in small amounts on fields where crops such as sunflowers grow. Blackbirds feeding on the sunflowers also pick up some of the cracked corn. When a bird eats a treated corn particle, it usually displays distress reactions (call and erratic flight) in from 10 to 30 minutes, causing the rest of the flock to become alarmed and leave the field. Although the adventurers that pick up the corn usually die, there is little impact on the population.[23]

Exclosures. Exclusion fences, or exclosures, have been used in Australia to keep rabbits out of crops and wild dogs away from livestock. In the United States, restriction placed on other damage-control methods have led to greater use of fences for the protection of crops and livestock from dogs, coyotes, and other animals. Electric fences have been used occasionally to protect agricultural areas from black bear and some wildlife species from carnivores.

Various other types of fences, including conventional, net, wire, and high-voltage electrical wire fences, have been used.[24] Fences appear to be most useful and cost effective for small, open pastures with intensive production and less effective for large pastures with low production and high vegetation cover, which restricts the removal of predators. But in large areas, fences may help direct predators to areas where other control methods can be used.

The disadvantages of exclusion fences are construction and maintenance costs, their inability to exclude some predators, and the difficulty of removal of predators from areas with thick cover. Regulations for the protection of some wildlife species, such as the pronghorn antelope, may prohibit the construction of exclusion fences, particularly on public lands. The cost of maintaining a fence is related to the terrain, the type of soil, the density of vegetation, damage from livestock and other animals, heavy snow, and flooding.

Predators can gain access through damaged fences, by jumping over or digging under fences, or through the fence when the electric power fails. Evidence indicates that when predators are accidentally contained within enclosures, fences sometimes increase losses from predation rather than reducing them.

Biological Control. Biological control involves using some living material to check a population and thereby prevent or reduce damage. In Chapter 3, we discussed some of the classic cases in which predator–prey interactions have been established to reduce damage. For example, gypsy moths, which defoliate trees, are kept in check by parasitic flies introduced into the moth's habitat. Biological control also involves the use of alternative food supplies. When the small, ground-feeding Oregon junco was found to eat a large proportion of seed in a Douglas fir clear-cut, food was placed at selected areas to divert the birds from the main clear-cut. This method was successful in reducing predation on seeds.[25]

Placing a variety of seed mixtures attractive to deer mice in adjacent habitats, biologists in British Columbia were able to divert small rodents from a regenerating forest. Waterfowl have been diverted from vegetable crops, meadow voles from orchards, rabbits from Christmas tree farms, and deer from agricultural areas by the use of alternative foods.[26]

Husbandry. Of the many techniques used to help ranchers control coyote, not all involve killing. Some of the more effective ones involve husbandry. In a study conducted in a Kansas sheep-raising region, biologists found that once losses had been analyzed for, among other things, the time of day and type of animal, husbandry techniques could be recommended, including confinement, light placed in the corral, a different method of disposal of sheep carcasses, and noise devices.[27] These techniques appear to be effective, though expensive. While husbandry techniques have little appeal to ranchers because of their avid dislike of coyote, the techniques hold promise. Killing coyote does not permanently reduce the population, and if other techniques can be effective, costs can be reduced (Figure 22–9).

Cultural Control. Crops can be planted in a number of ways to reduce damage. For example they can be planted before migratory birds arrive. The harvest can sometimes be timed in the light of animal activity patterns. Sunflower fields can be planted at least 400 m (440 yards) from small marshes and sloughs where blackbirds are found. The sunflower crop should not be planted too close to woody areas or shelter belts where birds roost. Crops that blackbirds do not harm, such as sugar beet, potatoes, soybeans, and pinto beans, can be planted where these birds are found.

Figure 22–9 Bringing sheep into a corral at night is a form of husbandry that can reduce loss due to predators.

Noisemakers. Noise devices have been used to keep nuisance animals away. Coyote deterrents have included explosive devices, recordings of Hitler's speeches (even the coyote can't stand him), and loud horns. Birds—particularly blackbirds—are frightened by gunfire, explosive devices, and recordings of distress and alarm calls.[28] Most of these devices, however, are expensive (Figure 22–10).

These are only some of the techniques that have been utilized in damage control. For all of them, information must be available on the biology of the animal. Also, the controversy surrounding damage control is wide and vocal. Thus, in dealing with damage control, managers must also deal with strong public attitudes, which themselves compound the difficulties (Figure 22–11).

Figure 22–10 Noise device used for controlling blackbirds. (Courtesy of the U.S. Fish and Wildlife Service.)

Figure 22–11 Blackbirds in a sunflower field.

Hey, That's My Dog!

An older couple stopped at a roadside rest to stretch and let their dog run. The little poodle ran a short way off into the grass and then became airborne in the talons of a golden eagle. The distraught couple shouted and ran around, to no avail. Their next step was to stop at the nearest wildlife agency. Upon entering the office, the lady hysterically demanded that the manager get her poodle back; she had no idea that her dog was a legitimate meal for a predator.

CASES OF MANAGING ANIMAL-DAMAGE PROBLEMS

Birds and Aircraft

For the past 60 years, birds have caused collision damage to aircraft, resulting in the loss of human lives. High-speed turbine-powered jet aircraft are more vulnerable to damage than are those with slower, piston-type engines. A number of things can be done to reduce the chances of such collisions. Studies show that habitat management, including the reduction of available food, cover, and water near airports, can lessen the attractiveness of airfields for birds.[13] Proper design can make airfield structures unusable by birds. Casual bird visitors can be driven from airfields by mobile equipment. Air-traffic personnel can be trained to monitor the movement of birds so that pilots can be alerted to their presence. Knowledge of the biology of birds and their times of migration is especially important for airfield personnel, since it enables them to direct aircraft evasively during heavy migration periods. Active measures, such as amplified tape-recorded stress calls of bird species, firing of shells or flares, and other forms of harassment have been used successfully in Canada.

Snowy owls move south from the arctic to airfields in the northern United States during some winters. They perch on fences and snowbanks near the edges of runways and hunt on the open runways. These owls create a serious safety problem, since most of the collisions they have been in have resulted in severe damage to high-speed engines. Ornithologists have discovered that the owls move south as a result of a food shortage and remain in the area they select for the winter. This movement is part of an arctic predator–prey cycle involving the owls and their rodent food supply. In the winter, following the three- to five-year decline in lemming population, the excess numbers of snowy owls migrate south in search of food. It appears that relatively few owls survive to return to their breeding grounds the following season, for the population of owls is generally very low the next spring. When they are concentrated around airfields, the only way to reduce the population seems to be to remove them by shooting. Since most of the snowy owls would die of natural causes during the winter, this form of management has, in any event, little impact on the population. Unfortunately, it does cause some concern among wildlife groups.

Using trained birds to drive others away has been tried in many countries—in Canada, for instance with the peregrine and gyrfalcon. If the birds are carefully trained, they can drive away certain other bird species during the daylight hours, but they are not effective in heavy rain or during strong winds. Also, their cost is high.

Rats and Coconuts

Small mammals species, particularly rats, reduce the coconut yield in the Philippines. Active both at ground level and in the crowns of trees, the rodents chew on the coconuts at the base and around the nut itself, about a third of the time penetrating the shell. Often, the coconut falls to the ground before it is ripe. After trying a number of techniques that proved too expensive—bending the trees, removing ground debris, lacing the ground with poison—biologists found that if they placed anticoagulant chemicals mixed with rice in plastic bags in about every fourth tree, rat-damaged coconuts decreased to nearly zero in some fields. Rats fed on this material and died soon thereafter. Rats on the crowns of trees obligingly move freely from one tree to another, making it unnecessary to bait each tree.[29]

Vampire Bats and Cattle

For centuries, vampire bats have been a source of damage and therefore economic loss to the cattle industry in the southwest, Mexico, and Central and South America. Removing bats was long a problem: They are difficult to find, and some carry rabies. Fortunately, biologists have found a chemical that reduces the bat population. A dose of diphenadione, an anticoagulant, is given orally to cattle in vampire-bat country. The chemical passes into the rumen compartment of the bat's stomach, where it is absorbed into the bloodstream. It is deadly to vampire bats that bite the cattle within 72 hours after ingestion.

The diphenadione does not affect the cattle, apparently because the green, leafy plants that cattle consume contain vitamin K, an antidote to the chemical. In addition, bacteria in cattle rumen synthesize vitamin K, thus preventing excessive bleeding. Bats, in contrast, do not have much vitamin K, and their body movement during flight usually causes blood capillaries to rupture in the wings. With diphenadione from cattle in their bloodstream, they bleed to death.[30]

SUMMARY

Animal-damage problems result when wildlife species destroy or alter commodities or other species that are desired by people. Damage from wildlife often occurs when people alter natural systems, and disturb the homeostatic relationships there. The federal government participates in developing management tools to prevent animal damage. In the past, most federal activity in this area was in behalf of ranching and farming interests. Now the government is also interested in preventing wildlife damage to buildings, gardens, cities, aircraft, and endangered species.

Animal-damage management programs should be part of broad management objectives. It is important that the biology of the animal population causing the damage be understood. It is important, too, to know precisely what role a damaging animal plays and has played in the natural system. Acting without that knowledge can do more harm than good. Computation of the damage costs and the costs of control measures should be part of the evaluation of any planned program that is to be instituted for purely economic reasons—that is, where loss of human lives is not involved. In essence, animal-damage control is the basic management concept of reducing the numbers of a population or preventing that population from gaining access to areas where the members are not wanted.

DISCUSSION QUESTIONS

1. Why are integrated damage-control techniques more effective than one technique alone?
2. Why should a manager study the biology of a predator before taking action against it?
3. Why is it recommended that predator reduction not be a separate objective of management?
4. Relate some of the theoretical predator–prey studies to controlling animal damage.
5. What forms of management should be considered when a rabbit population increases to the point of destroying range and forage?
6. Killing members of a predator population does not always reduce damage. Why not?
7. Do you feel that 1080 can be used selectively? Explain.
8. How can biological control be used to reduce the damage from animals?
9. How can the grizzly bear population be controlled so that the animals do not kill humans?
10. Why are habitat-manipulation measures generally more effective than population management in damage control?

LITERATURE CITED

1. McCabe, R. A. 1972. A Position on Predator Management. *Journal of Wildlife Management* 36:382–94.
2. Barryman, J. H. 1971. The Principles of Predator Control. *Journal of Wildlife Management* 36:395–400.
3. Connolly, G. E., R. E. Griffiths, and P. J. Savarie. 1978. Toxic Collar for Control of Sheep-Killing Coyotes: A Progress Report. *Proceedings, 8th Vertebrate Pest Control Conference,* pp. 197–205.
4. Pearson, E. W. 1981. *A Review of Predator Research.* Denver, CO: U.S. Fish and Wildlife Service, Wildlife Research Center.
5. Howard, W. E. 1974. Predator Control: Whose Responsibility? *BioScience* 24:360–63.
6. O'Gara, B. W. 1982. Let's Tell the Truth about Predation. *Transactions, 47th North American Wildlife Conference,* pp. 476–84. Washington, DC: Wildlife Management Institute.
7. Schaefer, J. M., R. D. Andrews, and J. J. Dinsmore. 1981. An Assessment of Coyote and Dog Predation on Sheep in Southern Iowa. *Journal of Wildlife Management* 45:883–93.

8. Nass, R. D. 1977. Mortality Associated with Sheep Operation in Idaho. *Journal of Range Management* 30:253–58.

9. Johnson, S. J., and D. E. Griffel. 1982. Sheep Losses on Grizzly Bear Range. *Journal of Wildlife Management* 46:786–90.

10. Jahnke, L. 1983. A Methodology for Assessing Indirect Costs of Predation Control: A Study of Wyoming Sheep Producers. Unpublished master's thesis, Department of Zoology, University of Wyoming.

11. Wyoming Crop and Livestock Report Service. 1981. *Wyoming Agricultural Statistics, 1981.* Cheyenne, WY: Wyoming Department of Agriculture.

12. Canover, M. R. 1988. Effect of Grazing by Canada Geese on Winter Growth of Rye. *Journal of Wildlife Management* 52:76–80.

13. Solman, V. E. F. 1973. Birds and Aircraft. *Biological Conservation* 5:79–86.

14. Leopold, A. 1933. *Game Management.* New York: Scribners.

15. Fall, M. W. 1980. Management Strategies for Rodent Damage Problems in Agriculture. In F. E. Sanchez (Ed.), *Proceedings, Symposium on Small Mammals: Problems and Control.* BIOTROP Spec. Publ 12:248.

16. Kelley, S. T., and M. E. Decapita. 1982. Cowbird Control and Its Effects on Kirtland's Warbler Reproductive Success. *Wilson Bulletin* 94:363–65.

17. Till, J. A. 1982. Efficiency of Denning in Alleviating Coyote Depredation upon Domestic Sheep. Unpublished master's thesis, Department of Wildlife Science, Utah State University.

18. Lynch, G. M. 1972. Effects of Strychnine Control on Nest Predators of Dabbling Ducks. *Journal of Wildlife Management* 36:436–40.

19. Schafer, E. W., and R. B. Brunters. 1971. Chemicals as Bird Repellents: Two Promising Agents. *Journal of Wildlife Management* 35:569–72.

20. Beason, S. L. 1974. Selectivity of Predator Control Techniques in South Texas. *Journal of Wildlife Management* 38:837–44.

21. Savarie, P. J., and R. T. Stermer. 1979. Evaluation of Toxic Collars for Selective Control of Coyotes That Attack Sheep. *Journal of Wildlife Management* 43:780–83.

22. Gilarino, J. L., W. F. Shake, and E. W. Schafer. 1974. Reducing Bird Damage to Ripening Cherries with Methocarb. *Journal of Wildlife Management* 38:338–42.

23. Besser, J. F. 1978. Birds and Sunflowers. In J. F. Carter (Ed.), *Sunflower Science and Technology.* Agronomy 19. Madison, WI: American Society of Agronomy, Crop Science Society of America, and Soil Science Society of America.

24. Wade, D. A. 1982. The Use of Fences for Predator Damage Control. *Proceedings of the Vertebrate Pest Conference* 10:24–33.

25. Hagar, D. C. 1960. The Interrelationships of Logging, Birds and Timber Regeneration in the Douglas Fir Regions of Northwestern California. *Ecology* 41:116–25.

26. Sullivan, T. P. 1979. The Use of Alternative Foods to Reduce Conifer Predation by Deer Mice. *Journal of Applied Ecology* 16:474–95.

27. Robel, R. J., A. J. Dayton, F. R. Henderson, R. L. Medvoa, and C. W. Spaeth. 1981. Relationships between Husbandry Methods and Sheep Losses to Canine Predators. *Journal of Wildlife Management* 45:894–911.

28. Conover, M. R. 1980. Comparative Effectiveness of Avitrol, Exploders, and Hawk-Kites in Reducing Blackbird Damage to Corn. *Journal of Wildlife Management* 48:109–16.

29. Fiedler, L. A., M. W. Fall, and R. F. Reidinger. 1982. Development and Evaluation of Methods to Reduce Rat Damage to Coconuts in the Philippines. *Proceedings, 10th Vertebrate Pest Conference,* pp. 73–79. Denver.

30. Bullard, R. W. and R. D. Thompson. 1977. Efficiency and Safety of the Systematic Method of Vampire Bat Control. *Interciencia* 2:149–52.

23
PUTTING IT ALL TOGETHER

A 5,000-hectare (12,350 acre) plot of land is made available as a wildlife management area in the country. This area contains mixed conifer forests, over 2,000 hectares of grasslands, numerous wetlands and water impoundments, and riparian communities consisting of cottonwood, willow, and mixed shrubs. During the last 50 years, the land has had extensive human impact. Logging was common in the forest. Large-scale mining operations occurred on the grasslands, leaving the landscape scarred and without much vegetation. Little effort had been expended to reclaim these areas. Some 75 oil and gas wells dot the landscape. Many of these have slurry ponds associated with them where excess oil waste is pumped. These areas are death traps for all forms of wildlife. Seepages drain into some streams, destroying fish, their food base, and furbearer habitat. Cattle grazing is extensive on the grassland areas.

Although the land is now available for wildlife management, there is local and national interest in the project. Some local people oppose the project because they feel that it will reduce income to the local community. Others are in favor of wildlife use of the area, but feel that sanctuary for several endangered species should be high priority. Those people who graze their cattle on the land are pushing for grazing rights at minimal costs. Those who operate the oil and gas wells feel that they should continue to obtain all the benefits from the wells, but should not be made to protect the oil slurry basins to prevent destruction of wildlife. Two small logging operations are lobbying their congressional delegates, stating that they will be out of business if they are not allowed to log each year.

"I now suspect that just as a deer herd lives in mortal fear of its wolves, so does a mountain live in mortal fear of its deer. And perhaps with better cause, for while a buck pulled down by wolves can be replaced in two or three years, a range pulled down by too many deer may fail of replacement in as many decades.

So also with cows. The cowman who cleans his range of wolves does not realize that he is taking over the wolf's job of trimming the herd to fit the range. He has not learned to think like a mountain. Hence we have dustbowls, and rivers washing the future into the seas.

We all strive for safety, prosperity, comfort, long life, and dullness. The deer strives with his supple legs, the cowman with trap and poison, the statesman with pen, the most us with machines, votes, and dollars, but it all comes to the same thing: peace in our time. A measure of success in this is all well enough, and perhaps this is behind Thoreau's dictum: In wildness is the salvation of the world. Perhaps this is the hidden meaning in the howl of the wolf, long known among the mountains, but seldom perceived among men."

Leopold, A. 1949. *A Sand County Almanac.* New York: Oxford University Press.

Congressional delegates also have their ideas, which are as diverse as the delegates themselves. Some advocate complete preservation for wildlife. Others want the area managed for hunters and fishermen and -women. There are strong feelings pro and con on endangered-species management. The U.S. Congress has a say on the funding, and different delegates indicate that their approval for future funds are contingent on specific regulations being enforced.

Backers of the wildlife area were able to convince their congressional delegations that a management unit should be established. Local public forums were held to discuss all of the interest groups' ideas. These hearings resulted in a bill which was passed by Congress indicating that the unit would be created and managed for wildlife diversity. Hunting and fishing would be permitted and would follow state laws. No further mining or oil or gas drilling would occur, but wells in place would operate in accordance with the initial environmental restrictions placed on them. Grazing and logging would be allowed to a limited extent, but must be part of the management plan. Congress appropriated $1,000,000 for the first year of operation. Twelve positions were allotted for the management and operation of the unit.

You arrive as director of this operation. You must hire, manage, and supervise the personnel and act as the principal administrative officer. You must make decisions about immediate, as well as long-term, efforts that affect wildlife. The public, private operators, your supervisors, congressional delegates, environmental organizations, and your personnel must all be considered in your decisions.

Following your becoming familiar with the physical area, which already contains the buildings to house your operations, you must hire 11 employees. This could be a lengthy process, requiring advertising and getting people onto appropriate lists. You need to determine which positions need to be filled, write a position description of duties, and then advertise after the regional personnel people approve. Some individuals might be available to transfer from other units. Others need to be hired new to the system. The process for each type of hiring varies. Also, as you develop short- and long-term program objectives, keep in mind the various qualifications required to complete different tasks. It will evolve that certain types of personnel are needed more immediately than others. Hiring these, then, is top priority for quick action.

Day-to-day and long-term plans need to be developed. The main program objective, designated when the unit was initially funded, is management for a diversity of wildlife. As new personnel arrive, brainstorming helps define ways of implementing plans. Planning for special situations needs to be incorporated into your program, or you will end up spending much time and effort applying Band-Aids for emergencies, including fires, special visitors, lost people, and injured animals. Management for endangered species is another special situation that may or may not arise.

LONG-TERM PLANNING

How much of an effort should be placed on waterbird management? Do you want to have a hunting season for big game? Should nature trails, backpack trails, picnic areas, and other recreational areas be established? Are there endangered species to be considered? An inventory of wildlife on the unit should be made. Since the area has undergone a heavy impact by humans, historic records should be searched and the literature examined to see which wildlife species are possible. Care should be given to note occasional wildlife users of the areas (Figure 23–1). Which birds or other animals use the area for migration? Are there winter users? Habitats might be inventoried by aerial photos superimposed on geological survey maps and then classified by "on ground" checks.

Once the staff has developed a tentative plan, a public meeting is advisable. The public should not be asked what they want, but rather, what their reactions are to the ideas of the professionals. Remember, there are those who come to such a meeting just to push their own ideas. At any rate, it is good to record all ideas (Figure 23–2).

Figure 23–1 Planning must consider special needs of some species, such as this island for nesting white pelicans.

Figure 23–2 Special wildlife trails so people can see nongame species such as this ruby-throated hummingbird are important. (Courtesy of the U.S. Fish and Wildlife Service; photo by R. H. Baetson.)

Now you must modify the planning document on the basis of ideas gleaned from the public meeting, to submit for your supervisor's review. This document should logically be written in the form of several program objectives, with strategies and evaluations. The planning document should show how local conditions are met. Thus, when one of the program objectives is to increase the number of nesting waterfowl from the present 500 pairs to 2,500 pairs, the plan should show how you think it can best be done. This might include plans for creating new impoundments, improving habitats, controlling predators, and cleaning up slurry ponds. Cost should be associated with each action item. A time frame should be set so that it is possible to measure progress.

In long-term planning, you finally decide on four program objectives: to provide (1) 2,500 fisherman-days, together with success rates which include standards for sizes of fish; (2) a harvest of 3,000 waterfowl; (3) a harvest of 210 deer; (4) facilities for 3,900 visitor-

days, so that people can enjoy and see wildlife and wildlife habitats. Strategies and problems standing in the way of accomplishing the objectives must be identified. It is necessary to determine what arrangements or agreements exist with the people involved in grazing cattle on the land. Are they to keep fences in repair, or do funds to do so come from the unit's budget? If the unit is responsible for fencing, that material must be purchased through the unit's procurement system, and personnel must be assigned the task of doing the job. If the lessees must repair the fences, you need to meet with them to determine how best to get them to meet their obligation.

Logging roads need to be closed in areas where harassment of wildlife is believed to occur. This could meet with opposition from fishermen, hunters, bird-watchers, or loggers. Special access might be made at the appropriate time. It is helpful to inform the public of such actions through announcement or news releases.

Slurry ponds are a real hazard for wildlife. Migratory-bird legislation can be used to force well operators to protect these areas. Their actions may or may not help other wildlife. Law enforcement agencies may need to be called in to assist. This may, however, have an adverse impact on relations with the operators and bring complaints to the unit and Congress. Various state laws can also help. Unit staff need to monitor and assist in protecting wildlife at these sites. Oiled or dead animals help by bringing forth public concern.

Soil reclamation at old mining sites requires evaluating soils and vegetation. Plants are needed that can grow in the local climate and soils and that can also attract wildlife and control soil erosion. In some cases, acid drainage or other forms of toxicants must be cleaned up first. The current human use of facilities and habitats should be determined by checking fishing and hunting success rates at exit sites (Figure 23–3). Having nonconsumptive users fill out questionnaires is also a good way of determining the public's needs and desires.

Figure 23–3 If overgrazing exists, methods need to be instituted to improve the area for wildlife. (Courtesy of the U.S. Soil Conservation Service.)

IMPLEMENTATION

To accomplish your long-term program objectives, you decide to institute the following selected practices:

1. Create additional wetlands (Chapter 17).
2. Promote waterfowl reproduction (Chapter 17).
3. Reclaim soil in mined areas (Chapter 9).
4. Increase sport fish (Chapter 19).
5. Maintain deer habitat (Chapter 15).
6. Create nature trails.
7. Create a visitors' center and increase visitor use.

Once improvements are made, plans for evaluation similar to the inventory phase must be made to judge whether the goals have been reached. To attract more visitors to the refuge, you plan to build a visitors' center with wildlife habitat around the facility. A site needs to be located with access for the public. Types of wildlife that might be attracted near the center must be considered. Impoundments with isolated islands and vegetated shorelines will attract waterfowl, waterbirds, some furbearers, mammals, and fish. If the size of the pond is large enough, the visitors' center can overlook the pond so that people can see wildlife all year.

Care should be taken to find the plants that best attract wildlife and also add to the landscape. Trees planted near the visitors' center will attract birds and small mammals. Feeders can be used to attract some wildlife species. Hummingbird feeders are popular in the summer. Brush piles might attract some birds and mammals. The buildings, as well as the area around the center, must be carefully designed. An architect is needed for this purpose. The agency may have such a person or may need to contract out the job. Wildlife biologists must have input into the design of the preserve and display area.

When the design is accepted, construction will probably be contracted through a bidding process. Unit people should assist with the construction of the facility and wildlife habitat. This will help assure that the project is coordinated and that wildlife is minimally disturbed. Efforts must be made to monitor the program.

Funds above and beyond the regular budget must be requested not only to construct, but also to maintain, the facility. Special budget requests must be prepared to obtain funds. If no maintenance funds are available for salaries for visitors' center personnel and cleanup crews, the center will not be able to function. Funds must also be requested to build, maintain, and change displays. The printing of pamphlets might require extra funds.

Upon completion of the center, schedules must be prepared for its operation and maintenance. People must be selected and trained who are knowledgeable and who enjoy working with the public. This schedule then becomes a basis for budget planning. You can show your supervisor how much is needed to hire more people or contractors and to buy equipment and supplies to accomplish this goal (Figure 23–4).

Figure 23–4 It is important to consider how to avoid animal damage, such as damage from nutria to dikes. (Courtesy of U.S. Department of Agriculture.)

PLANNING FOR SPECIAL SITUATIONS

At the same time you are developing long-term plans, you must plan for unexpected events. Do you put out a fire or control it? Are your people trained in firefighting techniques? Do you have people and equipment standing by during the major fire season? What assistance can you expect from local, regional, and agency firefighting units? What is the response time? Are you expected to help them? Your plan should cover different forms of fires at different regions in your management unit. Equipment and supplies should be available so that all areas of the unit are adequately covered. Similarly, suppose there is an accident involving injury, a lost person, or a possible drowning. How does your staff respond at different times of the day or season? What about tour groups, visits with your supervisor, or a visit from a U.S. senator?

Hazards of the Occupation

Northern goshawks are swift-flying falcons that nest in old-growth forests of northern and western North America. Some individuals migrate south, while others do not. Goshawks are usually very secretive birds that can maneuver quickly through forests without being seen, even though they are slightly larger than crows. The birds build stick nests 10 to 15 m (30–45 ft) high in the tree branches. They hatch two to five young, which both parents raise. The female usually stays with the young, while the male searches for food along the forest edge. Food items include squirrel, meadowlarks, and robins, all of which are brought to the nest in a ritualistic manner. Both adults defend the nest but the female, which is slightly bigger than the male, is more vicious toward intruders.

While working with a research group in the forest one day in 1977, I had occasion to walk near a nest. Since we knew that predators such as pine marten will climb trees and eat the goshawk's young, we were careful to put mothballs around the base of the tree so no predator could follow our scent. When I was within 30m (90 ft) of the nest tree, the female goshawk swooped down, causing me to duck. As she approached she made a loud crackling noise, and the air swished. I decided that it was time to leave the area and turned to go. Within seconds, a loud crack resounded throughout the woods as I was clobbered on the back of the head by this bird going at a speed of more than 80 km/hr (50 mi/hr). The goshawk took my cap in her talons and flew off. My headache, which lasted the rest of the day, reminded me of why I have such great respect for raptors in general and goshawks in particular.

ADMINISTRATION

Planning and astute delegation of work result in more effective administration. You must decide how your initial annual budget of $1 million is to be expended to operate the unit, accomplish its goals, and follow its long-term plans. Twelve employees' wages and benefits average out to $40,000 each per year, for a total of $480,000. Equipment and supply purchases total $72,000. Vehicle costs and replacements are estimated at $123,000. Building maintenance is projected to be $55,000, and you need a contingency fund of $20,000 for emergencies such as firefighting and rebuilding fences. An $8,000 utility charge is expected. Travel cost to see supervisors and attend meetings are projected at $10,000. You now have received estimates based on recommendations from your employees that costs for plantings to restore mined areas are $130,000 per year for three years. Cleanup of sludge ponds is going to cost $275,000 each year for five years. Construction of impoundments is $175,000 each for four years. You must now determine which, if any, of these management activities can be accomplished with the base budget and which must be funded by an additional budget request. You may also need to prioritize items so that some sludge ponds are cleaned, a few roads built, and a visitors' trail constructed.

As you can see, paperwork is and always will be an integral part of your job. Besides writing up planning and budget reports, you must make a monthly report, in writing, to your supervisor on visitor-use days, budget expenditures, planning objectives accomplished, meetings attended, and all wildlife–people incidents. All accidents must be reported in writing within five days.

Quarterly reports are required for mileage driven, gas usage, building temperature, and personnel activities. Annual reports must be prepared on the budget, personnel, wildlife trends, and administrative operations. Several state agencies require wildlife reports. Written evaluations of all permanent employees are required annually. Temporary and probationary employees must be evaluated semiannually.

Each year, you must prepare a proposed budget for the year after next. This budget is reviewed by your supervisor and generally requires several meetings and numerous responses in order to justify requests. You must constantly be aware of the plans you have and see that they are being followed. Evaluations are very important to determine whether goals are being met. Your daily activities must follow the plans that have been established, but allow for modifications to adapt to changing situations.

> "To sum up, wildlife once fed us and shaped our culture. It still yields us pleasure for leisure hours, but we try to reap that pleasure by modern machinery and thus destroy part of its value. Reaping it by modern mentality would yield not only pleasure, but wisdom as well."
>
> Leopold, A. 1949. *A Sand County Almanac*. New York: Oxford University Press.

APPENDIX 1
SELECTED FISH AND WILDLIFE MANAGEMENT AGENCIES

STATE AGENCIES

Alabama

Department of Conservation and Natural Resources
Director, Game and Fish
64 N. Union St.
Montgomery 36130

Alaska

Department of Fish and Game
P.O. Box 25526
Juneau 99802

Department of Public Safety
Division of Fish and Wildlife Protection
P.O. Box 111200
Juneau 99811

Arizona

Game and Fish Department
2221 W. Greenway Rd.
Phoenix 85023

Arkansas

Game and Fish Commission
#2 Natural Resources Dr.
Little Rock 72205

California

Department of Fish and Game
1416 Ninth St.
Sacramento 95814

Colorado

Department of Natural Resources
Division of Wildlife
6060 Broadway
Denver 80216

Connecticut

Department of Environmental Protection
Director of Wildlife Bureau or Director of Fisheries Bureau
79 Elm St.
Hartford 06106

Delaware

Department of Natural Resources and Environmental Control
Division of Fish and Wildlife
P.O. Box 1401
Dover 19901

Florida

Game and Fresh Water Fish Commission
620 S. Meridian St.
Tallahassee 32399

Georgia

Department of Natural Resources
Wildlife Resources Division
U.S. Highway 278, S.E.
Social Circle 30279

Hawaii

Department of Land and Natural Resources
Division of Forestry and Wildlife
1151 Punchbowl St.
Honolulu 96813

Idaho

Fish and Game Department
600 S. Walnut, Box 25
Boise 83707

Illinois

Department of Conservation
Lincoln Tower Plaza
524 S. Second St.
Springfield 62701

Indiana

Department of Natural Resources
Division of Fish and Wildlife
402 W. Washington St.
Indianapolis 46204

Iowa

Department of Natural Resources
Fish and Wildlife Division
Wallace State Office Bldg. (E. Ninth and Grand Aves.)
Des Moines 50319

Kansas

Department of Wildlife and Parks
900 Jackson St. (Suite 502)
Topeka 66612

Kentucky

Department of Fish and Wildlife Resources
#1 Game Farm Rd.
Frankfort 40601

Louisiana

Department of Wildlife and Fisheries
P.O. Box 98000
Baton Rouge 70898

Maine

Department of Inland Fisheries and Wildlife
284 State St., Station #41
Augusta 04333

Department of Marine Resources
State House, Station #21
Augusta 04333

Maryland

Department of Natural Resources
Forests, Parks and Wildlife
Tawes State Office Bldg.
Annapolis 21401

Massachusetts

Department of Fisheries,
    Wildlife and Recreational Vehicles
100 Cambridge St.
Boston 02202

Michigan

Department of Natural Resources
Division of Fish and Wildlife
Box 30028
Lansing 48909

Minnesota

Department of Natural Resources
Division of Fish and Wildlife
500 Lafayette Rd.
St. Paul 55155

Mississippi

Department of Wildlife Conservation
Southport Mall, P.O. Box 451
Jackson 39205

Missouri

Department of Conservation
P.O. Box 180
Jefferson City 65102

Montana

Department of Fish, Wildlife and Parks
1420 E. Sixth, P.O. Box 200701
Helena 59620

Nebraska

Game and Parks Commission
2200 N. 33rd St., P.O. Box 30370
Lincoln 68503

Nevada

Department of Wildlife
Box 10678
Reno 89520

New Hampshire

Fish and Game Department
2 Hazen Dr.
Concord 03301

New Jersey

Department of Environmental Protection
Division of Fish, Game, and Wildlife
CN 400
Trenton 08625

New Mexico

Game and Fish Department
P.O. Box 25112
Santa Fe 87504

New York

Department of Environmental Conservation
Division of Fish and Wildlife
50 Wolf Rd.
Albany 12233

North Carolina

Wildlife Resources Commission
Archdale Bldg.
512 N. Salisbury St.
Raleigh 27611

North Dakota

State Game and Fish Department
100 N. Bismark Exp.
Bismark 58505

Ohio

Department of Natural Resources
Division of Wildlife
Fountain Square
Columbus 43224

Oklahoma

Department of Wildlife Conservation
1801 N. Lincoln
P.O. Box 53465
Oklahoma City 73152

Oregon

Department of Fish and Wildlife
2501 S.W. 1st Ave.
Portland 97207

Pennsylvania

Fish and Boat Commission
P.O. Box 67000
Harrisburg 17106

Game Commission
2001 Elmerton Ave.
Harrisburg 17110

Rhode Island

Department of Environmental Management
Division of Fish and Wildlife
Stedman Government Center
Wakefield 02879

South Carolina

Department of Natural Resources
Division of Wildlife
Rembert C. Derr Bldg.
P.O. Box 167
Columbia 29202

South Dakota

Game, Fish and Parks Department
523 East Capitol
Pierre 57501

Tennessee

Wildlife Resources Agency
P.O. Box 40747
Ellington Agricultural Center
Nashville 37204

Texas

Parks and Wildlife Department
4200 Smith School Rd.
Austin 78744

Utah

State Department of Natural Resources and Energy
Division of Wildlife Resources
1596 West North Temple
Salt Lake City 84116

Vermont

Department of Fish and Wildlife
103 S. Main
Waterbury 05671

Virginia

Department of Game and Inland Fisheries
4010 West Broad St.
Box 11104
Richmond 23230

Washington

Department of Fish and Wildlife
600 North Capitol Way
Olympia 98501

West Virginia

Department of Natural Resources
1900 Kanawha Blvd., East
Charleston 25305

Wisconsin

Department of Natural Resources
Box 7921
Madison 53707

Wyoming

Game and Fish Department
5400 Bishop Blvd.
Cheyenne 82002

FEDERAL AGENCIES

Army Corps of Engineers
Dept. of the Army
20 Massachusetts Ave., N.W.
Washington, DC 20310

Bureau of Land Management
Dept. of Interior
Interior Bldg.
Washington, DC 20240

Bureau of Oceans and International, Environmental, and Scientific Affairs
Dept. of State
Washington, DC 20520

Bureau of Reclamation
Dept. of Interior
Interior Bldg.
Washington, DC 20240

Council on Environmental Quality
722 Jackson Pl., NW
Washington, DC 20006

Customs Service
Dept. of Treasury
1301 Constitution Ave., N.W.
Washington, DC 20229

Environmental Protection Agency
401 M St., S.W.
Washington, DC 20460

Fish and Wildlife Service
Dept of Interior
Interior Bldg.
Washington, DC 20240

Forest Service
Dept. of Agriculture
P.O. Box 2417
Washington, DC 20013

Geological Survey
12201 Sunrise Valley Dr.
M S-300
Reston, VA 20192

National Marine Fisheries Service
Dept. of Commerce
NOAA
Washington, DC 20235

National Oceanic and Atmospheric Administration
Dept of Commerce
Rockville, MD 20852

National Park Service
Dept. of Interior
Interior Bldg.
Washington, DC 20240

National Science Foundation
Washington, DC 20550

Office of Surface Mining
Dept. of Interior
1951 Constitution Ave., N.W.
Washington, DC 20240

Office of Water Research and Technology
Dept. of Interior
Interior Bldg.
Washington, DC 20240

Soil Conservation Service
Dept of Agriculture
P.O. Box 2890
Washington, DC 20013

PRIVATE AGENCIES (INCLUDES MANAGEMENT AND LOBBYING ORGANIZATIONS)

American Association for the Advancement of Science
1200 New York Ave, N.W.
Washington, DC 20005

American Conservation Association, Inc.
30 Rockefeller Plaza, Rm. 5402
New York, NY 10112

American Fisheries Society
5410 Grosvenor Ln.
Bethesda, MD 20814

American Forestry Assoc.
1516 P St., N.W.
Washington, DC 20005

American Institute of Biological Science
1444 I St., N.W.
Washington, DC 20005

Boone and Crockett Club
250 Station Dr.
Missoula, MT 59801

Defenders of Wildlife
1101 14th Street, N.W.
Washington, DC 20005

Ducks Unlimited, Inc.
One Waterfowl Way
Long Grove, IL 60047

Environmental Defense Fund, Inc.
257 Park Ave., S.
New York, NY 10010

International Council for Bird Preservation
219c Huntington Rd.
Cambridge CB3 ODL., England

International Game and Fish Assoc.
1301 E. Atlantic Blvd.
Pompano Beach, FL 33060

Izaak Walton League of America, Inc.
707 Conservation Ln.
Gaithersburg, MD 20879

National Audubon Society
700 Broadway
New York, NY 10003

National Wildlife Federation
1400 16th St., N.W.
Washington, DC 20036

National Wildlife Refuge Assoc.
1000 Thomas Jefferson St., N.W.
Washington, DC 20007

Nature Conservancy
1815 N. Lynn St.
Arlington, VA 22209

Safari Club International
4800 W. Gates Pass Rd.
Tucson, AZ 85745

Sierra Club
730 Polk St.
San Francisco, CA 94109

Wilderness Society
900 17th St., N.W.
Washington, DC 20006

Wildlife Management Institute
1101 14th St. N.W., Suite 801
Washington, DC 20005

Wildlife Society
5410 Grosvenor Ln.
Bethesda, MD 20814

World Wildlife Fund—U.S.
1250 24th St., N.W.
Washington, DC 20037

Worldwatch Institute
1776 Massachusetts Ave., N.W.
Washington, DC 20036

National or Regional Associations of Wildlife Administrators and Professionals

Association of Midwest Fish and Wildlife Agencies

International Association of Fish and Wildlife Agencies

Northeast Association of Fish and Wildlife Agencies

Southeastern Association of Fish and Wildlife Agencies

Western Association of Fish and Wildlife Agencies

Selected List of Professional Societies (journals published listed in parenthesis)

American Association for the Advancement of Science (*Science*). Oriented toward furthering the work of scientists from many disciplines and increasing public understanding of science in human progress.

American Fisheries Society (*Transactions of the American Fisheries Society; Fisheries; North American Journal of Fisheries Management; Progressive Fish Culturist; Journal of Aquatic Animal Health*). Promotes wise use of fisheries resources, both recreational and commercial.

American Forestry Association (*American Forests*). Promotes intelligent management in the use of forests, soil, water, wildlife, and all other natural resources.

American Institute of Fishery Research Biologists (*Briefs*). Concerned with the professional development and performance of its members in order to advance the science of fishery biology and to promote conservation.

American Ornithologists' Union (*The Auk, Ornithological Monographs*). Advances ornithological science through its publications and meetings.

American Society of Ichthyologists and Herpetologists (*Copeia*). Advances the study of fishes, amphibians, and reptiles.

American Society of Mammalogists (*Journal of Mammalogy, Mammalian Species, Special Publications of American Society of Mammalogists*). Encourages the study of all mammals and the dissemination of the knowledge obtained thereby.

Cooper Ornithological Society (*The Condor, Studies in Avian Biology*). Promotes the study and conservation of birds and wildlife in general and the dissemination of ornithological information.

Desert Fishes Council (*Proceedings of the Desert Fishes Council*). Synthesizes and disseminates information on the status and management of desert ecosystems in North America.

Ecological Society of America (*Ecology, Ecological Monographs, Bulletin of the Ecological Society of America*). Promotes the scientific studies of organisms and environment.

Society for Range Management (*Journal of Range Management, Rangelands*). Promotes professional development and the understanding of rangeland ecosystems and their management.

Society of American Foresters (*Journal of Forestry, Forest Science, Southern Journal of Applied Forestry*). Advances the science, technology, education, and practice of professional forestry; the society is the accrediting authority for professional forestry education in the United States.

Soil Conservation Society of America (*Journal of Soil and Water Conservation*). Promotes good land use and disseminates information from many specialized areas.

Wildlife Society (*Serial Publications: The Journal of Wildlife Management, Wildlife Monographs, Wildlife Society Bulletin, The Wildlifer*). Promotes sound management of wildlife resources and the environment.

Wilson Ornithological Society (*The Wilson Bulletin*). Advances the science of ornithology and cooperation between amateurs and professionals.

APPENDIX 2
SPECIALIZED SOCIETIES

Many organizations exist for the management or protection of a species, a group of species, or a habitat. Some of these organizations are:

Bass Anglers Sportsman Society
Bass Research Foundation
Bear Biology Association
Canvasback Society
Chihuahuan Desert Research Institute
Coastal Society
Cousteau Society
Desert Bighorn Council
Federation of Fly Fishers
Foundation for North American Wild Sheep
Game Conservation International
Hawk Migration Association of North America
International Atlantic Salmon Foundation
International Crane Foundation
International Quail Foundation
National Wild Turkey Federation

Nature Conservancy
North American Bluebird Society
North American Falconers Association
North American Gamebird Association
North American Native Fishes Association
North American Wolf Society
Pacific Seabird Group
Prairie Grouse Technical Council
Raptor Research Foundation
Rocky Mountain Elk Foundation
Ruffed Grouse Society
Trumpeter Swan Society
Urban Wildlife Research Center
Wildlife Disease Association
Wildlife Preservation Trust International (endangered species)

GLOSSARY

Abiotic factors: Nonliving entities—e.g., temperature, humidity, pH.

Accidental species: Species that have a low degree of fidelity to a type of community; not good species for use in defining a community.

Accuracy: The closeness of a sample to the true value.

Acid rain: Precipitation having a pH less than rainwater, which is assumed to be 5.6, but can vary.

Adapted: Suited to a particular condition; usually used of an organism in relation to its habitat.

Aestivation: Condition in which an organism passes hot or dry seasons and in which its normal activities are greatly curtailed or temporarily suspended.

Age structure: Number of individuals in each age category of a population.

Aggregation: Coming together of organisms into a group.

Allopatric: Different; usually used in reference to populations that occupy mutually exclusive (but usually adjacent) geographic areas.

Alluvial soil: Soil deposited by running water.

Alpha diversity: Diversity within a community.

Altrical: Born in a helpless state, so not able to move or support oneself; opposite of *precocial*.

Amino acid: Organic compound from which protein is formed.

Animal unit month (AUM):   The amount of forage required by an animal for one month of grazing; used to compare use of a range by cattle and ungulates.

Annual: Living one year.

Aquaculture: The rearing of plants or animals in water under controlled conditions.

Association: A major unit in community ecology, characterized by essential uniformity of the composition of species.

Autotrophic: Not requiring an exogenous factor for normal metabolism; refers to organisms, usually green plants, that are capable of converting solar energy to chemical energy (sugar) by photosynthesis.

Bag limit: Number of animals, usually birds, that can be taken in a unit of time, usually a day.

Beta diversity: Comparison of diversity between similar communities.

Biennial: A plant living two years, usually flowering the second year; occurring every two years.

Big game: Large animals hunted, or potentially hunted, for sport—for example, elk, mule deer, big-horned sheep, pronghorn, black bear.

Biodiversity: The total diversity of living organisms in an area. Can be considered from the genetic, species, or community point of view.

Biological amplification: The process by which organisms higher in the food chain accumulate and retain materials, such as organochlorines, from organisms lower in the food chain.

Biological clock: An internal mechanism which signals animals that it is time for some activity, such as migration or nesting.

Biological control: The use of a predator, parasite, or pathogen to hold a pest species below the level at which it can cause economic damage.

Biological potential: The maximum production of a population under optimal conditions.

Biomass: The total quantity of living organisms of a species per unit of space (called *species biomass*) or of all the species in a community (called *community biomass*).

Biome: A complex of communities with a distinct type of vegetation.

Biota: All the plants and animals within an area or region.

Biotic factors: Living entities—living plants and animals; opposite of *abiotic factors.*

Biotic potential: The number of births divided by the size of the population in a given area in a given time.

Birthrate: Proportion of a population newly born in a unit of time.

Bog: An extremely wet, poorly drained area characterized by a floating, spongy mat of vegetation often composed of sphagnum, sedges, and heaths.

Browse: Leaves, stems, twigs, bark, and wood of woody plants consumed by animals.

Brucellosis: A bacterial disease that affects mammals, often causing abortions in cattle and some ungulates.

Canopy: A network of the uppermost branches of a forest that partially or fully covers the understory.

Captive breeding: Breeding animals in a captive facility, usually done with endangered species for later release into the wild.

Carnivore: A flesh-eating animal.

Carrion: Dead animal flesh.

Carrying capacity: The maximal population a habitat can support without causing damage to the vegetation; may vary from year to year because of changes in productivity of vegetation.

Case law: Law based on court decisions or interpretations of legislative acts.

Cavity nester: Member of a wildlife species that nests in tree cavities—for example, a woodpecker.

Cementum annuli: Layers on the teeth of some animals that can be used for determining the animal's age.

Chaining: A vegetation-maintenance technique in which a heavy anchor chain is dragged between two tractors to break off or uproot plants.

Channelization: Straightening out of a stream or river; sometimes involves putting in a concrete bottom.

Chlorophyll: A complex of mainly green pigments in the chloroplasts, characteristic of green plants whose light-energy-transforming properties permit photosynthesis.

Clear-cut: Removal of all trees in an area in one cutting operation.

Climax community: A stable vegetative community reached as a result of a progression through successional seres.

Clumped distribution: An aggregated distribution pattern—for example, a herd of animals.

Cluster sampling: Simple random sampling applied to distinct groups of population numbers.

Cohort: A group of individuals in a population born during a particular period, such as a year.

Colonial nesters: Birds that nest in large groups.

Commensalism: The relationship between two populations living together when only one receives a benefit; the other population is neither harmed nor benefited.

Common law: The body of traditional laws based on custom and precedent.

Community: A group of interacting and interdependent (plant and animal) populations that live in the same area.

Compaction: The process whereby firm, concentrated soil is produced as a result of pressure on the top layers.

Competition: The active demand by two or more organisms for a commonly required resource that is limited.

Competitive exclusion principal (Gause's hypothesis):   No two species can occupy the same niche at the same time.

Conifer (coniferous):   A cone-bearing plant—for example, pine and fir.

Conservation: Wise maintenance and use of natural resources.

Consumptive use: Use of resources that involves their removal (for example, hunting and fishing).

Cover: Plants or other objects used by wildlife for protection from predators and adverse weather and for rearing young.

Cover type: The dominant type of plant covering an area.

Cycle: A regular pattern, such as cyclical change in the size of a population.

Death rate: The percentage of a population dying in a unit of time.

Deciduous: Falling off or shedding, as of leaves; descriptive of perennial plants that are normally leafless for some time during the year.

Decompose: To separate into component parts or elements; to decay or putrefy.

Deer yard: An area of heavy cover where deer congregate in the winter for food and shelter.

Density: Number of individuals per unit area.

Density dependent: More severely affecting a population as the size of the population increases.

Density independent: Having an impact on a population not related to the population's size.

Den tree: A tree, either hollow or having holes, that is used by various animals for cover and nesting.

Deterministic model: A mathematical model in which all the relationships are fixed and the concept of probability does not enter; a given input produces a predictable output; opposite of *stochastic model*.

Detritus food chain: A process in which dead organisms are decomposed by other organisms, such as worms, larvae, and bacteria.

Diameter breast high (dbh):   The standard measurement of diameter for standing trees, including bark, taken at 1.37 m (4.5 ft) above the ground.

Dispersion: The pattern of distribution of individuals in a population. In the mathematical sense, *dispersion* describes the probability of occurrence of such individuals in specified places.

Diversity: The total range of wildlife species, plant species, communities, and habitat features in an area.

Dragging: The use of a log or metal grate behind a tractor to loosen cattle manure in a field.

Easement: An access area across another's land.

Ecological characteristics: The basic features of a species related to its distribution, habitat, reproduction, growth characteristics and needs, and responses to changes in habitat.

Ecological longevity: The average length of life of individuals of a population under stated conditions.

Ecology: The study of the interrelationships of organisms with one another and their environment.

Ecosystem: Living and nonliving components in an environment functioning together.

Ecotone: The community formed where two other communities meet, sometimes called *edge*.

Edge: The area where two communities meet (*ecotone*).

Efficiency: Proportion of incoming solar energy converted to chemical energy.

Electroshock: A means of collecting fish by shocking them with electric currents.

Emigration: Movement out of a given area.

Endangered species: A species, subspecies, or race that is threatened with extinction throughout all or a significant portion of its range.

Endemic: Native to a region.

Environment: All the biotic and abiotic factors that affect an individual organism.

Environmental resistance: Factors that act to slow a population's growth.

Enzyme: A catalyst, generally a specific protein joined to some simple substance produced by cellular activity, that is essential to biological action.

Epilimnion: The upper layer of a lake.

Eurytopic: Able to withstand wide variations in environmental conditions.

Eutrophic: Late in lake succession stages; containing high amounts of nutrients.

Eutrophication: Process of succession in lakes by the addition of nutrients.

Evolution: The change in a population's genetic composition over time leading to adaptations to the environment.

Exotic species: A species not native to a geographical area in which it is found.

Exponential growth: Population growth that exceeds the carrying capacity of a habitat until the population saturates the habitat; growth characterized by a progressively increasing, nonlinear relationship between population and time.

Featured-species management: (see also *indicator species*) A wildlife-management strategy to produce relatively high numbers of selected wildlife species in specific places for specific purposes.

Fecundity: Capability of an organism to produce reproductive units, such as eggs, sperm, or asexual structures.

Feedback: The output of a given system that affects the state of that same system.

Feral: Having reverted to a wild state after being domesticated—for example, feral horses.

Fertility: The average number of births to each female in a unit of time.

Fluctuations: Irregular changes.

Food chain: The energy flow from green plants through consumer organisms at each trophic level. There are two general forms: the grazing and detritus food chains.

Food web: A complex food chain.

Forage: Vegetation used as food by wildlife, particularly ungulates, and domestic livestock.

Forb: Any herbaceous plant other than those in the Gramineae, Cyperaceae, and Juncaceae families; a fleshy-leaved plant.

Furbearer: A mammal commonly harvested for its hide (e.g., muskrat or mink).

Gamma diversity: Comparison of diversity of large heterogeneous areas.

Gap phase: Localized area of disturbance in a larger community.

Gause: Russian microbiologist who studied the mechanism of competitive exclusion between two species in the early 1930s. He formulated the competitive exclusion principle.

Gause's hypothesis: *See* competitive exclusion principle.

Gene pool: Narrowly, all the genes of a localized interbreeding population; broadly, all the genes of a species throughout its entire range.

Generality: The applicability of a model to appropriate situations.

Genetic composition: The total genetic makeup of a population.

Genotype: The entire genetic constitution of an organism; contrast with *phenotype*.

Gestation: The length of time from conception to birth.

Girdling: The act of encircling a tree with cuts through the cambium layer to kill the tree.

Gleaning: A process of feeding, particularly in birds, in which food items are gathered from the surface of the foraging substrate, usually plants.

Grazing food chain: The movement of energy from green plants to herbivores to carnivores, excluding decomposition.

Gross primary productivity: The total amount of energy available from the conversion of solar energy to chemical energy during photosynthesis by green plants.

Habitat: The environment that supplies the needs of a population.

Harvest: Removal of animals from a population.

Hawking: Capturing food while flying.

Herbivore: An animal that feeds on plants.

Heterotroph: An organism that utilizes chemical energy supplied by autotrophic organisms.

Homeostasis: A stable state or the tendency of a system to maintain a stable or balanced state.

Homeotherm: A warm-blooded animal that can regulate its body temperature physiologically.

Home range: The activity area used by an animal; usually refers to daily activity.

Host: An organism that furnishes food, shelter, or other benefits to another species.

Hypolimnion: The deepwater layer of a lake.

Immigration: Movement into a given area.

Impact (on population):   A change in a population's natality, growth, and/or survival caused by some disturbance.

Imprinting: Recognition fixed through a short-interval learning process in animals; young might imprint on parents, or animals might imprint on a nesting habitat.

Inbreeding: Breeding among genetically similar individuals in a population; leads to reduced genetic variability (*homozygosity*).

Indicator species: One of a few species that can be used to indicate the status of a natural system.

Indices: Indicators of population changes through repeated measurements; indices generally show population trends.

Innate capacity for increase (r_m):   A measure of the rate of increase of a population under "ideal" conditions.

Insectivorous: Insect eating.

Interspecific competition: Competition between individuals of different species.

Intraspecific competition: Competition between members of the same species.

Intrinsic rate of increase: Difference between the birth and death rates in a population.

K strategists: Populations that tend to be stable in size near the environmental carrying capacity.

Lacustrine: Relating to, or formed by, lakes

Law of the minimum: An ecological axiom which states that any factor a population requires that is present in the smallest amount limits the population's growth accordingly.

Legume: Any of a large group of the pea family that has five pods enclosing seeds.

Lek: A site where birds (primarily grouse) traditionally gather for sexual display and courtship.

Life tables: A table of population data based on a sample (often 1,000 individuals) of the population showing the age at which each member died. A **cohort** life table starts with a group of individuals, all born during the same period. A **static** life table has a sample of individuals from each age class in the population.

Limiting factor: Any factor that limits a population's growth when the factor is in short supply or prevents growth when it is absent.

Lincoln–Peterson index: A mark-recapture formula for estimating the abundance of animals in a habitat.

Litter: The uppermost layer of organic debris on a forest floor; essentially freshly fallen or slightly decomposed vegetable material, mainly foliate or leaf litter, but also bark fragments, flowers, and fruits.

Loam: A soil consisting of an easily crumbled mixture of clay, sand, and silt.

Logistic growth: Growth of a population that approaches and remains near the carrying capacity of a habitat.

Management: Manipulation of populations or habitats to achieve desired goals by people.

Management by objectives: Planning by ranking program objectives in order of priority.

Mark-recapture: Technique in which animals are caught, marked, and released; a second capture period is used to estimate the ratio of marked to unmarked animals and thereby estimate the size of a population.

Marsh: A low, treeless, wet area characterized by sedges, rushes, and cattails.

Maximum sustainable yield: The largest number of fish or wildlife that can be removed without destroying a population's capability of reproducing.

Metapopulation: A term used to denote isolated populations that can occasionally have members exchange genetic material between them.

Migration: Movement by a population on some regular basis (e.g., seasonally or yearly) away from and back to an area.

Migration corridor: A narrow region such as a band of land or belt of vegetation that animals follow during migration, since it provides a completely or partially suitable habitat.

Mineral cycling: The cycling of minerals throughout an ecosystem.

Mitigation: Reduction of the impact of a change on wildlife.

Models: Descriptors or formulas that show or predict changes in a system.

Mortality rate: The proportion of a population dying in a unit of time (death rate).

Multiple use: A concept of land management in which a number of products are produced from the same land base—for example, forests for timber, wildlife, recreational areas, and water retention.

Mutation: A chemical change in the genetic material of an individual.

Mutualism: The mutually beneficial association of different kinds of organisms.

Natality rate: The proportion of a population born in a unit of time (birth rate).

National Environmental Policy Act: (U.S. Laws, Statutes, etc., Public Law 91–190, 1970) Declares a national policy of encouraging productive and enjoyable harmony between humans and the environment, to promote the prevention or elimination of damage to the environment and biosphere, to stimulate the health and welfare of people, to enrich the understanding of ecological systems and natural resources of the United States, and to establish a Council of Environmental Quality.

Negative feedback: Feedback that inhibits or stops a system's progress.

Net primary productivity: Energy available in a plant following respiration. Gross primary productivity minus respiration equals net primary productivity.

Net reproductive rate: The average number of female offspring produced by the females in a population.

Niche: The functional role of an organism within its habitat.

Nitrogen fixation: The conversion of elemental, atmospheric nitrogen (N_2) to organic combinations or to forms readily usable in biological processes.

Nocturnal: Active at night.

Nonconsumptive use: Use of natural resources without removing them—for example, photography and watching wildlife.

Nongame wildlife: All wildlife not subject to harvesting regulations.

Old-growth stand: A forest stand that is past full maturity and that shows decadence; the last stage of forest succession.

Oligotrophic: Deficient in plant nutrients; said of a lake early in its succession stages with a low nutrient level.

Onmivore: An animal that feeds on both plants and animals.

Optimum yield: The amount of material that, when removed from a population, will maximize biomass (or numbers, or profit, or any other type of variable) on a sustained basis.

Organochloride: Member of a group of pesticides, including DDT, that are not biodegradable and therefore remain active for many years after application.

Overgrazing: A continued overuse, usually by ungulates, that creates deteriorated range condition.

Palustrine: Forested wetland, usually adjacent to a lake; tidal areas where salinity due to ocean salts is below 0.5%.

Parasite: An organism that benefits while feeding upon, securing shelter from, or otherwise injuring another organism (the *host*).

Parasitic: Growing on, and deriving nourishment from, another organism.

Parasitism: The interaction of two individuals in which one, the host, serves as a food source for the other, the parasite.

Pattern: Type of distribution (random, regular, or aggregate).

Permafrost: Ground that is frozen a few inches below the surface all year.

Phenotype: Expression of the characteristics of an organism, as determined by the interaction of its genetic constitution and the environment; contrast with *genotype*.

Photoperiod: Length of daylight.

Photoperiodism: Response of plants and animals to the relative duration of light and darkness: for example, some migration patterns are triggered by the length of the day.

Photosynthesis: Formation of chemical-bond energy (sugar) from solar energy by green plants.

Physiological longevity: Maximum life span of individuals in a population under specified conditions; the organisms die of senescence.

Phytoplankton: Minute plants that float in an aquatic system; the plant community in marine and fresh water that floats free in the water and contains many species of algae and diatoms.

Poikilotherm: A cold-blooded animal; the internal temperature remains similar to that of the environment. Some poikilotherms regulate their temperature behaviorally.

Polyandry: Mating of a female with two or more males.

Polygamy: Mating of a male with two or more females.

Population: A group of organisms of a single species that interact and interbreed in a common place.

Positive feedback: Return of output to a system that allows the system to continue in its direction; feedback that enhances or promotes a system's progress.

Precision (model):   The ability of a model to provide repeated results.

Precision (sample):   Repeatability of measurement.

Precocial: Able to move about at an early age.

Predation: The act of predators capturing prey.

Prescribed burn: Intentional burning of an area under selected fuel, moisture, and wind conditions.

Preservation: Leaving the natural system as it is.

Primary cavity nesters: Wildlife species that excavate spaces in trees or snags.

Primary production: Production by green plants.

Probability: The frequency, expressed as a proportion or percentage of the total number of occurrences, that will produce a specified value for a variable over a long series of trials.

Production: Amount of energy (or material) formed by an individual, population, or community in a specified period.

Program: A series of planned activities.

Promiscuous: Not restricted to one sexual partner.

Protein: Group of amino acids linked end to end in a specific order.

Put-and-take: Planting of hatchery fish for removal of fishing enthusiasts or planting of game-farm birds for removal by hunters.

Race: A geographic variant of a species; often considered the same as a subspecies.

Radio collar: A collar that contains a radio transmitter and is fastened to an animal; signals from the transmitter are used by wildlife biologists to gain information, usually about the position of the animal.

Random distribution: A distribution pattern in which an organism's position is independent of that of others.

Random sample: A sample in which the probability of selection of each individual in the population is known; a sample free from selection bias.

Raptor: Any predatory bird—such as a falcon, hawk, eagle, or owl—that has feet with sharp talons or claws adapted for seizing prey and a hooked beak for tearing flesh.

Realism: The accuracy with which a model conforms to the system it models.

Recruitment: Increment to a natural population, usually from young animals or plants entering the adult population.

Regular distribution: A distribution in which there is a repeated pattern—for example, trees in an orchard.

Regulating mechanisms: Factors that act to control the density of a population.

Replacement rate (net reproduction rate) (R_o): The average number of female offspring produced by the females in a population.

Respiration: The breakdown of sugar into usable energy by living organisms.

Riparian: Bordering a natural waterway.

Rookeries: In the United States, colonies of nesting (usually great blue) herons.

Rut: Breeding season of some ungulates.

Sample: A subset of the total number of units in a population.

Savanna: Lowland tropical and subtropical grassland with a scattering of trees and shrubs.

Scat: Animal fecal matter.

Schnabel method: A mark-recapture formula for estimating the abundance of animals in a habitat.

Schumacker & Eschmeyer method: A mark-recapture formula for estimating the abundance of animals in a habitat.

Secondary cavity nesters: Wildlife species that occupies a cavity in a tree or snag excavated by another species.

Secondary production: Production by herbivores, carnivores, or detritus feeders; contrast with *primary production.*

Self-regulation: The process of regulating population in which an increase is prevented by a deterioration in the quality of individuals that make up the population; regulating population by internal adjustments in behavior and physiology within the population, rather than by external forces such as predators.

Self-sustaining population: A wildlife population large enough to assure its continued existence without the introduction of other individuals from outside the area inhabited by the population.

Sere: Each community in a successional sequence.

Shelterbelt: A strip of trees or shrubs planted or left standing in prairie areas to help reduce wind and erosion of topsoil.

Sigmoid curve: *S*-shaped curve—for example, the logistic curve.

Slash: Woody material left after a cutting operation.

Snag: Standing dead tree.

Species: A group of similar-looking individuals reproductively isolated from other such groups under natural conditions.

Species richness: An indicator of the number of species of plants or animals present in an area; the more species present, the higher is the degree of species richness.

Statutory laws: Legislative acts.

Stenotopic: Having little ability to withstand the modification of environmental conditions.

Steppe: An extensive area of natural, dry grassland; usually used in reference to grasslands in southwestern Asia and southeastern Europe; equivalent to *prairie* in North American usage.

Stochastic model: A mathematical model based on probabilities; the prediction of the model is not a single, fixed number, but a range of possible numbers; opposite of *deterministic model.*

Stock: A group of fish or other aquatic animals that can be treated as a single unit for management purposes.

Strain: Type of fish having genetic adaptability to a specific set of physical conditions (e.g., temperature).

Strategies: Means of carrying out plans.

Stratified sampling: Sampling by groups: a method that increases precision.

Structure (community): Physical makeup of vegetation in a community.

Subspecies: Geographic variant of a species.

Substrate: Supporting material; in biology, usually refers to soil or soillike material, such as *community substrate.*

Succession: Changes in a community brought about by the species of the community; the changes generally result in the replacement of communities and continue until a climax community is established.

Surrogate species: A close relative of a rare or endangered species; used to determine capture and breeding techniques and rearing procedures; also used in physiological tests for endangered species.

Survivorship curve: Data from column ℓ_x in a life table (individuals alive at the beginning of each interval), plotted on semilog paper.

Sustained yield: Number of animals or plants continuously taken from a population without destroying the population.

Swamp: A wet area that usually has standing trees.

Symbiosis: The living together of two animals in a positive (mutualistic), neutral (commensal), or negative (parasitic) relationship.

Sympatric: Similar; usually used in reference to populations that occupy the same geographic region.

System: A collection of (living or nonliving) interacting parts that functions as a unit.

Tagia: Boreal forests of the north.

Territory: An area defended by individuals in some populations, usually against other members of their own species.

Threatened species: Species that are liable to become endangered in the near future.

Transect: A route that cuts across a study area; a term used in sampling.

Translocate: To move from one place to another.

Trends: Indications of changes in populations over time.

Tribe: A subdivision of subfamily based on structure, plumage, habits, or courtship behavior.

Trophic level: A structure of producer and consumer organisms superimposed on food chains to trace the flow of energy among the organisms.

Tundra: Northern, high-elevation biome with few sizable woody plants because of a short growing season.

Understory: Foliage, consisting of seedlings, shrubs, and herbs, that lies beneath, and is shaded by, canopy or taller plants.

Ungulate: A hooved animal.

Viable population: A population large enough to perpetuate itself over time in spite of normal fluctuations in population levels.

Wetland: Any area where the water table is near or above the surface of the land during a considerable part of the year.

Wildlife: All nondomesticated animals in a natural environment; sometimes construed as including captive animals in zoos.

Wildlife management: The manipulation of populations and habitats to achieve the desires of people.

Wolf tree: A tree of dominant size and position that usurps light and space from smaller understory, preventing its growth.

Xeric: Deficient in moisture for the support of life—said of a desert environment.

Xerophyte: Plant that can grow in dry places—for example, cactus.

Zero-based budget: A budget made by analyzing necessary functions and the implementation of programs in priority order, rather than by using earlier comparable budgets as a point of departure.

Zooplankton: Animal portion of plankton; the animal community in marine and fresh water that floats free in the water, independently of the shore and the bottom and moving passively with the currents.

INDEX

Boldface indicates major discussion. *Italics* indicate a photograph or illustration.

Winter Waterfowl Survey 91
Wolf:
    grey 35, 58, 59, 60, 102, 265, 337, *450,* 478
    red 337, 451
Wolverine 334
Woodcock 391, 392
Woodcock Call Count 91, 380
Woodpecker:
    black-backed three-toed 156
    downy 110
    Lewis 445
    red-cockaded 457
    yellow-bellied sapsucker 424
World Wildlife Fund 461, 462
Writers (wildlife) 245